Lecture Notes in Computer Science   11974

More information about this series at http://www.springer.com/series/7407

Yaroslav D. Sergeyev · Dmitri E. Kvasov (Eds.)

Numerical Computations: Theory and Algorithms

Third International Conference, NUMTA 2019
Crotone, Italy, June 15–21, 2019
Revised Selected Papers, Part II

 Springer

Editors
Yaroslav D. Sergeyev 🆔
University of Calabria
Rende, Italy

Lobachevsky University of Nizhny
Novgorod
Nizhny Novgorod, Russia

Dmitri E. Kvasov 🆔
University of Calabria
Rende, Italy

Lobachevsky University of Nizhny
Novgorod
Nizhny Novgorod, Russia

ISSN 0302-9743   ISSN 1611-3349   (electronic)
Lecture Notes in Computer Science
ISBN 978-3-030-40615-8   ISBN 978-3-030-40616-5   (eBook)
https://doi.org/10.1007/978-3-030-40616-5

LNCS Sublibrary: SL1 – Theoretical Computer Science and General Issues

This Springer imprint is published by the registered company Springer Nature Switzerland AG
The registered company address is: Gewerbestrasse 11, 6330 Cham, Switzerland

Preface

This volume, edited by Yaroslav D. Sergeyev and Dmitri E. Kvasov, contains selected peer-reviewed papers from the Third Triennial International Conference and Summer School on Numerical Computations: Theory and Algorithms (NUMTA 2019) held in Le Castella – Isola Capo Rizzuto (Crotone), Italy, during June 15–21, 2019. The NUMTA 2019 conference has continued the previous successful editions of NUMTA that took place in 2013 and 2016 in Italy in the beautiful Calabria region.

NUMTA 2019 was organized by the University of Calabria, Department of Computer Engineering, Modeling, Electronics and Systems Science, Italy, in cooperation with the Society for Industrial and Applied Mathematics (SIAM), USA. This edition had the high patronage of the municipality of Crotone – the city of Pythagoras and his followers, the Pythagoreans. In fact, Pythagoras established the first Pythagorean community in this city in the 6th century B.C. It was a very special feeling for the participants of NUMTA 2019 to visit these holy, for any mathematician, places with a conference dedicated to numerical mathematics.

The goal of the NUMTA series of conferences is to create a multidisciplinary round table for an open discussion on numerical modeling nature by using traditional and emerging computational paradigms. Participants of the NUMTA 2019 conference discussed multiple aspects of numerical computations and modeling starting from foundations and philosophy of mathematics and computer science to advanced numerical techniques. New technological challenges and fundamental ideas from theoretical computer science, machine learning, linguistic, logic, set theory, and philosophy met the requirements, as well as fresh, new applications from physics, chemistry, biology, and economy.

Researchers from both theoretical and applied sciences were invited to use this excellent opportunity to exchange ideas with leading scientists from different research fields. Papers discussing new computational paradigms, relations with foundations of mathematics, and their impact on natural sciences were particularly solicited. Special attention during the conference was dedicated to numerical optimization techniques and a variety of issues related to the theory and practice of the usage of infinities and infinitesimals in numerical computations. In particular, there were a substantial number of talks dedicated to a new promising methodology allowing one to execute numerical computations with finite, infinite, and infinitesimal numbers on a new type of a computational device – the Infinity Computer patented in the EU, Russia, and the USA.

This edition of the NUMTA conference was dedicated to the 80th birthday of Professor Roman Strongin. For the past 50 years Roman Strongin has been a leader and an innovator in Global Optimization, an important field of Numerical Analysis having numerous real-life applications. His book on Global Optimization, published in 1978, was one of the first in the world on this subject. Now it is a classic and has been used by many as their first introduction and continued inspiration for Global Optimization. Since that time, Roman has published numerous books and more than 400 papers in several scientific fields and has been rewarded with many national and international honors including the President of the Russian Federation Prize. For decades Roman served as Dean, First Vice-Rector, and Rector of the famous Lobachevsky State University of Nizhny Novgorod. Since 2008 he has been President of this university. He is also Chairman of the Council of Presidents of Russian Universities, Vice-President of the Union of the Rectors of Russian Universities, and Chairman of the Public Chamber of the Nizhny Novgorod Region.

We are proud to inform you that 200 researchers from the following 30 countries participated at the NUMTA 2019 conference: Argentina, Bulgaria, Canada, China, Czech Republic, Estonia, Finland, France, Germany, Greece, India, Iran, Italy, Japan, Kazakhstan, Latvia, Lithuania, the Netherlands, Philippines, Portugal, Romania, Russia, Saudi Arabia, South Korea, Spain, Switzerland, Thailand, Ukraine, the UK, and the USA.

The following plenary lecturers shared their achievements with the NUMTA 2019 participants:

- Louis D'Alotto, USA: "Infinite games on finite graphs using Grossone"
- Renato De Leone, Italy: "Recent advances on the use of Grossone in optimization and regularization problems"
- Kalyanmoy Deb, USA: "Karush-Kuhn-Tucker proximity measure for convergence of real-parameter single and multi-criterion optimization"
- Luca Formaggia, Italy: "Numerical modeling of flow in fractured porous media and fault reactivation"
- Jan Hesthaven, Switzerland: "Precision algorithms"
- Francesca Mazzia, Italy: "Numerical differentiation on the Infinity Computer and applications for solving ODEs and approximating functions"
- Michael Vrahatis, Greece: "Generalizations of the intermediate value theorem for approximations of fixed points and zeroes of continuous functions"
- Anatoly Zhigljavsky, UK: "Uniformly distributed sequences and space-filling"

Moreover, the following tutorials were presented during the conference:

- Roberto Natalini, Italy: "Vector kinetic approximations to fluid-dynamics equations"
- Yaroslav Sergeyev, Italy and Russia: "Grossone-based Infinity Computing with numerical infinities and infinitesimals"
- Vassili Toropov, UK: "Design optimization techniques for industrial applications: Challenges and progress"

These proceedings of NUMTA 2019 consist of two volumes: Part I and Part II. The book you have in your hands is the second part containing peer-reviewed papers chosen from the general stream, plenary lectures, and small special sessions of NUMTA 2019. Papers carefully selected from big special streams and sessions held during the conference have been collected in the Part I of the NUMTA 2019 proceedings.

This volume contains 19 long papers and 32 short papers that were accepted for publication after a thorough peer review process (required up to three review rounds for some manuscripts) by the members of the NUMTA 2019 Program Committee and independent reviewers. This volume also contains the paper of the winner (Lorenzo Fiaschi, Pisa, Italy) of the Springer Young Researcher Prize for the best NUMTA 2019 presentation made by a young scientist. The support of the Springer LNCS editorial staff and the sponsorship of the Young Researcher Prize by Springer are greatly appreciated.

The editors express their gratitude to institutions that have offered their generous support to the international conference NUMTA 2019. This support was essential for the success of this event:

- University of Calabria (Italy)
- Department of Computer Engineering, Modeling, Electronics and Systems Science of the University of Calabria (Italy)
- Italian National Group for Scientific Computation of the National Institute for Advanced Mathematics F. Severi (Italy)
- Institute of High Performance Computing and Networking of the National Research Council (Italy)
- International Association for Mathematics and Computers in Simulation
- International Society of Global Optimization

The editors thank all the participants for their dedication to the success of NUMTA 2019 and are grateful to the reviewers for their valuable work. Many thanks go to Maria Chiara Nasso from the University of Calabria, Italy, for her kind support in the technical editing of this volume.

The next Triennial International Conference and Summer School NUMTA "Numerical Computations: Theory and Algorithms" will take place in 2022 in Italy. The editors of this volume, who are chairs of the NUMTA Scientific and Organizing Committees, respectively, invite all the participants of NUMTA 2019, and readers of this book, to submit their high-quality results to the next edition of this wonderful event.

October 2019

Yaroslav D. Sergeyev
Dmitri E. Kvasov

Organization

General Chair

Yaroslav Sergeyev — University of Calabria, Italy, and Lobachevsky University of Nizhny Novgorod, Russia

Scientific Committee

Lidia Aceto, Italy
Andy Adamatzky, UK
Francesco Archetti, Italy
Thomas Bäck, The Netherlands
Roberto Battiti, Italy
Roman Belavkin, UK
Giancarlo Bigi, Italy
Paola Bonizzoni, Italy
Luigi Brugnano, Italy
Sergiy Butenko, USA
Sonia Cafieri, France
Tianxin Cai, China
Cristian Calude, New Zealand
Antonio Candelieri, Italy
Mario Cannataro, Italy
Giovanni Capobianco, Italy
Domingos Cardoso, Portugal
Francesco Carrabs, Italy
Emilio Carrizosa, Spain
Leocadio Casado, Spain
Carmine Cerrone, Italy
Raffaele Cerulli, Italy
Marco Cococcioni, Italy
Salvatore Cuomo, Italy
Louis D'Alotto, USA
Oleg Davydov, Germany
Renato De Leone, Italy
Alessandra De Rossi, Italy
Kalyanmoy Deb, USA
Francesco Dell'Accio, Italy
Branko Dragovich, Serbia
Gintautas Dzemyda, Lithuania
Yalchin Efendiev, USA
Michael Emmerich, The Netherlands
Adil Erzin, Russia
Yury Evtushenko, Russia
Giovanni Fasano, Italy
Şerife Faydaoğlu, Turkey
Luca Formaggia, Italy
Elisa Francomano, Italy
Masao Fukushima, Japan
David Gao, Australia
Manlio Gaudioso, Italy
Victor Gergel, Russia
Jonathan Gillard, UK
Daniele Gregori, Italy
Vladimir Grishagin, Russia
Mario Guarracino, Italy
Nicola Guglielmi, Italy
Jan Hesthaven, Switzerland
Felice Iavernaro, Italy
Mikhail Khachay, Russia
Oleg Khamisov, Russia
Timos Kipouros, UK
Lefteris Kirousis, Greece
Yury Kochetov, Russia
Olga Kostyukova, Belarus
Vladik Kreinovich, USA
Dmitri Kvasov, Italy
Hoai An Le Thi, France
Wah June Leong, Malaysia
Øystein Linnebo, Norway
Antonio Liotta, UK
Marco Locatelli, Italy
Stefano Lucidi, Italy
Maurice Margenstern, France

Vladimir Mazalov, Russia
Francesca Mazzia, Italy
Maria Mellone, Italy
Kaisa Miettinen, Finland
Edmondo Minisci, UK
Nenad Mladenovic, Serbia
Ganesan Narayanasamy, USA
Ivo Nowak, Germany
Donatella Occorsio, Italy
Marco Panza, USA
Massimo Pappalardo, Italy
Panos Pardalos, USA
Remigijus Paulavičius, Lithuania
Hoang Xuan Phu, Vietnam
Stefan Pickl, Germany
Raffaele Pisano, France
Yuri Popkov, Russia
Mikhail Posypkin, Russia
Oleg Prokopyev, USA
Davide Rizza, UK
Massimo Roma, Italy
Valeria Ruggiero, Italy
Maria Grazia Russo, Italy
Nick Sahinidis, USA
Leonidas Sakalauskas, Lithuania
Yaroslav Sergeyev, Italy

Khodr Shamseddine, Canada
Sameer Shende, USA
Vladimir Shylo, Ukraine
Theodore Simos, Greece
Vinai Singh, India
Majid Soleimani-Damaneh, Iran
William Spataro, Italy
Maria Grazia Speranza, Italy
Giandomenico Spezzano, Italy
Rosamaria Spitaleri, Italy
Alexander Strekalovskiy, Russia
Roman Strongin, Russia
Gopal Tadepalli, India
Tatiana Tchemisova, Portugal
Gerardo Toraldo, Italy
Vassili Toropov, UK
Hiroshi Umeo, Japan
Michael Vrahatis, Greece
Song Wang, Australia
Gerhard-Wilhelm Weber, Poland
Luca Zanni, Italy
Anatoly Zhigljavsky, UK
Antanas Žilinskas, Lithuania
Julius Žilinskas, Lithuania
Joseph Zyss, France

Organizing Committee

Francesco Dall'Accio — University of Calabria, Rende (CS), Italy
Alfredo Garro — University of Calabria, Rende (CS), Italy
Vladimir Grishagin — Lobachevsky University of Nizhny Novgorod, Nizhny Novgorod, Russia
Dmitri Kvasov (Chair) — University of Calabria, Rende (CS), Italy, and Lobachevsky University of Nizhny Novgorod, Nizhny Novgorod, Russia
Marat Mukhametzhanov — University of Calabria, Rende (CS), Italy, and Lobachevsky University of Nizhny Novgorod, Nizhny Novgorod, Russia
Maria Chiara Nasso — University of Calabria, Rende (CS), Italy
Clara Pizzuti — National Research Council of Italy (CNR), Institute for High Performance Computing and Networking (ICAR), Rende (CS), Italy
Davide Rizza — University of East Anglia, Norwich, UK
Yaroslav Sergeyev — University of Calabria, Rende (CS), Italy, and Lobachevsky University of Nizhny Novgorod, Nizhny Novgorod, Russia

Sponsors

 Springer

In cooperation with siam.

and under the high patronage of the municipality of Crotone, Italy.

Contents – Part II

Contents – Part I

First Order Methods in Optimization: Theory and Applications

High Performance Computing in Modelling and Simulation

Numbers, Algorithms, and Applications

Optimization and Management of Water Supply

Long Papers

Long Papers

Numerical Algorithms for the Parametric Continuation of Stiff ODEs Deriving from the Modeling of Combustion with Detailed Chemical Mechanisms

Luigi Acampora[iD] and Francesco S. Marra[✉][iD]

Istituto di Ricerche sulla Combustione - CNR, Naples, Italy
{acampora,marra}@irc.cnr.it

Abstract. The use of detailed chemical mechanisms is becoming increasingly necessary during the actual transition of energy production from fossil to renewable fuels. Indeed, the modern renewable fuels are characterized by a composition more complex than traditional fossil fuels due to the variability of the properties of the primary source, i.e. biomass. The parametric continuation can be a formidable tool to study the behavior of these new fuels allowing to promptly assess equilibrium conditions varying the main operative parameters. However, parametric continuation is a very computationally demanding procedure, both for the number of elementary operations needed and for the memory requirements. Actually, only very recently some approaches that allow affording this computation with chemical mechanisms composed of hundreds of chemical species and thousands of reactions have been proposed [1,2,37]. Starting from the procedure presented in [1], this paper illustrates further improvements of key steps that usually represents a bottleneck for the effective computation of parametric continuations and for the identification of bifurcation points.

Keywords: Bifurcation of combustion systems · Parametric continuation · Detailed mechanisms

Nomenclature

α	continuation parameter
λ	eigenvalues
$\mathbf{f}$	right-hand side of system given by Eqs. (1–2)
$\mathbf{J_f}$	Jacobian matrix
$\mathbf{x}$	state vector
ψ	test functions
ρ	density, $\mathrm{kg\,m^{-3}}$
τ	residence time, s
a	real part of eigenvalues
b	complex part of eigenvalues

© Springer Nature Switzerland AG 2020
Y. D. Sergeyev and D. E. Kvasov (Eds.): NUMTA 2019, LNCS 11974, pp. 3–16, 2020.
https://doi.org/10.1007/978-3-030-40616-5_1

c_p	constant pressure specific heat, $J\,kg^{-1}\,K^{-1}$
h	mass specific enthalpy, $J\,kg^{-1}$
N_s	number of chemical species
N_{nz}	number of non zero element in a matrix
r	net production rate, $kmol\,s^{-1}$
T	temperature, K
t	time, s
V	volume of the reactor, m^3
W	molecular weight, $kg\,kmol^{-1}$
Y	mass fraction

Subscripts

F	Fold Bifurcation
f	feed conditions
H	Hopf Bifurcation
j	species index

1   Introduction

The increasing energy needs and the current environmental challenges (such as actions to contrast the climate change and reduce pollutant emissions) are driving the growth of the demand for low-carbon (alternative to petroleum) fuels [10]. Bio-fuels are one of the most promising low-carbon energy source, but many aspects of these fuels must be deepened ranging from the economic and environmental impact of their production (f.i. the annual greenhouse gas reductions that can be obtained depends on how they are produced) [10,44] and combustion aspects such as ignition, oxidation, and pollutant emissions [44]. Computer simulations and chemical kinetic studies alongside experiments are fundamental tools to carry out investigations on these combustion aspects [44].

Particularly, the studies of the combustion processes in a Perfectly Stirred Reactor (PSR) are important topics in chemical kinetic because they are frequently employed to develop reliable kinetics mechanisms or model fuels (e.g. [23,30,43]), and to validate reduced kinetic mechanisms (e.g. [3,20,24–27,37,42]). The bifurcation analysis and the parametric continuation technique are the tools of choice for studying the dynamical behavior of these chemical reactive systems because they permit to understand the phenomenology of combustion chemistry and to identify reactor instabilities, multiple steady-states and optimum operating conditions [20,22].

These analysis tools are very popular since the pioneering work of Uppal, Ray and Poore [40,41]. Consequently, they are widely discussed in the scientific literature [5,14,15,21,35] and several parametric continuation and bifurcation analysis software are available such as AUTO [9] and MATCONT [8].

However, despite the popularity, these tools are rarely applied to systems involving large detailed chemical mechanisms (consisting of hundreds of species and thousands of reactions) because of the computational complexity and effort that arise in these models [1,22]. Indeed, in the past, complete bifurcation analyses (both bifurcation curves and bifurcation points are computed) have been successfully conducted only for some elementary fuels like hydrogen [17,31] or methane (31 species and 177 reactions) [32,33] by using ad-hoc software for thermodynamic, transport and kinetic data. Only in the past few years, some significant progress have been made. Shan and Lu [36] presented a bifurcation analysis tool based on computer code that automatically generate mechanism-specific subroutines for analytical Jacobian evaluation from mechanisms described in CHEMKIN format, and its application to the bifurcation analysis of methane (53 species and 325 reactions), DME (55 species and 290 reactions) combustion in PSR [36,37]. Acampora and coworkers developed a very efficient parametric continuation and bifurcation analysis module (rely on the fully numerical evaluation of the Jacobians) able to compute one-parameter bifurcation curves and identifying Fold and Hopf bifurcation points starting from a classical predictor-corrector continuation algorithm [1,4]. Some basic elements of this module have been specifically designed to overcome the difficulties in dealing with reaction mechanisms with several hundreds of species and thousands of chemical reactions [1]. Cantera [13] libraries have been integrated into the module for the management of kinetic, thermodynamic and transport data. The algorithm was applied to the study of methane (53 species and 325 reactions), Jet-A (482 species and 19,072 reactions) [1] and n-dodecane (451 species and 17,848) [3] combustion in a PSR. Kooshkbaghy et al. [20] have obtained a tool able to compute both one-parameter and two-parameter bifurcation curves by coupling AUTO-07p [9] and Chemkin [18]. It was used to the study of n-heptane oxidation describer by using a reduced mechanism consisting of 149 species in 669 reactions.

Despite, all of these significant achievements, there are also still aspects that must be deepened in this topic. Particularly, the detection of bifurcation points appears to be crucial. Therefore, the present work discusses possible improvements on the formulation of the test functions for the identification of saddle-node bifurcation points starting from the approach proposed in [1,4], Then, the proposed tool was applied to the complete bifurcation analysis of three different reaction mechanisms, with increasing number of species and reactions in order to illustrate the improvements suggested.

2 Mathematical Problem Setup

The problem that must be faced in this paper consists of the complete bifurcation analysis of a combustion process (modelled by using large detailed reaction mechanisms) in a PSR. The governing equations of an unsteady adiabatic constant pressure and constant volume PSR can be written as [11]:

$$\frac{dY_j}{dt} = \frac{Y_{j,f} - Y_j}{\tau} + \frac{W_j r_j}{\rho V}, \quad j = 1, 2, \ldots, N_s \tag{1}$$

$$\frac{dT}{dt} = \sum_{j=1}^{N_s} \left(\frac{Y_{j,f}(h_{j,f} - h_j)}{c_p \tau} - \frac{h_j W_j r_j}{\rho V c_p} \right) \tag{2}$$

The system of Eqs. 1–2 can be recast in the form:

$$\frac{d\mathbf{x}}{dt} = \mathbf{f}(\mathbf{x}, \alpha) \tag{3}$$

where $\mathbf{x} = [Y_1, \ldots, Y_{N_s}, T]$ is the state vector, $\mathbf{f}$ is the right-hand side of the system given by Eqs. (1–2) and α is a system parameter (e.g. residence time, pressure, temperature of feeding reactant mixture, etc.).

The bifurcation analysis of the system Eq. (3) consists of determining the steady state solutions (*equilibrium points* or simply the *equilibria*) as function of the parameter α and locating the bifurcation points, i.e. the equilibria in which qualitative changes in the dynamics of the system occur (for further details see [21]).

Computing equilibria is equivalent to find the curve, called *equilibrium curve*, implicitly defined by:

$$F(\mathbf{X}) = 0, \quad F : \mathbb{R}^{n+1} \to \mathbb{R}^n \tag{4}$$

with $n = N_S + 1$, $\mathbf{X} = (\mathbf{x}, \alpha)$ and $F(\mathbf{X}) = \mathbf{f}(\mathbf{x}, \alpha)$.

The Eq. (4) is an example of an algebraic *continuation problem*. Its numerical solution is a sequence of points:

$$\mathbf{X}_1, \mathbf{X}_2, \ldots, \tag{5}$$

approximating the equilibrium curve with desired accuracy [21]. This sequence starts from a known equilibrium point that can be found at some fixed parameter value by numerical integration. This continuation problem is solved by adopting the predictor-corrector (PCM) method introduced in [1, 4]:

$$\mathbf{X}_i \xrightarrow{predictor} \tilde{\mathbf{X}}_{i+1} \xrightarrow{corrector} \mathbf{X}_{i+1}$$

However, Eq. (4) does not define a well-posed mathematical problem because the number of equations is lower than the number of the unknowns. Therefore, an extra scalar equation is appended to the system (4) in order to obtain a well-posed problem:

$$\mathbf{F}(\mathbf{X}) = \begin{bmatrix} F(\mathbf{X}) \\ p(\mathbf{X}) \end{bmatrix} = 0 \tag{6}$$

p represents the equation of a hyperplane passing through $\tilde{\mathbf{X}}_{i+1}$ that is orthogonal to the normalized vector $\mathbf{v}_i$ tangent to the equilibrium curve in the point $\mathbf{X}_i$ [19, 21]:

$$\langle \mathbf{X} - \tilde{\mathbf{X}}_{i+1}, \mathbf{v}_i \rangle = 0 \tag{7}$$

Locating the bifurcation points is addressed by studying the eigenvalues $\lambda = a + ib$ of the Jacobian matrix:

$$\mathbf{J_f} = \frac{\partial \mathbf{f}(\mathbf{x}, \alpha)}{\partial \mathbf{x}} \tag{8}$$

For system (3) only two different bifurcations can be detected: the *Fold* bifurcation also knows as *Saddle-Node* bifurcation or *Turning Point*, associated with the existence of a null eigenvalue and the *Hopf* (also known as *Andronov-Hopf*) bifurcation, corresponding to the presence of a pair of purely imaginary eigenvalues [21].

The ordinary approach adopted to detect these bifurcations is based on the monitoring of two functions, one for Fold and one for Hopf bifurcation [35], called *test functions* (ψ), defined in such a way as to change sign across the bifurcation point [14].

3   Test Functions

The choice of the test functions is the main topic of this work. They can be simply formulated based on the definition of the bifurcation points [21,35] to be zero when:

– a real eigenvalue is zero - Fold:

$$\psi_F(\mathbf{X}) = \prod_{i=1}^{n}(\lambda_i(\mathbf{X})) \tag{9}$$

– a pair of eigenvalues is purely complex - Hopf:

$$\psi_H(\mathbf{X}) = \prod_{1 \leq i \leq j \leq n}(\lambda_i(\mathbf{X}) + \lambda_j(\mathbf{X})) \tag{10}$$

To avoid to directly deal with the eigenvalues of the Jacobian matrix Eq. (8), these functions are usually rewritten by considering that the product of the eigenvalues of a matrix is equal to its determinant [12] and the Stéphanos theorem [21]:

$$\psi_F(\mathbf{X}) = \det(\mathbf{J_f}(\mathbf{X})) \tag{11}$$

$$\psi_H(\mathbf{X}) = \det(2\mathbf{J_f}(\mathbf{X}) \odot \mathbf{I}) \tag{12}$$

where $\mathbf{I}$ is the identity matrix and $\odot$ is the symbol of the *bialternate matrix product* (for further details see [14]).

This form of the test functions Eqs. (11)–(12) is popular but it is not immediately suitable for the problem at hand. Indeed, as it is discussed in [1], two issues can arise when large reaction mechanisms are adopted. First of all, in both test functions the determinant may be too large to be represented by conventional floating-point values (overflow). Furthermore, the test function Eq. (12) introduces a limit on the number of species in the reaction mechanism because the resulting matrix of the bialternate product, a square matrix with size $m = \frac{1}{2}n(n-1)$ and $N_{nz} = \mathcal{O}(n^3)$ nonzero elements [14] ($n = N_s + 1$), may require more memory than available to be stored. For instance, using Matlab in a 64-bit system environment, $(8+8)N_{nz} + 8(m+1)$ bytes are needed to store this matrix in memory as sparse while only $8n^2$ bytes are required to store the matrix $\mathbf{J_f}$ in memory as dense.

Despite the first issue can be faced by rescaling the elements of the Jacobian matrix, the most effective approach to overcome both issues is to introduce new test functions.

The test function for the fold bifurcation Eq. (11) can be rewritten by considering the LUP decomposition of the Jacobian matrix $\mathbf{J_f} = \mathbf{P}^{-1}\mathbf{LU}$, as:

$$\psi_F(\mathbf{X}) = (-1)^s \prod_{i=1}^{n} u_{ii} \tag{13}$$

Indeed, it results that:

- $\det(\mathbf{P}^{-1}) = (-1)^s$, where s is the number of row permutations performed during LUP factorization;
- $\det(\mathbf{L}) = 1$ if the Doolittle's factorization algorithm [34] is adopted;
- $\det(\mathbf{U}) = \prod_{i=1}^{n} u_{ii}$, being $\det(\mathbf{U})$ a upper triangular matrix (u_{ii} are the diagonal elements of the matrix $\mathbf{U}$).

Since a bifurcation point is detected if:

$$\psi(\mathbf{X}_i)\,\psi(\mathbf{X}_{i+1}) < 0 \tag{14}$$

only the sign of the test function is important. To locate the bifurcation point with the desired accuracy, the procedure suggested in [1] is considered. Therefore, the test function Eq. (13) can be modified as suggested in [2,35]:

$$\psi_F(\mathbf{X}) = (-1)^s \prod_{i=1}^{n} \text{sign}(u_{ii}) \tag{15}$$

The procedure to compute this test function is described in Algorithm 1. The pseudocode is written by following the conventions reported in [6]. The matrix operations are not explicitly described by *for* cycles but they have interpretations similar to those in Matlab.

Algorithm 1. Test function for Fold bifurcations based on LUP decomposition

```
1:  procedure SGNDETLUP(A)
2:      n ← rows[A]
3:      sgndet ← 1
4:      for i ← 1 to n − 1 do
5:          cp ← 1
6:          ndx ← FINDMAX(ABS(A[i..n, i]))
7:          if ndx ≠ 1 then
8:              cp ← −1
9:              pivot ← A[i, i..n]
10:             A[i, i..n] ← A[ndx + (i − 1), i..n]
11:             A[ndx + (i − 1), i..n] ← pivot
12:         end if
13:         j ← i + 1
14:         AV ← A[j..n, i]/A[i, i]
15:         A[j..n, j..n] ← A[j..n, j..n] − AV · A[i, j..n]
16:         sgndet ← sgndet · cp· SGN(A[i, i])
17:     end for
18:     sgndet ← sgndet· SGN(A[n, n])
19:     return sgndet
20: end procedure
```

In the Algorithm 1, FINDMAX, ABS and SGN are the functions used to find the index of the greatest values in a vector, the absolute value and the sign function, respectively.

As suggested by [16] to mitigate the memory requirements of the test functions for computing Hopf bifurcations (Eq. 12), it is possible to exploit the sparsity pattern of the bialternate product matrix. The pattern of the bialternate product matrix obtained by considering the reduction of $\mathbf{J_f}$ to Hessenberg form appears particularly interesting to this purpose (see Fig. 1) [16].

The structure in Fig. 1 is exploited to compute directly the sign of the determinant of the bialternate product matrix in Eq. (12) without needing to store in memory the whole matrix. The procedure here proposed is described in Algorithm 2, where the functions HESS, SQRT and FINDMIN are used for: computing the Hessenberg form of the input matrix, for computing the square root and for finding the index of the lowest value in a vector, respectively. NZSDH finds the row index of the non-zero element below the first sub-diagonal of the bialternate product matrix (see Fig. 1). From the pattern in Fig. 1 it is deduced that the index of the row containing this element of the matrix follows the numerical sequence defined in [38] (2, 4, 5, 7, 8, 9, 11, 12, 13, 14, 16 . . .). PERSEQ(i, p) is a procedure based on the numerical sequence defined in [39] that returns the element i-th of the *simple periodic sequence* with period p (formula $x = 1 + mod(i, p)$). The function BPROD2AI is described in the Algorithm 3. In the Algorithm 3, the function ROUND rounds to nearest decimal or integer and the function TIMES performs the element-wise multiplication (Hadamard product).

Algorithm 2. Test function for Hopf bifurcations based on LUP decomposition

---

1: **procedure** SGNDETBALUP(A)
2:     $H \leftarrow$ HESS(A)
3:     $nH \leftarrow rows[H]$
4:     $nBH \leftarrow nH \cdot (nH - 1)/2$
5:     $Ak \leftarrow$ BPROD2AI($H, 1, 1..nBH$)
6:     $nzv \leftarrow$ NZSDH($\langle 1, 2, \ldots, nBH - 1 \rangle$)
7:     $maxd \leftarrow$ FINDMIN(ABS($nzv - nBH$))
8:     $Mnr \leftarrow nzv[maxd] - maxd$
9:     Allocate rectangular matrix M of size $Mnr \times nBH$
10:     $nrr \leftarrow 0$
11:     $nfr \leftarrow 1$
12:     $ndxnz\_old \leftarrow 0$
13:     $sgndet \leftarrow 1$
14:     **for** $k \leftarrow 1$ **to** $nBH - 1$ **do**
15:         $cp \leftarrow 1$
16:         $ndxbnda \leftarrow$ MAX($\langle k + 1, ndxnza\_old + 1 \rangle$)
17:         $nzr \leftarrow$ MIN($\langle nzv[k], nBH \rangle$)
18:         $ndxbndb \leftarrow$ MAX($\langle k + 1, nzr \rangle$)
19:         **if** $ndxbnda \leq ndxbndb$ **then**
20:             $Akadd \leftarrow$ BPROD2AI($H, ndxbnda : ndxbndb, k : nBH$)
21:             $nr\_ad \leftarrow rows[Akadd]$
22:             $vec \leftarrow \langle nfr - nrr, \ldots, nfr + nr\_ad - 1 \rangle$
23:             $nfr \leftarrow nfr + nr\_ad$
24:             $insrows \leftarrow$ PERSEQ($vec[nrr + 1..length(vec)] - 1, Mnr$)
25:             $subM[insrows, k..nBH] = Akadd$
26:             $nrr \leftarrow nrr + nr\_ad$
27:             $ndxnza\_old \leftarrow ndxbndb$
28:         **else**
29:             $vec \leftarrow \langle nfr - nrr, \ldots, nfr - 1 \rangle$
30:         **end if**
31:         $actrows \leftarrow$ PERSEQ($vec - 1, Mnr$)
32:         $Ap \leftarrow \langle Ak[1], M[actrows, k] \rangle$
33:         $ndxp \leftarrow$ FINDMAX(ABS(Ap))
34:         **if** $ndxp \neq 1$ **then**
35:             $cp \leftarrow -1$
36:             $pivot \leftarrow Ak$
37:             $Ak \leftarrow M[actrows[ndxp - 1], k..nBH]$
38:             $M[actrows[ndxp - 1], k..nBH] \leftarrow pivot$
39:         **end if**
40:         **if** $Ak[1] \neq 0$ **then**
41:             $MM \leftarrow M[actrows, k]/Ak[1]$
42:             $ndxnz \leftarrow$ FINDNONZERO(Ak)
43:             $M[actrows, k + ndxnz - 1] \leftarrow M[actrows, k + ndxnz - 1] - MM \cdot Ak[1..length(Ak), ndxnz]$
44:         **end if**
45:         $sgndet \leftarrow sgndet \cdot cp \cdot$ SGN($Ak[1]$)
46:         $Ak \leftarrow M[actrows[1], k + 1..length(M)]$
47:         $nrr \leftarrow nrr - 1$
48:     **end for**
49:     $sgndet \leftarrow sgndet \cdot$ SGN(Ak)
50:     **return** $sgndet$
51: **end procedure**

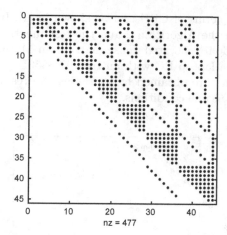

Fig. 1. Sparsity pattern in the bialternate product matrix produced by the Hessenberg form of a dense square matrix of dimension n = 10.

Algorithm 3. Compute the bialternate matrix product $2A \odot I$

1: **procedure** BPROD2AI(H, r, c)
2:     $A \leftarrow 2 \cdot H$
3:     Define I Identity matrix of the same size of A
4:     $i \leftarrow$ ROUND(SQRT($2 \cdot r$) + 1)
5:     $k \leftarrow r -$ TIMES(($i - 2$), ($i - 1$)/2)
6:     $j \leftarrow$ ROUND(SQRT($2 \cdot c$) + 1)
7:     $l \leftarrow c -$ TIMES(($j - 2$), ($j - 1$)/2)
8:     $crc \leftarrow$ (TIMES($A[i,j], I[k,l]$) $-$ TIMES($A[i,l], I[k,j]$) + TIMES($I[i,j], A[k,l]$) $-$
   TIMES($I[i,l], A[k,j]$))/2
9: **end procedure**

The proposed method reduces drastically the memory requirement of the test function Eq. 12. For example, a Jacobian matrix of size 812 leads to a memory requirement to store the result of the bialternate product of about 8 GB (storing it as a sparse matrix) while it requires only 2 GB if the test function Eq. 12 is computed by using the Algorithm 2. Despite this remarkable result, this approach is not effective in terms of CPU cost.

To find a computationally attractive algorithm in terms of computational cost for chemical systems involving large detailed reaction mechanisms, it is needed to exploit the eigenvalues. Therefore, a different Hopf test function can be defined as done in [1]:

$$\psi_H(\mathbf{X}) = \prod_{1 \leq i \leq j \leq n} \text{sign}(\lambda_i(\mathbf{X}) + \lambda_j(\mathbf{X})) \qquad (16)$$

The test function Eq. 16 force to compute the eigenvalues of the Jacobian matrix but it has low memory requirements (only the vector containing the eigenvalues must be stored in addition to the Jacobian matrix) and it has

better performance in terms of CPU cost as shown in [1]. In Fig. 2 is reported
a comparison between the Algorithm 2 and the test function Eq. 16. The tests
are conducted on random generated dense matrices in Matlab computing environment by adopting a laptop PC with an Intel i7-8550U with 16 GB of RAM
memory. All the results are obtained by averaging the results coming from 10
runs of the code in the same conditions.

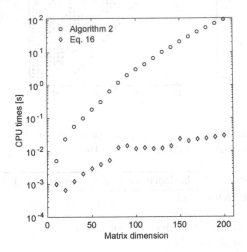

Fig. 2. CPU times (in seconds) required for computing Hopf test function by using
Algorithm 2 and Eq. 16.

The results reported in Fig. 2 clearly show that the function based on Eq. 16
increase significantly the performance of the bifurcation analysis. A further
advantage of this approach is that the same eigenvalues can be promptly adopted
for the location of the Fold bifurcations [1]:

$$\psi_F(\mathbf{X}) = \prod_{i=1}^{n} \mathrm{sign}(\lambda_i(\mathbf{X})) \tag{17}$$

4   Study Case

This study case consists in the computation and analysis of the equilibrium
curve, the stability of equilibrium points and the bifurcation behavior of a stoi-
chiometric premixed n-heptane/air mixture in PSR (Eqs. 1 and 2). This case was
similar to that studied by Kooshkbaghi et al. [20] but in their paper the authors
performed the location of bifurcation point only by using a reduced reaction
mechanism. Therefore, the purpose of this study is to demonstrate the ability of
the algorithm to deal with a reaction mechanism with hundreds of species and
thousands of reactions.

The chemical kinetic mechanism for the study of the oxidation of n-heptane in flow reactors, shock tubes and rapid compression machines developed by Curran et al. [7,28,29] (Version 3) is used in order to stress the method in conditions relevant for practical applications. This mechanism consists of 654 species and 2827 reactions.

The residence time is chosen as the bifurcation parameter. The reactor operates at 1013250 Pa (10 atm) and it is fed with a fuel-air mixture at 700 K. Air is defined as 21% O_2 and 79% N_2. The complete analysis is obtained in about 31 min of execution time on the same laptop described in the previous paragraph and the results are reported in Fig. 3.

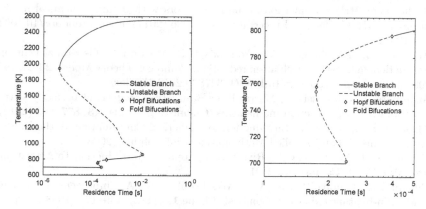

Fig. 3. Reactor temperature versus residence time. The dashed lines identify the unstable branches. The plot on the right is a magnification of the region of first ignition.

The bifurcation diagram in figure Fig. 3 shows three stable branches: a weakly or non-reacting branch, a *cool flame* branch, and a strongly burning branch. The *cool flame* branch is typical of hydrocarbons displaying multi-stage ignitions and the Negative Temperature Coefficient (NTC) regime [20]. This branch is broken by an unstable branch between two Hopf bifurcations revealing the existence of oscillatory states. Similar results are reported in [20] for the pressure of 101325 Pa (1 atm).

5   Conclusions

This work deepens some aspects of locating the bifurcation points in systems of equations arising from the adoption of large, detailed chemical mechanisms. It introduces a new algorithm able to reduce the memory requirement of the bialternate product matrix by exploiting the sparse structure of this matrix. However, it is shown that the execution time of this algorithm is not affordable when it is adopted repetitively in the analysis of equilibrium points. Instead, by dealing with the direct evaluation of eigenvalues when a large detailed reaction mechanism is considered becomes effective both in terms of computational effort and memory requirements. Finally, the bifurcation analysis method proposed is used to study the bifurcation behavior of a n-heptane/air mixture in a PSR with an entire detailed mechanism, demonstrating the effectiveness of the approach.

References

1. Acampora, L., Marra, F.: Numerical strategies for the bifurcation analysis of perfectly stirred reactors with detailed combustion mechanisms. Comput. Chem. Eng. **82**, 273–282 (2015). https://doi.org/10.1016/j.compchemeng.2015.07.008
2. Acampora, L., Marra, F.: Numerical strategies for detection of bifurcation points in the parametric continuation of model reactors with detailed chemical mechanisms. In: AIP Conference Proceedings, vol. 1906 (2017). https://doi.org/10.1063/1.5012382
3. Acampora, L., Kooshkbaghi, M., Frouzakis, C.E., Marra, F.S.: Generalized entropy production analysis for mechanism reduction. Combust. Theory Model. **23**(2), 197–209 (2019). https://doi.org/10.1080/13647830.2018.1504990
4. Acampora, L., Mancusi, E., Marra, F.S.: Bifurcation analysis of perfectly stirred reactors with large reaction mechanisms. Chem. Eng. Trans. **43**, 877–882 (2015)
5. Allgower, E.L., Georg, K.: Introduction to Numerical Continuation Methods. Society for Industrial and Applied Mathematics, Philadelphia (2003)
6. Cormen, T.H., Leiserson, C.E., Rivest, R.L., Stein, C.: Introduction To Algorithms. The MIT Press, Cambridge (2001)
7. Curran, H.J., Gaffuri, P., Pitz, W.J., Westbrook, C.K.: A comprehensive modeling study of n-heptane oxidation. Combust. Flame **114**(1–2), 149–177 (1998). https://doi.org/10.1016/S0010-2180(97)00282-4
8. Dhooge, A., Govaerts, W., Kuznetsov, Y.A.: MATCONT: a MATLAB package for numerical bifurcation analysis of ODEs. ACM Trans. Math. Softw. **29**(2), 141–164 (2003). https://doi.org/10.1145/779359.779362
9. Doedel, E.J.: AUTO: a program for the automatic bifurcation analysis of autonomous systems. Congr. Numer. **30**, 265–284 (1981)
10. Fargione, J., Hill, J., Tilman, D., Polasky, S., Hawthorne, P.: Land clearing and the biofuel carbon debt. Science **319**(5867), 1235–1238 (2008). https://doi.org/10.1126/science.1152747
11. Glarborg, P., Miller, J.A., Kee, R.J.: Kinetic modeling and sensitivity analysis of nitrogen oxide formation in well-stirred reactors. Combust. Flame **65**(2), 177–202 (1986). https://doi.org/10.1016/0010-2180(86)90018-0
12. Golub, G., Van Loan, C.: Matrix Computations. Johns Hopkins Studies in the Mathematical Sciences. Johns Hopkins University Press, Baltimore (2013)
13. Goodwin, D.G.: An open-source, extensible software suite for CVD process simulation. In: Allendorf, M., Maury, F., Teyssandier, F. (eds.) Chemical Vapor Deposition XVI and EUROCVD 14, vol. 2003-08, pp. 155–162. The Electrochemical Society ECS, Pennington (2003)
14. Govaerts, W.: Numerical bifurcation analysis for ODEs. J. Comput. Appl. Math. **125**(1–2), 57–68 (2000). https://doi.org/10.1016/S0377-0427(00)00458-1
15. Govaerts, W., Kuznetsov, Y.A., Dhooge, A.: Numerical continuation of bifurcations of limit cycles in MATLAB. SIAM J. Sci. Comput. **27**(1), 231–252 (2005). https://doi.org/10.1137/030600746

16. Guckenheimer, J., Myers, M., Sturmfels, B.: Computing hopf bifurcations I. SIAM J. Numer. Anal. **34**(1), 1–21 (1997)
17. Kalamatianos, S., Vlachos, D.G.: Bifurcation behavior of premixed hydrogen/air mixtures in a continuous stirred tank reactor. Combust. Sci. Technol. **109**(1–6), 347–371 (1995). https://doi.org/10.1080/00102209508951909
18. Kee, R.J., Rupley, F.M., Meeks, E., Miller, J.A.: CHEMKIN-III: a FORTRAN chemical kinetics package for the analysis of gas-phase chemical and plasma kinetics. Technical report, SAND-96-8216, Sandia National Laboratories (1996)
19. Keller, H.: Lectures on Numerical Methods in Bifurcation Problems. Lectures on Mathematics and Physics. Springer, Berlin (1987). Published for the TATA Institute of Fundamental Research
20. Kooshkbaghi, M., Frouzakis, C.E., Boulouchos, K., Karlin, I.V.: n-Heptane/air combustion in perfectly stirred reactors: dynamics, bifurcations and dominant reactions at critical conditions. Combust. Flame **162**(9), 3166–3179 (2015). https://doi.org/10.1016/j.combustflame.2015.05.002
21. Kuznetsov, Y.A.: Elements of Applied Bifurcation Theory. Applied Mathematical Sciences. Springer, Berlin (1998). https://doi.org/10.1007/b98848
22. Lengyel, I., West, D.H.: Numerical bifurcation analysis of large-scale detailed kinetics mechanisms. Curr. Opin. Chem. Eng. **21**, 41–47 (2018). https://doi.org/10.1016/j.coche.2018.02.013
23. Lindstedt, R.P., Maurice, L.Q.: Detailed kinetic modelling of n-Heptane combustion. Combust. Sci. Technol. **107**(4–6), 317–353 (1995). https://doi.org/10.1080/00102209508907810
24. Lu, T., Ju, Y., Law, C.K.: Complex CSP for chemistry reduction and analysis. Combust. Flame **126**(1–2), 1445–1455 (2001). https://doi.org/10.1016/S0010-2180(01)00252-8
25. Lu, T., Law, C.K.: A directed relation graph method for mechanism reduction. Proc. Combust. Inst. **30**(1), 1333–1341 (2005). https://doi.org/10.1016/j.proci.2004.08.145
26. Lu, T., Law, C.K.: A criterion based on computational singular perturbation for the identification of quasi steady state species: a reduced mechanism for methane oxidation with NO chemistry. Combust. Flame **154**(4), 761–774 (2008). https://doi.org/10.1016/j.combustflame.2008.04.025
27. Lu, T., Law, C.K.: Toward accommodating realistic fuel chemistry in large-scale computations. Prog. Energy Combust. Sci. **35**(2), 192–215 (2009). https://doi.org/10.1016/j.pecs.2008.10.002
28. Mehl, M., Pitz, W.J., Sjöberg, M., Dec, J.E.: Detailed kinetic modeling of low-temperature heat release for PRF fuels in an HCCI engine.Technical Paper 2009-01-1806, SAE International, June 2009. https://doi.org/10.4271/2009-01-1806
29. Mehl, M., Pitz, W.J., Westbrook, C.K., Curran, H.J.: Kinetic modeling of gasoline surrogate components and mixtures under engine conditions. Proc. Combust. Inst. **33**(1), 193–200 (2011). https://doi.org/10.1016/j.proci.2010.05.027
30. Metcalfe, W.K., Dooley, S., Dryer, F.L.: Comprehensive detailed chemical kinetic modeling study of toluene oxidation. Energy Fuels **25**(11), 4915–4936 (2011). https://doi.org/10.1021/ef200900q
31. Olsen, R.J., Vlachos, D.G.: A complete pressure-temperature diagram for air oxidation of hydrogen in a continuous-flow stirred tank reactor. J. Phys. Chem. A **103**(40), 7990–7999 (1999). https://doi.org/10.1021/jp991148b
32. Park, Y.K., Vlachos, D.G.: Isothermal chain-branching, reaction exothermicity, and transport interactions in the stability of methane/air mixtures*. Combust. Flame **114**(1–2), 214–230 (1998). https://doi.org/10.1016/S0010-2180(97)00285-X

33. Park, Y.K., Vlachos, D.G.: Kinetically driven instabilities and selectivities in methane oxidation. AIChE J. **43**(8), 2083–2095 (1997). https://doi.org/10.1002/aic.690430816

34. Quarteroni, A., Sacco, R., Saleri, F.: Numerical Mathematics. Texts in Applied Mathematics. Springer, New York (2017)

35. Seydel, R.: Practical Bifurcation and Stability Analysis, Interdisciplinary Applied Mathematics, vol. 5. Springer, New York (2010). https://doi.org/10.1007/978-1-4419-1740-9

36. Shan, R., Lu, T.: Ignition and extinction in perfectly stirred reactors with detailed chemistry. Combust. Flame **159**(6), 2069–2076 (2012). https://doi.org/10.1016/j.combustflame.2012.01.023

37. Shan, R., Lu, T.: A bifurcation analysis for limit flame phenomena of DME/air in perfectly stirred reactors. Combust. Flame **161**(7), 1716–1723 (2014). https://doi.org/10.1016/j.combustflame.2013.12.025

38. Sloane, N.J.A.: The on-line encyclopedia of integer sequences (2019). https://oeis.org/A014132

39. Sloane, N.J.A.: The on-line encyclopedia of integer sequences (2019). http://oeis.org/A010883

40. Uppal, A., Ray, W.H., Poore, A.B.: On the dynamic behavior of continuous stirred tank reactors. Chem. Eng. Sci. **29**(4), 967–985 (1974). https://doi.org/10.1016/0009-2509(74)80089-8

41. Uppal, A., Ray, W.H., Poore, A.B.: The classification of the dynamic behavior of continuous stirred tank reactors—influence of reactor residence time. Chem. Eng. Sci. **31**(3), 205–214 (1976). https://doi.org/10.1016/0009-2509(76)85058-0

42. Valorani, M., Creta, F., Goussis, D.A., Lee, J.C., Najm, H.N.: An automatic procedure for the simplification of chemical kinetic mechanisms based on CSP. Combust. Flame **146**(1–2), 29–51 (2006). https://doi.org/10.1016/j.combustflame.2006.03.011

43. Westbrook, C.K., Pitz, W.J., Herbinet, O., Curran, H.J., Silke, E.J.: A comprehensive detailed chemical kinetic reaction mechanism for combustion of n-alkane hydrocarbons from n-octane to n-hexadecane. Combust. Flame **156**(1), 181–199 (2009). https://doi.org/10.1016/j.combustflame.2008.07.014

44. Westbrook, C.K., et al.: Detailed chemical kinetic reaction mechanisms for soy and rapeseed biodiesel fuels. Combust. Flame **158**(4), 742–755 (2011). https://doi.org/10.1016/j.combustflame.2010.10.020

Stability Analysis of DESA Optimization Algorithm

Rizavel C. Addawe[1]([✉]) [iD] and Joselito C. Magadia[2] [iD]

[1] University of the Philippines Baguio, Baguio, Philippines
rcaddawe@up.edu.ph
[2] University of the Philippines School of Statistics,
Diliman, Quezon City, Philippines
jcmagadia@up.edu.ph

Abstract. This paper investigates the dynamics of the hybrid evolutionary optimization algorithm, Differential Evolution-Simulated Annealing (DESA) algorithm with the binomial crossover and SA-like selection operators. A detailed mathematical framework of the operators of the DESA/rand/1/bin algorithm is provided to characterize the behavior of the DESA-population system. In DESA, the SA-like selection operation provides a nonzero probability of accepting a deteriorated solution that decreases with a sufficient number of generations. This paper shows that the system defined by the DESA-population is stable. Moreover, the DESA-population system time constant, learning and momentum rates are dependent on the value of the crossover constant and the probability of accepting deterioration in the quality of the objective function.

Keywords: Differential evolution - simulated annealing · Stability analysis · Lyapunov's theorem

1 Introduction

Differential Evolution (DE) [1] is one of the best genetic types of Evolutionary Algorithms (EAs) [2,3] for solving problems with the real-valued variables from diverse areas of science and technology, engineering, and economic [4]. From a theoretical point of view, the combination of the search properties of stochastic algorithms in the development of a hybrid algorithm that is equally applicable and has searching ability and power to reach the optimal solution is of importance to many in the growing population of machine-learning researchers. Also, the analysis of increasingly many areas of application generate data requires theory and robust methods that consistently find the optimal fit to the data. Furthermore, there is also a great need for methods that can run with minimal need for manual human input. Only a number of theoretical studies on the DE-population include time complexity [5], dynamical behavior [6–10] and convergence properties [11–14].

Whether or not an annealed version of DE or a combination of Simulated Annealing (SA) [15] and DE is of practical use can contribute to the growth of

© Springer Nature Switzerland AG 2020
Y. D. Sergeyev and D. E. Kvasov (Eds.): NUMTA 2019, LNCS 11974, pp. 17–31, 2020.
https://doi.org/10.1007/978-3-030-40616-5_2

DE algorithm [16]. Indeed, developing the DESA algorithm [17] for fitting high-dimensional functions was a worthy topic for several reasons. Despite already promising results, DE is still in its infancy and can most probably be improved. Hence, theoretical analysis of DE and its hybrid is also necessary to understand its search process, to detect the allowable ranges of its control parameters, and to find problem classes in which a given set of parameters will perform successfully or will fail [4].

There is therefore a need for methods that robustly obtain the optimal fit to the data. Stochastic nature-inspired meta-heuristics optimization algorithms [1,18–21] have been the interest of a community of engineers due to their simplicity and adaptability in solving real-life problems. On the other hand, deterministic mathematical programming methods [22–26] are actively studied in the academe due to their interesting theoretical convergence properties. In a study done in [24], results show that both stochastic nature-inspired meta-heuristics and deterministic global optimization methods are competitive and surpass each other in dependence on the available budget of function evaluations.

Theoretical analysis of the EA and its convergence analysis has been an important research topic in the evolutionary community. Markovian stochastic process [27–31] and global random search [32,33] are two theoretical frameworks that have been used to model the evolutionary process. Despite some theoretical analysis of the DE behavior, the theory of DE is still behind the empirical studies.

In this paper, we investigate the stability of the DESA algorithm, a DE based algorithm with the SA-like selection operator. Section 2 gives a detailed algorithmic framework of the binomial crossover and SA-like selection components of the DESA algorithm. Section 3 provides the computations of the mean and variance of the DESA-population as it goes through a number of mutations, crossovers and selections. Section 4 shows that the system defined by the DESA-population is stable. We also establish that an estimate of the DESA-population system time constant can be expressed as γ_β^{-1}. Finally, the learning rate and momentum rate of DESA is compared with those of the classical DE algorithm.

2   The DESA Algorithm

Differential Evolution-Simulated Annealing (DESA) optimization algorithm [17] was developed as an optimization algorithm for high dimensional functions. The combination is done by incorporating a SA-like selection criterion in a DE framework to form the DESA algorithm. DESA has been used in several applications [34–36]. The objective function to be minimized is $f(\vec{x}), \vec{x} = (x_1, \dots, x_D) \in \mathcal{R}^D$ with D number of parameters, and the feasible solution $\Phi = \Pi_{j=1}^{D}[x_{j_{min}}, x_{j_{max}}]$, where $x_{j_{min}}$ and $x_{j_{max}}$ are the lower and upper bounds of the parameter values, respectively.

2.1   The Implementation of the DESA Algorithms

The mathematical framework of the DESA algorithm are detailed as follows:

(1) **Initialization**: Generate an *initial population* denoted by $\vec{x}(0) = (x_{i,j,0})$ for $i = 1, \ldots, Np$ individuals; $j = 1, \ldots, D$ parameters; and, let the number of generation $g \leftarrow 0$. This initial population is generated by assigning random values in the search space to the variables of every solution.

(2) **Reproduction**: Generate a *trial population* $\vec{y}(g)$ from the current population $\vec{x}(g)$.

  (a) *Mutation*: Generate a *mutant population* from $\vec{x}(g)$ by a mutation operator, denoted by $\vec{v}(g)$. The binomial mutation operator, *DE/rand/1/bin*, is given by

$$\vec{v}(g) = x_{r1} + F(x_{r2} - x_{r3}) \tag{1}$$

where $F \in [0,2]$, a constant; while x_{r1}, x_{r2} and x_{r3} are distinct and randomly chosen from the current population $\vec{x}(g) = (x_{i,j,g})$, for $i = 1, \ldots, Np$ individuals, and $j = 1, \ldots, D$ parameters.

  (b) *Crossover*: Generate a *trial population* $\vec{y}(g)$ from $\vec{x}(g)$ and $\vec{v}(g)$ by a crossover operator, denoted by $\vec{u}(g)$. Let $\vec{y}(g) \leftarrow \vec{u}(g)$. The crossover operator is given by,

$$\vec{u}_{i,g} = \begin{cases} \vec{v}_{i,g} = x_{r1} + F(x_{r2} - x_{r3}) \ \text{if} \ rand(0,1) \leq CR, \\ \vec{x}_{i,g} \qquad\qquad\qquad\qquad\qquad\qquad \text{otherwise} \end{cases} \tag{2}$$

where $F \in [0,2]$, and $CR \in [0,1]$.

For the *binomial crossover* we let $\vec{u}_{i,g} = \text{binomial\_crossover}(\vec{v}_{i,g}, \vec{x}_{i,g})$, for $i = 1, \ldots, NP$.

(3) **Selection**: Generate a new population $\vec{z}(g)$ from $\vec{y}(g)$ by a selection operator, denoted by $\vec{x}_{i,g+1}$. Let $\vec{z}(g) \leftarrow \vec{x}_{i,g+1}$. The *SA-like selection* operator is implemented as

$$\vec{x}_{i,g+1} = \begin{cases} \vec{u}_{i,g} \ \text{if} \ f(\vec{u}_{i,g}) < f(\vec{x}_{i,g}) \\ \qquad \text{or} \ [f(\vec{u}_{i,g}) \geq f(\vec{x}_{i,g})] \wedge [rand(0,1) \leq \beta_{i,g}]; \\ \vec{x}_{i,g} \ \text{otherwise} \end{cases} \tag{3}$$

where

$$\beta_{i,g} = \exp\left[-\frac{f(\vec{u}_{i,g}) - f(\vec{x}_{i,g})}{kT} \right] \tag{4}$$

is expected to attain a smaller value in $(0,1)$. $T > 0$ starts high and kT gradually decreases according to the parameter $0 < k < 1$. For the *SA-like selection* we let $\vec{x}_{i,g+1} = \text{SA-like\_selection}(\vec{u}_{i,g}, \vec{x}_{i,g})$, for $i = 1, \ldots, NP$.

(4) **Termination**: If the termination condition is satisfied, then stop; else, let $g + 1 \leftarrow g$ and $\vec{x}(g) \leftarrow \vec{z}(g)$; then go to Step 2.

2.2   The Binomial Crossover Operator of DESA Algorithm

For the DESA algorithm, we use the $DE/rand/1/bin$ [16] mutation. Let x_{r1}, x_{r2} and x_{r3} be three trial solutions picked-up randomly from population $x_{ri}, i = 1,\dots,Np$. Here, we assume that the trial solutions are drawn with replacement such that x_{r1}, x_{r2}, and x_{r3} are independent of each other. Thus, $P[(x_{ri} = x_l) \cap (x_{rj} = x_k)] = P(x_{ri} = x_l)P(x_{rj} = x_k)$ where $i,j = 1,2,3$ and $k,l = 1,\dots,Np$ and $i \neq j$. Hence, from the $DE/rand/1/bin$ mutation operator in (1), DESA may choose $\vec{v}_{i,g}$ with probability (wp) CR. Consequently, we have

$$\vec{u}_{i,g} = \begin{cases} \vec{v}_{i,g} = x_i + F(x_j - x_k) \text{ if } rand(0,1) \leq CR, & \text{wp CR} \\ \vec{x}_{i,g} & \text{otherwise,} \quad \text{wp 1-CR.} \end{cases} \tag{5}$$

2.3   The SA-like Selection Operator of DESA Algorithm

With the SA-like selection of DESA algorithm, we now compute the expected value of the trial solution u_m corresponding to the target solution x_m. Using (3), let α be equal to the true probability of the event that $f(\vec{u}_m) < f(\vec{x}_m)$ which is mutually exclusive from the event that $f(\vec{u}_m) \geq f(\vec{x}_m)$. with probability equal to $1 - \alpha$. Also, $P(rand[0,1] < \beta) = \beta$. Hence, $P[f(\vec{u}_m) \geq f(\vec{x}_m) \cap (rand[0,1] < \beta)] = (1-\alpha)\beta$. So, the probabilities to the mutually exclusive and exhaustive events as follows:

$$\vec{x}_{i,g+1} = \begin{cases} \vec{u}_{i,g} \text{ if } f(\vec{u}_{i,g}) < f(\vec{x}_{i,g}) & \text{wp } \alpha \\ \quad \text{or } f(\vec{u}_{i,g}) \geq f(\vec{x}_{i,g}) \wedge rand[0,1] \leq \beta_i, & \text{wp } (1-\alpha)\beta \\ \vec{x}_{i,g} \text{ otherwise,} & \text{wp } 1 - [\alpha + \beta - \alpha\beta]. \end{cases} \tag{6}$$

If the trial vector $\vec{u}_{i,g}$ has the lower value of the fitness function f, it will survive to the population of the next generation; otherwise, it is subjected to the *Metropolis Criterion* as done in SA [15].

DESA algorithm incorporates a typical *SA-like selection* mechanism that conditionally accepts an inferior solution to the next generation. When $\beta = 0$, (6) assumes the DE selection process such that

$$\vec{x}_{i,g+1} = \begin{cases} \vec{u}_{i,g} & \text{wp } \alpha \\ \vec{x}_{i,g} \text{ otherwise, wp } 1 - \alpha. \end{cases} \tag{7}$$

By the independence of the binomial crossover and SA-like selection operators of DESA, we have the following probability assignments:

$$\vec{x}_{i,g+1} = \begin{cases} \vec{u}_{i,g} & \text{wp } (\alpha + \beta - \alpha\beta)CR \\ \vec{x}_{i,g} & \text{wp } 1 - (\alpha + \beta - \alpha\beta)CR. \end{cases} \tag{8}$$

For ease of notation, let $\gamma_\beta = (\alpha + \beta - \alpha\beta)CR$ denotes the probability of accepting poor solutions in DESA where β is non-zero. Thus,

$$\vec{x}_{i,g+1} = \begin{cases} \vec{u}_{i,g} & \text{wp } \gamma_\beta \\ \vec{x}_{i,g} & \text{wp } 1 - \gamma_\beta. \end{cases} \tag{9}$$

2.4   The Mean and Variance of the DESA-population

The *mean* or the *expected value* of the mth individual in the next DESA-population, denoted as $E(x_{m,g+1})_{DESA}$, is computed as

$$E\left[(x_{m,g+1})_{DESA}\right] = (1 - \gamma_\beta)x_m + \gamma_\beta \mu_x \qquad (10)$$

where $\mu_x = \frac{1}{Np}\sum_{m=1}^{Np} x_m$, the mean of the current population with $m = 1, \ldots, Np$. Similarly, $E(x_{m,g+1}^2)_{DESA}$ is given by

$$E\left[(x_{m,g+1}^2)_{DESA}\right] = (1 - \gamma_\beta)x_m^2 + \gamma_\beta \left[(2F^2 + 1)\sigma_x^2 + \mu_x^2\right]. \qquad (11)$$

Therefore, the *variance* of the DESA-population is given by

$$Var(x_{m,g+1})_{DESA} = \gamma_\beta(1 - \gamma_\beta)(x_m - \mu_x)^2 + \gamma_\beta \left[(2F^2 + 1)\sigma_x^2\right]. \qquad (12)$$

Again, with $\beta = 0$, the probabilities to events of the DESA-population in (6) becomes a classical DE-population. Hence, we have

$$E(x_{m,g+1})_{DE} = (1 - \gamma_0)x_m + \gamma_0 \mu_x, \qquad (13)$$

$$E(x_{m,g+1}^2)_{DE} = (1 - \gamma_0)x_m^2 + \gamma_0(2F^2 + 1)Var(x) + \gamma_0 \mu_x^2, \qquad (14)$$

and

$$Var(x_{m,g+1})_{DE} = \gamma_0(1 - \gamma_0)(x_m - \mu_x)^2 + \gamma_0(2F^2 + 1)\sigma_x^2 \qquad (15)$$

where $\gamma_0 = \alpha CR$.

In DESA, the probability of accepting a set of bad solution, β, decreases exponentially with the badness of the move, which is the amount $f(g+1) - f(g)$. For some sufficient number of generations, or when $x_m = x_{g_{max}}$ for $m = 1, 2, \ldots, Np$, or $x_1 = \ldots = x_m = \mu_x$, then $\sigma_x^2 = 0$, the DESA algorithm converges to the optimal solution. Hence, $f(x_m) - f(x_j)$ becomes negligible $\forall m, j \in \{1, \ldots, Np\}$. DESA-population attains convergence to the optimal solution.

3   Stability Analysis of the DESA Population System

For this section, we show the stability of the DESA population system as shown in [8,9]. In order to validate the analysis, we make certain assumptions, which are enumerated below:

(i) The objective function $f(x)$ is assumed to be of class C^2, derivatives $f^{(1)}$, $f^{(2)}, \ldots, f^{(k)}$ exist and are continuous [37]. In addition, let $f(x)$ be Lipschitz continuous [38], and unimodal in the region of interest.

(ii) The population of Np individual trial solutions are located very close to each other. That is, the parameter vectors gather in a compact cluster around the global optimum during the later stages of the search and especially when the scaling factor $F = 0.5$ [39,40].

(iii) Dynamics is modeled assuming the vectors as search-agents moving in continuous time.

(iv) Assume that the mutation and crossover of DESA occur in unit time to give rise to offsprings. In the SA-like selection of individuals, $\vec{x}_m$ is replaced by $\vec{u}_m$ if the $f(\vec{u}_m) < f(\vec{x}_m)$ or $f(\vec{u}_m) \geq f(\vec{x}_m) \land rand[0,1] < \beta$.

Theorem 1 *Velocity of an individual point, X_m, in* **DESA** *system. If DESA-population may be modeled as a continuous-time dynamic system, then the expected value of the velocity of an individual point on the fitness landscape may be given as:*

$$E\left(\frac{dx_m}{dt}\right) = -\frac{k}{8}\gamma_\beta \left\{(2F^2 + 1)\sigma_x^2 + (\mu_x - x_m)^2\right\} f'(x_m) + \frac{1}{2}\gamma_\beta(\mu_x - x_m) \quad (16)$$

where $\gamma_\beta = (\alpha + \beta - \alpha\beta)CR$, $\alpha = P[f(\vec{u}_m) < f(\vec{x}_m)]$, $\beta = P[rand(0,1) \leq \beta]$, $0 < CR < 1$ is a crossover constant operator, μ_x and σ_x^2 are the mean and variance of $x_m, m = 1, 2, ..., Np$, respectively.

Proof. Let us assume that mutation and crossover occur in unit time to give rise to offspring. In DESA's selection, x_m is replaced by $\vec{u}_m$ if $f(\vec{x}_m) > f(\vec{u}_m)$ or $[f(\vec{x}_m) \leq f(\vec{u}_m) \land rand[0,1] < \beta]$. This decision making is performed using *Heaviside's unit step function* [41], which is defined as follows:

$$u(p) = \begin{cases} 1 & p \geq 0 \\ 0 & otherwise. \end{cases} \quad (17)$$

Hence,

$$\frac{dx_m}{dt} = u\left[-f'(x_m)\frac{dx_m}{dt}\right](u_m - x_m). \quad (18)$$

Now, we replace the unit step function by logistic function: $u(p) = \lim_{k\to\infty} \frac{1}{1+e^{-kp}}$. For the analysis, take a reasonable value of k to get an approximate value of $u(p)$ which is given by $u(p) \approx \frac{1}{1+e^{-kp}}$. Now, with a very small value of p and by neglecting higher order terms, we obtain $e^{-kp} \approx 1 - kp$. Therefore,

$$u(p) \approx \frac{1}{1 + e^{-kp}} \approx \frac{1}{2 - kp} = \frac{1}{2}\left(1 - \frac{kp}{2}\right)^{-1}. \quad (19)$$

Also implying that,

$$u(p) \approx \frac{1}{2} + \frac{k}{4}p. \quad (20)$$

Now, the DESA-population has a small divergence [8]. Therefore, $u_m - x_m$ is not very large, and as $\left|\frac{dx_m}{dt}\right|$ is either 0 or $|u_m - x_m|$. This ensures that $\left|\frac{dx_m}{dt}\right|$ is small. Also, we have assumed that fitness landscape has a moderate slope. That is, $f'(x_m)$ is also small which in turn suggests that $\left|f'(x_m)\frac{dx_m}{dt}\right|$ is small. Thus, from (18) we get, $\frac{dx_m}{dt} = \left[\frac{1}{2} - \frac{k}{4}f'(x_m)\right](u_m - x_m)$. Hence,

$$\frac{dx_m}{dt} = \frac{\frac{1}{2}(u_m - x_m)}{1 + \frac{k}{4}f'(x_m)(u_m - x_m)}. \quad (21)$$

Since $\left|\frac{k}{4}f'(x_m)(u_m - x_m)\right|$ is small, we have

$$\left[1 + \frac{k}{4}f'(x_m)(u_m - x_m)\right]^{-1} \approx 1 - \frac{k}{4}f'(x_m)(u_m - x_m). \tag{22}$$

From (21) we get

$$\frac{dx_m}{dt} = -\frac{k}{8}(u_m - x_m)^2 f'(x_m) + \frac{(u_m - x_m)}{2}. \tag{23}$$

Now, u_m is a random variable. Therefore, $\frac{dx_m}{dt}$ which is a function of u_m is also a random variable. So, we may compute its expected value as follows:

$$E\left(\frac{dx_m}{dt}\right) = -\frac{k}{8}f'(x_m)E(u_m - x_m)^2 + \frac{1}{2}E(u_m - x_m). \tag{24}$$

This implies that

$$E\left(\frac{dx_m}{dt}\right) = -\frac{k}{8}f'(x_m)\left[E(u_m^2) + x_m^2 - 2x_m E(u_m)\right] + \frac{1}{2}\left[E(u_m) - x_m\right]. \tag{25}$$

Now, substitute the values of $E(u_m)$ and $E(u_m^2)$ from (10) and (11), respectively. Then, set $\mu_x = \frac{1}{Np}\sum_{m=1}^{NP} x_m$ and $\sigma_x^2 = Var(x)$ to obtain the expression for the expected value given in (16). This completes the proof.

Theorem 2 *Velocity of the Centroid of DESA system. Let $\mu_x = \frac{1}{Np}\sum_{m=1}^{Np} x_m$ denotes the centroid of all points of the current population, $\mu_{f'} = \frac{1}{Np}\sum_{m=1}^{Np} f'(x_m)$ be the average slope of the fitness landscape, and $\varepsilon_m = \mu_x - x_m$ denotes the deviation of an individual from this centroid. Then, the expected velocity of the centroid of the population is given by*

$$E\left(\frac{d\mu_x}{dt}\right) = -\frac{k}{8}\gamma_\beta(2F^2 + 1)\sigma_x^2\mu_{f'} - \frac{k}{8}\gamma_\beta\left(\frac{1}{Np}\sum_{m=1}^{Np}\varepsilon_m^2 f'(\mu_x + \varepsilon_m)\right) \tag{26}$$

Proof. Since,

$$\mu_x = \frac{1}{Np}\sum_{i=1}^{Np} x_i = \frac{1}{Np}\sum_{m=1}^{Np} x_m, \tag{27}$$

we may now solve for $\frac{d\mu_x}{dt}$ as follows:

$$\frac{d\mu_x}{dt} = \frac{d}{dt}\left(\frac{1}{Np}\sum_{m=1}^{Np} x_m\right) = \frac{1}{Np}\sum_{m=1}^{Np}\frac{dx_m}{dt}. \tag{28}$$

Getting expected value of its derivative with respect to time t (generation g) gives

$$E\left(\frac{d\mu_x}{dt}\right) = E\left(\frac{1}{Np}\sum_{m=1}^{Np}\frac{dx_m}{dt}\right) = \frac{1}{Np}\sum_{m=1}^{Np} E\left(\frac{dx_m}{dt}\right). \tag{29}$$

Substituting the value of $E\left(\frac{dx_m}{dt}\right)$ from (16), we obtain

$$E\left(\frac{d\mu_x}{dt}\right) = \frac{1}{Np}\sum_{m=1}^{Np}\left(-\frac{k}{8}\gamma_\beta\left\{(2F^2+1)\sigma_x^2+\varepsilon^2\right\}f'(x_m)+\frac{1}{2}\gamma_\beta\cdot\varepsilon\right) \quad (30)$$

where $\varepsilon_m = \mu_x - x_m$ denotes the deviation of an individual from the centroid. Hence,

$$E\left(\frac{d\mu_x}{dt}\right) = -\frac{k}{8}\gamma_\beta\left(\frac{1}{Np}\sum_{m=1}^{Np}f'(x_m)\left[(2F^2+1)\sigma_x^2+\varepsilon^2\right]\right)+0 \quad (31)$$

since the sum of all deviations from the mean is zero. Hence,

$$E\left(\frac{d\mu_x}{dt}\right) = -\frac{k}{8}\gamma_\beta\cdot f_{\mu'}\left[(2F^2+1)\sigma_x^2\right]-\frac{k}{8}\gamma_\beta\left(\frac{1}{Np}\sum_{m=1}^{Np}\varepsilon^2 f'(x_m)\right) \quad (32)$$

where $f_{\mu'} = \frac{1}{Np}\sum_{m=1}^{Np}f'(x_m)$. Resetting $x_m = \mu_x - \varepsilon_m$ gives the desired expression in (26). This completes the proof.

As done in [8], to study the stability of DESA algorithm, we model DESA algorithm as an autonomous control system. Here, each population member x_m is a state variable of the control system. For the DESA-population, the expected value of the velocity of an individual point on the fitness landscape is given in (16).

Assuming the DESA-population to be concentrated into a small neighborhood around an optimum in a flatter portion of the function, we have $|f'(x_m)| \ll 1$. Hence, the equation can be written as,

$$E\left(\frac{dx_m}{dt}\right) = \frac{1}{2}\gamma_\beta(\mu_x - x_m) \text{ for } m = 1, 2, \ldots, Np. \quad (33)$$

Using (27) we now rewrite the above expectation as

$$E\left(\frac{dx_m}{dt}\right) = \frac{1}{2}\gamma_\beta\left(\frac{1}{Np}\sum_{i=1}^{Np}x_i - x_m\right) \text{ for } i = 1, 2, \ldots, Np. \quad (34)$$

Hence, (34) represents Np number of simultaneous equations:

$$E\left(\frac{dx_1}{dt}\right) = \frac{1}{2}\gamma_\beta\left(\frac{1}{Np}\sum_{i=1}^{Np}x_i - x_1\right); \quad (35)$$

$$\vdots$$

$$E\left(\frac{dx_{Np}}{dt}\right) = \frac{1}{2}\gamma_\beta\left(\frac{1}{Np}\sum_{i=1}^{Np}x_i - x_2\right) \text{ for } i = 1, 2, \ldots, Np. \quad (36)$$

So, we may now represent them using this matrix notation:

$$
\begin{bmatrix}
E\left(\frac{dx_1}{dt}\right) \\
E\left(\frac{dx_2}{dt}\right) \\
\vdots \\
E\left(\frac{dx_{Np}}{dt}\right)
\end{bmatrix}
= \frac{1}{2}\gamma_\beta
\begin{bmatrix}
\frac{1}{Np}-1 & \frac{1}{Np} & \cdots & \frac{1}{Np} \\
\frac{1}{Np} & \frac{1}{Np}-1 & \cdots & \frac{1}{Np} \\
\vdots & \vdots & \ddots & \vdots \\
\frac{1}{Np} & \frac{1}{Np} & \cdots & \frac{1}{Np}-1
\end{bmatrix}
\cdot
\begin{bmatrix}
x_1 \\
x_2 \\
\vdots \\
x_{Np}
\end{bmatrix}
\tag{37}
$$

The above matrix equation is of the form $\left[E\left(\frac{d\vec{x}}{dt}\right)\right] = A\left[\vec{x}\right]$, where $\vec{x} = [x_1, x_2, \ldots, x_{Np}]^T$ is the set of state variables and

$$
A = \frac{1}{2}\gamma_\beta
\begin{bmatrix}
\frac{1}{Np}-1 & \frac{1}{Np} & \cdots & \frac{1}{Np} \\
\frac{1}{Np} & \frac{1}{Np}-1 & \cdots & \frac{1}{Np} \\
\vdots & \vdots & \ddots & \vdots \\
\frac{1}{Np} & \frac{1}{Np} & \cdots & \frac{1}{Np}-1
\end{bmatrix}.
\tag{38}
$$

The eigenvalues of the system-matrix A are those of λ satifying $det\left[\lambda I - A\right] = 0$, where I is the identity matrix of order Np. Now, $det\left[\lambda I - A\right] = 0$ implies that

$$
det
\begin{bmatrix}
\frac{2}{\gamma_\beta}\lambda - \frac{1}{Np} + 1 & -\frac{1}{Np} & \cdots & -\frac{1}{Np} \\
-\frac{1}{Np} & \frac{2}{\gamma_\beta}\lambda - \frac{1}{Np} + 1 & \cdots & -\frac{1}{Np} \\
\vdots & \vdots & \ddots & \vdots \\
-\frac{1}{Np} & -\frac{1}{Np} & \cdots & \frac{2}{\gamma_\beta}\lambda - \frac{1}{Np} + 1
\end{bmatrix}
= 0.
\tag{39}
$$

After doing simple algebraic operations on the rows of the determinant in LHS of (39) we get,

$$
\lambda\left(\lambda + \frac{\gamma_\beta}{2}\right)^{Np-1} = 0,
\tag{40}
$$

which is the characteristic equation of matrix A. Hence, we get the system eigenvalues as: $\lambda = 0, -\frac{\gamma_\beta}{2}, \ldots, -\frac{\gamma_\beta}{2}$. Since one eigenvalue is zero, the system is not asymptotically stable and must have a *DC* component in the output.

Theorem 3. *The DESA-population dynamics system is asymptotically stable.*

Proof. Recall that in the SA-like selection in DESA algorithm, $\gamma_\beta = (\alpha + \beta - \alpha\beta)CR > 0$ since the probability of accepting bad solutions β is greater than zero. Now, in (40), $\lambda + \frac{\gamma_\beta}{2} = 0$. This implies that $\gamma_\beta = -2\lambda > 0$, only if $\lambda < 0$. Hence, we may argue that the system eigenvalues: $\lambda < 0, -\frac{\gamma_\beta}{2}, \ldots, -\frac{\gamma_\beta}{2}$ satisfy (40). This completes the proof.

Theorem 4. *The DESA-population system,*

$$
E\left(\frac{dx_m}{dt}\right) = \frac{1}{2}\gamma_\beta\left(\frac{1}{Np}\sum_{i-1}^{Np} x_i - x_m\right) \quad for\ i = 1, 2, \ldots, Np,
\tag{41}
$$

is stable in the sense of Lyapunov.

Proof. We are assuming the DESA-population is located very close to optima. So, (41) holds true in such a region where the value of the gradient is negligibly small. Hence, a condition for an equilibrium point is $E\left(\frac{dx_m}{dt}\right) = 0$ [42]. We consider the case where the DESA-population is confined within a small neighborhood of an isolated optimum and over the entire population value of the gradient is minimal. In this case, the preferred equilibrium point should be the optimum itself. With time there is no change in values of state variables after they hit the optimum. Now, since $E\left(\frac{dx_m}{dt}\right) = 0$, then $x_m = \frac{1}{Np}\sum_{j=1}^{Np} x_j = \mu_x$ for $j = 1, \ldots, Np$. This completes the proof.

This is possible only if all of the state variables are equal in value. In case of a smooth, unimodal fitness landscape, the solution vectors generally crowd into a small neighborhood surrounding the optimum. Thus, during the later stages of the search, the equilibrium point x_e is identical to the optimum, and population members are expected not to change any further, and thus this point should satisfy the condition $x_1 = \ldots = x_{Np} = x_e$.

Now, we examine the stability of the solution vectors very near to such an optimum point of the search space. First, we define the *Lyapunov's Energy function* V as follows:

$$V(x_1, \ldots, x_{Np}, t) = \sum_{i=1}^{Np} (x_i - \mu_x)^2. \tag{42}$$

Observe that V, the sum of the squared differences of each individual from the centroid, is always positive except the equilibrium, where it becomes zero. V is positive definite with respect to equilibrium [42]. In addition, we note that $V = Np(\sigma_x^2)$. Differentiating (42) with respect to time and getting expectations, we get

$$E\left(\frac{dV}{dt}\right) = 2\sum_{m=1}^{Np} (x_m - \mu_x)\left[E\left(\frac{dx_m}{dt}\right) - E\left(\frac{d\mu_x}{dt}\right)\right] \tag{43}$$

From (25) we get, $E\left(\frac{dx_m}{dt}\right) = \frac{1}{2}\gamma_\beta(\mu_x - x_m)$ and

$$E\left(\frac{d\mu_x}{dt}\right) = E\left[\frac{d}{dt}\left(\frac{1}{Np}\sum_{i=1}^{Np} x_i\right)\right] = \frac{1}{Np}E\left(\sum_{i=1}^{Np}\frac{dx_i}{dt}\right) = 0. \tag{44}$$

Putting these expected values in (43), we get

$$E\left(\frac{dV}{dt}\right) = \mu_{V'} = -\gamma_\beta\sum_{i=1}^{Np} (x_i - \mu_x)^2. \tag{45}$$

It is clear that $\mu_{V'} = 0$ when $x_1 = x_2 = \ldots = x_{Np} = x_e$ and is negative otherwise, since $\gamma_\beta > 0$. Hence, $\mu_{V'}$ is a negative definite with respect to equilibrium point. Therefore, V is positive definite and $\mu_{V'}$ is negative definite, satisfying Lyapunov's stability theorem.

Theorem 5. *An estimate of the system time-constant of DESA is γ_β^{-1}.*

Proof. Using determinant of the population system in (39) the Lyapunov's energy function in (42) is written as $\frac{V}{-\mu_{V'}} = \frac{1}{\gamma_\beta}$ where $\frac{V}{-\mu_{V'}}$ is the average value of time rate of change of energy function. Let μ_V denotes the average of the energy function, V, when the process be carried out repeatedly for same initial conditions and parameter values. Similarly, the time rate of change of the average is denoted as $\mu_{V'}$. We assume that the runs of the algorithm that the process is time invariant. Hence, $-E(\mu_{V'})\mu_V = \gamma_\beta^{-1}$ which implies that $\mu_V = V_0 \exp\left(-t\,\gamma_\beta^{-1}\right)$ where V_0 is the initial value of V. Now, define a time-constant for the system as the time interval in which the energy function reduces to e^{-1} part of its initial value. Denoting this time-constant by T and putting $\mu_V = e^{-1}V_0$, and $t = T$ in (45) gives time-constant $T = \gamma_\beta^{-1}$.

4 The Learning Rate and Momentum of DESA-population

From (16) we may write,

$$E\left(\frac{dx_m}{dt}\right) = -\eta_{DESA}f'(x_m) + \tau_{DESA}, \tag{46}$$

where $\eta_{DESA} = \frac{k}{8}\gamma_\beta\left[(2F^2+1)Var(x) + (\mu_x - x_m)^2\right]$ and $\tau_{DESA} = \frac{1}{2}\gamma_\beta(\mu_x - x_m)$. The classical gradient descent search algorithm is given by the following continuous-time dynamics in single dimension [43]:

$$\frac{d\theta}{dt} = -\eta G + \tau \tag{47}$$

where η is the *learning rate* and τ is the *momentum*.

$$E\left(\frac{d\mu_x}{dt}\right) = -\frac{k}{8}\gamma_\beta(2F^2+1)Var(x)f'_{av} - \gamma_\beta\left(\frac{1}{N}\sum_{m=1}^{Np}\varepsilon_m^2 f(\mu_x + \varepsilon_m)\right) \tag{48}$$

The resemblance of (46) and (47) suggests that, the dynamics of DESA uses some kind of estimation for the gradient of the objective function, $f'(x_m)$. In Eq. (48), $-\eta_{DESA}f'(x_m)$ term on the RHS is responsible for moving along the direction of the *negative gradient*, whereas τ_{DESA} represents a component of *velocity of a trial solution towards the centroid* of the population. Clearly, this individual x_m is very near to an optimum, when $f'(x_m) \to 0$,

$$E\left(\frac{dx_m}{dt}\right) \approx \tau_{DESA} = \frac{1}{2}\gamma_\beta(\mu_x - x_m). \tag{49}$$

Now, if the population converges towards the optimum, $(\mu_x - x_m)$ tends to zero and $E\left(\frac{dx_m}{dt}\right) \to 0$. Thus, once reaching the optimum, the average velocity of the population members approaches zero, $Var(x) \to 0$, $\mu_x - x_m \to 0$, and $\varepsilon_m \to 0$. We get $E\left(\frac{dx_m}{dt}\right) \to 0$ in (49), and $E\left(\frac{d\mu_x}{dt}\right) \to 0$.

4.1   Comparison of Learning Rate and Momentum Rate of de and DESA

We compare the classical DE and DESA algorithms with respect to the learning rate η and the momentum rate τ.

We compute the relative difference of DE and DESA as: $\left|\frac{\eta_{DE}-\eta_{DESA}}{\eta_{DE}}\right|$. With $\gamma_\beta = (\alpha + \beta - \alpha\beta)CR$. The learning rates are given by

$$\eta_{DESA} = \frac{k}{8}(\alpha + \beta - \alpha\beta)CR\left[(2F^2 + 1)Var(x) + (\mu_x - x_m)^2\right] \qquad (50)$$

and

$$\eta_{DE} = \frac{k}{8}CR\left[(2F^2 + 1)Var(x) + (\mu_x - x_m)^2\right], \qquad (51)$$

respectively. Hence,

$$\left|\frac{\eta_{DE} - \eta_{DESA}}{\eta_{DE}}\right| = \left|\frac{CR - (\alpha + \beta - \alpha\beta)CR}{CR}\right| = (1 - \alpha)(1 - \beta).$$

Thus, the relative difference in learning rate of DE and DESA approaches 0, if either α or β approaches 1. It approaches 1, otherwise.

Similarly, we compute the relative difference in momentum rates of DE and DESA as

$$\left|\frac{\tau_{DE} - \tau_{DESA}}{\tau_{DE}}\right| = \left|\frac{\frac{1}{2}CR(\mu_x - x_m) - \frac{1}{2}(\alpha + \beta - \alpha\beta)CR(\mu_x - x_m)}{\frac{1}{2}CR(\mu_x - x_m)}\right|. \qquad (52)$$

Therefore, the relative difference (change) in momentum rates is also $(1 - \alpha)$ $(1 - \beta)$ which also approaches 0, if either α or β approaches 1. It approaches 1, otherwise.

4.2   On the Stability of DESA in the Sense of Lyapunov

Lyapunov stability analysis is based on the idea that if the total energy in the system continually decreases, then the system will asymptotically reach the zero energy state associated with an equilibrium point of the system. A system is said to be asymptotically stable if all the states approach the equilibrium state with time [4].

It can be noted that the centroid (μ_x) of the population system does not change with time since

$$E\left(\frac{d\mu_x}{dt}\right) = E\left(\frac{d}{dt}\left(\frac{1}{Np}\sum_{i=1}^{Np} x_i\right)\right) = \frac{1}{Np}E\left(\sum_{i=1}^{Np} \frac{dx_i}{dt}\right) = 0. \qquad (53)$$

The condition for equilibrium is $x_1 = \ldots = x_{Np} = x_e$, where x_e the equilibrium position is. If all population members are equal, then $x_i = \mu_x, i = 1, \ldots, Np$

of the population. Hence, $x_e \equiv \mu_x$. Initially, the population spread within a small region around the optima. So, the centroid was also very close to the actual optima. Lyapunov's function, in this case, is directly proportional to the population variance. With time, the initially dispersed populations gather at the center of the system, and eventually, population variance diminishes to 0, which leads to convergence of the population system. Average velocity of mth population member is $E\left(\frac{dx_m}{dt}\right) = \frac{1}{2}\gamma_\beta(\mu_x - x_m)$ and the average acceleration is $\frac{1}{2}\gamma_\beta\left(\frac{d\mu_x}{dt} - \frac{dx_m}{dt}\right) = -\frac{1}{2}\gamma_\beta\frac{dx_m}{dt}$. So, acceleration is directly proportional to velocity, and the negative sign indicates that it acts in opposite direction.

5   Conclusions

From mathematical framework of the operators of the DESA/rand/1/bin algorithm, we described the characteristics of the DESA-population as it goes through mutation and selection operators. This allows a comparison of the characteristics and convergence of the classical DE and its hybrid DESA. An analysis using basic concepts and interpretations of nonlinear control theory, showed that the DESA-population dynamics system is asymptotically stable. The dynamics of DE and DESA uses some kind of estimation for the gradient of the objective function. The SA-like selection operator of DESA increased its learning rates and momentum, which results to convergence faster than the classical DE. In addition, the time-constant of DESA algorithm is given by γ_β^{-1}.

Acknowledgements. The author RCA would like to thank the University of the Philippines Baguio, Baguio City, Philippines through the Ph.D. Incentive Grant and Research Dissemination Grant.

References

1. Storn, R., Price, K.: Differential evolution - a simple and efficient adaptive scheme for global optimization over continuous spaces. International Computer Science Institute, Berkeley, TR-95-012 (1995)
2. Back, T.: Evolution Strategies, Evolutionary Programming. Genetic Algorithms. Oxford University Press, New York (1996)
3. Eiben, A., Rudolph, G.: Theory of evolutionary algorithms: a bird's eye view. Theor. Comput. Sci. **229**(1–2), 3–9 (1999)
4. Das, S., Mullick, S.S., Suganthan, P.N.: Recent advances in differential evolution - an updated survey. Swarm Evol. Comput. **27**, 1–30 (2016)
5. Zielinski, K., Peters, D., Laur, R.: Run time analysis regarding stopping criteria for differential evolution and particle swarm optimization. In: Proceedings of the 1st International Conference on Process/System Modelling/Simulation/Optimization, vol. 1, pp. 235–243 (2005)
6. Zaharie, D.: Critical values for the control parameters of differential evolution algorithms. In: Proceedings 8th International Mendel Conference Soft Computing, pp. 62–67 (2002)

7. Zaharie, D.: Influence of crossover on the behavior of differential evolution algorithms. Appl. Soft Comput. **9**(3), 1126–1138 (2009)
8. Dasgupta, S., Das, S., Biswas, A., Abraham, A.: On stability and convergence of the population-dynamics in differential evolution. AI Commun. **22**(1), 1–20 (2009)
9. Das, S., Abraham, A., Konar, A.: Modeling and analysis of the population-dynamics of differential evolution algorithm. In: Das, s, Abraham, A., Konar, A. (eds.) Metaheuristic Clustering. SCI, vol. 178, pp. 111–135. Springer, Berlin (2009). https://doi.org/10.1007/978-3-540-93964-1_3
10. Kvasov, D.E., Mukhametzhanov, M.S.: Metaheuristic vs. deterministic global optimization algorithms: the univariate case. Appl. Math. Comput. **318**, 245–259 (2018)
11. Xue, F., Sanderson, C., Graves, J.: Modeling and convergence analysis of a continuous multi-objective differential evolution algorithm. IEEE Press, Edinburgh, UK (2005)
12. Zhao, Y., Wang, J., Song, Y.: An improved differential evolution to continuous domains and its convergence. In: Proceedings of the 1st ACM/SIGEVO Summit on Genetic and Evolutionary Computation (GEC 2009), pp. 1061–1064 (2009)
13. Ghosh, S., Das, S., Vasilakos, A.V., Suresh, K.: On convergence of differential evolution over a class of continuous functions with unique global optimum. IEEE Trans. Syst. Man Cybern. B **42**(1), 107–124 (2012)
14. Zhan, Z., Zhang, J.: Enhanced differential evolution with random walk. In: Proceedings of the 14th International Conference on Genetic and Evolutionary Computation Conference Companion (GECCO 2012), pp. 1513–1514 (2012)
15. Kirkpatrick, S., Gelatt Jr., C.D., Vecchi, M.P.: Optimization by simulated annealing. Science **220**, 671–680 (1983)
16. Storn, R., Price, K.: Differential evolution: a simple and efficient heuristic for global optimization over continuous spaces. J. Glob. Optim. **11**(4), 341–359 (1997)
17. Addawe, R., Adorio, E., Addawe, J., Magadia, J.: DESA: a hybrid optimization algorithm for high dimensional functions. In: Proceedings of the Eight IASTED International Conference on Control and Applications, pp. 316–321, ACTA Press, Montreal, Canada (2006)
18. Holland, J.H.: Adaptation in Natural and Artificial Systems: An Introductory Analysis with Application to Biology, Control, and Artificial Intelligence. University of Michigan Press, Ann Arbor (1975)
19. Deb, K., Kumar, A.: Real-coded genetic algorithms with simulated binary crossover: studies on multimodal and multiobjective problems. Complex Syst. **9**, 431–454 (1995)
20. Price, K.V., Storn, R.M., Lampinen, J.A.: Differential Evolution. NCS. Springer, Heidelberg (2005). https://doi.org/10.1007/3-540-31306-0
21. Yang, X.-S., He, X.: Firefly algorithm: recent advances and applications. Int. J. Swarm Intell. **1**, 36–50 (2013)
22. Sergeyev, Y.D., Strongin, R.G., Lera, D.: Introduction to Global Optimization Exploiting Space-Filling Curves. Springer, New York (2013). https://doi.org/10.1007/978-1-4614-8042-6
23. Strongin, R.G., Sergeyev, Y.D.: Global Optimization with Non-Convex Constraints: Sequential and Parallel Algorithms. Kluwer Academic Publishers, Dordrecht (2010)
24. Sergeyev, Y.D., Kvasov, D.E., Mukhametzhanov, M.S.: On the efficiency of nature-inspired metaheuristics in expensive global optimization with limited budget. Sci. Rep. **8**, article 453 (2018)

25. Horst, R., Pardalos, P.M. (eds.): Handbook of Global Optimization, vol. 1. Kluwer Academic Publishers, Dordrecht (1995)
26. Sergeyev, Y.D., Kvasov, D.E.: Global search based on efficient diagonal partitions and a set of lipschitz constants. SIAM J. Optim. **16**(3), 910–937 (2006)
27. Suzuki, J.: A Markov chain analysis on simple genetic algorithms. IEEE Trans. Syst. Man Cybern. **25**, 655–659 (1995)
28. Eiben, A.E., Aarts, E.H.L., Van Hee, K.M.: Global convergence of genetic algorithms: a Markov chain analysis. In: Schwefel, H.-P., Männer, R. (eds.) PPSN 1990. LNCS, vol. 496, pp. 3–12. Springer, Heidelberg (1991). https://doi.org/10.1007/BFb0029725
29. De Jong, K.A., Spears, W.M., Gordon, D.F.: Using Markov chains to analyze GAFOS. In: Proceedings Foundation of Genetic Algorithm, pp. 115–137 (1994)
30. Rudolph, G.: Convergence analysis of canonical genetic algorithms. IEEE Trans. Neural Netw. **5**(1), 96–101 (1994)
31. Vose, M.: Modeling simple genetic algorithms. Evol. Comput. **3**(4), 453–472 (1996)
32. Qi, X., Palmieri, F.: Theoretical analysis of evolutionary algorithms with an infinite population size in continuous space part I: basic properties of selection and mutation. IEEE Trans. Neural Netw. **5**(1), 102–119 (1994)
33. Peck, C.C., Dhawan, A.P.: Genetic algorithms as global random search methods: an alternative perspective. Evol. Comput. **3**(1), 39–80 (1995)
34. Addawe, R., Addawe, J., Magadia, J.: Optimization of seasonal ARIMA models using differential evolution - simulated annealing (DESA) algorithm in forecasting dengue cases in Baguio City. In: AIP Conference Proceedings, vol. 1776, pp. 090021–090028 (2016)
35. Addawe, R., Magadia, J.: Differential Evolution-Simulated Annealing (DESA) algorithm for fitting autoregressive models to data. In: OPT-i 2014 International Conference on Engineering and Applied Sciences Optimization. National Technical University, Kos Island, Greece (2014)
36. Addawe, R., Addawe, J., Sueno, M., Magadia, J.: Differential evolution-simulated annealing for multiple sequence alignment. IOP Conf. Ser. J. Phys. Conf. Ser. **893**, 012061 (2016)
37. Kirk, W., Sims, B. (eds.): Handbook of Metric Fixed Point Theory. Kluwer Academic, London (2001)
38. Fletcher, R.: Practical Methods of Optimization, 2nd edn. Wiley, Chichester (1987)
39. Zaharie, D.: On the explorative power of differential evolution. In: 3rd International Workshop on Symbolic and Numerical Algorithms on Scientific Computing (SYNASC 2001) (2001)
40. Das, S., Konar, A., Chakraborty, U.: Two improved differential evolution schemes for faster global search. In: ACM-SIGEVO Proceedings of GECCO Washington D.C., pp. 991–998 (2005)
41. Anwal, P.: Generalized Functions: Theory and Technique, 2nd edn. Birkher, Boston MA (1998)
42. Hahn, W.: Theory and Application of Lyapunov's Direct Method. Prentice-Hall, Englewood Cliffs (1963)
43. Snyman, J.A.: Practical Mathematical Optimization: An Introduction to Basic Optimization Theory and Classical and New Gradient-Based Algorithms. Springer, Berlin (2005). https://doi.org/10.1007/b105200

Are Humans Bayesian in the Optimization of Black-Box Functions?

Antonio Candelieri[✉] [ID], Riccardo Perego [ID], Ilaria Giordani [ID],
and Francesco Archetti [ID]

University of Milano-Bicocca, 20126 Milan, Italy
antonio.candelieri@unimib.it

Abstract. Many real-world problems have complicated objective functions whose optimization requires sophisticated sequential decision-making strategies. Modelling human function learning has been the subject of intense research in cognitive sciences. The topic is relevant in black-box optimization where information about the objective and/or constraints is not available and must be learned through function evaluations. The Gaussian Process based Bayesian learning paradigm is central in the development of active learning approaches balancing exploration/exploitation in uncertain conditions towards effective generalization in large decision spaces. In this paper we focus on Bayesian Optimization and analyse experimentally how it compares to humans while searching for the maximum of an unknown 2D function. A set of controlled experiments with 53 subjects confirm that Gaussian Processes provide a general model to explain different patterns of learning enabled search and optimization in humans.

Keywords: Bayesian Optimization · Cognitive models · Search strategy

1 Introduction

We consider as a reference problem the black-box optimization problem: the objective function and/or constraints are analytically unknown and evaluating the function might be very expensive and noisy. In black-box situations as we cannot assume any prior knowledge about the objective function $f(x)$, any functional form is a priori admissible and the value of the function at a point says nothing about the value at other points, as postulated by the No Free Lunch theorems (Adam et al. 2019). The only way to develop a problem specific algorithm is through a sample of function evaluations.

An algorithm fitting for such applications should have global properties and be sample efficient, because the cost of function evaluations is the dominating cost. This problem has been addressed in several fields under different names, including active learning (Kruschke et al. 2008), Bayesian Optimization (Jones et al. 1998), (Zhigljavsky and Zilinskas 2007), (Candelieri et al. 2018), optimal search, optimal experimental design, hyperparameter optimization (Eggensperger et al. 2019) and others.

Efficient sampling is active sampling in which a surrogated model of the objective function is built upon the observations already performed and the next sampled point is

© Springer Nature Switzerland AG 2020
Y. D. Sergeyev and D. E. Kvasov (Eds.): NUMTA 2019, LNCS 11974, pp. 32–42, 2020.
https://doi.org/10.1007/978-3-030-40616-5_3

chosen on the basis of its informative value: this choice brings up the so called "exploration vs exploitation dilemma", where exploration means devoting resources to know more about possible solutions while exploitation devotes resources to improve on solutions already identified in previous phases. The search for the new point must strike an effective balance between the needs of exploration and exploitation.

To do this, the surrogate model must sum up our a priori beliefs: the Gaussian Process (GP) is the best probabilistic framework to update our beliefs as new data arrives and provide an estimate of the expected value of the objective function and the uncertainty in this estimate.

Psychologists have extensively studied how humans balance exploration and exploitation (Krusche et al. 2008), (Mehlhorn et al. 2015), with a recent attention on the links between modern machine learning algorithms and psychological processes. (Gershman 2018; Schulz et al. 2017; (Gopnik et al. 2017). Psychological research has mostly focused on how people learn functions according to a protocol in which an input is presented to participants and they are asked to predict the corresponding output value. Then they observe the true output value (usually noisy) in order to improve their own "predictive model". Through this outcome feedback, people are thought to adjust their internal representation of the underlying function. The pioneering work of (Wilson et al. 2014) demonstrated that humans use both random and directed exploration. The issue of GP regression, kernel composition for different degrees of smoothness and safe optimization in their relation to cognition is also studied in a recent survey by (Shultz et al. 2018). Directed exploration is realized by adding *uncertainty bonuses* (Auer et al. 2002). to estimated values obtaining the *upper confidence bound* (UCB) algorithm (Srinivas et al. 2010). In (Wu et al. 2018) the human search strategy is analysed for rewards under limited search horizons, concluding that GP offers the best model for generalization and UCB the best solution of the exploration/exploitation dilemma.

This paper considers optimization, a task related with function learning, but with its own specific features. Contrary to function learning, optimization is not yet widely considered in the literature; as a reference, in (Borji and Itti 2013) a simple 1-D optimization problem has been considered. Specifically, the aim of this paper is how humans choose the next x to be queried when attempting to locate the maximum of an unknown 2D function. We'll focus on the questions: do humans follow a Bayesian approach, and if so, how do humans balance between exploration and exploitation during optimization? Which space representation do they use? Can GPs offer a unifying theory of human function optimization?

The structure of the paper is as follows: Sect. 2 outlines the methodological background of GP based optimization linking them to the issue of learning including the temporal trade-off over uncertain rewards (this is also a central topic in cognitive science). Section 3 is devoted to the experimental set-up and Sect. 4 reports the experimental results on how humans behave in black-box Bayesian Optimization (BO).

2   Methodological Background

This section provides the underlying methodological framework of the study.

2.1 Gaussian Processes

GPs are a powerful non-parametric model for implementing both regression and classification. One way to interpret a GP is as a distribution over functions, with inference taking
place directly in the space of functions (Williams and Rasmussen 2006). A GP, therefore,
is a collection of random variables, any finite number of which have a joint Gaussian
distribution. A GP is completely specified by its mean function $\mu(x)$ and covariance
function $cov(f(x), f(x')) = k(x, x')$:

$$\mu(x) = \mathbb{E}[f(x)]$$

$$cov(f(x), f(x')) = k(x, x') = \mathbb{E}[(f(x) - \mu(x))(f(x') - \mu(x'))]$$

and will be denoted by: $f(x) \sim GP(\mu(x), k(x, x'))$.

Usually, for notational simplicity we will take the prior of the mean function to be
zero, although this is not necessary. The covariance function assumes a critical role int
the GP modelling, as it specifies the distribution over functions, depending on a sample
$X_{1:n}$ with $f(X_{1:n}) \sim \mathcal{N}(0, K(X_{1:n}, X_{1:n}))$.

We usually have access only to noisy function values, denoted by $y = f(x) + \varepsilon$.
Assuming additive independent identically distributed Gaussian noise ε with variance
λ^2, cov $y = (y_1, \ldots, y_n) = K(X_{1:n}, X_{1:n}) + \lambda^2 I$.

Therefore, the predictive equations for GP regression, that are $\mu(x)$ and $k(x, x')$,
can be easily updated, by conditioning the joint Gaussian prior distribution on the
observations:

$$\mu(x) = \mathbb{E}[f(x)|D_{1:n}, x] = k(x, X_{1:n})\Big[K(X_{1:n}, X_{1:n}) + \lambda^2 I\Big]^{-1} y$$

$$\sigma^2(x) = k(x, x) - k(x, X_{1:n})\Big[K(X_{1:n}, X_{1:n}) + \lambda^2 I\Big]^{-1} k(X_{1:n}, x)$$

The covariance function is the crucial ingredient in a GP predictor, as it encodes
assumptions about the function to approximate: function evaluations that are near to a
given point should be informative about the prediction at that point. Under the GP view
it is the covariance function that defines nearness or similarity. Examples of covariance
(aka kernel) functions:

Squared Exponential (SE) kernel:

$$k_{SE}(x, x') = e^{-\frac{\|x-x'\|^2}{2\ell^2}}$$

with ℓ known as *characteristic length-scale.*

Exponential kernel:

$$k_{Exp}(x, x') = e^{-\frac{|x-x'|}{\ell}}$$

Power Exponential kernel:

$$k_{PowExp}(x, x') = e^{-\left(\frac{|x-x'|}{\ell}\right)^p}$$

Matérn kernels:

$$k_{Mat}(x, x') = \frac{2^{1-\nu}}{\Gamma(\nu)} \left(\frac{|x - x'|\sqrt{2\nu}}{\ell} \right)^{\nu} K_\nu \left(\frac{|x - x'|\sqrt{2\nu}}{\ell} \right)$$

With two hyperparameters ν and ℓ, and where K_ν is a modified Bessel function, that is a product of an exponential and a polynomial of order r. The most widely adopted versions, specifically in the Machine Learning community, are $\nu = 3/2$ and $\nu = 5/2$.

2.2 GP-Based Optimization

The acquisition function is the mechanism to implement the trade-off between *exploration* and *exploitation* in BO. More precisely, any acquisition function aims to guide the search of the optimum towards points with potential low values of objective function either because the prediction of $f(x)$, based on the probabilistic surrogate model, is low or the uncertainty, also based on the same model, is high (or both). Indeed, *exploiting* means to consider the area providing more chance to improve the current solution (with respect to the current surrogate model), while *exploring* means to move towards less explored regions of the search space where predictions based on the surrogate model are more uncertain, with higher variance.

Probability of Improvement (PI) was the first acquisition function proposed in the literature (Kushner 1964):

$$PI(x) = P(f(x) \le f(x^+)) = \Phi \left(\frac{f(x^+) - \mu(x)}{\sigma(x)} \right).$$

where $f(x^+)$ is the best value of the objective function observed so far, $\mu(x)$ and $\sigma(x)$ are mean and standard deviation of the probabilistic surrogate model, such as a GP, and $\Phi(\cdot)$ is the normal distribution function. The next point to evaluate is chosen according to:

$$x_{n+1} = \underset{x \in X}{\operatorname{argmax}} \, PI(x)$$

Expected Improvement (EI) was initially proposed in (Mockus et al. 1978) and then made popular in (Jones et al. 1998). It measures the expectation of the improvement on $f(x)$ with respect to the predictive distribution of the probabilistic surrogate model.

$$EI(x) = \begin{cases} (f(x^+) - \mu(x))\Phi(Z) + \sigma(x)\phi(Z) \, if \, \sigma(x) > 0 \\ 0 \, if \, \sigma(x) = 0 \end{cases},$$

where $\phi(Z)$ and $\Phi(Z)$ are the probability distribution and the cumulative distribution of the standardized normal, respectively, where

$$Z = \begin{cases} \frac{f(x^+) - \mu(x)}{\sigma(x)} \, if \, \sigma(x) > 0 \\ 0 \, if \, \sigma(x) = 0 \end{cases}.$$

The EI is made up of 2 terms: the first is increased by decreasing the predictive mean; the second by increasing the predictive uncertainty. The next point to evaluate is chosen according to:

$$x_{n+1} = \underset{x \in X}{\mathrm{argmax}} \, EI(x)$$

Upper/Lower Confidence Bound, where Upper and Lower are used, respectively, for maximization and minimization problems, is an acquisition function that manages exploration-exploitation by being optimistic in the face of uncertainty.

In the case of a minimization problem, LCB (Lower Confidence Bound) is given by:

$$LCB(x) = \mu(x) - \xi\sigma(x)$$

while in the case of a maximization problem the UCB acquisition function is used:

$$UCB(x) = \mu(x) + \xi\sigma(x)$$

where $\xi \geq 0$ is the parameter to manage the trade-off between exploration and exploitation ($\xi = 0$ is for pure exploitation; on the contrary, higher values of ξ emphasizes exploration by inflating the model uncertainty). In (Srinivas et al. 2010), a policy is provided for updating the value of ξ along function evaluations, with also a proof of convergence of such a policy.

In the case of a minimization problem the next point is chosen as

$$x_{n+1} = \underset{x \in X}{\mathrm{argmin}} \, LCB(x)$$

while, in the case of a maximization problem the next point is selected as

$$x_{n+1} = \underset{x \in X}{\mathrm{argmax}} \, UCB(x)$$

2.3 Bayesian Optimization

The following algorithm summarizes a general Bayesian Optimization process where the acquisition function, whichever it is, is denoted by $\alpha(x, D_{1:n})$. This function is generally maximized, but in the case of $\alpha = LCB$.

With respect to the probabilistic surrogate model, the summarized algorithm does not specify the probabilistic surrogate model, as well as the kernel in the case of a GP. This is basically done in order to maintain the algorithm as general as possible.

In this study we have used a GP as a surrogate probabilistic model, considering all the five different kernels presented in the previous section, and the three different acquisition functions previously described.

General Bayesian Optimization Algorithm

1 **for** $n = 1, 2, \ldots$ **do**
2   select a new x_{n+1} by optimizing an acquisition function α, such that
$$x_{n+1} = \text{argmax}_x \, \alpha(x, D_{1:n})$$
3   evaluate the objective function to obtain $y_{n+1} = f(x_{n+1})$
4   update the dataset of observations $D_{1:n+1} = D_{1:n} \cup \{(x_{n+1}, y_{n+1})\}$
5   update the probabilistic surrogate model, $\mu(x)$ and $\sigma(x)$, as in section 2.1
6 **endfor**
7 Output: the best f value observed over the entire optimization process

---

3 Experimental Setup

3.1 Test Function as a "Task"

In this study, we have selected the Styblinski-Tang test function, as defined in https://www.sfu.ca/~ssurjano/optimization.html.

$$f(x) = \frac{1}{2} \sum_{i=1}^{d} \left(x_i^4 - 16 x_i^2 + 5 x_i \right)$$

where d is the number of dimensions ($d = 2$, in this study) and $f(x)$ is minimized in the hypercube $x_i \in [-5; 5] \, \forall i = 1, \ldots, d$.

Since the optimization performed by humans was defined as a black-box maximization task, this means that the optimization problem considered is:

$$\max_{x \in [-5;5]^d} -f(x)$$

3.2 Experiment: Optimization by Humans

3.2.1 Participants Fifty-three participants (14 female), with an average age of 26 (standard deviation: 5.82) were recruited. The experiment took around 15 min to complete the task, on average. The experimental procedure is defined in the following.

3.2.2 Procedure In order to conduct the test, each one of the participants was sat in front of a personal computer, asked to play for a game with the following rules:

- *In front of the player there is a white panel: the goal of the game is to click on it and find a point with maximum score, within 15 clicks*
- *Everytime the player clicks on the panel a score is shown for that selected point: higher the score, better the choice. Points are also colored according to the associated score, providing a visual feedback about the distribution of the scores collected so far.*

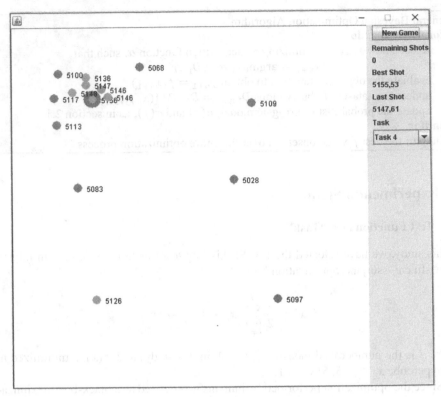

Fig. 1. An example of a game play

Figure 1 shows a frame of the game, with points selected by one of the participants.

For each player, at each iteration, a GP is fitted on the observed points and the GP next point is compared, via Euclidean distance, to the choice made by the player.

To fit the GP the five kernels described in Sect. 2.1 have been considered, along with the three acquisition functions: PI, EI and UCB.

The two strategies, human player and BO, are considered compliant, pointwise, if the distance between the point chosen by the human player and the algorithmic player is less than a given "threshold". The specific procedure is summarized in the following pseudo-code.

Let denote by:
- p a participant
- k a kernel
- α an acquisition function
- n a generic iteration
- $s^{p,k,n}$ a search strategy (i.e., an acquisition fuction, under a GP with a given kernl, at a specific iteration)

1 **foreach** p, k and n
2   fit a Gaussian Process $GP^{p,k,n}$
3   **foreach** α
4   select $x_{n+1}^{GP_{p,k,n}}$
5   compute $d^{p,k,n,\alpha} = \left\| x_{n+1}^{GP^{p,k,n}} - x_{n+1}^{p} \right\|$, where

           $x_{n+1}^{GP^{p,k,n}} = \text{argmax}_x \, \alpha(x) | GP^{p,k,n}$ is the next point according to acquisition function α under $GP^{p,k,n}$

           and x_{n+1}^{p} is the next point chosen by participant p at iteration $n+1$
6   **endforeach**
7     $\bar{d}^{p,k,n} = \min_\alpha \{d^{p,k,n,\alpha}\}$
8   **if** $\bar{d}^{p,k,n} \leq threshold$ **then**
9        $s^{p,k,n} = \text{argmin}_\alpha \{d^{p,k,n,\alpha}\}$
10 **else**
11     $s^{p,k,n} = \emptyset$

Finally, for a given kernel $\bar{k}$, the search strategy of the participant $\bar{p}$ is compliant to the most frequent acquisition function in the series $s^{p,k} = \left\{ s^{p,k,n} \right\}_{n=1:N}$.

According to the mentioned procedure, the following figures summarize the main results of the study (Figs. 2, 3, 4, 5 and 6).

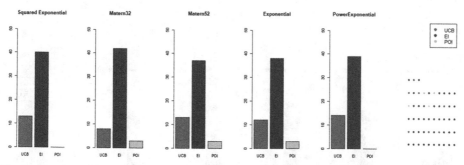

Fig. 2. Number of human players whose strategy is compliant with respect to kernel type and acquisition functions, with "threshold" set to 0.10, and $\beta = 1$ in UCB

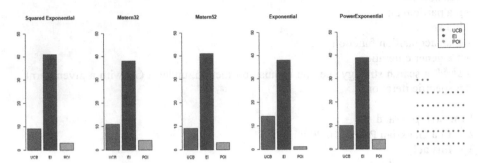

Fig. 3. Number of human players whose strategy is compliant with respect to kernel type and acquisition functions, with "threshold" set to 0.15, and $\beta = 1$ in UCB

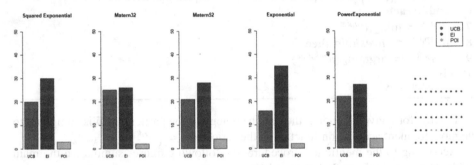

Fig. 4. Number of human players whose strategy is compliant with respect to kernel type and acquisition functions, with "threshold" set to 0.15, and $\beta = 0.5$ in UCB

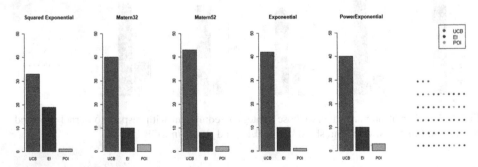

Fig. 5. Number of human players whose strategy is compliant with respect to kernel type and acquisition functions, with "threshold" set to 0.15, and $\beta = 0$ in UCB

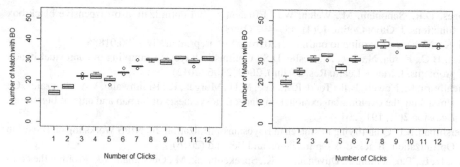

Fig. 6. Number of players, depending on the number of clicks, whose strategy is compliant with respect to anyone of the BO implementations considered (i.e., pair "kernel type – acquisition function"). This figure refers to threshold = 0.10 (left) and threshold = 0.15 (right)

4 Conclusions

In conclusion, 40 out of 53 participants (75%) shown search patterns compliant with Bayesian Optimization. This percentage is increasing with the number of iterations (i.e., clicks) and the "threshold" value. Also interesting is the analysis of which space model, that is kernel, and which exploitation-exploration balance, that is the acquisition function, are implied by human search. Contrary to previous results, kernel is not a major factor in determining compliance, while acquisition functions, and specifically the balancing parameter β in UCB, are the main determinants.

References

Adam, S.P., Alexandropoulos, S.A.N., Pardalos, P.M., Vrahatis, M.N.: No free lunch theorem: a review. In: Demetriou, I.C., Pardalos, P.M. (eds.) Approximation and Optimization. SOIA, vol. 145, pp. 57–82. Springer, Cham (2019). https://doi.org/10.1007/978-3-030-12767-1_5

Auer, P., Cesa-Bianchi, N., Fischer, P.: Finite-time analysis of the multiarmed bandit problem. Mach. Learn. **47**(2–3), 235–256 (2002)

Borji, A., Itti, L.: Bayesian optimization explains human active search. In: Advances in Neural Information Processing System 26 (NIPS 2013), pp. 55–63 (2013)

Candelieri, A., Perego, R., Archetti, F.: Bayesian optimization of pump operations in water distribution systems. J. Glob. Optim. **71**, 213–235 (2018)

Chapelle, O., Li, L.: An empirical evaluation of thompson sampling. In: Advances in Neural Information Processing Systems, pp. 2249–2257 (2011)

Eggensperger, K., Lindauer, M., Hutter, F.: Pitfalls and best practices in algorithm configuration. J. Artif. Intell. Res. **64**, 861–893 (2019)

Gershman, S.J.: Uncertainty and exploration. bioRxiv 265504 (2018). https://doi.org/10.1101/265504

Gopnik, A., O'Grady, S., Lucas, C.G., Griffiths, T.L., Wente, A., Bridgers, S., Dahl, R.E.: Changes in cognitive flexibility and hypothesis search across human life history from childhood to adolescence to adulthood. Proc. Nat. Acad. Sci. **114**(30), 7892–7899 (2017)

Kruschke, J.K.: Bayesian approaches to associative learning: from passive to active learning. Learn. Behav. **36**(3), 210–226 (2008)

Jones, D.R., Schonlau, M., Welch, W.J.: Efficient global optimization of expensive black-box functions. J. Glob. Optim. **13**(4), 455–492 (1998)

Li, K., Malik, J.: Learning to optimize. (2016) arXiv preprint arXiv:1606.01885

May, B.C., Korda, N., Lee, A., Leslie, D.S.: Optimistic Bayesian sampling in contextual-bandit problems. J. Mach. Learn. Res. **13**(Jun), 2069–2106 (2012)

Mehlhorn, K., Newell, B.R., Todd, P.M., Lee, M.D., Morgan, K., Braithwaite, V.A., Gonzalez, C.: Unpacking the exploration–exploitation tradeoff: a synthesis of human and animal literatures. Decision **2**(3), 191 (2015)

Gershman, S.J.: Quantifying mismatch in Bayesian optimization. In: NIPS Workshop on Bayesian Optimization: Black-Box Optimization and Beyond (2016)

Schulz, E., Tenenbaum, J., Duvenaud, D.K., Speekenbrink, M., Gershman, S.J.: Probing the compositionality of intuitive functions. In: Advances in Neural Information Processing Systems, pp. 3729–3737 (2016)

Schulz, E., Speekenbrink, M., Krause, A.: A tutorial on Gaussian process regression: modelling, exploring, and exploiting functions. J. Math. Psychol. **85**, 1–16 (2018)

Srinivas, N., Krause, A., Kakade, S., Seeger, M.: Gaussian process optimization in the bandit setting: no regret and experimental design. In: Proceedings of the 27th International Conference on Machine Learning, pp. 1015–1022. Omnipress, June 2010

Thompson, W.R.: On the likelihood that one unknown probability exceeds another in view of the evidence of two samples. Biometrika **25**(3/4), 285–294 (1933)

Wilson, R.C., Geana, A., White, J.M., Ludvig, E.A., Cohen, J.D.: Humans use directed and random exploration to solve the explore–exploit dilemma. J. Exp. Psychol. Gen. **143**(6), 2074 (2014)

Wu, C.M., Schulz, E., Speekenbrink, M., Nelson, J.D., Meder, B.: Generalization guides human exploration in vast decision spaces. Nat. Hum. Behav. **2**(12), 915 (2018)

Zhigljavsky, A., Zilinskas, A.: Stochastic Global Optimization, vol. 9. Springer, Berlin (2007). https://doi.org/10.1007/978-0-387-74740-8

A New Syntax for Diagrammatic Logic: A Generic Figures Approach

Gianluca Caterina$^{(\boxtimes)}$ and Rocco Gangle

Endicott College, Beverly, MA 01915, USA
gcaterin@endicott.edu

Abstract. In this paper we propose a new syntactical representation of C.S. Peirce's diagrammatic systems for propositional and predicate logic. In particular, we use the categorical notion of *generic figures* to represent the syntax of the diagrammatic language as a category of functors from a suitable, simple category into the category of sets, highlighting the relational nature of Peirce's diagrammatic logic.

1 Introduction

In this paper we present a new syntactical representation of the Alpha and Beta Existential Graphs (from now on denoted by EG_α and EG_β) – introduced by C.S. Peirce towards the end of the 19th century – which are diagrammatic systems that have been proven to be equivalent to propositional and predicate logic respectively.

These are representations of the syntax and semantics for propositional and predicate logic solely based on simple diagrams and some basic topological relations that such diagrams entail. For too long these graphs have been considered just a curious variation of the standard linear notation in logic. The renewed interest in diagrammatic reasoning via category theory, however (see the work in diagrammatic quantum computation done by Coecke [5], Spivak's wiring diagrams [11] and Ahti Pietarinen's diagrammatic proof analysis [8]), motivated us to look deeper into the structure of Peirce's graphs, and we believe that the categorical notion of *generic figures* developed by Reyes [10] is especially apt for highlighting some interesting characteristics of these logical systems.

Among other things, category theory is a powerful and efficient framework to model logic. At a very basic level, *categorical logic* is based on the idea that, in a suitable category, objects can model propositions, whereas morphisms model proofs. In our work we study a variation on this theme, as we aim to model propositions as functors from a suitable "base" category into the category of sets. In this way, some of the lurking problems that arise when trying to formalize in a rigorous way Peirce's Existential Graphs are solved in a single stroke by introducing an intuitively structured category that "generates" the graphs. In some sense, ours is a hybrid model that aims to build a link between the categorical and the classical approaches to logic.

© Springer Nature Switzerland AG 2020
Y. D. Sergeyev and D. E. Kvasov (Eds.): NUMTA 2019, LNCS 11974, pp. 43–58, 2020.
https://doi.org/10.1007/978-3-030-40616-5_4

The aim of this paper is modest, as we are only concerned, at this level, to represent the syntax of these systems in a categorical framework. This, however, is meant to be the first of a series of papers with a much broader scope, in which a thorough investigation of the inference rules and the semantics of the graphs will be provided in terms of this very general framework.

In order to keep the paper self-contained, in Sect. 2 below we present a brief introduction to the syntax of EG_α and EG_α. In Sect. 3 we provide the necessary basic categorical background needed to introduce the notion of generic figures, and in Sect. 4 we present a full account of the syntax of such systems in terms of functor categories.

2   Peirce's Existential Graphs

In this section we introduce Peirce's diagrammatic syntax for formal logic that he named Existential Graphs. Peirce himself developed three increasingly sophisticated levels of the graphs, which he denoted alpha, beta and gamma. The alpha level (EG_α) corresponds to classical propositional logic, the beta level (EG_β) to first-order predicate logic, and the gamma level (EG_γ) to modal and higher-order logics. We summarize here only the basic syntax of Peirce's EG_α and EG_β.

2.1   Existential Graphs: Alpha

We now discuss the basic elements of Peirce's Existential Graphs and present a schematic introduction for those unfamiliar with the graphs, following the basic outline presented in [2] and using some of that material in a slightly modified form. Besides the rigorous mathematical presentation in [1], detailed but more accessible treatments may be found in [12]. A categorical approach to the graphs can be found in [3] and [4].

Every EG_α graph is understood by Peirce to represent a propositional assertion according to a syntax consisting of three types of elements: the Sheet of Assertion, characters and seps (or cuts). An EG_α graph is composed of characters and seps "scribed" on the Sheet of Assertion. The syntactical relations among these elements may be described as follows:

- The blank sheet (called the *Sheet of Assertion* or SA) is both the site on which graphs are *scribed* and is itself a graph (called the empty graph).
- A *character* is any reproducible image (an iterable symbol, typically letters of the Roman alphabet) scribed on part of the SA.
- Characters may be enclosed, along with a local area surrounding them (a neighborhood on the SA) that may or may not include other characters, by a closed curve called a *sep* (or *cut*). These curves are usually drawn as ovals or circles. These seps may not intersect characters, nor may they intersect one another. They may, however, be nested with any number of characters and seps scribed in the areas or enclosures they distinguish.

The class of EG_α graphs may be characterized through the following recursive definition:

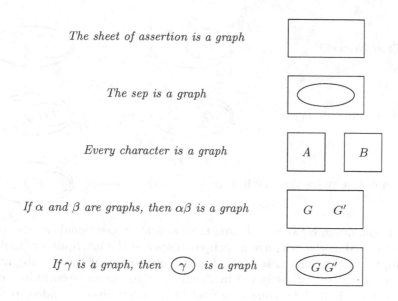

The sheet of assertion is a graph

The sep is a graph

Every character is a graph

If α and β are graphs, then $\alpha\beta$ is a graph

If γ is a graph, then (γ) is a graph

Let us first notice that the EG_α graphs constructed in this way separate regions in the sheet of assertion into evenly and oddly enclosed areas, as the following picture shows.

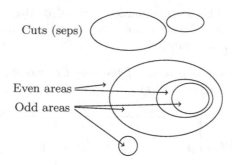

Cuts (seps)

Even areas

Odd areas

We omit a detailed discussion of Peirce's transformation rules for $\mathcal{EG}_\alpha$. A full treatment of these rules may be found in [12]. As a reminder to the reader, examples of all of the rules are provided in an informal, iconic way below.

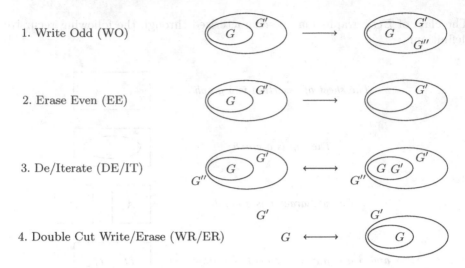

1. Write Odd (WO)

2. Erase Even (EE)

3. De/Iterate (DE/IT)

4. Double Cut Write/Erase (WR/ER)

The correspondence between EG_α and the classical propositional calculus (with seps interpreted as the *negation* of their contents and the inscription of multiple (sub)graphs on the same area as the logical *conjunction* of those (sub)graphs) has been exhaustively investigated by Zeman [16] and more recently in a much broader context in [1]. The iconic quality of Peirce's calculus does indeed possess, as shown by several authors, including the knot theorist Kauffman [7], a truly topological nature. The derivation rules themselves may be characterized in a combinatorial-topological nature, in such a way that the dual nature of a graph as equally syntactical and semantical emerges naturally. In fact, two of the three basic elements of the EG_α syntax, the Sheet of Assertion and the seps, are defined topologically. This feature induces several of the aspects of the EG_α system that distinguish it from the more common linear notation, such as the obviation of axioms of commutativity.

Example 1. *Here below is a proof of the classical* modus ponens *in Peirce's style: from A and $A \to B$ we can infer B.*

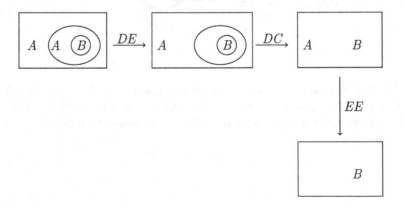

2.2   Existential Graphs: Beta

In order to extend diagrammatic propositional calculus to predicate logic, Peirce introduced what he called the *line of identity*. Essentially, the line of identity is an existential quantifier that augments the language of EG_α to generate a complete and sound first order logic (without free variables). We refer to the excellent work by Dau [6] for details. In what follows we only sketch the main traits of the system, in order to offer the reader an idea of what the graphs in EG_β look like (the structure of their syntax) along with a flavor of their interpretation (their semantics).

First, we assert the existence of an object by drawing a dot on the *SA*.

•

A *line of identity* asserts the equality of all the points on the line.

To express that an object has a certain property P, write the predicate symbol next to the object:

To express that two objects are not the same (notice that the cut is still interpreted as a negation, as in EG_α):

The following examples should help the reader to clarify the interpretation of the EG_β graphs.

Example 2. *The graphs below represent the classical Aristotelean square of opposition. The translation into the classic linear notation of the graphs, starting from the top left in the counterclockwise direction, is the following:*

- $\forall x(B(x) \rightarrow A(x))$: *every B is A;*
- $\forall x(B(x) \rightarrow \neg A(x))$: *no B is A;*
- $\exists x(B(x) \wedge A(x))$: *some B is A;*
- $\exists x(B(x) \wedge \neg A(x))$: *some B is not A.*

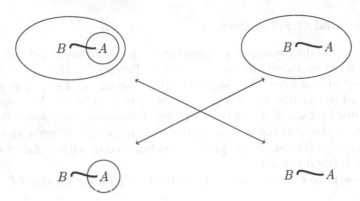

Example 3. *The graph below encloses the property (i.e. the unary relation) "is a woman" together with its single hook within a cut. The same line of identity is connected to the first hook of the triadic relation "gives". Thus, the graph may be understood to assert that* someone who is not a woman gives something to someone.

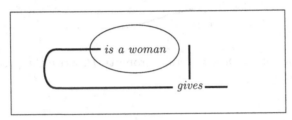

3   Presheaf Categories and Their Generic Figures

In this section we present a brief outline of the mathematics needed to represent the Existential Graphs in a categorical fashion. This presentation follows that given in [2], Appendix A.

3.1   Category Theory: Basic Notions

A mathematical *category* consists of a class of *objects* together with *morphisms* or arrows between objects subject to axioms of *identity* (every object is equipped with an identity morphism that composes inertly), *composition* (head-to-tail morphisms compose to a unique morphism) and *associativity* (paths of morphisms compose uniquely). For a comprehensive introduction and details filling out this rough characterization, see [9].

A bit more formally, a category $\mathcal{C}$ consists of

- a collection of objects $Ob(\mathcal{C})$ and
- a collection of arrows $Ar(\mathcal{C})$ between objects

subject to the following axioms:

- *(A1) Axiom of Composition*: Given any two arrows ordered such that the target of one is the source of the other, the composition of the first followed by the second exists in the category as a unique and definite arrow.

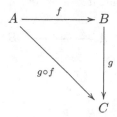

- *(A2) Axiom of Associativity*: Given any three arrows ordered such that the target of the first is the source of the second and the target of the second is the source of the third, the composite of the first two arrows composed with the third is the same as the first arrow composed with the composite of the second and third.

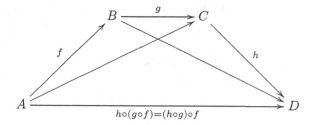

- *(A3) Axiom of Identity*: There exists at least one arrow from any object into itself, called the *identity* arrow.

Example 4. *A few example of categories (objects and arrows):*

- **Set** *(sets and functions)*
- **Mon** *(monoids and monoid homomorphisms)*
- **Grp** *(groups and group homomorphisms)*
- **Vect**$_k$ *(vector spaces over a field k, and linear maps)*
- **Pos** *(partially ordered sets and monotone functions)*
- **Top** *(topological spaces and continuous functions).*

3.2 Functors Between Categories

Categories may be related to one another via mappings called *functors*, which may be understood at a first approach on analogy with functions between sets.

Given two categories $\mathcal{C}$ and $\mathcal{D}$, a functor is in the first place a map F from objects of $\mathcal{C}$ into objects of $\mathcal{D}$ and from arrows of $\mathcal{C}$ into arrows of $\mathcal{D}$, that is, roughly, a function from the set of the objects of C to the set of the objects of D together with a function from the set of arrows of C to the set of arrows of D.[1]

The condition that this map has to satisfy is that (a) relations linking arrows to their source and target objects are preserved and (b) composition relations between arrows are preserved across the mapping.

This amounts to saying that, if

$$A \xrightarrow{\quad f \quad} B$$

is an arrow in C, then

$$F(A) \xrightarrow{\quad F(f) \quad} F(B)$$

is an arrow in D.

Furthermore, the following diagram must commute in D for all suitable objects A, B and C and arrows f, g and $g \circ f$ in C:

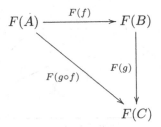

Finally, identity arrows must "track" with their objects across the mapping. Formally, for any object A in C,

$$F(Id_A) = Id_{F(A)}$$

Such a mapping F is called a *covariant functor*. There is a dual notion of *contravariant functor*, which, essentially, instead of "preserving" arrows it "reverses" them. That is, $F : C \longrightarrow D$ is said to be *contravariant* if, given two object A and B in C we have that[2]

$$(F : A \longrightarrow B) \Rightarrow F(f) : F(B) \longrightarrow F(A)$$

---

[1] This formulation only causes difficulties in the (not infrequent) cases when the objects and/or arrows of either C or D cannot be gathered into a set, for instance when one of these is the category **Set** of sets and functions. The ensuing problems and the various strategies for resolving them are readily located in the standard literature on categories.

[2] Contravariant functors also reverse the direction of composition.

3.3   Presheaf Categories and Generic Figures

The approach we follow in this paper was introduced in the book *Generic Figures and their Glueings* by Reyes et al. [10]. In a nutshell, the main idea is to consider classes of structures, defined set-theoretically yet intuitively pictured as types of diagrams (sets, bisets, directed graphs, etc.) and, given that each element in the class is itself built up from determinate *kinds* of ingredients (set-elements represented as dots or points, pairs of such points, vertices and arrows, etc.) a category of *generic figures* may be defined in which those very ingredients are the objects. Such a construction is very general, and lends itself to application to a large class of diagrammatic structures. In fact, in Sect. 4 we will apply this notion to EG_α and EG_β. Here below we outline the steps of this approach.

Given a (small) category $\mathcal{C}$, a functor F from $\mathcal{C}^{op}$ into **Set**, the category of sets and functions, is a *presheaf*. Here, $\mathcal{C}^{op}$ represents the category generated by reversing all the arrows in $\mathcal{C}$ while preserving objects, identity morphisms and composition of morphisms. Schematically, a morphism $f : A \to B$ in $\mathcal{C}$ becomes a morphism $f' : B \to A$ in $\mathcal{C}^{op}$. A natural transformation from one such functor F to another F' is then a family of functions in **Set** that maps one presheaf into another.

More generally, given any two categories $\mathcal{C}$ and $\mathcal{D}$, the *functor category* $\mathcal{D}^{\mathcal{C}}$ is defined as follows:

- The objects of $\mathcal{D}^{\mathcal{C}}$ are all *functors* $\mathcal{C} \to \mathcal{D}$.
- The arrows of $\mathcal{D}^{\mathcal{C}}$ are all *natural transformations* between functors $\mathcal{C} \to \mathcal{D}$.

Natural transformations are morphisms between functors: given two functors $F \in \mathcal{D}^{\mathcal{C}}$ and $G \in \mathcal{D}^{\mathcal{C}}$, a natural transformation between F and G is a family of morphisms η_O parametrized by the objects $O \in \mathcal{C}$ such that the following diagram commutes for any two objects A and B that are connected by a morphism f in $\mathcal{C}$:

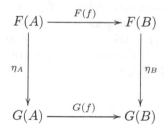

In the case of the functor category of all presheaves over $\mathcal{C}$, the collection of all functors $\mathcal{C}^{op} \to$ **Set** may be taken to be the objects of the new category $\mathbf{Set}^{\mathcal{C}^{op}}$, the morphisms of which are all the natural transformations between these functors. Each functor may be associated with an identity natural transformation. Also, natural transformations compose in the appropriate way and this composition is associative, so the axioms of a category are satisfied. Such a category is called a *presheaf category*. For details, see [10] and [9] ch. 2.

Example: The Category Set$^{\mathcal{C}^{op}}$ of Directed Graphs

Example 5. *Consider the category $\mathcal{C}$ consisting of exactly two objects V and A and two non-identity arrows $s, t : V \to A$. This category is pictured here, with the identity arrows on V and A not shown:*

$$V \underset{t}{\overset{s}{\rightrightarrows}} A$$

Contravariant functors from $\mathcal{C}$ into **Set** *may be thought of as directed graphs. Given such a functor $D: \mathcal{C}^{op} \to$ **Set**, where $D(V)$ is the set of graph vertices and $D(A)$ is the set of graph arrows, $D(s)$ and $D(t)$ are then two functions $D(A) \to D(V)$ assigning a source-vertex and a target-vertex, respectively, to each arrow, that is, each element of $D(A)$.*

The example above illustrates clearly how the structural properties of a given class of diagrams are completely recast into the structure of $\mathcal{C}^{op}$: an arbitrary directed graph G can be encoded/reconstructed by the data provided by a unique associated functor $G : \mathcal{C}^{op} \to$ **Set**. Notice that the structure of $\mathcal{C}^{op}$ in this case is very simple, since it has only two objects and two arrows. This motivates the following definition:

Definition 1. *Given a class of structures $\mathcal{F}$ whose elements are in one-to-one correspondence with the elements of the collection of functors from $\mathcal{C}^{op}$ into* **Set**, *we call* generic figures *the elements of $\mathcal{C}^{op}$.*

4   Existential Graphs as Functor Categories

At this point we are in the position to represent EG_α and EG_β in terms of their generic figures. We will proceed in steps. We first look at EG_{α^*}, defined as the unlabeled version of EG_α, then we consider $EG_{\beta+}$, defined as EG_β without the cuts, and finally we will present the generic figures for the full EG_β system. We will use the italic fonts to refer to the categorical version of each of these systems (so, for instance, we use $\mathcal{EG}_{\alpha^*}$ to denote the category of functors correspondent to the collection of graphs EG_α).

4.1 EG_{α^*}

The iconic syntax of Peirce's EG_{α^*} is given by the class of contravariant functors from the category pictured below, which we notate $\mathcal{A}^*$, into the category **FinSet** of finite sets and functions between these.[3] We denote such a functor category by $\mathcal{EG}_{\alpha^*}$.

$$A_1 \longrightarrow A_2 \longrightarrow A_3 \ldots$$

---

[3] Restriction: for some n, $F(A_n) = \emptyset$.

Notice how the complexity of the nesting of the cuts is represented in an entirely straightforward and natural way by functors from the category of the ordered natural numbers into **FinSet**.

A few examples of cuts-only graphs are given below:

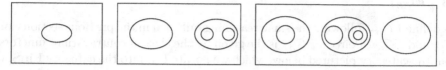

As indicated in the footnote on the previous page, we do require $F(A_i)$ to be definitively equal to the empty set for i greater or equal to a given, finite n. It is only natural to conceive a system in which such a restriction is dropped, and therefore study graphs with infinite depth, so to speak. The notion of *grossone* – a groundbreaking idea developed by Sergeyev (see for instance [13], [14] and [15]) – provides a fundamentally finer structure underlying the infinite, discrete nature of the natural numbers, and we believe that representing graphs of infinite depth in such a framework will set the basis for a progressive research program. Needless to say, a category that captures the nature of grossone would need to be defined, and functors from such a category into the category of sets investigated.

4.2 EG_α

In order to add labeled variables to EG_{α^*}, we need to add the generic figures correspondent to tokens and types of variables, along with their structural (syntactical) relations with the cuts. Here the iconic syntax is given by the category of contravariant functors from the category pictured below, which we notate $\mathcal{A}$, into the category **FinSet** of finite sets and functions between these. We indicate such a functor category by $\mathcal{EG}_\alpha$.

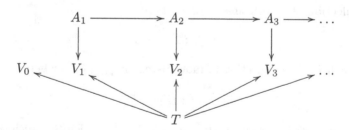

Examples of objects of $\mathcal{EG}_\alpha$ (expressed diagrammatically):

(A, B, C, D) (A, B, C, D, X, Y, Z)

Some readers may find the appearance of the letters at the bottom of each graph not very elegant: we could get rid of them to the cost of adding some conditions on the arrows, which we prefer not to do.

4.3 $EG_{\beta+}$

We define $EG_{\beta+}$ following the same reasoning outlined in the previous sections for EG_α and EG_{α^*}, Peirce's $EG_{\beta+}$ is the given by the class of contravariant functors from the category pictured below, which we notate $\mathcal{B}_+$, into the category **FinSet** of finite sets and functions between these. We call this functor category $\mathcal{EG}_{\beta_+}$.

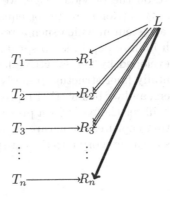

More precisely, the category $\mathcal{B}_+$ consists of objects and morphisms specified as follows:

- Objects: $\{T_i\}_{i\in\mathbb{N}}, \{R_i\}_{i\in\mathbb{N}}, L$
- Morphisms:
  - identities;
  - a collection of morphisms $\{t_i\}_{i\in\mathbb{N}}$ where

$$t_i : T_i \longrightarrow R_i$$

  - for each $i \in \mathbb{N}$ a collection of morphisms $\{r_i^j\}_{j=0,1,\dots,i}$ where

$$r_i^j : L \longrightarrow R_i$$

Formally, then, a $EG_{\beta+}$ graph is a functor $G : \mathcal{B}_+{}^{op} \longrightarrow$ **FinSet** such that there is some n such that for all $m > n$ $G(T_m) = \emptyset$. This latter condition simply ensures for the sake of tidiness that every graph has a maximal relation arity.

Example 6. *The graph below represents the same situation illustrated in Example 3 but without the cut. It may be read as saying that someone who is a woman gives something to someone. Without cuts and, in particular, without nested cuts, neither negation nor universal quantification may be expressed in the reduced system.*

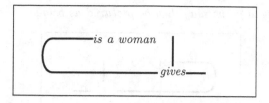

For the sake of clarity, here below we present an example in which the correspondence between a given graph and the correspondent functor is illustrated in detail.

Example 7. *Consider the diagram below, which represents a functor from* $\mathcal{B}_+$ *into* **FinSet**.

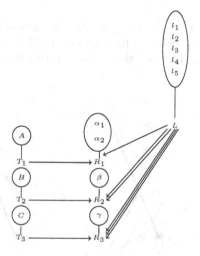

The ovals above each of the category objects represent the sets to which the functor G sends those objects. All objects that are not shown, such as T_4 are understood to be sent to the empty set. For instance, $G(R_1)$ is the two-element set containing α_1 and α_2. The three functions $G(t_1)$, $G(t_2)$ and $G(t_3)$ are completely determined since their codomains are singletons (the reader should keep in mind the contravariance of the functor).

We may stipulate that the remaining functions are defined as follows (the functions are listed in the top row and their argument in the leftmost column):

	$G(r_1^1)$	$G(r_2^1)$	$G(r_2^2)$	$G(r_3^1)$	$G(r_3^2)$	$G(r_3^3)$
α_1	l_1					
α_2	l_5					
β		l_4	l_5			
γ				l_1	l_2	l_3

The resulting beta graph may then be pictured as below:

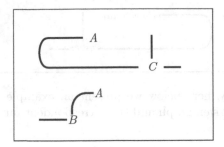

4.4 EG_β

Finally, the full-fledged EG_β system is given by the class of contravariant functors from the category pictured below, which we notate $\mathcal{B}$, into the category **FinSet** of finite sets and functions between these. We call this functor category $\mathcal{EG}_\beta$.

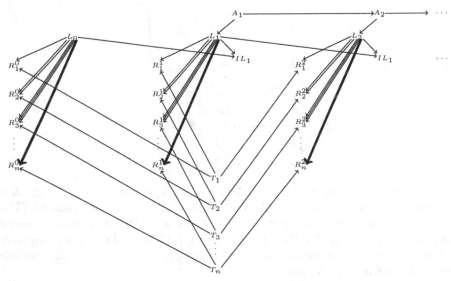

Without specifying all the details, we note only that this category is constructed by, roughly speaking, gluing copies of $\mathcal{B}_+$ below each object of $\mathcal{A}^*$, with one additional copy added at the beginning of the sequence. The objects labeled IL_n represent "lines-in", that is, lines of identity that extend across a cut and are glued to lines of identity on the area of that cut, functioning in this respect like "soldering points". In this way, the structures presented at the *beta* level by lines of identity and n-ary relations are embedded in the nested cut structures formalized at the *alpha* level.

Example 8. *The diagram below presents the graph discussed in Example 7 with two cuts added to it. For the sake of brevity, we omit the details describing the structure of its associated functor. The attentive reader should at this point be able to construct this functor as a useful exercise.*

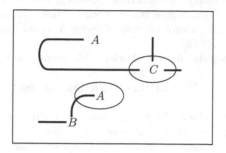

5   Conclusion

We have shown how the generic figures approach within categories of presheaf functors provides an appropriate mathematical setting for representing the diagrammatic syntax of Peirce's Existential Graphs. The same general mathematical setting captures the relevant syntactical structure of diagrams in both *alpha* and *beta* systems, the only difference being the complexity of the base category over which the presheafs are constructed. We remarked above that the grossone research program might find interesting possibilities for syntactical representation in proximity to Peirce's *alpha* system (with the controlled lifting of the restriction on finite nesting of cuts). This suggests one possible path for future research that may very well find additional, perhaps fruitful, connections between the grossone program and Peirce's logic. Another potential line of development would examine the semantics for Peirce's graphs and aim to formulate Peirce's transformation rules in the generic figures framework. Finally, the generic figures approach suggests itself as a natural medium for rigorously formulating a variety of otherwise quite different diagrammatic systems. Perhaps the basis for a general theory of diagrammatic representation may eventually be worked out on this categorical mathematical terrain.

References

1. Brady, G., Trimble, T.H.: A categorical interpretation of C.S Peirce's propositional logic alpha. J. Pure Appl. Algebra **149**, 213–239 (2000)
2. Caterina, G., Gangle, R.: Iconicity and Abduction. SAPERE, vol. 29. Springer, Cham (2016). https://doi.org/10.1007/978-3-319-44245-7
3. Caterina, G., Gangle, R.: The sheet of indication: a diagrammatic semantics for Peirce's EG-alpha. Synthese **192**(4), 923–940 (2015)
4. Caterina, G., Tohmè, F., Gangle, R.: Abduction: a categorical characterization. J. Appl. Logic **13**(1), 78–90 (2015)

5. Coecke, B., Kissinger, A.: Picturing Quantum Processes: A First Course in Quantum Theory and Diagrammatic Reasoning. Cambridge University Press, Cambridge (2017)
6. Dau, F.: The Logic System of Concept Graphs with Negation. LNCS (LNAI), vol. 2892. Springer, Heidelberg (2003). https://doi.org/10.1007/b94030
7. Kauffman, L.: Peirce's existential graphs. Cybern. Hum. Knowing **18**, 49–81 (2001)
8. Ma, M., Pietarinen, A.: Proof analysis of Peirce's alpha system of graphs. Stud. Logica **105**(3), 625–647 (2017)
9. Mac Lane, S.: Categories for the Working Mathematician. Springer, New York (1998)
10. Reyes, M., Reyes, G., Zolfaghari, H.: Generic Figures and their Glueings. Polimetrica, Milan (2004)
11. Spivak, D.: Category Theory for the Sciences. MIT Press, Cambridge (2014)
12. Roberts, D.D.: The Existential Graphs of C.S. Peirce. Mouton, The Hague (1973)
13. Sergeyev, Y.D.: Arithmetic of Infinity. Edizioni Orizzonti Meridionali, Cosenza (2003)
14. Sergeyev, Y.D.: New applied approach for executing computations with infinite and infinitesimal quantities. Informatica **19**(4), 567–596 (2008)
15. Sergeyev, Y.D.: Numerical infinities and infinitesimals: methodology, applications, and repercussions on two Hilbert problems. EMS Surv. Math. Sci. **4**, 219–320 (2017)
16. Zeman, J.: Peirce's logical graphs. Semiotica **12**, 239–256 (1974)

Objective and Violation Upper Bounds on a DIRECT-Filter Method for Global Optimization

M. Fernanda P. Costa[1]([⊠]) [iD], Ana Maria A. C. Rocha[2] [iD],
and Edite M. G. P. Fernandes[2] [iD]

[1] Centre of Mathematics, University of Minho, 4710-057 Braga, Portugal
mfc@math.uminho.pt
[2] ALGORITMI Center, University of Minho, 4710-057 Braga, Portugal
{arocha,emgpf}@dps.uminho.pt

Abstract. This paper addresses the problem of solving a constrained global optimization problem using a modification of the DIRECT method that incorporates the filter methodology to simultaneously minimize the objective function and the constraints violation. Thus, in the "Selection" step of the herein proposed DIRECT-filter algorithm, the hyperrectangles are classified in four categories and subsequently handled separately. The new algorithm also imposes upper bounds on the objective function and constraints violation aiming to discard some hyperrectangles from the process of identifying the potentially optimal ones. A heuristic to avoid the exploration of the hyperrectangles that have been mostly divided is also implemented. Preliminary numerical experiments are carried out to show the effectiveness of the imposed upper bounds on the objective and violation as well as the goodness of the heuristic.

Keywords: Global optimization · DIRECT · Filter method · Heuristic

1 Introduction

The paper aims to address the use of the filter methodology [8] combined with a DIRECT-type method [12] to globally solve non-smooth and non-convex constrained optimization problems. The constrained global optimization (CGO) problem has the form:

$$\min_{x \in \Omega} f(x)$$
$$\text{subject to} \quad h(x) = 0 \tag{1}$$
$$g(x) \leq 0,$$

where $f : \mathbb{R}^n \to \mathbb{R}$, $h : \mathbb{R}^n \to \mathbb{R}^m$ and $g : \mathbb{R}^n \to \mathbb{R}^p$ are nonlinear continuous functions and $\Omega = \{x \in \mathbb{R}^n : -\infty < l_i \leq x_i \leq u_i < \infty, i = 1, \dots, n\}$. Since convexity is not assumed, many local minima may exist in the feasible region, although we require only a global solution. For non-smooth problems, the derivative-free methods are the most appropriate. Popular methods to solve

© Springer Nature Switzerland AG 2020
Y. D. Sergeyev and D. E. Kvasov (Eds.): NUMTA 2019, LNCS 11974, pp. 59–71, 2020.
https://doi.org/10.1007/978-3-030-40616-5_5

problem (1) combine a penalty term, which depends on a constraint violation measure, with the objective function to give the so-called penalty function. The penalty term aims to penalize f whenever an approximation point is found that is infeasible. Penalty functions within a DIRECT-type framework are proposed in [5,21]. An auxiliary function that combines in a special manner information on the objective and constraints is presented in [11]. Other techniques that involve the handling of the objective function and constraints violation separately can be found in [3,4,13].

This paper addresses the exploration of the DIRECT method in order to solve CGO problems. It uses the filter methodology [8] to handle the constraints. The objective function and the constraints violation measure are separately handled and both simultaneously minimized. The main differences relative to the work reported in [3] are the following:

1. four categories of hyperrectangles (according to the violation measure and the non-dominance *vs* dominance feature of their center points) are defined instead of three;
2. upper bounds on the objective and violation values are imposed during the selection step in order to reduce the number of explored hyperrectangles;
3. a heuristic is used to prevent the mostly divided hyperrectangles to be selected and identified as potentially optimal.

The paper is organized as follows. Section 2 briefly presents some ideas and the main steps of the DIRECT method. Section 3 describes the proposed extension to handle CGO problems, in particular, the use of a filter method to classify each hyperrectangle according to its non-dominance/dominance feature and constraints violation magnitude. Further, the strategy that imposes upper bounds on f and violation values, as well as the heuristic are exposed. Finally, Sect. 4 contains the results of our preliminary numerical experiments and we conclude the paper with the Sect. 5.

2   Features About DIRECT Method

The DIRECT (DIviding RECTangles) algorithm, originally proposed to solve bound constrained global optimization problems, assumes that the objective function, f, is a continuous function and creates finer and finer partitions of the hyperrectangles generated from the set Ω [6,7,12]. The algorithm is a modification of the standard Lipschitzian approach, in which f is assumed to satisfy the Lipschitz condition

$$|f(x_1) - f(x_2)| \leq K\|x_1 - x_2\| \text{ for all } x_1, x_2 \in \Omega,$$

where the Lipschitz constant $K > 0$ is viewed as a weighting parameter that indicates how much emphasis to place on global versus local search.

DIRECT is a deterministic and derivative-free method that is able to explore optimal regions aiming to converge to the global optimum and at the same time avoiding being trapped in a local optimum.

DIRECT is described by six main steps: "Initialization", "Selection", "Sampling", "Division", "Iteration" and "Termination" [9, 10, 12].

The "Selection" step serves the purpose of identifying the set of indices of hyperrectangles that are the most promising, denoted by potentially optimal hyperrectangles (POH), based on the current partition of Ω. In the "Sampling" steps, the set of dimensions with the maximum size in each POH is identified to define points where the objective function is evaluated. For the "Division" step, DIRECT uses two measures: (i) the *size* of the hyperrectangle to favor the global search feature of the algorithm; (ii) the *value* of the hyperrectangle to give preference to the local search feature. The *value* corresponds to the objective function value alone at the center, for bound constrained problems (and to the objective function and constraint violation values, when problem (1) is addressed).

For further details on the original DIRECT and other recent interesting modifications, we refer the reader to [14–20, 22].

3   DIRECT-Filter Method

In this section, we reveal how the DIRECT algorithm is extended to incorporate the filter methodology in order to minimize both the objective function and constraints violation. First, we briefly present the filter methodology and the proposed extensions to be incorporated in the main steps of DIRECT. Second, the strategy that uses the upper bounds on objective and violation values and the heuristic to avoid the selection of the mostly divided hyperrectangles are presented.

3.1   Filter Methodology

Based on the filter methodology [1, 8], the problem (1) is reformulated into the following bound constrained bi-objective optimization problem:

$$\min_{x \in \Omega} \ (\theta(x), f(x)), \tag{2}$$

where $\theta(x) = \|h(x)\|_1 + \|g(x)_+\|_1$ is a non-negative function to measure equality and inequality constraints violation, and $g_+ \in \mathbb{R}^p$ is defined componentwise by $\max\{0, g_i\}$, $i = 1, \ldots, p$. A point x is feasible when $\theta(x) = 0$ and is infeasible when $\theta(x) > 0$. While minimizing the constraints violation, θ, and the objective function, f, the filter method builds a region of dominated points that will not be accepted as new approximations to the solution. The concept of dominance arises from the multi-objective optimization area:

Definition 1. *A point x, or the corresponding pair $(\theta(x), f(x))$, is said to dominate y, or the corresponding pair $(\theta(y), f(y))$, denoted by $x \prec y$, if and only if*

$$\theta(x) \le \theta(y) \ \ and \ \ f(x) \le f(y),$$

with at least one inequality being strict.

The filter $\mathcal{F}$ contains a finite set of pairs $(\theta(x), f(x))$, none of which is dominated by any of the others, and the corresponding points x are known as the non-dominated points [8].

Let $x^{(k)}$ be a trial point (approximation to the optimal solution of the CGO problem (1)) and $\mathcal{F}_k$ be the filter, at iteration k, of the algorithm. To avoid the acceptance of the trial point, or the corresponding pair $(\theta(x^{(k)}), f(x^{(k)}))$, that is arbitrary close to the boundary of the filter, the conditions of acceptability define an envelope around the filter and are as follows:

$$\theta(x^{(k)}) \leq (1 - \gamma)\theta(x^l) \ \text{ or } \ f(x^{(k)}) \leq f(x^l) - \gamma\theta(x^l) \tag{3}$$

for all points x^l that correspond to pairs $(\theta(x^l), f(x^l))$ in the filter $\mathcal{F}_k$, where $\gamma \in (0, 1)$ is fixed. When the point is acceptable to the filter, the filter is updated and whenever a point is added to the filter, all the dominated points are removed from the filter.

We note that the filter contains only infeasible points. However, the feasible point with the least function value, denoted by f_{best}, is saved and is used to filter other feasible points.

3.2   Identifying POH in the DIRECT-Filter Method

In the context of solving a CGO problem, the herein proposed algorithm defines two separate regions within the usually called infeasible region. One is denoted by "infeasible" region (identified by I) and contains hyperrectangles with center points c_j that satisfy $\theta(c_j) > \theta_{feas}$, for a sufficiently small positive tolerance θ_{feas}. The other is called "feasible-band" region (identified by FB) and contains the hyperrectangles with center points that satisfy $0 < \theta(c_j) \leq \theta_{feas}$. On the other hand, the herein coined "feasible" region (identified with F) contains hyperrectangles with $\theta(c_j) = 0$.

When applying a DIRECT-type method, in the partition of $\{H^i : i \in I_k\}$ of iteration k, using the filter methodology and the three regions above defined, the identification of POH (in the "Selection" step) is implemented separately for the following four sets of indices:

- the set $I_k^{FB/ND+b}$, contains indices of hyperrectangles with center points in the "feasible-band" region that are non-dominated (FB/ND), appended with the index of the hyperrectangle that corresponds to f_{best} (+b);
- the set $I_k^{FB/D+F\backslash b}$ contains the indices of hyperrectangles with center points in the "feasible-band" region that are dominated (FB/D), appended with the indices of the hyperrectangles with centers in the region F except b (+$F \backslash$ b);
- the set $I_k^{I/ND}$ contains the indices of hyperrectangles with non-dominated center points that are in the "infeasible" region (I/ND);
- the set $I_k^{I/D}$ contains the indices of hyperrectangles with dominated center points that belong to the "infeasible" region (I/D).

As usual, the hyperrectangles are organized by groups of the same *size*. The proposed strategy aims to identify, from each hull, indices of promising hyperrectangles, in terms of the

- optimality measure f, when the indices for exploration belong to the sets $I^{FB/ND+b}$ and $I^{FB/D+F\backslash b}$;
- feasibility measure θ, when the indices belong to the sets $I^{I/ND}$ and $I^{I/D}$.

Thus, in this filter-type method context, the algorithm identifies POH with respect to (w.r.t.) f, using the following definition [12]:

Definition 2. *Given the partition $\{H^i : i \in I\}$ of Ω, let ϵ be a positive constant and let $f_{\min}$ be the current best function value among center points in the regions "feasible" and "feasible-band". A hyperrectangle j is said to be potentially optimal w.r.t. f if there exists some rate-of-change constant $\hat{K} > 0$ such that*

$$f(c_j) - \hat{K}d_j \leq f(c_i) - \hat{K}d_i, \text{ for all } i \in I$$
$$f(c_j) - \hat{K}d_j \leq f_{\min} - \epsilon|f_{\min}| \qquad (4)$$

where c_j is the center, d_j is a measure of the size of the hyperrectangle j (for instance, the distance from c_j to its vertices) and I is $I^{FB/ND+b}$ or $I^{FB/D+F\backslash b}$.

The value of $f_{\min}$ coincides with f_{best} if there are center points with $\theta = 0$; otherwise $f_{\min}$ is set to the least function value of the center points in the region FB/ND.

On the other hand, for the remaining sets of indices (hyperrectangles) where θ is used to define the hull, the algorithm identifies POH w.r.t θ, by adopting the following definition [3,4]:

Definition 3. *Given the partition $\{H^i : i \in I\}$ of Ω, let ϵ be a positive constant. A hyperrectangle j is said to be potentially optimal w.r.t. the function θ if there exists some constant $\hat{K} > 0$ such that*

$$\theta(c_j) - \hat{K}d_j \leq \theta(c_i) - \hat{K}d_i, \text{ for all } i \in I$$
$$\theta(c_j) - \hat{K}d_j \leq \theta_{\min} - \epsilon\theta_{\min} \qquad (5)$$

where $\theta_{\min} > 0$ is the θ value that corresponds to $f_{\min}$ if the "feasible-band" region is non-empty; otherwise is the least value of θ reached by a point in the "infeasible" region. The set I is $I^{I/ND}$ or $I^{I/D}$.

3.3 Objective and Violation Upper Bounds

We now show how upper bounds on objective function and constraints violation, denoted by f^U and θ^U respectively, are imposed in a way that hyperrectangles with f and/or θ values greater than the corresponding upper bounds are not considered in the "Selection" step to identify POH. The bounds f^U and θ^U are defined at each iteration and depend on the information available at that moment.

Thus, the bound on f to apply to the set $I^{FB/D+F\backslash b}$ is defined by

$$f^U_{FB} = f_{FB} + \beta_f|f_{FB}| \text{ with } f_{FB} = \max\{f^{FB/ND}_{\max}, f_{best}\},$$

where $f_{\max}^{FB/ND}$ - directly identified from the filter $\mathcal{F}$ - is the f value of the center of the hyperrectangle with the lowest θ value among the hyperrectangles with center in the region FB/ND, and $\beta_f \geq 0$ is a constant factor.

On the other hand, the bound on f to apply to the set $I^{I/D}$ is defined by

$$f_I^U = f_I + \beta_f |f_I|$$

where f_I - directly identified from the filter - is the f value of the center of the hyperrectangle with the lowest θ value among the hyperrectangles with center in the region I/ND. This θ value will be denoted by $\theta_{\min} > \theta_{feas}$.

Moreover, θ^U is computed using $\theta_{\min}$ as follows:

$$\theta^U = \theta_{\min} + \beta_\theta \theta_{\min}$$

where $\beta_\theta > 0$ is a constant factor. This upper bound on θ is applied only to the sets $I^{I/ND}$ and $I^{I/D}$, since the other two are naturally bounded by θ_{feas}.

From hereafter, we denote the basic DIRECT-filter method (as described in the previous subsection) by "DIRECT-f" and the variant that incorporates the upper bounds on f and θ (as reported here in this subsection) by "UB-DIRECT-f".

3.4   Heuristic

Besides using the above described upper bounds, the "UB-DIRECT-f" algorithm can be enhanced with a heuristic that aims to avoid identifying POH among those hyperrectangles that were mostly divided [17].

The heuristic is applied only to the two sets of indices $I^{FB/D+F\setminus \mathsf{b}}$ and $I^{I/D}$. Thus, hyperrectangles with indices based on *size* that are larger than $\lfloor i_b/4 \rfloor$ are discarded, where $\lfloor t \rfloor$ gives the greatest integer less than or equal to t, and i_b is the index based on the *size* of the hyperrectangle that corresponds to

- $f_{\min}$, when the hull from the set $I^{FB/D+F\setminus \mathsf{b}}$ is explored;
- $\theta_{\min}$, when the hull from the set $I^{I/D}$ is explored.

(We note that the larger the *size*, the smaller is the index based on *size*.)

This heuristic runs for a cycle of 10 iterations and aims to potentiate the exploration of hyperrectangles of larger *sizes* in order to identify POH. With this selection, global information during the search is reinforced and the likelihood is that $f_{\min}$ and/or $\theta_{\min}$ may be improved. This cycle of iterations is implemented every 10 iterations of the original "UB-DIRECT-f". While the heuristic is active, the upper bounds on f and θ are disabled. This variant is denoted by "UB-DIRECT-f+Heur" in the subsequent tables of results.

4   Numerical Experiments

During the preliminary numerical experiments, a set of benchmark problems is used. The MATLAB® (MATLAB is a registered trademark of the MathWorks, Inc.) programming language is used to code the algorithm and the tested problems.

Unless otherwise stated, the stopping conditions for the algorithm are the following. We consider that a good approximate solution $x^{(k)}$, at iteration k, is found, if the conditions

$$\theta(x^{(k)}) \leq \eta_1 \text{ and } p_{error} \equiv \frac{|f(x^{(k)}) - f^*|}{\max\{1, |f^*|\}} \leq \eta_2 \tag{6}$$

are satisfied, for sufficiently small tolerances $\eta_1, \eta_2 > 0$, where f^* is the best known solution to the problem. However, if conditions (6) are not satisfied, the algorithm runs until a maximum number of function evaluations, nfe_{max}, is reached.

The parameter values for the algorithm are set as follows: $\gamma = 1E{-}05$, $\theta_{feas} = 1E{-}04$, $\epsilon = 1E{-}04$, $\beta_f = 1.1$, $\beta_\theta = 1E{+}04$, $\eta_1 = 1E{-}04$, $\eta_2 = 1E{-}04$ and $nfe_{max} = 1E{+}06$. (We note that a smaller value of β_f was also tested but the reported choice gave better results specially for the larger problems.)

Our goal is to reveal the effectiveness of the proposed objective function and constraint violation upper bounds in reducing the computational burden without affecting the robustness of the DIRECT-filter method.

Table 1 presents a comparison of our solutions with others reported in the literature, when solving the problem "Gomez #3" [11], with global optimum value $f^* = -0.9711$, occurring at $(0.109, -0.623)$:

$$\min_{x \in \Omega} \left(4 - 2.1x_1^2 + \frac{x_1^4}{3}\right) x_1^2 + x_1 x_2 + (-4 + 4x_2^2)x_2^2$$
$$\text{subject to } -\sin(4\pi x_1) + 2\sin^2(2\pi x_2) \leq 0$$

with $\Omega = \{x \in \mathbb{R}^2 : -1 \leq x_i \leq 1, i = 1, 2\}$. The solutions reported in the table have 1% and 0.01% error relative to the known global solution. The results are compared to those available in [11] and to another filter-based DIRECT algorithm (in [3]). We can see that the implementation of the upper bounds on f and θ as well as the heuristic make the DIRECT-filter method more efficient.

To compare the results to those in [21] (variants DIRECT-GLc and DIRECT-GLce), problem "T1" (with several instances depending on n) is used:

$$\min_{x \in \Omega} \sum_{i=1}^{n} x_i$$
$$\text{subject to } \sum_{i=1}^{n} x_i^2 \leq 6$$

with $\Omega = \{x \in \mathbb{R}^n : -1 \leq x_i \leq 1, i = 1, \ldots, n\}$. The algorithms stop with the condition $p_{error} \leq 1E{-}04$ alone (or a maximum of $1E{+}06$ function evaluations). See Table 2. Although we are not yet able to achieve convergence before $1E{+}06$ function evaluations on the larger instances, $n = 5$ and $n = 6$ of the problem

Table 1. Comparison results when solving problem "Gomez #3".

Algorithm	p_{error}	$f(x^{(k)})$	$\theta(x^{(k)})$	k	nfe	f^*
"DIRECT-f"	1%	−0.961782	0.00E+00	9	185	−0.9711
"UB-DIRECT-f"		−0.961782	0.00E+00	9	225	
"UB-DIRECT-f+Heur"		−0.961782	0.00E+00	10	149	
In [3]		–	–	9	219	
In [11]		–	–	–	89	
"DIRECT-f"	0.01%	−0.971006	6.00E−05	17	615	
"UB-DIRECT-f"		−0.971006	6.00E−05	17	683	
"UB-DIRECT-f+Heur"		−0.971041	3.17E−05	18	555	
In [3]		–	–	18	733	
In [11]		–	–	–	513	

"T1", the results obtained by "UB-DIRECT-f+Heur" for the other instances outperform the others in comparison.

To analyze the quality of the obtained solutions we use problem "5" (available in [2]):

$$\min_{x \in \Omega} x_3$$
$$\text{subject to} \quad 30x_1 - 6x_1^2 - x_3 = -250$$
$$20x_2 - 12x_2^2 - x_3 = -300$$
$$0.5(x_1 + x_2)^2 - x_3 = -150$$

with $\Omega = \{x \in \mathbb{R}^3 : 0 \le x_1 \le 9.422, 0 \le x_2 \le 5.903, 0 \le x_3 \le 267.42\}$ and problem "8" [2]:

$$\min_{x \in \Omega} x_1^4 - 14x_1^2 + 24x_1 - x_2^2$$
$$\text{subject to} \quad x_2 - x_1^2 - 2x_1 \le -2$$
$$-x_1 + x_2 \le 8$$

with $\Omega = \{x \in \mathbb{R}^2 : -8 \le x_1 \le 10, 0 \le x_2 \le 10\}$ and stop the algorithm after $k_{max} = 20$ iterations and then after $k_{max} = 50$ iterations. The results are compared to those obtained previously in [3], and are shown in Table 3. On the other hand, to analyze the gain in efficiency of the present algorithm variants, Table 4 reports the best f and θ values obtained by the algorithms when the stopping conditions in (6) are used. The gain in quality and efficiency of the proposed DIRECT-filter method, in particular when the upper bounds on f and θ, and the heuristic are implemented, have been once more demonstrated with the problems "5" and "8". The results reported in [3] and those obtained by variants DIRECT-GLc and DIRECT-GLce in [21] are also used in the comparison.

Figures 1(a), (b) and (c) show the center points generated by the three variants of the DIRECT-filter method when solving the problem "8". Feasible points are marked with '+' (blue) and infeasible points with '×' (red). It can be seen that the variant "UB-DIRECT-f+Heur" is more effective in reaching the

Table 2. Comparison results when solving problem "T1".

	Algorithm	$f(x^{(k)})$	$\theta(x^{(k)})$	k	nfe	f^*
$n = 2$	"DIRECT-f"	-3.464106	9.29E$-$05	14	1395	-3.4641
	"UB-DIRECT-f"	-3.464106	9.29E$-$05	14	893	
	"UB-DIRECT-f+Heur"	-3.464106	5.72E$-$05	13	335	
	DIRECT-GLc	$-$	$-$	$-$	1373	
	DIRECT-GLce	$-$	$-$	$-$	2933	
$n = 3$	"DIRECT-f"	-4.242443	0.00E+00	28	16885	-4.2426
	"UB-DIRECT-f"	-4.242443	0.00E+00	35	37977	
	"UB-DIRECT-f+Heur"	-4.242443	9.17E$-$05	29	3233	
	DIRECT-GLc	$-$	$-$	$-$	26643	
	DIRECT-GLce	$-$	$-$	$-$	8297	
$n = 4$	"DIRECT-f"	-4.898847	0.00E+00	42	151753	-4.899
	"UB-DIRECT-f"	-4.898847	3.42E$-$05	39	78859	
	"UB-DIRECT-f+Heur"	-4.898440	3.30E$-$05	51	36219	
	DIRECT-GLc	$-$	$-$	$-$	192951	
	DIRECT-GLce	$-$	$-$	$-$	47431	
$n = 5$	"DIRECT-f"	(-5.470982)	$(6.65\mathrm{E}-05)$	(61)	$>$1E+06	-5.4772
	"UB-DIRECT-f"	(-5.470711)	$(0.00\mathrm{E}+00)$	(63)	$>$1E+06	
	"UB-DIRECT-f+Heur"	(-5.474293)	$(1.00\mathrm{E}-04)$	(117)	$>$1E+06	
	DIRECT-GLc	$-$	$-$	$-$	253805	
	DIRECT-GLce	$-$	$-$	$-$	78257	
$n = 6$	"DIRECT-f"	(-5.991770)	$(0.00\mathrm{E}+00)$	(45)	$>$1E+06	-6.0000
	"UB-DIRECT-f"	(-5.996647)	$(0.00\mathrm{E}+00)$	(50)	$>$1E+06	
	"UB-DIRECT-f+Heur"	(-5.988112)	$(0.00\mathrm{E}+00)$	(79)	$>$1E+06	
	DIRECT-GLc	$-$	$-$	$-$	239697	
	DIRECT-GLce	$-$	$-$	$-$	135843	

In parentheses, the achieved values when the algorithm stops due to $nfe > 1E+06$

solution. The points cluster around the global solution, being "UB-DIRECT-f+Heur" the one that concentrates the search the most. Figure 1(d) shows the pairs (θ, f) corresponding to the center points of all the hyperrectangles generated by variant "UB-DIRECT-f+Heur". Dominated points are marked with 'circle' (red) and non-dominated points (or filter points) are marked with 'full circle' (blue). The smaller plot shows an overview of the filter points.

Table 3. Quality of the results when solving problems "5" and "8".

Prob.	Algorithm	$k = k_{\max}$	$|f(x^{(k)}) - f^*|$	$\theta(x^{(k)})$	nfe	f^*
"5"	"DIRECT-f"	20	2.512E−04	5.92E−03	471	201.16
	"UB-DIRECT-f"		2.512E−04	5.92E−03	471	
	"UB-DIRECT-f+Heur"		2.512E−04	5.92E−03	379	
	In [3]		2.512E−04	5.92E−03	(471)	
	"DIRECT-f"	50	6.819E−04	9.44E−05	3307	
	"UB-DIRECT-f"		6.819E−04	9.44E−05	2653	
	"UB-DIRECT-f+Heur"		6.819E−04	9.44E−05	2167	
	In [3]		6.819E−04	9.55E−05	(2827)	
"8"	"DIRECT-f"	20	9.756E−04	0.00E+00	881	−118.70
	"UB-DIRECT-f"		9.756E−04	0.00E+00	873	
	"UB-DIRECT-f+Heur"		7.611E−02	5.08E−05	587	
	In [3]		5.372E−02	0.00E+00	(717)	
	"DIRECT-f"	50	3.724E−03	9.85E−05	3363	
	"UB-DIRECT-f"		3.724E−03	9.85E−05	2715	
	"UB-DIRECT-f+Heur"		2.993E−03	9.82E−05	1971	
	In [3]		3.623E−03	9.62E−05	(3333)	

In parentheses, values computed for the comparison, but not reported in [3]

Table 4. Efficiency when solving problems "5" and "8".

Prob.	Algorithm	k	$f(x^{(k)})$	$\theta(x^{(k)})$	nfe	f^*
"5"	"DIRECT-f"	30	201.159343	7.83E−05	1015	201.16
	"UB-DIRECT-f"	30	201.159343	7.83E−05	883	
	"UB-DIRECT-f+Heur"	30	201.159343	7.83E−05	769	
	In [3]	30	201.159343	7.83E−05	1009	
	DIRECT-GLc	−	201.1593	−	819	
	DIRECT-GLce	−	201.1593	−	819	
"8"	"DIRECT-f"	19	−118.700976	0.00E+00	823	−118.70
	"UB-DIRECT-f"	19	−118.700976	0.00E+00	797	
	"UB-DIRECT-f+Heur"	23	−118.692210	0.00E+00	689	
	In [3]	23	−118.700976	0.00E+00	881	
	DIRECT-GLc	−	−118.6892	−	1197	
	DIRECT-GLce	−	−118.6898	−	1947	

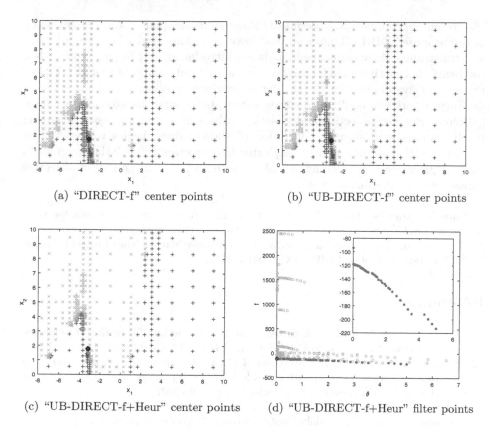

(a) "DIRECT-f" center points   (b) "UB-DIRECT-f" center points

(c) "UB-DIRECT-f+Heur" center points   (d) "UB-DIRECT-f+Heur" filter points

Fig. 1. Plots of center points in problem "8": '+' (blue) - feasible points and '×' (red) - infeasible points; 'circle' (red) - dominated points and 'full circle' (blue) - non-dominated points (Color figure online)

5   Conclusions

In this paper, we present an extension of the DIRECT method for solving equality and inequality constrained global optimization problems. The extension integrates the filter methodology into the DIRECT and aims to minimize both the objective function and the constraints violation simultaneously. The use of the filter method allows the classification of the hyperrectangles, through the objective and violation values of their center points, in four categories. Features like non-dominance/dominance and almost feasible/infeasibility are used to classify and separately handle the hyperrectangles. Furthermore, upper bounds on the objective function and on the constraints violation are imposed to identify the hyperrectangles that should be avoided from the process of selecting the most promising hyperrectangles. Furthermore, a heuristic that avoids the identification of potentially optimal hyperrectangles, among those that were mostly divided, has been cyclically (every 10 iterations) implemented.

Preliminary numerical experiments show that the quality and the efficiency of the proposed DIRECT-filter method have been improved when the objective and constraints violation upper bounds are introduced, and in particular, when the heuristic is activated. The comparison carried out with other DIRECT-type methods is encouraging for the smaller dimensional problems.

Future work will be directed to generate upper bounds based on information gathered from the objective and violation values from each category, resorting to the average and standard deviation of those values. Issues related to the extension of the heuristic to avoid exploring hyperrectangles with the larger sizes, while focusing on hyperrectangles with very small violation and lower objective values, will require further work.

Acknowledgments. The authors wish to thank two anonymous referees for their comments and suggestions to improve the paper. This work has been supported by FCT – Fundação para a Ciência e Tecnologia within the Projects Scope: UID/CEC/00319/2019 and UID/MAT/00013/2013.

References

1. Audet, C., Dennis Jr., J.E.: A pattern search filter method for nonlinear programming without derivatives. SIAM J. Optim. **14**(4), 980–1010 (2004)
2. Birgin, E.G., Floudas, C.A., Martínez, J.M.: Global minimization using an Augmented Lagrangian method with variable lower-level constraints. Technical report MCDO121206, University of São Paulo, SP, Brazil (2007)
3. Costa, M.F.P., Rocha, A.M.A.C., Fernandes, E.M.G.P.: Filter-based DIRECT method for constrained global optimization. J. Global Optim. **71**(3), 517–536 (2018)
4. DiPillo, G., Liuzzi, G., Lucidi, S., Piccialli, V., Rinaldi, F.: A DIRECT-type approach for derivative-free constrained global optimization. Comput. Optim. Appl. **65**(2), 361–397 (2016)
5. DiPillo, G., Lucidi, S., Rinaldi, F.: A derivative-free algorithm for constrained global optimization based on exact penalty functions. J. Optim. Theory Appl. **164**(3), 862–882 (2015)
6. Finkel, D.E., Kelley, C.T.: Convergence analysis of the DIRECT algorithm. Technical report TR04-28, North Carolina State University, NC, USA (2004)
7. Finkel, D.E., Kelley, C.T.: Additive scaling and the DIRECT algorithm. J. Global Optim. **36**(4), 597–608 (2006)
8. Fletcher, R., Leyffer, S.: Nonlinear programming without a penalty function. Math. Program. Ser. A **91**(2), 239–269 (2002)
9. Gablonsky, J.M.: DIRECT version 2.0 user guide. Technical report TR-01-08, North Carolina State University, NC, USA (2001)
10. Gablonsky, J.M., Kelley, C.T.: A locally-biased form of the DIRECT algorithm. J. Global Optim. **21**(1), 27–37 (2001)
11. Jones, D.R.: Direct global optimization algorithm. In: Floudas, C., Pardalos, P. (eds.) Encyclopedia of Optimization, pp. 431–440. Springer, Boston (2008). https://doi.org/10.1007/0-306-48332-7_93
12. Jones, D.R., Perttunen, C.D., Stuckman, B.E.: Lipschitzian optimization without the Lipschitz constant. J. Optim. Theory Appl. **79**(1), 157–181 (1993)

13. Liu, H., Xu, S., Chen, X., Wang, X., Ma, Q.: Constrained global optimization via a DIRECT-type constraint-handling technique and an adaptive metamodeling strategy. Struct. Multidiscip. Optim. **55**(1), 155–177 (2017)
14. Liu, Q., Cheng, W.: A modified DIRECT algorithm with bilevel partition. J. Global Optim. **60**(3), 483–499 (2014)
15. Liuzzi, G., Lucidi, S., Piccialli, V.: Exploiting derivative-free local searches in DIRECT-type algorithms for global optimization. Comput. Optim. Appl. **65**(2), 449–475 (2016)
16. Paulavičius, R., Žilinskas, J.: Simplicial Lipschitz optimization without the Lipschitz constant. J. Global Optim. **59**(1), 23–40 (2014)
17. Paulavičius, R., Sergeyev, Y.D., Kvasov, D.E., Žilinskas, J.: Globally-biased DIS-IMPL algorithm for expensive global optimization. J. Global Optim. **59**(2–3), 545–567 (2014)
18. Sergeyev, Y.D., Kvasov, D.E.: Global search based on efficient diagonal partitions and a set of Lipschitz constants. SIAM J. Optim. **16**(3), 910–937 (2006)
19. Sergeyev, Y.D., Kvasov, D.E.: Deterministic Global Optimization: An Introduction to the Diagonal Approach. SpringerBriefs in Optimization Series. Springer, New York (2017). https://doi.org/10.1007/978-1-4939-7199-2
20. Stripinis, L., Paulavičius, R., Žilinskas, J.: Improved scheme for selection of potentially optimal hyper-rectangles in DIRECT. Optim. Lett. **12**(7), 1699–1712 (2018)
21. Stripinis, L., Paulavičius, R., Žilinskas, J.: Penalty functions and two-step selection procedure based DIRECT-type algorithm for constrained global optimization. Struct. Multidiscip. Optim. **59**(6), 2155–2175 (2019)
22. Wu, Y., Ozdamar, L., Kumar, A.: TRIOPT: a triangular-based partitioning algorithm for global optimization. J. Comput. Appl. Math. **177**, 35–53 (2005)

The Approximate Synthesis of Optimal Control for Heterogeneous Discrete Systems with Intermediate Criteria

Olga Danilenko[1]([✉])(iD) and Irina Rasina[2](iD)

[1] Trapeznikov Institute of Control Sciences of RAS,
65 Profsoyuznaya Street, Moscow 117997, Russia
olga@danilenko.org
[2] Ailamazyan Program Systems Institute of RAS,
4a Petra Pervogo Street, Perslavl-Zalessky 152020, Russia
irinarasina@gmail.com

Abstract. We consider one of the classes of hybrid systems, heterogeneous discrete systems (HDSs). The mathematical model of an HDS is a two-level model, where the lower level represents descriptions of homogeneous discrete processes at separate stages and the upper (discrete) level connects these descriptions into a single process and controls the functioning of the entire system to ensure a minimum of functionality. In addition, each homogeneous subsystem has its own goal. A method of the approximate synthesis of optimal control is constructed on the basis of Krotov-type sufficient optimality conditions obtained for such a model in two forms. A theorem on the convergence of the method with respect to a function is proved, and an illustrative example is given.

Keywords: Heterogeneous discrete system · Intermediate criteria · Approximate synthesis · Optimal control

1 Introduction

The direct use of the optimal control theory's theoretical results is associated with insurmountable difficulties regarding the solvability of practical problems in analytical form. Therefore, theoretical results have always been accompanied by the construction and development of various iterative methods. It is nearly impossible to track the many works that represent various scientific schools and areas. Therefore, generalization and analogs of Krotov's sufficient optimality conditions [1] in two forms will be used substantially in this paper. Some insight into this field is given via an overview [2] and several publications [3–5].

The approach that is proposed in [6] is based on an interpretation of the abstract model of multi-step controlled processes [7] as a discrete-continuous system and extended to heterogeneous discrete systems (HDS) [8]. This method has essentially allowed the decomposition of the inhomogeneous system into homogeneous subsystems by constructing a two-level hierarchical model and

© Springer Nature Switzerland AG 2020
Y. D. Sergeyev and D. E. Kvasov (Eds.): NUMTA 2019, LNCS 11974, pp. 72–83, 2020.
https://doi.org/10.1007/978-3-030-40616-5_6

generalizing optimality conditions and optimization algorithms that were developed for homogeneous systems. This refers to systems with a fixed structure that are studied within the classical theory of optimal control.

Notably, with this approach, all homogeneous subsystems are linked by a common goal and represented by a function in the model. However, each homogeneous subsystem can also have its own goal. Such a generalization of the HDS model was carried out in [11], where sufficient conditions for optimal control in two forms were obtained.

In this paper a method of approximate synthesis of optimal control is constructed, and an illustrative example is considered.

Previously, the authors proposed a more sophisticated improvement method [12] for another class of heterogeneous systems, discrete-continuous systems, that requires searching for a global extremum in control variables at both levels of the hierarchical model. For the class of heterogeneous discrete systems considered in the present paper, the derivation of its analogue is not possible due to the structural features of the discrete models and the construction of sufficient optimality conditions.

2   Heterogeneous Discrete Processes with Intermediate Criteria

Let us consider a two-level model where the lower level consists of discrete dynamic systems of homogeneous structure. A discrete model of general form appears on the top level.

$$x(k+1) = f(k, x(k), u(k)),$$
$$k \in \mathbf{K} = \{k_I, k_I + 1, ..., k_F\}, \quad u \in \mathbf{U}(k, x), \tag{1}$$

where k is the number of the step, x and u are respectively variables of state and control of arbitrary nature (possibly different) for different k, and $\mathbf{U}(k, x)$ is the set given for each k and x. On some subset $\mathbf{K}' \subset \mathbf{K}$, $k_F \notin \mathbf{K}'$, $u(k)$ is interpreted as a pair $\left(u^v(k), m^d(k)\right)$, where $m^d(k)$ is a process $(x^d(k, t), u^d(k, t))$, $t \in \mathbf{T}(k, z(k))$, $m^d(k) \in \mathbf{D}^d(k, z(k))$, and $\mathbf{D}^d$ is the set of admissible processes m^d, complying with the system

$$x^d(k, t+1) = f^d\left(k, z, t, x^d(k, t), u^d(k, t)\right),$$
$$t \in \mathbf{T} = \{t_I(z), t_I(z) + 1, ... t_F(z)\}, \tag{2}$$

$$x^d \in \mathbf{X}^d(k, z, t), \quad u^d \in \mathbf{U}^d\left(k, z, t, x^d\right), \quad z = (k, x, u^v).$$

For this system an intermediate goal is defined on the set $\mathbf{T}$ in the form of a functional that needs to be minimized:

$$I^k = \sum_{\mathbf{T}(z) \backslash t_F(z)} f^k(t, x^d(k, t), u^d(k, t)) \to \inf.$$

Here $\mathbf{X}^d(k, z, t)$, $\mathbf{U}^d\left(k, z, t, x^d\right)$ are given sets for each t, z, and x^d. The right-hand side operator of the 1 is the following on the set $\mathbf{K}'$:

$$f(k, x, u) = \theta\left(z, \gamma^d(z)\right), \quad \gamma^d = \left(t_I, x_I^d, t_F, x_F^d\right) \in \mathbf{\Gamma}^d(k, z),$$

$$\mathbf{\Gamma}^d(z) = \{\gamma^d \colon t_I = \tau(k, z), t_F = \vartheta(k, z),$$
$$x_I^d = \xi(k, z), \quad x_F^d \in \mathbf{\Gamma}_F^d(k, z)\}.$$

On the set $\mathbf{D}$ of the processes

$$m = \left(x(k), u(k), x^d(k, t), u^d(k, t)\right),$$

satisfying 1, 2, the optimal control problem on minimization of a terminal functional $I = F\left(x\left(k_F\right)\right)$ is considered. Here $k_I = 0$, k_F, $x\left(k_I\right)$ are fixed and $x(k) \in \mathbf{X}(k)$.

3 Sufficient Optimality Conditions

The following theorems are valid [11]:

Theorem 1. *Let there be a sequence of processes* $\{m_s\} \subset \mathbf{D}$ *and functions* φ, φ^d *such that:*

(1) $R\left(k, x_s\left(k\right), u_s\left(k\right)\right) \to \mu\left(k\right)$, $k \in \mathbf{K}$;
(2) $R^d\left(z_s, t, x_s^d\left(t\right), u_s^d\left(t\right)\right) - \mu^d\left(z_s, t\right) \to 0$, $k \in \mathbf{K}'$, $t \in \mathbf{T}\left(z_s\right)$;
(3) $G^d\left(z_s, \gamma_s^d\right) - l^d\left(z_s\right) \to 0$, $k \in \mathbf{K}'$;
(4) $G\left(x_s\left(t_F\right)\right) \to l$.

Then the sequence $\{m_s\}$ *is a minimizing sequence for* I *on the set* $\mathbf{D}$.

Theorem 2. *For each element* $m \in \mathbf{D}$ *and any functionals* φ, φ^d *the estimate is*

$$I(m) - \inf_{\mathbf{D}} I \leq \Delta = I(m) - l.$$

Let there be two processes $m^{\mathrm{I}} \in \mathbf{D}$ *and* $m^{\mathrm{II}} \in \mathbf{E}$ *and functionals* φ *and* φ^d *such that* $L\left(m^{\mathrm{II}}\right) < L\left(m^{\mathrm{I}}\right) = I\left(m^{\mathrm{I}}\right)$, *and* $m^{\mathrm{II}} \in \mathbf{D}$.
Then $I(m^{\mathrm{II}}) < I(m^{\mathrm{I}})$.

Here:

$$L = G\left(x\left(k_F\right)\right) - \sum_{\mathbf{K} \setminus \mathbf{K}' \setminus k_F} R(k, x(k), u(k))$$

$$+ \sum_{\mathbf{K}'} \left(G^d(z) - \sum_{\mathbf{T}(z) \setminus t_F} R^d(z, t, x^d(k, t), u^d(k, t))\right),$$

$$G\left(x\right) = F\left(x(k_F)\right) + \varphi\left(k_F, x\right) - \varphi\left(k_I, x\left(k_I\right)\right),$$

$$R\left(k, x, u\right) = \varphi\left(k + 1, f\left(k, x, u\right)\right) - \varphi\left(k, x\right),$$

$$G^d\left(k, z, \gamma^d\right) = -\varphi\left(k + 1, \theta\left(k, z, \gamma^d\right)\right) + \varphi\left(k, x\left(k\right)\right)$$
$$+ \varphi^d\left(k, z, t_F, x_F^d\right) - \varphi^d\left(k, z, t_I, x_I^d\right),$$

$$R^d\left(k, z, t, x^d, u^d\right) = \varphi^d(k, z, t + 1, f^d\left(k, z, t, x^d, u^d\right))$$
$$- f^k(t, x^d(k, t), u^d(k, t)) - \varphi^d(k, z, t, x^d),$$

$$\mu^d\left(k,z,t\right) = \sup\ \{R^d\left(k,z,t,x^d,u^d\right) : x^d \in \mathbf{X}^d(k,z,t), u^d \in \mathbf{U}^d\left(k,z,t,x^d\right)\},$$

$$l^d\left(k,z\right) = \inf\ \{G^d\left(k,z,\gamma^d\right) : \left(\gamma^d\right) \in \mathbf{\Gamma}^d(k,z), x^d \in \mathbf{X}^d(k,z,t_F)\},$$

$$\mu\left(k\right) = \begin{cases} \sup\{R\left(k,x,u\right) : x \in \mathbf{X}(k), u \in \mathbf{U}\left(k,x\right)\}, & t \in \mathbf{K}\backslash\mathbf{K}', \\ -\inf\{l^d\left(z\right) : x \in \mathbf{X}\left(k\right),\ u^v \in \mathbf{U}^v\left(k,x\right)\}, & k \in \mathbf{K}', \end{cases}$$

$$l = \inf\{G\left(x\right) : x \in \mathbf{\Gamma} \cap \mathbf{X}\left(k_F\right)\}.$$

Here $\varphi\left(k,x\right)$ is an arbitrary functional and $\varphi^d(k,z,t,x^d)$ is an arbitrary parametric family of functionals with parameters k and z.

We note that $L(m)$ and $I(m)$ coincide for $m \in \mathbf{D}$.

Theorem 1 allows us to reduce the solution of the optimal control problem posed to an extremum study of the constructions R, G and R^d, G^d by the arguments for each k and t, respectively. Theorem 2 indicates a way to construct improvement methods. One of the variants of these methods is implemented below.

4   Sufficient Conditions in the Bellman Form

One of the possible ways to set a pair $(\varphi,\ \tilde{\varphi}^d)$ is to require fulfillment of condition $\inf\limits_{\{m_u\}} L = 0$ for any m_x. Here $m_u = (u(k),\ u^v(k),\ u^d(k,t))$ is a set of control functions from the sets $\mathbf{U}$, $\mathbf{U}^v$, and $\mathbf{U}^d$, respectively, $m_x = (x(k),\ \tilde{x}^d(k,t))$ is a set of state variables of upper and lower levels. Such a requirement leads directly to concrete optimality conditions of the Bellman type that can also be used to construct effective iterations of process improvement. Let $\mathbf{\Gamma}_F^d\left(z\right) = \mathbb{R}^{n(k)}$, $\theta\left(z,\gamma^d\right) = \theta\left(z,x_F^d\right)$. There are no other restrictions on the state variables.

The following recurrent chain is obtained with respect to the Krotov-Bellman functionals φ and $\varphi^d\left(z\right)$ of two levels:

$$\varphi\left(k,x\right) = \sup_{u\in\mathbf{U}(k,x)} \varphi\left(k+1, f\left(k, x\left(k\right), u\right)\right), \quad k \in \mathbf{K}\backslash\mathbf{K}'\backslash k_F,$$

$$\varphi\left(k_F, x\right) = -F\left(x\right),$$

$$\varphi^d(k,t) = \sup_{u^d\in\mathbf{U}^d(z,t,x^d)} \left(\varphi^d\left(k,t+1, f^d\left(k,t,x^d\left(k,t\right),u^d\right)\right)\right. \tag{3}$$

$$\left. - f^k(t, x^d(k,t), u^d(k,t))\right),$$

$$\varphi^d\left(z, t_F, x_F^d\right) = \varphi\left(k+1, \theta\left(z, x_F^d\right)\right),$$

$$\varphi\left(k,x\right) = \sup_{u^v\in\mathbf{U}^v(t,x)} \varphi^d\left(z, \tau\left(z\right), \xi\left(z\right)\right), \quad k \in \mathbf{K}',$$

which is resolved in the order from k_F to k_I. Suppose that a solution to this chain $\left(\varphi\left(k, x\left(k\right)\right),\ \varphi^d\left(z,t,x^d\right)\right)$ exists and, moreover, that there are controls corresponding to this solution $\tilde{u}\left(k,x\right)$, $\tilde{u}^v\left(k,x\right)$, $\tilde{u}^d\left(z,t,x^d\right)$, obtained from the maximum operations in 3. Substituting the found controls in the right parts of the given discrete formulas, we obtain

$$x(k+1) = f(k, x(t), \tilde{u}(k, x(t))), \quad x(k_I) = x_I, \quad k \in \mathbf{K} \backslash \mathbf{K}' \backslash k_F,$$

$$x(k+1) = \theta\left(k, x(k), \tilde{u}^v(k, x(k)), \gamma^d(\tilde{z})\right),$$

$$x^d(k, t+1) = f^d\left(k, x(k), \tilde{u}^v(k, x(k)), t, x^d, \tilde{u}^d(\tilde{z}(k), t, x^d)\right),$$

$$t_I = \tau(\tilde{z}(k)), \quad x^d(t_I) = \xi(\tilde{z}(k)), \quad \tilde{z}(k) = (k, x(k), \tilde{u}^v(k, x(k)))$$

for $k \in \mathbf{K}'$. The solution of this chain is

$$(x(k), u(k))_*, \quad k \in \mathbf{K} \backslash \mathbf{K}',$$

$$\left(x(k), \hat{u}(k), x^d(k, t), u^d(k, t)\right)_*, \quad k \in \mathbf{K}', \quad t \in \mathbf{T}(z_*(k)).$$

If this solution exists, it sets the optimal heterogeneous discrete process m_*. We note that the functional $\varphi^d(z, t, x^d)$ in this case can be considered independent of x, because it "serves" a family of problems for different initial conditions.

The first variant of these conditions is obtained in [8,11].

5   The Approximate Synthesis of Optimal Control

Suppose that $\mathbf{X}(k) = \mathbb{R}^{m(k)}$, $\mathbf{X}^d(z, t) = \mathbb{R}^{n(k)}$, $x_I^d = \xi(z)$, $k_I, x_I, k_F, t_I(k), t_F(k)$ are given, $x_F^d \in \mathbb{R}^{n(k)}$, and lower-level systems do not depend on control u^v.

We will develop the method based on the principles of expansion [9] and localization [10]. The task of improvement is to build an operator $\eta(m)$, $\eta : \mathbf{D} \to \mathbf{D}$, such that $I(\eta(m)) \leq I(m)$. For some given initial element, such an operator generates improving, specifically a minimizing sequence $\{m_s\} : m_{s+1} = \eta(m_s)$.

According to the localization principle, the task of improving an element m^{I} resolves itself into the problem of the minimum of the intermediary functional

$$I_\alpha(m) = \alpha I(m) + (1 - \alpha)J(m^{\mathrm{I}}, m), \quad \alpha \in [0, 1], \tag{4}$$

where $J(m^{\mathrm{I}}, m)$ is the functional of a metric type. By varying α from 0 to 1, we can achieve the necessary degree of proximity m_α to m^{I} and effectively use the approximations of the constructions of sufficient conditions in the neighbourhood of m^{I}. As a result, we obtain an algorithm with the parameter α, which is a regulator configurable for a specific application. This parameter is chosen so that the difference $I(m^{\mathrm{I}}) - I(m_\alpha)$ is the largest; then the corresponding element m_α is taken as m^{II}. We consider the intermediary functional of the form

$$I_\alpha = \alpha I + (1 - \alpha)\left(\sum_{\mathbf{K} \backslash \mathbf{K}' \backslash k_F} \frac{1}{2}|\Delta u(k)|^2 + \sum_{\mathbf{K}'} \sum_{\mathbf{T}(z) \backslash t_F} \frac{1}{2}|\Delta u^d(k, t)|^2\right),$$

where $\alpha \in [0, 1]$, $\Delta u = u - u^{\mathrm{I}}$, $\Delta u^d = u^d - u^{d\mathrm{I}}$.

According to said extension principle for the given element $m^{\mathrm{I}} \in \mathbf{D}$, we need to find an element $m^{\mathrm{II}} \in \mathbf{D}$ for which $I_\alpha(m^{\mathrm{II}}) = L_\alpha\left(m^{\mathrm{II}}\right) < I_\alpha(m^{\mathrm{I}}) = L_\alpha\left(m^{\mathrm{I}}\right)$, or $L_\alpha\left(m^{\mathrm{II}}\right) - L_\alpha\left(m^{\mathrm{I}}\right) < 0$. We consider the increment of the functional $L_\alpha(m)$:

$$\Delta L_\alpha \approx G_{x_F}^{\mathrm{T}} \Delta x_F + \frac{1}{2} \Delta x_F^{\mathrm{T}} G_{x_F x_F}^{\mathrm{T}} \Delta x_F$$

$$- \sum_{\mathbf{K} \backslash \mathbf{K}' \backslash k_F} \left(R_x^{\mathrm{T}} \Delta x + R_u^{\mathrm{T}} \Delta u + \frac{1}{2} \Delta u^{\mathrm{T}} R_{uu} \Delta u \right.$$

$$+ \frac{1}{2} \Delta x^{\mathrm{T}} R_{xx} \Delta x + \Delta u^{\mathrm{T}} R_{ux} \Delta x \bigg) + \sum_{\mathbf{K}' \backslash k_F} \left(G_{x_F^d}^{d\mathrm{T}} \Delta x_F^d + G_x^{d\mathrm{T}} \Delta x \right.$$

$$+ \frac{1}{2} \Delta x_F^{d\mathrm{T}} G_{x_F^d x_F^d}^{d\mathrm{T}} \Delta x_F^d + \frac{1}{2} \Delta x^{\mathrm{T}} G_{xx}^{d\mathrm{T}} \Delta x + \Delta x_F^{d\mathrm{T}} G_{x_F^d x}^{d\mathrm{T}} \Delta x \bigg)$$

$$- \sum_{\mathbf{T}(z) \backslash t_F} \left(R_{x^d}^{d\mathrm{T}} \Delta x^d + R_x^{d\mathrm{T}} \Delta x + R_{u^d}^{d\mathrm{T}} \Delta u^d + \frac{1}{2} \Delta u^{d\mathrm{T}} R_{u^d u^d}^d \Delta u^d \right.$$

$$+ \frac{1}{2} \Delta x^{d\mathrm{T}} R_{x^d x^d}^d \Delta x^d + \Delta x^{\mathrm{T}} R_{xx^d}^d \Delta x^d + \Delta x^{\mathrm{T}} R_{xu^d}^d \Delta u^d + \Delta u^{d\mathrm{T}} R_{u^d x^d}^d \Delta x^d \bigg),$$

where $\Delta u = u - u^{\mathrm{I}}$, $\Delta x = x - x^{\mathrm{I}}$, $\Delta u^d = u^d - u^{d\mathrm{I}}$, $\Delta x^d = x^d - x^{d\mathrm{I}}$, $\Delta x_F^d = x_F^d - x_F^{d\mathrm{I}}$, and $x_F = x(k_F)$. Here the functions R, G, R^d, and G^d are defined for the functional I_α, and their first and second derivatives are calculated at $u = u^{\mathrm{I}}(k)$, $x = x^{\mathrm{I}}(k)$, $x^d = x^{d\mathrm{I}}(k,t)$, and $u^d = u^{d\mathrm{I}}(k,t)$. We suppose that matrices R_{uu} and $R_{u^d u^d}^d$ are negative definite (this can always be achieved by choosing a parameter α [10]). We find $\Delta u, \Delta u^d$ such that $\sum_{\mathbf{K} \backslash \mathbf{K}' \backslash k_F}, \sum_{\mathbf{T}(z) \backslash t_F}$ reach their respective maximum values. It is easy to see that

$$\Delta u = -(R_{uu})^{-1}(R_u + R_{ux} \Delta x(k)),$$

$$\Delta u^d = -(R_{u^d u^d}^d)^{-1}(R_{u^d}^d + R_{u^d x}^d \Delta x(k) + R_{u^d x^d}^d \Delta x^d(k,t)).$$

We substitute the found formulas for the control increments into the formula for the increment of the functional ΔL_α. Then we perform the necessary transformations and denote the result by ΔM_α. We obtain

$$\Delta M_\alpha \approx G_x^{\mathrm{T}} \Delta x + \frac{1}{2} \Delta x^{\mathrm{T}} G_{xx} \Delta x - \sum_{\mathbf{K} \backslash \mathbf{K}' \backslash k_F} \left((R_x - R_{xu} R_{uu}^{-1} R_u^{\mathrm{T}}) \Delta x \right.$$

$$+ \frac{1}{2} \Delta x^{\mathrm{T}} \left(R_{xx} - R_{xu} R_{uu}^{-1} R_{xu}^{\mathrm{T}} \right) \Delta x - \frac{1}{2} R_u^{\mathrm{T}} R_{uu}^{-1} R_u \bigg)$$

$$+ \sum_{\mathbf{K}' \backslash k_F} \left(G_{x_F^d}^{d\mathrm{T}} \Delta x_F^d + \frac{1}{2} \Delta x_F^{d\mathrm{T}} G_{x_F^d x_F^d}^d \Delta x_F^d + G_x^{d\mathrm{T}} \Delta x + \frac{1}{2} \Delta x^{\mathrm{T}} G_{xx}^d \Delta x \right.$$

$$+ \Delta x_F^{d\mathrm{T}} G_{x_F^d x}^d \Delta x \bigg) - \sum_{\mathbf{T}(z) \backslash t_F} \left(R_{x^d}^d - R_{x^d u^d}^d (R_{u^d u^d}^d)^{-1} R_{u^d}^{d\mathrm{T}} \right) \Delta x^d$$

$$+ \frac{1}{2} \Delta x^{d\mathrm{T}} \left(R^d_{x^d x^d} - R^d_{x^d u^d} (R^d_{u^d u^d})^{-1} R^{d\mathrm{T}}_{x^d u^d} \right) \Delta x^d$$

$$+ \left(R^{d\mathrm{T}}_x - R^d_{x u^d} (R^d_{u^d u^d})^{-1} R^d_{u^d} \right) \Delta x + \frac{1}{2} \Delta x^{\mathrm{T}} \left(R^d_{xx} - R^d_{x u^d} (R^d_{u^d u^d})^{-1} R^d \mathrm{T}_{x u^d} \right) \Delta x$$

$$+ \left(\Delta x^{\mathrm{T}} \left(R^d_{x x^d} - R^d_{x u^d} (R^d_{u^d u^d})^{-1} R^{d\mathrm{T}}_{x^d u^d} \right) \Delta x^d - \frac{1}{2} R^{d\mathrm{T}}_{u^d} (R^d_{u^d u^d})^{-1} R^d_{u^d} \right).$$

We define the functions φ, φ^d as $\varphi = \psi^{\mathrm{T}}(k) x(k) + \frac{1}{2} \Delta x^{\mathrm{T}}(k) \sigma(k) \Delta x(k)$,

$$\varphi^d = \lambda^{\mathrm{T}}(k,t) x(k) + \psi^{d\mathrm{T}}(k,t) x^d(k,t) + \frac{1}{2} \Delta x^{d\mathrm{T}}(k,t) \sigma^d(k,t) \Delta x^d(k,t)$$

$$+ \Delta x^{\mathrm{T}}(k) \Lambda(k,t) \Delta x^d(k,t) + \frac{1}{2} \Delta x^{\mathrm{T}}(k) S(k,t) \Delta x(k),$$

where ψ, ψ^d, λ are vector functions and $\sigma, \sigma^d, S, \Lambda$ are matrices, and so that the increment of the functional ΔM_α does not depend on $\Delta x, \Delta x_F, \Delta x^d, \Delta x^d_F$. The last requirement will be achieved if

$$R_x - R_{xu} R_{uu}^{-1} R_u^{\mathrm{T}} = 0,$$

$$R_{xx} - R_{xu} R_{uu}^{-1} R_{xu}^{\mathrm{T}} = 0,$$

$$R^d_x - R^d_{x u^d} \left(R^d_{u^d u^d} \right)^{-1} R^{d\mathrm{T}}_{u^d} = 0,$$

$$R^d_{x^d} - R^d_{x^d u^d} (R^d_{u^d u^d})^{-1} R^{d\mathrm{T}}_{u^d} = 0,$$

$$R^d_{x^d x^d} - R^d_{x^d u^d} (R^d_{u^d u^d})^{-1} R^{\mathrm{T}}_{x^d u^d} = 0,$$

$$R^d_{xx} - R^d_{x u^d} (R^d_{u^d u^d})^{-1} R^{\mathrm{T}}_{x u^d} = 0,$$

$$R^d_{x x^d} - R^d_{x u^d} (R^d_{u^d u^d})^{-1} R^{\mathrm{T}}_{x^d u^d} = 0,$$

$$G_x = 0, \quad G^d_x = 0, \quad G^d_{x^d_F} = 0, \quad G_{xx} = 0, \quad G^d_{x^d_F x^d_F} = 0, \quad G^d_{x^d_F x} = 0, \quad G^d_{xx} = 0.$$

Transformation of these conditions leads to a Cauchy problem for HDS with respect to ψ, $\psi^d, \lambda, \sigma, \sigma^d, S,$ and Λ, with initial conditions on the right end:

$$\psi(k_F) = -\alpha F_x, \quad \sigma(k_F) = -\alpha F_{xx},$$

$$\psi(k) = H_x - \left(f_x^T \sigma(k+1) f_u + H_{xu} \right) \left(f_u^T \sigma(k+1) f_u + H_{uu} \right)^{-1} H_u,$$

$$\sigma(k) = f_x^{\mathrm{T}} \sigma(k+1) f_x + H_{xx}$$

$$- \left(f_x^{\mathrm{T}} \sigma(k+1) f_u + H_{xu} \right) \left(f_u^{\mathrm{T}} \sigma(k+1) f_u + H_{uu} \right)^{-1}$$

$$\left(f_x^{\mathrm{T}} \sigma(k+1) f_u + H_{xu} \right)^{\mathrm{T}}, \quad k \in \mathbf{K} \backslash \mathbf{K}' \backslash k_F,$$

$$\psi(k) = H_x + \xi_x^{\mathrm{T}} H_{x^d} + \xi_x^{\mathrm{T}} \psi^d(k, t_I) + \lambda(t_I) - \lambda(t_F), \quad k \in \mathbf{K}',$$

$$\sigma(k) = \theta_x^{\mathrm{T}} \sigma(k+1) \theta_x + H_{xx} + \xi_x^{\mathrm{T}} \theta_{x^d}(t_I) \sigma(k+1) \theta_x + \theta_x^{\mathrm{T}} \sigma(k+1) \theta_{x^d}(t_I) \xi_x$$

$$+ \xi_x^{\mathrm{T}} \theta_{x^d}^{\mathrm{T}}(t_I) \sigma(k+1) \theta_{x^d}(t_I) \xi_x + \xi_x^{\mathrm{T}} \sigma^d(k, t_I) \xi_x + S(k, t_I)$$

$$+\xi_{xx}\psi^d(t_I) + \xi_x^T \sigma^d(k,t_I)\xi_x + \xi_x^T \Lambda(t_I), \ k \in \mathbf{K}',$$

$$\psi^d = H_{x^d}^d - \left(f_{x^d}^{dT}\sigma^d(k,t+1)f_{u^d}^d + H_{x^d u^d}^d\right)\left(f_{u^d}^{dT}\sigma^d(k,t+1)f_{u^d}^d + H_{u^d u^d}^d\right)^{-1}H_{u^d}^d,$$

$$\psi^d(k,t_F) = H_{x_F^d},$$

$$\lambda(k,t) = \lambda(k,t+1) + H_x^d - \left(\Lambda(k,t+1)f_{u^d}^d + f_x^{dT}\sigma^d(k,t+1)f_{u^d}^d + H_{xu^d}^d\right)$$

$$\left(f_{u^d}^{dT}\sigma(k,t+1)f_{u^d}^d + H_{u^d u^d}^d\right)^{-1}H_{u^d}^d, \quad \lambda(k,t_F) = 0,$$

$$\sigma^d(k,t) = f_{x^d}^{dT}\sigma^d(k,t+1)f_{x^d}^d + H_{x^d x^d}^d - \left(f_{x^d}^{dT}\sigma^d(k,t+1)f_{u^d}^d + H_{x^d u^d}^d\right)$$

$$\left(f_{u^d}^{dT}\sigma(k,t+1)f_{u^d}^d + H_{u^d u^d}^d\right)^{-1}\left(f_{x^d}^{dT}\sigma^d(k,t+1)f_{u^d}^d + H_{x^d u^d}^d\right)^T,$$

$$\sigma^d(k,t_F) = \theta_{x_F^d}^T\sigma(k+1)\theta_{x_F^d} + H_{x_F^d x_F^d},$$

$$\Lambda(k,t) = f_x^{dT}\Lambda(k,t+1)f_{x^d}^d + H_{xx^d}^d - \left(f_x^{dT}\Lambda(k,t+1)f_{u^d}^d + H_{xu^d}^d\right)$$

$$\left(f_{u^d}^{dT}\sigma^d(k+1)f_{u^d}^d + H_{u^d u^d}^d\right)^{-1}\left(f_{x^d}^{dT}\sigma^d(k,t+1)f_{u^d}^d + H_{x^d u^d}^d\right)^T,$$

$$\Lambda(k,t_F) = \theta_x^T\sigma(k+1)\theta_{x^d} + H_{xx^d},$$

$$S(k,t) = S(k,t+1) + f_x^{dT}\Lambda^{dT}(k,t+1) + \Lambda(k,t+1)f_x^d + H_{xx}^d + f_x^{dT}\sigma^d(k,t+1)f_x^d$$

$$- \left(f_x^{dT}\Lambda(k,t+1)f_{u^d}^d + H_{xu^d}^d\right)\left(f_{u^d}^{dT}\sigma^d(k+1)f_{u^d}^d + H_{u^d u^d}^d\right)^{-1}$$

$$\left(\left(f_x^{dT}\Lambda(k,t+1)f_{u^d}^d + H_{xu^d}^d\right)\right)^T, \ S(k,t_F) = 0,$$

where

$$H = \psi^T(k+1)f(k,x(k),u(k)) - \frac{1}{2}(1-\alpha)\,|\Delta u(k)|^2, \ k \in \mathbf{K}\backslash\mathbf{K}'\backslash k_F$$

and

$$H = \psi^T(k+1)\theta(k,x(k),x_I^c,x_F^c) \ k \in \mathbf{K}',$$

$$H^d = \psi^{dT}(k,t+1)f^d(k,t,x(k),x^d,u^d) - f^k(t,x^d,u^d) - \frac{1}{2}(1-\alpha)\,|\Delta u^d(k)|^2,$$

$$x(k_I) = x_I, \ x(k_F) = x_F, \ x^d(k,t_I) = x_I^d, \ x^d(k,t_F) = x_F^d.$$

Wherein

$$\Delta u(k) = -\left(f_u^T\sigma(k+1)f_u + H_{uu}\right)^{-1}\left(H_u + \left(f_x^T\sigma(k+1)f_u + H_{xu}\right)^T\Delta x(k)\right),$$

$$\Delta u^d(k,t) = -(H_{u^d u^d}^d)^{-1}\left(H_{u^d}^d + (\Lambda^T f_{u^d}^d + H_{xu^d}^d)^T\Delta x(k)\right.$$

$$\left. + \left(\sigma^d f_{u^d}^d + H_{x^d u^d}^d\right)^T\Delta x^d(k,t)\right).$$

We note that the formulas obtained for the control increments of the upper and lower levels depend on the state increments of the same levels. The method then gives a solution to the problem in the form of approximately optimal linear synthesis.

6   Iterative Procedure

Based on the formulas obtained, we can formulate the following iterative procedure:

1. We calculate the initial HDS from left to right for $u = u_s(k)$, $u^d = u_s^d(k,t)$ with the given initial conditions to obtain the corresponding trajectory $(x_s(k),\ x_s^d(k,t))$.
2. We resolve the HDS from right to left with respect to $\psi(k)$, $\psi^d(k,t)$, $\lambda(k,t)$, $\sigma(k)$, $\sigma^d(k,t)$, $\Lambda(k,t)$, and $S(k,t)$.
3. We find Δu, Δu^d and new controls $u = u_s(k) + \Delta u$, $u^d = u_s^d(k,t) + \Delta u^d$.
4. With the controls found and the initial condition $x(k_I) = x_I$, we calculate the initial HDS from left to right. This defines a new element m_{s+1}.

The iteration process ends when $|I_{s+1} - I_s| \approx 0$ with a specified accuracy.

Theorem 3. *Suppose that the indicated iteration procedure is developed for a given HDS and the functional I is bounded from below. Then it generates an improving sequence of elements $\{m_s\} \in \mathbf{D}$, convergent in terms of the functional, i.e., there is a number I^* such that $I^* \leq I(m_s)$, $I(m_s) \to I^*$.*

Proof. The proof follows directly from the monotonicity property with respect to the functional of the improvement operator under consideration. Thus, we obtain a monotonic numerical sequence

$$\{I_s\} = \{I(m_s)\}, \quad I_{s+1} \leq I_s,$$

bounded from below, which according to the well-known analysis theorem converges to a certain limit: $I_s \to I_*$.

Remark 1. The equations for the matrices σ, σ^d are analogs of the matrix Riccati equations and can therefore have singular points. Points $k^* \in \mathbf{K}$, $t^* \in \mathbf{T}(k)$ are called singular if there are changes in the sign of definiteness of matrices R_{uu}, $R_{u^d u^d}^d$. In these cases, by analogy with homogeneous discrete processes, singular points can be shifted to the points k_I, $t_I(k)$ due to the special choice of the parameter α, and we can find the control increments by the modified formulas [13]. In the particular case when the discrete process of the lower level does not depend on x and u^d, these formulas have the simplest form:

$$R_{uu}(k_I)\Delta u(k_I) = 0, \quad R_{u^d u^d}^d(k,t_I)\Delta u^d(k,t_I) = 0.$$

The last equalities are systems of linear homogeneous algebraic equations with degenerate matrices $R_{uu}(k_I)$, $R_{u^d u^d}^d(k,t_I)$ and therefore always have non-zero solutions.

Remark 2. If $\sigma = 0$, $\sigma^d = 0$, $\Lambda = 0$ in the resulting algorithm, then we obtain the first-order improvement method. In this case, the formulas Δu, Δu^d will still depend on the state increments. Consequently, the resulting solution, as before, is an approximate synthesis of optimal control.

7 Example

We illustrate the work of the method with an example. Let the HDS be given:

$$x^d(t+1) = -2x^d(t) + (u_1^d - 1)^2, \; x^d(0) = 1, t = 0, 1, 2, 3,$$

$$I^0 = \frac{1}{2}(x^d(t))^2 + \frac{1}{3}(u_1^d)^3,$$

$$x^d(t+1) = (t - u_2^d)^2, \; t = 4, 5, 6, \; I^1 = \frac{1}{2}(x^d)^2 + u_2^d,$$

$$I = x^d(7) \to \min.$$

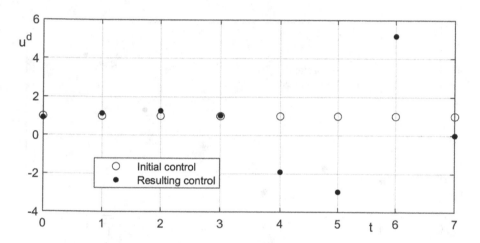

Fig. 1. Control variables in different iterations

It is easy to see that $K = 0, 1, 2$. Since x^d is a linking variable in the two periods under consideration, we can write the process of the upper level in terms of this variable:

$$x(0) = x^d(0,0), \; x(1) = x^d(0,4), \; x(2) = x^d(1,7), \; x^d(1,4) = x(1).$$

Then $\theta = x^d(0,4), \; \xi = x(1), \; I = x(2)$.

Since at both stages the process of the lower level does not depend on the state variables of the upper level, then $\lambda(0,t) = \lambda(1,t) = 0, \; \Lambda(0,t) = \Lambda(1,t) = 0, \; S(0,t) = S(1,t) = 0$.

We obtain

$$\psi(2) = -\alpha, \; \sigma(2) = 0, \; \psi(1) = \psi(2) + \psi^d(1,4), \; \sigma(1) = 2\sigma^d(1,4)$$

$$H^d(0,t) = \psi^d(0,t+1)(-2x^d + (u_1^d - 1)^2) - \frac{1}{2}(x^d(t))^2 - \frac{1}{3}(u_1^d)^3 - \frac{1}{2}(1-\alpha)(\Delta u_1^d)^2,$$

$$H^d(1,t) = \psi^d(1,t+1)(t-u_2^d)^2 - \frac{1}{2}(x^d)^2 - u_2^d - \frac{1}{2}(1-\alpha)(\Delta u_2^d)^2,$$

$$\psi^d(0,t) = -2\psi^d(0,t+1) - x^d - 4\sigma^d(0,t+1)(1-u_1^d)(4(u_1^d-1)^2)\sigma^d(0,t+1)$$
$$+ 2\psi^d(0,t+1) - 2u_1^d - (1-\alpha))^{-1}(2\psi^d(0,t+1)(u_1^d-1) - (u_1^d)^2), \ \ \psi^d(0,4) = \psi(2),$$

$$\sigma^d(0,t) = 4\sigma^d(0,t+1) - 1 - (4\sigma^d(0,t+1)(1-u_1^d))^2)(4(u_1^d-1)^2)\sigma^d(0,t+1)$$
$$+ 2\psi^d(0,t+1) - 2u_1^d - (1-\alpha))^{-1},$$

$$\psi^d(1,t) = -x^d, \ \ \psi^d(1,7) = 0, \ \ \sigma^d(1,t) = -1, \ \ \sigma^d(1,7) = 0,$$

$$\Delta u_1^d = (2\psi^d(0,t+1) - 2u_1^d - (1-\alpha))^{-1}(2\psi^d(0,t+1)(u_1^d-1) - (u_1^d)^2$$
$$+ 2\sigma^d(0,t+1)(u_1^d-1)\Delta x^d(0,t),$$

$$\Delta u_2^d = -(2\psi^d(1,t+1) - (1-\alpha))^{-1}(2\psi^d(1,t+1)(t-u_2^d) + 1).$$

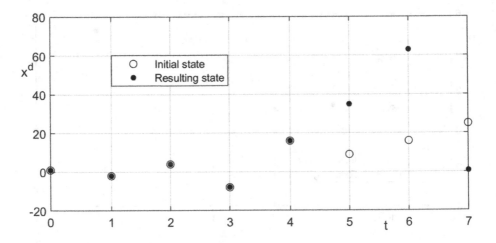

Fig. 2. State variables in different iterations

Numerical experiments show that the improvement of the functional does not depend significantly on the choice of the parameter α and occurs in almost one iteration. The result of calculations is shown for $\alpha = 0.76$ and $u(t) = 1, t = 0,..,6$. The functional value is improved from 25 to 0.64 in one iteration. Initial and resulting controls and states are shown in Figs. 1 and 2.

For comparison, calculations using the gradient method were also performed. The result is obtained in six iterations, while the value of the functional is 2.87. This indicates the efficiency of the proposed method.

8   Conclusion

This paper considers HDS with intermediate criteria. On the basis of an analogue of Krotov's sufficient optimality conditions, a method for the approximate synthesis of optimal control is constructed, its algorithm formulated, and an illustrative example given to demonstrate the efficiency of the proposed method.

References

1. Krotov, V.F., Gurman, V.I.: Methods and Problems of Optimal Control. Nauka, Moscow (1973). (in Russian)
2. Gurman, V.I., Rasina, I.V., Blinov, A.O.: Evolution and prospects of approximate methods of optimal control. Program Syst. Theor. Appl. **2**(2), 11–29 (2011). (in Russian)
3. Rasina, I.V.: Iterative optimization algorithms for discrete-continuous processes. Autom. Remote Control **73**(10), 1591–1603 (2012)
4. Rasina, I.V.: Hierarchical Models of Control Systems of Heterogeneous Structure. Fizmatlit, Moscow (2014). (in Russian)
5. Gurman, V.I., Rasina, I.V.: Global control improvement method for non-homogeneous discrete systems. Program Syst. Theor. Appl. **7**(1), 171–186 (2016). (in Russian)
6. Gurman, V.I.: To the theory of optimal discrete processes. Autom. Remote Control **34**(7), 1082–1087 (1973)
7. Krotov, V.F.: Sufficient optimality conditions for discrete controlled systems. DAN USSR **172**(1), 18–21 (1967)
8. Rasina, I.V.: Discrete heterogeneous systems and sufficient optimality conditions. Bull. Irkutsk State Univ. Ser. Math. **19**(1), 62–73 (2016). (in Russian)
9. Gurman, V.I.: The Expansion Principle in Control Problems. Nauka, Moscow (1985). (in Russian)
10. Gurman, V.I., Rasina, I.V.: Practical applications of sufficient conditions for a strong relative minimum. Autom. Remote Control **40**(10), 1410–1415 (1980)
11. Rasina, I.V., Guseva, I.S.: Control improvement method for non-homogeneous discrete systems with intermediate criterions. Program Syst. Theor. Appl. **9**(2), 23–38 (2018). (in Russian)
12. Rasina, I., Danilenko, O.: Second-order improvement method for discrete-continuous systems with intermediate criteria. IFAC-PapersOnLine **51**(32), 184–188 (2018)
13. Gurman, V.I., Baturin, V.A., Rasina, I.V.: Approximate methods of optimal control. Irkutsk (1983) (in Russian)

Modelling Climate Changes with Stationary Models: Is It Possible or Is It a Paradox?

Davide Luciano De Luca[1]([✉]) [iD], Andrea Petroselli[2] [iD], and Luciano Galasso[1]

[1] Department of Informatics, Modelling, Electronics and System Engineering,
University of Calabria, 87036 Arcavacata di Rende, CS, Italy
davide.deluca@unical.it
[2] DEIM Department - Water Engineering Section, Tuscia University, 01100 Viterbo, VT, Italy

Abstract. Climate is changing; many studies of time series confirm this sentence, but this does not imply that the past is no more representative of the future, and then that "stationarity is dead".

In fact, "stationarity" and "change" are not mutually exclusive. As examples: (1) according to Newton's first law, without an external force, the position of a body in motion changes in time but the velocity is unchanged; (2) according to Newton's second law, a constant force implies a constant acceleration and a changing velocity.

Consequently, "non-stationarity" is not synonymous with change; change is a general notion applicable everywhere, including the real (material) world, while stationarity and non-stationarity only regard the adopted models. Thus, stationary models can be also adopted for environmental changes.

With this aim, in this work Authors show some numerical experiments concerning rainfall processes. In detail, a Neymann Scott Rectangular Pulse model (NRSP), with some changing temporal scenarios for its parameters, is adopted, and the derived Annual Maximum Rainfall (AMR) time series are investigated for several temporal resolutions (sub-hourly and hourly scales). The goal is to analyze if there are some particular scales in which the assumed temporal changes in parameters could be "hidden" when AMR series (which are nowadays more available and longer than high-resolution continuous time series for many sites in the world) are studied, and then stationary models for Extreme Value distributions could be adopted.

The results confirm what is obtained from analysis of AMR series in some parts of Italy, for which it is not essential to remove the hypothesis of stationary parameters: significant trends could not appear only from the observed AMR data, as a relevant rate of outlier events also occurred in the central part of the last century.

Keywords: Rainfall processes · Climate changes · Stationary models

1 Introduction

Climate changes are widely described in many technical reports and scientific papers (e.g. [1, 2]). Concerning Europe, and in particular Italy, the report of the European

© Springer Nature Switzerland AG 2020
Y. D. Sergeyev and D. E. Kvasov (Eds.): NUMTA 2019, LNCS 11974, pp. 84–96, 2020.
https://doi.org/10.1007/978-3-030-40616-5_7

Environmental Agency (EEA) [3] and the publication of the Italian Institute for Environmental Protection and Research (ISPRA) [4] can be mentioned. In these reports, projections of future climate were derived by four Regional Climate Models (RCMs), named as ALADIN, GUF, CMCC, LMD. For each RCM, four scenarios of Representative Concentration Pathways (RCPs) of total radiative forcing (i.e., cumulative measure of human emissions of greenhouse gasses from all sources expressed in W/m^2) were used as input: RCP 2.6, RCP 4.5, RCP 6 and RCP 8.5. Moreover, an ensemble mean projection from all the RCMs was also derived for each RCP. Focusing the attention on RCP 4.5 (intermediate emissions) and RCP 8.5 (high emissions), with respect to the reference period 1971–2000, the results of the simulations related to a future period until 2090 for Italy are:

- a decrease of annual precipitation. In details, the ensemble mean of reduction is 13 mm for RCP 4.5 and 71 mm for RCP 8.5;
- a modest increase for the annual maximum daily rainfall. The ensemble mean is, for both RCP 4.5 and RCP 8.5, up to 5–7 mm;
- a significant increase for the waiting time between two consecutive rainfall events. The ensemble mean is up to 8 days for RCP 4.5, and up to 16 days for RCP 8.5.

However, RCMs usually underestimate intensity of extreme rainfall, due to structure of the adopted numerical schemes and to their temporal and spatial resolutions. Concerning these last aspects, spatial resolutions between 10 and 30 km, typically used in RCMs for climate change studies, are still too coarse to well reproduce sub-daily localized heavy precipitation events [5, 6].

Moreover, from analysis of some time series in southern Italy, related to Annual Maximum Rainfall (AMR) for daily and sub-daily resolutions, significant trends do not appear from the observed data, as a relevant number of heavy events also occurred in the central part of the last century [7].

Consequently, the contrast between the perception of climate changes by people (also supported by RCMs projections) and the evidence from some time data series implies a more in-depth analysis of rainfall processes, mainly for extremes from high-resolution data. In this context, only an analysis of observed records may not be sufficient, as high-resolution rainfall data usually present a very short sample size; this aspect, together with the need to obtain perturbed time series which are representative of future rainfall fields, makes preferable the use of stochastic rainfall generators.

In this paper, a modified version of the Neymann Scott Rectangular Pulse (NSRP, [8–10]) was implemented (Sect. 2.1), in which:

- the parameters were estimated by considering the AMR time series at sub-daily durations, that are usually more lengthy with respect to high-resolution continuous time series. Moreover, with this choice, a better reconstruction of AMR time series can be obtained, which are of main interest in this study;
- a simple goniometric scheme was introduced for the mean value of inter-arrivals between two consecutive storms, in order to better reproduce seasonality and annual precipitation, without over parameterizing the model by using monthly or seasonal parameter sets.

Once the NSRP model has been calibrated, some parameters trends were hypothesized, which provided compatible results with RCM scenarios in terms of variation for maximum daily rainfall and cumulative annual precipitation. Time series of Viterbo raingauge (central Italy) were used as case study (Sect. 2.2). The obtained results are discussed in Sect. 3.

2 Methods and Materials

2.1 Brief Overview of NSRP Model

The single-site Neymann Scott Rectangular Pulses (NSRP, [8–10]) model presents a flexible structure, in which the meaning of model parameters is strictly related to the underlying physical features observed in rainfall events. In details, the basic formulation is:

1. it is assumed that the inter-arrivals Ts between the origins of two consecutive storms are independent and identically distributed, and follow an exponential distribution:

$$P_{T_S}(t_S) = 1 - e^{-\lambda t_S},\tag{1}$$

   where $1/\lambda$ represents the mean value for inter-arrivals;
2. for each origin, a number M of rain cells (also named bursts) is associated. M is usually considered as Geometric or Poisson distributed. In the following, a geometric distribution is assumed and, with the aim of having at least one burst for a storm, the random variable $C = M - 1$ is used, with $E[C] = \theta - 1$, so that $E[M] = \theta$ and:

$$P_C(c) = \frac{1}{\theta} \cdot \left(1 - \frac{1}{\theta}\right)^c\tag{2}$$

3. the starting time of each rain cell Tc, measured from the origin of the associated storm, is exponentially distributed with parameter β:

$$P_{T_c}(\tau_c) = 1 - e^{-\beta \tau_c}\tag{3}$$

4. a rectangular pulse is then related to each burst, with an Intensity I and a duration D, which are both exponential distributed with parameters η and ξ, respectively:

$$P_I(i) = 1 - e^{-\eta i}\tag{4}$$
$$P_D(d) = 1 - e^{-\xi d}\tag{5}$$

5. the total precipitation intensity at time t, $Y(t)$, is then calculated as sum of the intensities related to the active bursts at time t:

$$Y(t) = \int_0^{+\infty} I_{t-u}(u)dM(t-u), \tag{6}$$

where $I_{t-u}(u)$ is the intensity of a single rectangular pulse at time u and $M(t)$ is the counting process of the burst occurrences. Then, the aggregated process, i.e. the rainfall height $H_j^{(\tau)}$ cumulated on the temporal τ - resolution and related to the time interval j is:

$$H_j^{(\tau)} = \int_{(j-1)\tau}^{j\tau} Y(t)dt \tag{7}$$

Therefore, the basic version of a NSRP model has 5 parameters that can be estimated by minimizing an objective function, evaluated as sum of residuals (normalized or not) between the considered (by users) statistical properties of the observed data at chosen resolutions and their theoretical expressions. The statistical properties are typically referred to high-resolution continuous time series (e.g. 5-min rainfall time series), for example: mean, variance, k-lag autocorrelation for $H_j^{(\tau)}$ at several values of τ [11].

However, the sample size of these datasets is usually short (at most 15–20 years of records) and then not very suitable for obtaining robust estimations, even more so when a specific 5-parameter set is considered for each month or season, in order to take into account the seasonality of the process.

To overcome this problem, in this work Authors considered statistical properties from the annual maximum time series of rainfall heights at hourly and sub-hourly resolutions, which are usually longer than continuous series. This choice also allows for a better reproduction of extreme events, which are generally underestimated if continuous time series are used for model calibration [12].

Moreover, due to the fact that the information about the seasonality of the rainfall process during the year is lost with the use of Block Maxima (BM) series, a very simple schematization was introduced for modelling the seasonality. In detail, with the goal to not over parameterize the model (i.e. introducing monthly or seasonal parameter sets), if no specific information can be derived from BM series, the following goniometric function was introduced for the mean value of the inter-arrivals between two consecutive storm origins (i.e. $1/\lambda$):

$$\frac{1}{\lambda(t)} = \frac{1}{\lambda_{min}} + A \cdot (1 + cos(\varphi(t))), \tag{8}$$

where $1/\lambda_{min}$ is the assumed minimum value of mean waiting time between two storms, A is the difference $(1/\lambda - 1/\lambda_{min})$, with λ from Eq. (1), and

$$\varphi(t) = \frac{2\pi t}{T_y} + \pi, \tag{9}$$

in which T_y is the total number of minutes in one year, and $0 \le t \le T_y$. With adoption of Eq. (8), Authors hypothesize that the mean waiting time assumes the minimum value

$(1/\lambda_{min})$ when t is very close to 0 and T_y (i.e. in the winter period), the maximum value during summer season, and the mean value along the whole interval $[0; T_y]$ is equal to $1/\lambda$.

The use of this goniometric function only introduces one more parameter, λ_{min}, with respect to the basic formulation for NSRP, and it also allows for a better reconstruction of the annual precipitation, as explained in Sect. 3.1. In the following, the standard model version (without goniometric function) is indicated as NSRP1, while the version with Eq. (8) is NSRP2.

Once the model calibration was completed, some parameters trends were hypothe-sized, from which Authors considered those that provided compatible results with RCM scenarios, in terms of variation for maximum daily rainfall and annual precipitation. In this paper, as a representative example, Authors discussed the obtained outcomes (Sect. 3.2) for the following scenario, which is characterized by:

- a linear increasing trend of 50% in 100 years concerning Intensity of Bursts;
- a linear decreasing trend of 25% in 100 years concerning Duration of Bursts;
- a linear increasing trend of 50% in 100 years concerning the mean waiting time between two consecutive storms.

2.2 The Case Study of Viterbo Raingauge

Authors focused attention on Viterbo rain gauge (central Italy), characterized by a Mean Annual Precipitation (MAP) equal to 746 mm. The analyzed data, provided by 'Agenzia Regionale di Protezione Civile – Centro Funzionale Regionale' of Lazio region, were: (i) Annual Maximum Rainfall (AMR) series related to sub-daily durations (1–24 h), with a sample size $N = 71$ years; (ii) continuous high-resolution (5-min) rainfall series from 1994 to 2015, from which the AMR series related to 5, 15 and 30 min were obtained, with $N = 22$ years.

As explained in Sect. 2.1, because of the relatively short sample size for the continu-ous rainfall series, Authors preferred to calibrate (Sect. 3) the model by using the longer hourly AMR series, in order to better reproduce the extreme rainfall events, which are of main interest for specific topics, like analysis of induced events (floods and landslides) and related strategies of disaster risk reduction. However, the statistics of sub-hourly AMR data (derived from continuous series) were considered in the validation step.

3   Results and Discussion

3.1   Calibration of NSRP Models Without Parameter Trend

The calibration for NSRP1 and NSRP2 was carried out by optimizing with respect to the mean sample values of hourly AMR series (1–24 h) and annual precipitation. According with [11, 13], the ranges of variation for the 5 parameters related to NSRP1 were: $[0.002 \text{ h}^{-1}; 0.01 \text{ h}^{-1}]$ for λ; $[2, 10]$ for θ; $[0.02 \text{ h}^{-1}; 0.5 \text{ h}^{-1}]$ for β; $[0.05 \text{ h/mm};$ $0.2 \text{ h/mm}]$ for η and $[1 \text{ h}^{-1}; 10 \text{ h}^{-1}]$ for ξ. Moreover, the range $[2 \text{ days}; 5 \text{ days}]$ was adopted for $1/\lambda_{min}$, introduced in NSRP2 (see Eq. 8). For each hypothesized parameter set, concerning both NSRP1 and NSRP2, a single 500-year realization of continuous 1-min rainfall heights was generated with the usual Monte Carlo technique [14], according to the property of ergodicity [15, 16] related to a stationary process.

The calibration results are reported in Table 1. From Table 2 (where there are also indicated the mean values for 5, 15 and 30-min AMR series, not adopted for calibration), it is clear that there is not a significant difference between NSRP1 and NRSP2 in terms of mean values for AMR (there is in general a slight overestimation from NSRP2 when sample data are compared). The more noticeable difference is concerned with the reconstruction of MAP: NSRP1 provided a value of 552.0 mm, while 739.3 mm (closer to the sample value of 746 mm) were obtained from NSRP2, which is clearly preferable for the successive elaborations.

Table 1. NSRP calibration results

	$1/\lambda$ (h)	θ (-)	$1/\beta$ (h)	$1/\eta$ (mm/h)	$1/\xi$ (h)	$1/\lambda_{min}$ (h)
NSRP1 model	214.5	5.7	11.9	14.2	0.16	
NSRP2 model						72.0

Table 2. NSRP performances, in terms of reconstruction of mean values for Annual Maximum Rainfall (AMR) and Mean Annual Precipitation (MAP)

	5-min AMR (mm)	15-min AMR (mm)	30-min AMR (mm)	1-h AMR (mm)	3-h AMR (mm)
NSRP1 model	6.9	17.6	26.9	32.9	38.2
NSRP2 model	7.1	18.3	28.5	35.4	41.1
Sample data	8.3	15.7	22.1	31.0	40.3
	6-h AMR (mm)	12-h AMR (mm)	24-h AMR (mm)	MAP (mm)	
NSRP1 model	44.4	53.3	64.0	552.0	
NSRP2 model	47.9	60.2	72.8	739.3	
Sample data	47.3	55.6	65.5	746.0	

In Fig. 1, the comparison among Amount-Duration-Frequency (ADF) curves obtained from the sample AMR series and those derived from the simulated continuous NSRP2 process is illustrated. For values of return period from 10 to 1000 years, we can observe at most a difference of about ±2.5 mm for longer durations, ranging from −2% to 4% with respect to sample data. Consequently, the proposed calibration for NSRP, carried out by considering sample statistics from observed AMR series, seems useful in order to reproduce extreme rainfall events along several sub-daily durations.

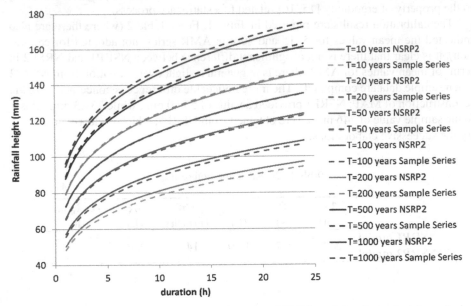

Fig. 1. Comparison among ADF curves obtained from NSRP2 model and sample data analysis

3.2   Results from the Assumed Parameter Trends

Starting from the hypothesized scenario described in Sect. 2.1 for parameter trends, 500 realizations, each one concerning 101 years of continuous 1-min rainfall heights, were generated with the usual Monte Carlo technique [14]. Obviously, the first generated year, denoted as "0", has all the features of the calibrated stationary process in Sect. 3.1.

Focusing on Annual Precipitation (AP) and annual maximum 24-h rainfall height, the temporal evolution of their mean values (calculated for each year from the correspondent 500 realizations) is compatible with RCM projections, reported in Sect. 1. In fact (see also Table 3):

- a mean reduction of 82.5 mm in 100 years is obtained for AP (well-matched with 71 mm in 90 years from RCP 8.5);
- there is a slight increase for 24-h AMR, of about 4 mm in 100 years (the ensemble mean is, for both RCP 4.5 and RCP 8.5, up to 5–7 mm in 90 years, related to daily duration).

Moreover, interesting comments can be made from analysis of temporal evolution for the generated distributions regarding AMR series at sub-hourly and hourly time scales. A specific AMR distribution, related to a fixed year, is obviously derived from the

Table 3. Evolution of mean values for Annual Maximum Rainfall (AMR) and Mean Annual Precipitation (MAP): from now (t0) to 25 (t25), 50 (t50), 75 (t75) and 100 (t100) years

	5-min AMR (mm)	15-min AMR (mm)	30-min AMR (mm)	1-h AMR (mm)	3-h AMR (mm)
t0	7.1	18.3	28.5	35.4	41.1
t25	7.8	19.9	30.7	37.5	43.4
t50	8.4	20.8	31.0	36.9	43.2
t75	9.0	22.5	33.1	38.6	44.3
t100	9.7	23.5	33.4	37.9	44.0

	6-h AMR (mm)	12-h AMR (mm)	24-h AMR (mm)	MAP (mm)	
t0	47.9	60.2	72.8	739.3	
t25	50.8	61.4	74.8	732.6	
t50	51.0	61.6	74.9	700.8	
t75	51.8	62.7	75.3	687.0	
t100	51.5	63.1	76.7	656.8	

correspondent 500 annual extremes associated to the generated realizations. From Fig. 2 it is clear that there is a significant difference among AMR distributions (represented in EV1 probabilistic plots, [14]) at higher resolutions (5–15 min); from Table 3 it can be noted that there is an increase in mean values of about 37% (5-min AMR), 28% (15-min AMR) and 17% (30-min AMR) in 100 years.

These differences are less and less evident for hourly time scales (Fig. 3); the increase in mean values is at most about 5–7% in 100 years (Table 3).

These results can be easily justified from the assumed trend scenario:

- an increase in burst intensity induces a clear increase in rainfall height for finer time scales (5–30 min), which are less influenced by a contemporary reduction of burst duration;
- on the contrary, for coarser resolutions (from 1 h), the simultaneous presence of an increase for intensity and a reduction in duration for bursts produces a sort of balance for rainfall heights, and then it is not possible to highlight a significant trend for AMR series;
- the increase of mean waiting time between two consecutive storms mainly influences the reduction of annual precipitation, as expected from RCM projections.

Such considerations are also confirmed by Figs. 4 and 5. For finer time scales (5 and 15 min), the temporal evolutions of 2.5%, 97.5% quantiles and mean value (derived from all the 500 realizations) clearly show an increasing trend more significant than those associated to the coarser ones (1–24 h), for which the temporal slopes could be also considered as horizontal.

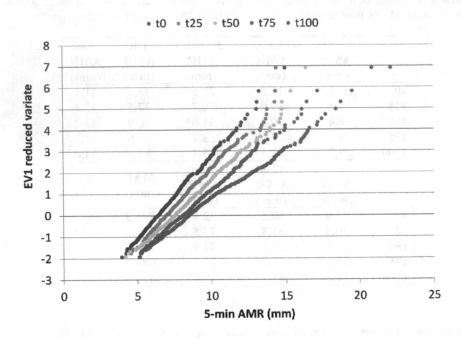

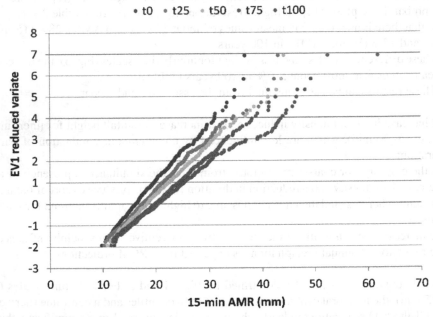

Fig. 2. Evolution of distributions for 5-min AMR (top) and 15-min AMR (bottom): from now (t0) to 25 (t25), 50 (t50), 75 (t75) and 100 (t100) years.

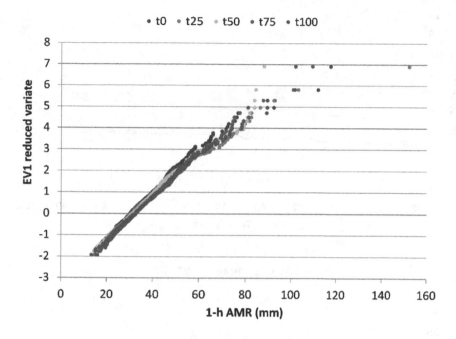

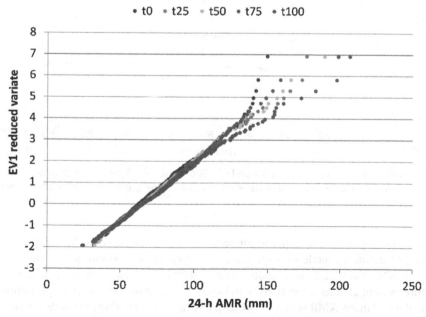

Fig. 3. Evolution of distributions for 1-h AMR (top) and 24-h AMR (bottom): from now (t0) to 25 (t25), 50 (t50), 75 (t75) and 100 (t100) years.

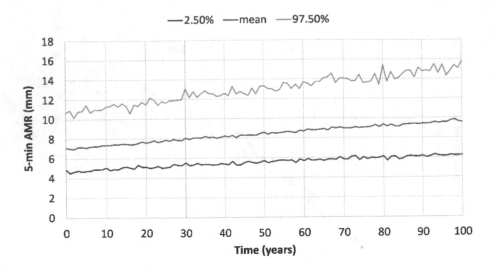

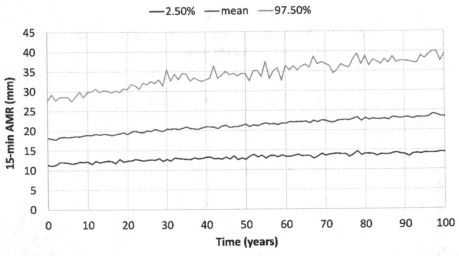

Fig. 4. Temporal evolution of mean, 2.5% and 97.5% quantiles for 5-min AMR (top) and 15-min AMR (bottom), from Monte Carlo simulation carried out for the adopted NSRP2 with parameter trend.

Consequently, from the exposed numerical experiments it could be affirmed that adoption of stationary models for rainfall extreme value distributions still remains as a valid tool, when the specific time scale is such as to "hide" climatic change effects which are more evident at finer resolutions. Obviously, this is also confirmed by preliminary data analysis of many AMR series, for which a relevant rate of outlier events also occurred in the last century.

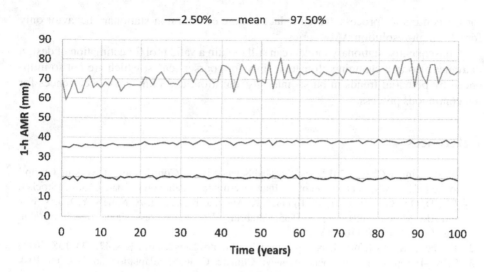

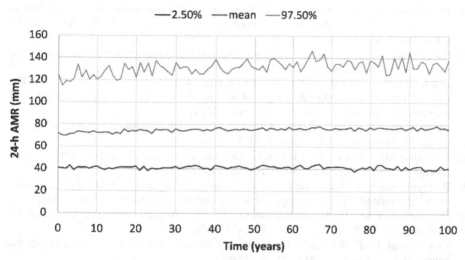

Fig. 5. Temporal evolution of mean, 2.5% and 97.5% quantiles for 1-h AMR (top) and 24-h AMR (bottom), from Monte Carlo simulation carried out for the adopted NSRP2 with parameter trend.

4   Conclusions

From the proposed numerical experiments, the obtained results can clearly constitute an interesting contribution for discussion about climate change effects on several time resolutions for rainfall time series. As well-known by preliminary analysis of many observed Annual Maximum Rainfall (AMR) time series, significant trends do not appear from the sample data at specific resolutions in some parts of Italy [7], and this is confirmed by simulations carried out in this work, although some parameter trends were imposed for the continuous process, which induce a more evident non-stationary behavior only for very high-resolution AMR series.

Consequently, stationary models can still remain a valid tool for estimation of design extremes also in a changing climate, mainly for coarser scales, which are not so influenced by potential trends in bursts intensity, duration, and number of occurrences for the continuous process.

References

1. IPCC: Climate Change 2013: The Physical Science Basis. Contribution of Working Group I to the Fifth Assessment Report of the Intergovernmental Panel on Climate Change [Stocker, T.F., D. Qin, G.K. Plattner, M. Tignor, S.K. Allen, J. Boschung, A. Nauels, Y. Xia, V. Bex and P.M. Midgley (eds.)], pp. 1–1535. Cambridge University Press, Cambridge. https://doi.org/10.1017/cbo9781107415324
2. Trenberth, K.E.: Changes in precipitation with climate change. Clim. Res. **47**, 123–138 (2011)
3. EEA—European Environment Agency: Climate Change Adaptation ad Disaster Risk Reduction in Europe, vol. 15, pp. 1–171. EEA, Kongens Nytorv (2017)
4. ISPRA—Istituto Superiore per la Protezione e la Ricerca Ambientale: Il Clima Futuro in Italia: Analisi delle proiezioni dei modelli regionali, vol. 58, pp. 1–64. ISPRA, Rome (2015). ISBN 978-88-448-0723-8. (in Italian)
5. Ban, N., Schmidli, J., Schär, C.: Heavy precipitation in a changing climate: does short-term summer precipitation increase faster? Geophys. Res. Lett. **42**, 1165–1172 (2015)
6. Chan, S.C., Kendon, E.J., Fowler, H.J., Blenkinsop, S., Roberts, N.M., Ferro, C.A.T.: The value of high-resolution Met Office regional climate models in the simulation of multi-hourly precipitation extremes. J. Clim. **27**, 6155–6174 (2014)
7. De Luca, D.L., Galasso, L.: Stationary and non-stationary frameworks for extreme rainfall time series in Southern Italy. Water **10**, 1477 (2018)
8. Rodriguez-Iturbe, I., Cox, D.R., Isham, V.: Some models for rainfall based on stochastic point processes. Proc. R. Soc. Lond. A **410**, 269–288 (1987)
9. Rodriguez-Iturbe, I., Febres De Power, B., Valdes, J.B.: Rectangular pulses point process models for rainfall: analysis of empirical data. J. Geophys. Res **92**, 9645–9656 (1987)
10. Cowpertwait, P.S.P.: Further developments of the Neyman-Scott clustered point process for modeling rainfall. Water Resour. Res. **27**, 1431–1438 (1991)
11. Cowpertwait, P.S.P.: A Poisson-cluster model of rainfall: high-order moments and extreme values. Proc. R. Soc. Lond. **454**, 885–898 (1998)
12. Cross, D., Onof, C., Winter, H., Bernardara, P.: Censored rainfall modelling for estimation of fine-scale extremes. Hydrol. Earth Syst. Sci. **22**, 727–756 (2018). https://doi.org/10.5194/hess-22-727-2018
13. Calenda, G., Napolitano, F.: Parameter estimation of Neyman-Scott processes for temporal point rainfall simulation. J. Hydrol. **225**(1–2), 45–66 (1999)
14. Kottegoda, N.T., Rosso, R.: Applied Statistics for Civil and Environmental Engineers, p. 736. Wiley-Blackwell, Hoboken (2008). ISBN 978-1-405-17917-1
15. Koutsoyiannis, D., Montanari, A.: Negligent killing of scientific concepts: the stationarity case. Hydrol. Sci. J. **60**, 1174–1183 (2014)
16. Serinaldi, F., Kilsby, C.G.: Stationarity is undead: uncertainty dominates the distribution of extremes. Adv. Water Resour. **77**, 17–36 (2015)

Differential Equations and Uniqueness Theorems for the Generalized Attenuated Ray Transforms of Tensor Fields

Evgeny Yu. Derevtsov[1,2] , Yuriy S. Volkov[1,2(✉)] , and Thomas Schuster[3]

[1] Sobolev Institute of Mathematics SB RAS, 630090 Novosibirsk, Russia
{dert,volkov}@math.nsc.ru
[2] Novosibirsk State University, 630090 Novosibirsk, Russia
[3] Saarland University, 66041 Saarbrücken, Germany
thomas.schuster@num.uni-sb.de

Abstract. Properties of operators of generalized attenuated ray transforms (ART) are investigated. Starting with Radon transform in the mathematical model of computer tomography, attenuated ray transform in emission tomography and longitudinal ray transform in tensor tomography, we come to the operators of ART of order k over symmetric m-tensor fields, depending on spatial and temporal variables. The operators of ART of order k over tensor fields contain complex-valued absorption, different weights, and depend on time. Connections between ART of various orders are established by means of application of linear part of transport equation. This connections lead to the inhomogeneous k-th order differential equations for the ART of order k over symmetric m-tensor field. The right hand parts of such equations are m-homogeneous polynomials containing the components of the tensor field as the coefficients. The polynomial variables are the components ξ^j of direction vector ξ participating in differential part of transport equation. Uniqueness theorems of boundary-value and initial boundary-value problems for the obtained equations are proved, with significant application of Gauss-Ostrogradsky theorem. The connections of specified operators with integral geometry of tensor fields, emission tomography, photometry and wave optics allow to treat the problem of inversion of the ART of order k as the inverse problem of determining the right hand part of certain differential equation.

Keywords: Tensor tomography · Attenuated ray transform · Transport equation · Boundary-value problem · Uniqueness theorem

1 Preliminaries

Two well-known tomography and integral geometry problems underlay in a notion of generalized attenuated ray transform (ART) of tensor fields. The first

The reported study was funded by Russian Foundation for Basic Research (RFBR) and German Science Foundation (DFG) according to the joint German-Russian research project 19-51-12008 and by German Science Foundation (DFG) under project Lo 310/17-1.

© Springer Nature Switzerland AG 2020
Y. D. Sergeyev and D. E. Kvasov (Eds.): NUMTA 2019, LNCS 11974, pp. 97–111, 2020.
https://doi.org/10.1007/978-3-030-40616-5_8

is the problem of emission computer tomography, and the second is the problem of integral geometry, consisting in reconstruction of a tensor field by its known longitudinal ray transform.

A progress of emission tomography in biology and medicine diagnostics is well known [1,2]. In contradistinction to the transmission computer tomography, a setting of emission tomography problem contains, in general, two unknown functions that should be reconstructed. First function (absorption coefficient) is responsible for a medium absorption, and the second describes a distribution of internal sources, which radiation is fixed by detectors. A purpose is to find distribution of internal sources f and/or absorption coefficient ε by given values of attenuated ray transform

$$I = \int_L f(q) \exp\left\{-\int_{L(q)} \varepsilon(p)\, dp\right\} dq, \tag{1}$$

where $L(q)$ is a segment of straight line L between point q and detector. At the most part of settings of emission tomography problem the absorption coefficient supposed to be known. Later a phenomenon of absorption arises in the models of vector tomography [3–8]. The authors of listed articles develop as approaches to some aspects of applications of vector tomography, so investigate certain theoretical questions.

Tensor tomography has traditional applications to the problems of photoelasticity and fiber optics [9,10], new approaches and achievements in diffractive tomography of strains [11], polarization tomography of quantum radiation [12], diffusion MRI-tomography and cross-polarization optic coherent tomography [13–16]. A success of tensor tomography in studying of anisotropic objects and materials in physics, geophysics, biology and medicine makes a deep impression and closely connected with progress in integral geometry of tensor fields, wherein many types of ray transforms are suggested and investigated [17,18] as in 2D-case [19–22] so in case of arbitrary dimension of Euclidean space and Riemannian surface or manifold [23,24].

Initial data for well-known problem of integral geometry for tensor fields represent, in particular, the longitudinal ray transform

$$\mathcal{P}w(x,\xi) = \int_{-\infty}^{\infty} w_{i_1\ldots i_m}(x - s\xi)\xi^{i_1}\ldots\xi^{i_m}\,ds, \tag{2}$$

where $w_{i_1\ldots i_m}(x)$ is a symmetric tensor field of rank m (m-tensor field), ξ is a unit direction vector, $|\xi| = 1$, for a straight line L along which the integration is carried out. Here and below the Einstein rule consisting in that by repeating super- and subscripts in a monomial a summation from 1 to n is meant (n is a dimension of Euclidean space). The purpose of this integral geometry problem is to find tensor field w by given values of the longitudinal ray transform (2).

The operators of generalized ART for tensor fields are defined and studied in the article. A generalization of the operators of attenuated ray transform for tensor fields is implemented in three directions. At first, the attenuation function $\exp\left\{-\int_{L(q)} \varepsilon(p)\,dp\right\}$ becomes complex-valued similar to those arising in inverse scattering problem at Rytov's approach [25,26], and then within diffraction tomography, see [2] for instance. Secondly we take into account a concept of generalized ray transform of tensor fields (integral moments of generalized tensor fields) considered in [17,18]. A third direction of generalization is connected with settings of dynamic tomography and consists in consideration of depending upon a time internal sources [27–29].

We would like to point out some connections between partial cases of generalized ART for tensor fields and certain tomography problems, photometry, and wave optics.

Let rectangular Cartesian coordinates system be given in Euclidean space R^3 with inner product $\langle x, y\rangle$ of elements $x = (x^1, x^2, x^3)$, $y = (y^1, y^2, y^3) \in R^3$. Let in R^3 a bounded convex domain D, with smooth boundary ∂D, be given. The domain D contains distribution $f(x)$, $x \in D$, of sources of monochromatic scalar wave field. Usage of a notion of *the optical system* which is mathematical formalization of device like a camera [30] leads to a formulation of direct problem of wave optics consisting in solving the boundary-value problem for the Helmholtz equation satisfying to discontinuous boundary conditions of Kirchhoff [31] and to radiation condition of Sommerfeld on the infinity.

An application of the Green's function for a half-space gives a solution of direct problem in a form of Kirchhoff integral [31]. Usually in optics the Fraunhofer approach is exploited [32] allowing to simplify a solution of direct problem significantly. The obtained approximate solution can be represented as the convolution $\widetilde{u}_\delta * \Delta$ with known kernel Δ. Function $\widetilde{u}_\delta$ is known as *the ideal wave image* [33],

$$\widetilde{u}_\delta(x, \xi) = \int_0^\infty s e^{iks} f(x - s\xi)\,ds, \tag{3}$$

where k is wave number, $k = \text{const}$. A similar setting of direct problem with incoherent sources in a medium with constant absorption coefficient $\varepsilon > 0$ leads to so called notion of *ideal photometric image*,

$$u_\delta(x, \xi) = \int_0^\infty e^{-\varepsilon s} f(x - s\xi)\,ds. \tag{4}$$

The inverse problems of wave optics and photometry are formulated as a problem of determination of a function f which describes distributions of sources of monochromatic wave field or distributions of incoherent sources.

The attenuated ray transform (ART) of order k for m-tensor fields, k, m is integer, $k, m \geq 0$, is defined by a formula

$$u_m^k(x, \xi) = \int_0^\infty s^k \exp\left\{-\int_0^s \left(\varepsilon(x - \sigma\xi, \xi) + i\rho(x - \sigma\xi, \xi)\right)d\sigma\right\}$$
$$\times w_{i_1\ldots i_m}(x - s\xi)\xi^{i_1}\ldots\xi^{i_m}\,ds. \tag{5}$$

Functions $\varepsilon \geq 0$, ρ, and symmetric m-tensor field $w = w_{i_1 \ldots i_m}$ (i.e. every component of w) are finite and bounded.

The attenuated ray transform $u_0^0(x, \xi)$ of order 0 for scalar fields at $\rho \equiv 0$ and $\varepsilon = \text{const}$ may be treated as ideal photometric image (4), and for order $k = 1$ and rank $m = 0$ the operator $u_0^1(x, \xi)$ coincides with ideal wave transform (3). The generalized ART for scalar fields is connected with tomographic transforms also, see [2]. Thus at $k, m = 0$, $\rho \equiv 0$ and $\varepsilon \equiv 0$, the operator (5) describes fan-beam (in R^2) or cone-beam (in R^3) transforms. The same operator with limits from $-\infty$ to $+\infty$ is ray transform defined in the space R^n of any dimension $n \geq 2$; for $n = 2$ this operator may be treated as the Radon transform. At $k, m = 0$, $\rho \equiv 0$, $n \geq 2$ the operator (5) is standard attenuated ray transform. For $\rho \equiv 0$, $\varepsilon \equiv 0$, $k = 0$, m integer, $m \geq 1$, $n \geq 2$, we obtain well-known longitudinal ray transform for tensor fields [18]. Besides, if $k \geq 1$ then we have integral k-moments of generalized tensor fields. Thus we make sure that, at first, the operators of ART of order k for m-tensor fields are connected with tomographic transforms. Secondly, settings of inverse problems consisting in determination of m-tensor field w by its known ART of order k arise naturally. At last, the notion has an obvious potential to further generalizations and new settings of inverse problems. For example we may consider the Riemannian domain instead of Euclidean space.

The formula (5) defines the stationary ART of order k for m-tensor fields. We assume now that a symmetric tensor field w depends and of time also. A speed of propagation of perturbation is equal, for simplicity, to unit. Then a function $u_m^k(t, x, \xi)$ of a form

$$u_m^k(t, x, \xi) = \int\limits_0^\infty s^k \exp\left\{ -\int\limits_0^s \left(\varepsilon(x - \sigma\xi, \xi) + \mathrm{i}\rho(x - \sigma\xi, \xi) \right) d\sigma \right\}$$
$$\times w_{i_1 \ldots i_m}(t - s, x - s\xi)\xi^{i_1} \ldots \xi^{i_m} ds \qquad (6)$$

is called the *non-stationary ART of order k for m-tensor fields*.

First section of the paper contains certain connections between generalized ART for tensor fields of different orders. We obtain differential equations which solutions are the ART of order k for m-tensor fields. Differential equations of the first order coincides with stationary and non-stationary transport equations with right-hand part of the form $w_{i_1 \ldots i_m}(t - s, x - s\xi)\xi^{i_1} \ldots \xi^{i_m}$, complex-valued absorption coefficient, and without integral part, describing the scattering [34]. Next section is devoted to a proof of uniqueness theorems for the boundary-value and initial boundary-value problems for obtained equations.

2   Main Equations

At first we derive the equations which solutions are the ART of order k for m-tensor fields. It should be remarked that all constructions are valid and for the space R^2 of dimension 2.

We denote a set of pairs (x, ξ) as $SR = R^3 \times R^3 = \{(x, \xi) \mid x, \xi \in R^3, |\xi| = 1\}$. The set of pairs $(x, \xi) \in SR$ with fixed x is denoted as S_x^2. A set given in D symmetric m-tensor fields $w(x) = (w_{i_1 \ldots i_m}(x))$, $i_1, \ldots, i_m = 1, 2, 3$, is designated by $S^m(D)$. Below we often omit the letter "D" in the designation $S^m(D) = S^m$. The scalar product in S^m is defined by the formula

$$\langle u(x), v(x) \rangle = u_{i_1 \ldots i_m}(x) v^{i_1 \ldots i_m}(x). \tag{7}$$

We recall that in Euclidean spaces with rectangular Cartesian coordinate system there is no difference between contravariant and covariant components of tensors. Below the covariant components of tensors are used usually.

An operator $\mathcal{H}$, acting on differentiable on SR functions $\psi(x, \xi)$, is defined by relation

$$(\mathcal{H}\psi)(x, \xi) = \frac{d}{d\tau} \psi(x + \tau \xi, \xi) \Big|_{\tau=0}. \tag{8}$$

In particular the function ψ may be depending on x only, but $\mathcal{H}\psi$ depends on the pair (x, ψ) always. In Cartesian coordinates the operator is represented as

$$(\mathcal{H}\psi)(x, \xi) = \frac{d}{d\tau} \psi(x + \tau \xi, \xi) \Big|_{\tau=0} = \xi^1 \frac{\partial \psi}{\partial x^1} + \xi^2 \frac{\partial \psi}{\partial x^2} + \xi^3 \frac{\partial \psi}{\partial x^3}, \tag{9}$$

and hence

$$(\mathcal{H}\psi)(x, \xi) = \langle \xi, \nabla \psi \rangle = \mathrm{div}(\psi \xi). \tag{10}$$

Remark 1. For differentiability of a function $\psi(x, \xi)$ on SR its differentiability upon coordinates of spatial variables $x \in R^3$ is enough.

We use a short transcription $\langle w, \xi^m \rangle$ for a sum $w_{i_1 \ldots i_m} \xi^{i_1} \ldots \xi^{i_m}$, and denotation $\alpha(x, \xi) = \varepsilon(x, \xi) + i\rho(x, \xi)$ below.

Lemma 1. *Let $\varepsilon(x, \xi) \geq 0$, $\rho(x, \xi)$ be the elements of $C^1(SR)$, the components $w_{i_1 \ldots i_m}$, $i_1, \ldots, i_m = 1, 2, 3$, of a finite symmetric m-tensor field $w(x)$ be the elements of $C^1(R^3)$, m integer, $m \geq 0$. Then for integer k, $k \geq 1$, the formula*

$$((\mathcal{H} + \alpha)u_m^k)(x, \xi) = k\, u_m^{k-1}(x, \xi) \tag{11}$$

is valid with u_m^k determined by (5).

Proof. We set a designation $f(x) := w_{i_1 \ldots i_m}(x)$ for a fixed indexes $i_1, \ldots, i_m$, and prove relation (11) for the function f. For this purpose we change the sum $\langle w(x - s\xi), \xi^m \rangle$ in $u_m^k(x, \xi)$ by $f(x - s\xi)$. Taking in account this replacement we prove the relation (11) directly. By definition (8)

$$(\mathcal{H}u_m^k)(x, \xi) = \frac{d}{d\tau} u_m^k(x + \tau \xi, \xi)\big|_{\tau=0}$$

$$= \int_0^\infty s^k \frac{d}{d\tau} \exp\left\{ -\int_0^s \alpha(x + (\tau - \sigma)\xi, \xi)d\sigma \right\}\Big|_{\tau=0} f(x - s\xi)ds$$

$$+ \int_0^\infty s^k \exp\left\{ -\int_0^s \alpha(x - \sigma\xi, \xi)d\sigma \right\} \left(\frac{d}{d\tau} f(x + (\tau - s)\xi) \right)\Big|_{\tau=0} ds.$$

For fixed point x and unit vector ξ the functions $f(x + v\xi)$ and $\alpha(x + v\xi, \xi)$ can be treated as functions depending on a real variable v only,

$$f(x + v\xi) = \varphi(v), \quad \alpha(x + v\xi, \xi) = \psi(v).$$

Then

$$\frac{d}{d\tau}\varphi(\tau - s) = -\frac{d}{ds}\varphi(\tau - s), \quad \frac{d}{d\tau}\psi(\tau - \sigma) = -\frac{d}{d\sigma}\psi(\tau - \sigma),$$

and hence

$$(\mathcal{H}u_m^k)(x, \xi) = \int_0^\infty s^k \varphi(-s) \exp\left\{ -\int_0^s \psi(-\sigma)d\sigma\right\} \int_0^s \frac{d}{d\sigma}\psi(\tau - \sigma)d\sigma\bigg|_{\tau=0} ds$$

$$- \int_0^\infty s^k \exp\left\{ -\int_0^s \psi(-\sigma)d\sigma\right\}\left(\frac{d\varphi(\tau - s)}{ds}\right)\bigg|_{\tau=0} ds.$$

Passing to the limit at $\tau \to 0$ and integrating by parts the second term of right-hand part, we obtain

$$(\mathcal{H}u_m^k)(x, \xi) = \int_0^\infty s^k \exp\left\{ -\int_0^s \psi(-\sigma)d\sigma\right\}\psi(-s)\Big(\psi(-s) - \psi(0)\Big)ds$$

$$- s^k \exp\left\{ -\int_0^s \psi(-\sigma)d\sigma\right\}\varphi(-s)\bigg|_{s=0}^\infty$$

$$+ k \int_0^\infty s^{k-1} \exp\left\{ -\int_0^s \psi(-\sigma)d\sigma\right\}\varphi(-s)ds$$

$$- \int_0^\infty s^k \exp\left\{ -\int_0^s \psi(-\sigma)d\sigma\right\}\varphi(-s)\left(\frac{d}{ds}\int_0^s \psi(-\sigma)d\sigma\right)ds. \tag{12}$$

As the function $f(x)$ is finite, $s^k = 0$ at $k \geq 1$, $s = 0$, then the second term at $s \to \infty$, and at $s = 0$ vanishes. The third term on the right is equal to $ku_m^{k-1}(x, \xi)$ according to (5). Summing up the first and the last items we obtain $-\alpha(x, \xi)u_m^k(x, \xi)$. Thus

$$(\mathcal{H}u_m^k)(x, \xi) = ku_m^{k-1}(x, \xi) - \alpha(x, \xi)u_m^k(x, \xi),$$

and the formula (11) for the component $w_{i_1 \ldots i_m}(x)$ of m-tensor field $w(x)$ is proved. We take into account now the linearity property

$$\int_0^\infty s^k \exp\left\{ -\int_0^s \alpha(x - \sigma\xi, \xi)d\sigma\right\}\langle w(x - s\xi), \xi^m\rangle ds$$

$$= \left\langle \xi^m, \int_0^\infty s^k \exp\left\{ -\int_0^s \alpha(x - \sigma\xi, \xi)d\sigma\right\}w(x - s\xi)ds\right\rangle \tag{13}$$

of inner product and remind that Einstein rule works here again. We prove the lemma for each component $w_{i_1...i_m}$ of m-tensor field w. Consequently as (13) fulfilled so (11) is proved for the tensor field w.

Lemma 2. *For $\varepsilon \geq 0$, $\rho \in C^1(SR)$, $w_{i_1...i_m} \in C^1(R^3)$, suppose that symmetric m-tensor field w is finite, $m \geq 0$. Then for $k = 0$ the formula*

$$((\mathcal{H} + \alpha)u_m^0)(x, \xi) = \langle w(x), \xi^m \rangle \tag{14}$$

is valid.

Proof. Coming back to the proof of Lemma 1 we obtain the formula (12) for the fixed component $w_{i_1...i_m} =: f$ of the field w. As $k = 0$ then the third term vanishes, and the second at $s = 0$ is equal to $\varphi(0) = f(x)$. The first and the last terms give $-\alpha(x, \xi)\, u_m^0(x, \xi)$. Accordingly $\mathcal{H}u_m^0(x, \xi) = w_{i_1...i_m}(x) - \alpha(x, \xi)u_m^0(x, \xi)$. Turning to (13) and taking it into account, we come to the statement of Lemma 2.

An operator $\mathcal{L}_k : C^k(SR) \to C(SR)$ is defined by relations

$$(\mathcal{L}_1\psi)(x, \xi) = ((\mathcal{H} + \alpha)\psi)(x, \xi),$$

$$(\mathcal{L}_k\psi)(x, \xi) = \frac{1}{k-1}(\mathcal{H} + \alpha)(\mathcal{L}_{k-1}\psi)(x, \xi),$$

for k integer, $k > 1$.

Theorem 1. *For ε, $\rho \in C^{k+1}(SR)$, $w \in C^{k+1}(S^k)$, suppose that ε is non-negative, w is finite symmetric m-tensor field. Then for integer k, m $(k, m \geq 0)$, the formula*

$$(\mathcal{L}_{k+1}u_m^k)(x, \xi) = \langle w(x), \xi^m \rangle \tag{15}$$

is valid with $u_m^k(x, \xi)$ defined by (5).

Proof. Under assumption of Lemma 1, we act on the both parts of (11) $(k - 1)$ times by the operator $(\mathcal{H}+\alpha)$. The formula $(\mathcal{H}+\alpha)^k u_m^k(x, \xi) = k!\, u_m^0(x, \xi)$ arises as a result. Applying the operator $\mathcal{H} + \alpha$ once more and using Lemma 2, we get the statement of the theorem.

Thus we have a differential equation (15), connecting ART of order k for m-tensor fields u_m^k with symmetric m-tensor fields.

We consider non-stationary case now.

Lemma 3. *For functions ε, $\rho \in C^1(SR)$, symmetric m-tensor field $w(t, x) \in C^1(R \times S^m)$, suppose that ε is non-negative, w is finite. Then for $u_m^k(t, x, \xi)$ determined by the formula (6), $k, m \geq 0$, the relations*

$$\left(\frac{\partial}{\partial t} + \mathcal{H} + \alpha\right)u_m^k = k\, u_m^{k-1}, \quad k \geq 1, \tag{16}$$

$$\left(\frac{\partial}{\partial t} + \mathcal{H} + \alpha\right)u_m^0 = \langle w, \xi^m \rangle \tag{17}$$

are valid.

Proof. As it was done above, we choose one of the components $w_{i_1...i_m}(t, x)$ of symmetric m-tensor field $w(t, x)$, with fixed indexes $i_1, \ldots, i_m$, and prove formulas (16), (17) for this any component $w_{i_1...i_m}(t, x)$ of w, denoted by (a function) $f(t, x)$. We fix also new designation $v^k(t, x, \xi)$ for non-stationary ART of order k, acting on the function f.

For calculation of derivative $\dfrac{\partial v^k}{\partial t}(t, x, \xi)$ we need to find $\dfrac{\partial f(t - s, x - s\xi)}{\partial t}$ only, as merely function f depends on t,

$$\frac{\partial v^k}{\partial t} = \int_0^\infty s^k \exp\left\{ - \int_0^s \alpha(x - \sigma\xi, \xi)d\sigma \right\} \frac{\partial f(t - s, x - s\xi)}{\partial t} ds. \qquad (18)$$

We calculate a total derivative $\dfrac{df}{ds}$ at first,

$$\frac{df}{ds} = \left\langle \frac{\partial f(t - s, x - s\xi)}{\partial(x - s\xi)}, \frac{\partial(x - s\xi)}{\partial s} \right\rangle + \frac{\partial f(t - s, x - s\xi)}{\partial(t - s)} \frac{\partial(t - s)}{\partial s}.$$

With usage of

$$\frac{\partial f(\theta, y)}{\partial y} = \nabla_y f(\theta, y)$$

for $y = x - s\xi$, $\theta = t - s$, and $\dfrac{\partial y}{\partial s} = -\xi$ the derivative can be represented in a form

$$\frac{df}{ds} = \langle \nabla_y f(\theta, y), -\xi \rangle + \frac{\partial f(\theta, y)}{\partial \theta} \frac{\partial \theta}{\partial s}.$$

Next step consists in calculation of a result of the operator $\mathcal{H}$ action on the function f,

$$\mathcal{H}f \equiv \frac{df(\theta, y + \tau\xi)}{d\tau}\bigg|_{\tau=0} = \left\langle \frac{\partial f(\theta, y + \tau\xi)}{\partial(y + \tau\xi)}, \frac{\partial(y + \tau\xi)}{\partial \tau} \right\rangle\bigg|_{\tau=0}$$

$$= \langle \nabla_y f(\theta, y), \xi \rangle.$$

Noting that $\dfrac{\partial \theta}{\partial s} = -\dfrac{\partial \theta}{\partial t}$, we obtain

$$\frac{\partial f(\theta, y)}{\partial t} = -\frac{df(\theta, y + \tau\xi)}{d\tau}\bigg|_{\tau=0} - \frac{df(\theta, y)}{ds}.$$

This implies

$$\frac{\partial v^k}{\partial t}(t, x, \xi) = -\int_0^\infty s^k \exp\left\{ -\int_0^s \alpha(x - \sigma\xi, \xi)d\sigma \right\} \frac{df(t - s, x + (\tau - s)\xi)}{d\tau}\bigg|_{\tau=0} ds$$

$$- \int_0^\infty s^k \exp\left\{ -\int_0^s \alpha(x - \sigma\xi, \xi)d\sigma \right\} \left(\frac{df(t - s, x - s\xi)}{ds} \right) ds.$$

Based on the definition of operator $\mathcal{H}$, we integrate the second term to the right-hand part of the last expression,

$$
\left(\frac{\partial v^k}{\partial t} + \mathcal{H} v^k \right)(t, x, \xi) = \int_0^\infty s^k \frac{d}{d\tau} \exp\left\{ - \int_0^s \alpha(x + (\tau - \sigma)\xi, \xi) d\sigma \right\} \Big|_{\tau=0}
$$
$$
\times f(t - s, x - s\xi) ds
$$
$$
- s^k \exp\left\{ - \int_0^s \alpha(x - \sigma\xi, \xi) d\sigma \right\} f(t - s, x - s\xi) \Big|_{s=0}^\infty
$$
$$
+ \int_0^\infty f(t - s, x - s\xi) \frac{d}{ds} \left(s^k \exp\left\{ - \int_0^s \alpha(x - \sigma\xi, \xi) d\sigma \right\} \right) ds.
$$

We refer now to the reasonings and calculations similar to those in the proofs of Lemmas 1 (see (12), (13) as samples) and 2. As it was done above, taking into account distinctions in final results depending on the $k \geq 1$ or $k = 0$, we obtain the statement of Lemma 3 for the non-stationary ART v^k of order k for the function $f := w_{i_1 \ldots i_m}$ i.e. for a fixed component of the field w. A choice of the component was arbitrary so the result is valid for every component. Now we consider the expression $\langle w, \xi^m \rangle$ (a linear combination of components) and, accordingly, the ART u_m^k. Taking into account the property (13), we obtain the Eqs. (16) and (17).

We define an operator $\mathcal{L}_k^t : C^k(R \times SR) \to C(R \times SR)$ by induction on m,

$$
(\mathcal{L}_1^t \psi)(t, x, \xi) = \left(\left(\frac{\partial}{\partial t} + \mathcal{H} + \alpha \right) \psi \right)(t, x, \xi),
$$
$$
(\mathcal{L}_k^t \psi)(t, x, \xi) = \left(\frac{\partial}{\partial t} + \mathcal{H} + \alpha \right)(\mathcal{L}_{k-1}^t \psi)(t, x, \xi), \quad k > 1.
$$

Theorem 2. *For ε, $\rho \in C^{k+1}(SR)$, $w \in C^{k+1}(R \times S^m)$, suppose that ε is non-negative, w is finite m-tensor field. Then*

$$
(\mathcal{L}_{k+1}^t u_m^k)(t, x, \xi) = \langle w(t, x), \xi^m \rangle. \tag{19}
$$

Proof. The proof of formulated theorem is based just on the same reasons as Theorem 1. We use Lemma 3 significantly.

3   Uniqueness Theorems

We prove uniqueness theorems for boundary-value and initial boundary-value problems of the Eqs. (15) and (19), respectively. We remind that D is a bounded convex domain in R^3 with smooth boundary ∂D.

Theorem 3. *For given functions ε, $\rho \in C^k(D \times S_x^2)$, $\varepsilon(x, \xi) \geq 0$ at $x \in D$, $\xi \in S_x^2$, a function $\varphi(x, \xi) \in C^{k+1}(D \times S_x^2)$, $x \in D$, satisfies in D to the Eq. (15) with zero right-hand part,*

$$
\frac{1}{k!}(\mathcal{H} + \alpha)^{k+1} \varphi \equiv \mathcal{L}_{k+1} \varphi = 0, \tag{20}
$$

and boundary-value conditions

$$\varphi(x,\xi) = (\mathcal{H}\varphi)(x,\xi) = \cdots = (\mathcal{H}^k\varphi)(x,\xi) = 0, \quad x \in \partial D, \quad \langle n_x, \xi \rangle < 0, \quad (21)$$

where n_x is outer normal to the surface ∂D at the point x. Then $\varphi(x,\xi) = 0$ for all $x \in D$, $\xi \in S_x^2$.

Proof. We prove the theorem for $k = 0$ at first. Then (20) looks like $(\mathcal{H}+\alpha)\varphi = 0$. As the coefficient $\alpha(x,\xi) = \varepsilon(x,\xi) + i\rho(x,\xi)$ in (21) is complex-valued, then the function $\varphi(x,\xi)$ is complex-valued also, and thus it can be represented in a form $\varphi = \varphi_1 + i\varphi_2$. Let's write $(\mathcal{H} + \alpha)\varphi$ in more details,

$$(\mathcal{H} + \varepsilon + i\rho)(\varphi_1 + i\varphi_2) = (\mathcal{H}\varphi_1 + \varepsilon\varphi_1 - \rho\varphi_2) + i(\mathcal{H}\varphi_2 + \varepsilon\varphi_2 + \rho\varphi_1) = 0,$$

and multiple it on $\bar{\varphi} = \varphi_1 - i\varphi_2$. Here the designations $\bar{\varphi}$ for complex conjugate and $|\varphi|$ for modulus of complex-valued function φ are used. Then

$$Re\,\{\bar{\varphi}(\mathcal{H} + \bar{\alpha})\varphi\} = \frac{1}{2}\mathcal{H}(|\varphi|^2) + \varepsilon|\varphi|^2 = 0.$$

After integration of last equation over D and unit sphere S_x^2, we get

$$\frac{1}{2}\int\limits_{D}\int\limits_{S_x^2}\mathcal{H}(|\varphi|^2)dxd\lambda_x(\xi) + \int\limits_{D}\int\limits_{S_x^2}\varepsilon|\varphi|^2dxd\lambda_x(\xi) = 0,$$

where $d\lambda_x(\xi)$ is angular measure on S_x^2, $x \in D$.

As $\mathcal{H}(|\varphi|^2) = \mathrm{div}(|\varphi|^2\xi)$ (see (10)), then Gauss-Ostrogradsky formula can be applied to the first integral of last expression. We have as a result

$$\frac{1}{2}\int\limits_{\partial D}\int\limits_{S_x^2}\langle n_x, \xi\rangle|\varphi|^2dsd\lambda_x(\xi) + \int\limits_{D}\int\limits_{S_x^2}\varepsilon|\varphi|^2d\lambda_x(\xi)dx = 0. \quad (22)$$

The condition (21), for $k = 0$, implies $\varphi(x,\xi)$ vanishes at $\langle n_x,\xi\rangle < 0$, $x \in \partial D$. Hence the first integral at the left-hand part of (22) is equal to zero. Hereof and from nonnegativity of $\varepsilon(x,\xi)$ it follows that (22) is performed if and only if $\varphi(x,\xi) = 0$ for all $x \in D$, $\xi \in S_x^2$. We proved the theorem for $k = 0$.

We assume now that the theorem is true for some $j = k-1$, $j \geq 1$, and prove it for $j = k$ (namely for the equation of order $k+1$). Let's consider the equation

$$\mathcal{L}_{k+1}\varphi = \frac{1}{k!}(\mathcal{H} + \alpha)^{k+1}\varphi = \frac{1}{k}(\mathcal{H} + \alpha)(\mathcal{L}_k\varphi) = 0,$$

and denote $\mathcal{L}_k\varphi$ as $A + iB$. Then

$$\frac{1}{k}(\mathcal{H} + \alpha)(A + iB) = \frac{1}{k}(\mathcal{H}A + \varepsilon A - \rho B) + \frac{i}{k}(\mathcal{H}B + \varepsilon B + \rho A) = 0.$$

Let's multiply obtained expression on complex conjugate $\overline{\mathcal{L}_k\varphi}$ to $\mathcal{L}_k\varphi$. The multiplication is possible by virtue induction assumption, so

$$(A - iB)\frac{1}{k}(\mathcal{H} + \alpha)(A + iB) = \frac{1}{k}\left(A\mathcal{H}A + B\mathcal{H}B + \varepsilon A^2 + \varepsilon B^2\right)$$

$$+ \frac{i}{k}\left(A\mathcal{H}B - B\mathcal{H}A + \rho A^2 + \rho B^2\right) = 0.$$

Hence

$$Re\{(\overline{\mathcal{L}_k\varphi})\tfrac{1}{k}(\mathcal{H}+\alpha)(\mathcal{L}_k\varphi)\} = \frac{1}{2k}\mathcal{H}(A^2+B^2) + \frac{1}{k}\varepsilon(A^2+B^2)$$

$$= \frac{1}{2k}\mathcal{H}\big(|\mathcal{L}_k\varphi|^2\big) + \frac{1}{k}\varepsilon|\mathcal{L}_k\varphi|^2 = 0.$$

After integration of last expression over D and S_x^2, $x \in D$, and application of Gauss-Ostrogradsky formula we obtain expression of a form (22), where instead of $|\varphi|^2$ the term $|\mathcal{L}_j\varphi|^2$ appears. The term contains degrees $\mathcal{H}^j\varphi$ of the operator $\mathcal{H}$ no more then k, so at $x \in \partial D$, $\langle n_x, \xi\rangle < 0$, it follows that $\mathcal{L}_k\varphi = 0$. We can conclude now, as in the case for $k = 0$, that $\varphi(x,\xi) = 0$ for all $x \in D$, $\xi \in S_x^2$. The theorem is proved.

Theorem 4. *For ε, $\rho \in C^k(D \times S_x^2)$, $x \in D$, $\varepsilon(x,\xi) \geq 0$ at $x \in D$, $\xi \in S_x^2$. If $\varphi(t,x,\xi) \in C^{k+1}(R_+ \times D \times S_x^2)$ satisfies the equation*

$$\frac{1}{k!}\Big(\frac{\partial}{\partial t} + \mathcal{H} + \alpha\Big)^{k+1}\varphi \equiv \mathcal{L}_{k+1}^t\varphi = 0 \tag{23}$$

in D, initial conditions

$$\varphi(0,x,\xi) = \frac{\partial\varphi}{\partial t}(0,x,\xi) = \ldots = \frac{\partial^k\varphi}{\partial t^k}(0,x,\xi) = 0, \tag{24}$$

and, for $x \in \partial D$, $\langle n_x, \xi\rangle < 0$, $t \geq 0$, boundary conditions

$$\varphi(t,x,\xi) = (\mathcal{H}\varphi)(t,x,\xi) = \ldots = (\mathcal{H}^k\varphi)(t,x,\xi) = 0, \tag{25}$$

where n_x is outer normal to the surface ∂D at the point x, then $\varphi(t,x,\xi) = 0$ for $t > 0$, $x \in D$, $\xi \in S_x^2$.

Proof. We check a rightness of the theorem for $k = 0$ at first, i.e. for the equation of the first order.

Considering $\varphi = \varphi_1 + i\varphi_2$ and multiplying both parts of the equality

$$\Big(\frac{\partial}{\partial t} + \mathcal{H} + \alpha\Big)(\varphi_1 + i\varphi_2) = 0$$

on $\overline{\varphi} = \varphi_1 - i\varphi_2$, we obtain

$$\frac{1}{2}\frac{\partial}{\partial t}|\varphi|^2 + \frac{1}{2}\mathcal{H}\big(|\varphi|^2\big) + \varepsilon|\varphi|^2 = 0.$$

We integrate obtained expression by t from 0 until $T \in (0,\infty)$, domain D and sphere S_x^2, then use Gauss-Ostrogradsky formula,

$$\frac{1}{2}\int\limits_{D}\int\limits_{S_x^2}\Big(|\varphi(T,x,\xi)|^2 - |\varphi(0,x,\xi)|^2\Big)d\lambda_x(\xi)dx$$

$$+ \frac{1}{2} \int_0^T \int_{\partial D} \int_{S_x^2} \langle n_x, \xi \rangle |\varphi|^2 d\lambda_x(\xi) dS dt + \int_0^T \int_D \int_{S_x^2} \varepsilon |\varphi|^2 d\lambda_x(\xi) dx dt = 0.$$

Taking into account arbitrariness of T, the initial (24) and boundary (25) conditions (for $k = 0$), we make sure that the last formula is correct if $\varphi(t, x, \xi) = 0$ for $t \in (0, \infty)$, $x \in D$, $\xi \in S_x^2$.

Remaining part of proof of this theorem is quite similar to the proof of the second part of the Theorem 3.

4   Conclusion

In the article the generalized attenuated ray transforms (ART) for tensor fields are considered and investigated. The transforms are connected with attenuated ray transform arising in emission tomography problem and some ray transforms of the other types. The generalization is implemented in three directions. Namely, a function of attenuation $\exp\{- \int_{L(x)} \varepsilon(y) \, dy\}$ becomes complex-valued, the weight have more general form, and mathematical model contains internal sources (in scalar case) or symmetric tensor fields depending on time.

The generalization of ART operator leads to stationary $u_m^k(x, \xi)$ and non-stationary $u_m^k(t, x, \xi)$ ART of order k for m-tensor fields. They may be treated as the integral moments of a source distribution f or of a symmetric tensor field w with components $w_{i_1 \ldots i_m}$ with a weight generated by exponential function. Connections between ART of different orders are established. Differential equations which solutions are the generalized ART-operators of order k for m-tensor fields are derived. In particular, for 0-tensor field (scalar field) the differential equations of the first order coincide with stationary and non-stationary transport equations with complex-valued absorption coefficient, but without integral part responsible for the scattering phenomenon [34]. Uniqueness theorems for boundary-valued problems in stationary case, and initial boundary-value problems in non-stationary case are proved.

There exist close connections of ART of order k for m-tensor fields with different problems of integral geometry, tomography and optics. According to optical terminology it can be seen easily that $u_0^1(x, \xi)$ for $\varepsilon \equiv 0$ and $\rho(x, \xi) = $ const is the ideal wave image, and $u_0^0(x, \xi)$ for $\rho \equiv 0$ and $\varepsilon = $ const is the ideal photometric image [30, 31, 33]. Concerning notions and terms of computerized tomography the operator (5), for $m = 0$, $\rho = 0$, $\varepsilon = 0$, may be treated as fan-beam or cone-beam transforms, and as well as well-known Radon or ray transforms. In more complicated mathematical models, for example in emission tomography, the operator (5) is standard attenuated ray transform, and certain natural generalization of the integrand leads to a notion of longitudinal ray transform of symmetric tensor fields [18] and to integral moments of generalized tensor fields [17].

We make sure that introduced in the article notion of ART of order k is connected in its partial cases with various transforms of tomographic types. So it arises naturally settings of inverse problems of determination of a scalar, vector or tensor fields by its known generalized ART of order k. This inverse problem can be treated and as the inverse problem for generalized transport equation by determining of its right-hand part. Considered in the article notions may be treated as the first step of investigations towards this direction, and have good potential for further generalization and settings of inverse problems. In particular, the generalized ART of order k can be extended in natural way onto the case of Riemannian metric, including stationary and non-stationary settings. The operator $\mathcal{H}$ then turns into the operator $H = \xi^j \dfrac{\partial}{\partial x^j} - \Gamma^j_{kl} \xi^k \xi^l \dfrac{\partial}{\partial \xi^j}$ known in differential geometry as *geodesic vector field*, and as before the differential equations can contain or not contain the variable t. This way leads to the construction and subsequent investigations of mathematical models for dynamic refractive tensor tomography.

References

1. Budinger, T., Gullberg, G., Huesman, R.: Emission computed tomography. In: Herman, G. (ed.) Image Reconstruction from Projections: Implementation and Applications, pp. 147–246. Springer, Heidelberg (1979). https://doi.org/10.1007/3-540-09417-2_5

2. Natterer, F.: The Mathematics of Computerized Tomography. Wiley, Chichester (1986)

3. Natterer, F.: Inverting the attenuated vectorial Radon transform. J. Inverse Ill Posed Probl. **13**(1), 93–101 (2005). https://doi.org/10.1515/1569394053583720

4. Kazantsev, S., Bukhgeim, A.: Inversion of the scalar and vector attenuated X-ray transforms in a unit disc. J. Inverse Ill Posed Probl. **15**(7), 735–765 (2007). https://doi.org/10.1515/jiip.2007.040

5. Tamasan, A.: Tomographic reconstruction of vector fields in variable background media. Inverse Probl. **23**(5), 2197–2205 (2007). https://doi.org/10.1088/0266-5611/23/5/022

6. Ainsworth, G.: The attenuated magnetic ray transform on surfaces. Inverse Probl. Imaging **7**(1), 27–46 (2013). https://doi.org/10.3934/ipi.2013.7.27

7. Sadiq, K., Tamasan, A.: On the range characterization of the two-dimensional attenuated doppler transform. SIAM J. Math. Anal. **47**(3), 2001–2021 (2015). https://doi.org/10.1137/140984282

8. Monard, F.: Inversion of the attenuated geodesic X-ray transform over functions and vector fields on simple surfaces. SIAM J. Math. Anal. **48**(2), 1155–1177 (2016). https://doi.org/10.1137/15M1016412

9. Aben, H., Puro, A.: Photoelastic tomography for three-dimensional flow birefringence studies. Inverse Probl. **13**(2), 215–221 (1997). https://doi.org/10.1088/0266-5611/13/2/002

10. Ainola, L., Aben, H.: Principal formulas of integrated photoelasticity of characteristic parameters. J. Opt. Soc. Am. A **22**(6), 1181–1186 (2005). https://doi.org/10.1364/JOSAA.22.001181

11. Lionheart, W.R.B., Withers, P.J.: Diffraction tomography of strain. Inverse Probl. **31**(4), 045005 (2015). https://doi.org/10.1088/0266-5611/31/4/045005

12. Karassiov, V.P.: Polarization tomography of quantum radiation: theoretical aspects and operator approach. Theor. Math. Phys. **145**(3), 1666–1677 (2005). https://doi.org/10.1007/s11232-005-0189-4

13. Panin, V.Y., Zeng, G.L., Defrise, M., Gullberg, G.T.: Diffusion tensor MR imaging of principal directions: a tensor tomography approach. Phys. Med. Biol. **47**(15), 2737–2757 (2002). https://doi.org/10.1088/0031-9155/47/15/314

14. Schmitt, J.M., Xiang, S.H.: Cross-polarized backscatter in optical coherence tomography of biological tissue. Opt. Lett. **23**(13), 1060–1062 (1998). https://doi.org/10.1364/OL.23.001060

15. Kuranov, R.V., Sapozhnikova, V.V., et al.: Complementary use of cross-polarization and standard OCT for differential diagnosis of pathological tissues. Opt. Express **10**(15), 707–713 (2002). https://doi.org/10.1364/OE.10.000707

16. Gelikonov, V.M., Gelikonov, G.V.: New approach to cross-polarized optical coherence tomography based on orthogonal arbitrarily polarized modes. Laser Phys. Lett. **3**(9), 445–451 (2006). https://doi.org/10.1002/lapl.200610030

17. Sharafutdinov, V.: A problem of integral geometry for generalized tensor fields on R^n. Sov. Math. Dokl. **33**(1), 100–102 (1986)

18. Sharafutdinov, V.: Integral Geometry of Tensor Fields. VSP, Utrecht (1994)

19. Derevtsov, E.Yu., Polyakova, A.P.: Solution of the integral geometry problem for 2-tensor fields by the singular value decomposition method. J. Math. Sci. **202**(1), 50–71 (2014). https://doi.org/10.1007/s10958-014-2033-6

20. Svetov, I.E., Derevtsov, E.Yu., Volkov, Yu.S, Schuster, T.: A numerical solver based on B-splines for 2D vector field tomography in a refracting medium. Math. Comput. Simul. **97**, 207–223 (2014). https://doi.org/10.1016/j.matcom.2013.10.002

21. Derevtsov, E., Svetov, I.: Tomography of tensor fields in the plane. Eurasian J. Math. Comput. Appl. **3**(2), 24–68 (2015)

22. Derevtsov, E.Yu., Maltseva, S.V.: Reconstruction of the singular support of a tensor field given in a refracting medium by its ray transform. J. Appl. Ind. Math. **9**(4), 447–460 (2015). https://doi.org/10.1134/S1990478915040018

23. Monard, F.: Efficient tensor tomography in fan-beam coordinates. Inverse Probl. Imaging **10**(2), 433–459 (2016). https://doi.org/10.3934/ipi.2016007

24. Monard, F.: Efficient tensor tomography in fan-beam coordinates. II: attenuated transforms. Inverse Probl. Imaging **12**(2), 433–460 (2018). https://doi.org/10.3934/ipi.2018019

25. Mueller, R.K., Kaveh, M., Wade, G.: Reconstructive tomography and applications to ultrasonic. Proc. IEEE **67**(4), 567–587 (1979). https://doi.org/10.1109/PROC.1979.11284

26. Ball, J., Johnson, S.A., Stenger, F.: Explicit inversion of the Helmholtz equation for ultrasound insonification and spherical detection. In: Wang, K. (ed.) Acoustical Imaging, vol. 9. Springer, Boston (1980). https://doi.org/10.1007/978-1-4684-3755-3_26

27. Schmitt, U., Louis, A.K.: Efficient algorithms for the regularization of dynamic inverse problems: I. Theory. Inverse Probl. **18**(3), 645–658 (2002). https://doi.org/10.1088/0266-5611/18/3/308

28. Schmitt, U., Louis, A.K., Wolters, C., Vauhkonen, M.: Efficient algorithms for the regularization of dynamic inverse problems: II. Applications. Inverse Probl. **18**(3), 659–676 (2002). https://doi.org/10.1088/0266-5611/18/3/309

29. Hahn, B., Louis, A.K.: Reconstruction in the three-dimensional parallel scanning geometry with application in synchrotron-based X-ray tomography. Inverse Probl. **28**(4), 045013 (2012). https://doi.org/10.1088/0266-5611/28/4/045013

30. Kireitov, V.R.: On the problem of determining an optical surface by its reflections. Funct. Anal. Appl. **10**(3), 201–209 (1976). https://doi.org/10.1007/BF01075526

31. Born, M., Wolf, E.: Principles of Optics. Cambridge University Press, Cambridge (1999)

32. Goodman, J.: Introduction to Fourier optics. McGraw-Hill Book Company, New York (1968)

33. Kireitov, V.R.: Inverse Problems of the Photometry. Computing Center of the USSR Acad. Sci., Novosibirsk (1983). (in Russian)

34. Case, K., Zweifel, P.: Linear Transport Theory. Addison-Wesley Publishing Company, Boston (1967)

Multiextremal Optimization in Feasible Regions with Computable Boundaries on the Base of the Adaptive Nested Scheme

Victor Gergel[ID], Vladimir Grishagin[✉][ID], and Ruslan Israfilov[ID]

Lobachevsky State University, Gagarin Avenue 23, 603950 Nizhni Novgorod, Russia
{gergel,vagris}@unn.ru, ruslan@israfilov.com

Abstract. The paper is devoted to consideration of multidimensional optimization problems with multiextremal objective functions over search domains determined by constraints, which form a special type of domain boundaries called computable ones, which, in general case, are non-linear and multiextremal. The regions of this class can be very complicated, in particular, non-convex, non-simply connected, and even disconnected. For solving such problems, a new global optimization technique based on the adaptive nested scheme developed recently for unconstrained optimization is proposed. The novelty consists in combination of the adaptive scheme with a technique for reducing the constraints to an explicit form of feasible subregions in internal subproblems of the nested scheme that allows one to evaluate the objective function at the feasible points only. For efficiency estimation of the proposed adaptive nested algorithm in comparison with the classical nested optimization and the penalty function method, a representative numerical experiment on the test classes of multidimensional multiextremal functions has been carried out. The results of the experiment demonstrate a significant advantage of the adaptive scheme over its competitors.

Keywords: Multiextremal optimization · Dimensionality reduction · Computable boundaries

1 Introduction

Many important applied problems of decision making can be stated as problems of searching the global minimum of a multidimensional multiextremal function subject to complicated constraints [1, 6, 13, 22, 28, 33, 38]. The property of multiextremality generates significant complexity of these problems because analytical methods are not almost applicable to solve them and numerical algorithms in general case require essential computational expenditures. This feature is explained by the fact that the global minimizer is an integral characteristic of the objective function, i.e., in order to confirm that a point is the global minimizer, it is necessary to compare the objective function value at this point with function values at all points in the region of the search. As a consequence,

© Springer Nature Switzerland AG 2020
Y. D. Sergeyev and D. E. Kvasov (Eds.): NUMTA 2019, LNCS 11974, pp. 112–123, 2020.
https://doi.org/10.1007/978-3-030-40616-5_9

the global optimization method is obliged to build in the search domain a grid of trial points (the term trial means the evaluation of objective function value at a point). Such the grids can be simple enough, for example, regular rectangular or random Monte-Carlo ones [39], but efficient methods build non-uniform grids which adapt to the behavior of the objective function placing trials densely in subregions with low function values and rarely in subdomains where the function has high values. For essentially multiextremal functions like Lipschitzian ones the number of grid nodes grows exponentially when increasing the problem dimension. Just this circumstance explains the high complexity of global optimization problems.

As the main approaches to designing efficient and theoretically substantiated methods one can consider the paradigm of component methods and the idea of reducing the dimensionality of optimization problems.

The component methods [4,21,23,27,28,30] partition the search region into several subdomains and introduce a criterion that evaluates numerically each subdomain from the point of view of its efficiency for search continuation and after that a new iteration is executed in the subdomain with the best criterion value. The methods of this class differ in the strategies of partitioning and criteria of efficiency of subdomains.

The algorithms based on the idea of dimensionality reduction can be divided into two groups. The methods of the first group replace the multidimensional problem with an equivalent univariate one applying a continuous mapping of the multidimensional search domain onto a subregion of the real axis by means of the Peano space-filling curves, or evolvents [3,14,24–26,32,36].

The second group of optimization algorithms is based on the known scheme of nested optimization [4]. According to this approach the initial multidimensional problems is reduced to a family one-dimensional subproblems connected recursively [5,9–12,17,18,29,34,36]. In the paper [9], a generalization of the classical scheme called adaptive nested optimization has been proposed and the research [18] has demonstrated that this version of the nested scheme in combination with information-statistic algorithm of univariate global search has the high efficiency being better significantly than the classical prototype and one of the most qualitative popular method DIRECT [21].

For solving relatively simple problems of global optimization characterized by a small number of local optima with regions of attraction being large enough, a so called multi-start approach [2,5,35] can be used when a local optimization method is launched from several starting points. This approach is clear geometrically, but, unfortunately, the methods of this type are semiheuristical and are not efficient for complicated multiextremal problems.

Another challenge in global optimization refers to problems with complicated constraints. The traditional way to solve such the problems consists in transforming the constrained problem to an equivalent problem either without constraints or, as a rule, in a simple region like a box.

There exist two main approaches in this way. The first transformation is classical in optimization and is connected with the penalty function method [7,20,37].

This method is sufficiently universal but it requires a tuning of its parameters (penalty constant, equalizing coefficients for constraints). In many cases this tuning is not simple and a wrong choice of parameters does not allow obtaining the correct solution of the constrained problem. For example, if the penalty constant is small the solution of the unconstrained problem can differ significantly from the solution of the initial problem. At the same time, if the penalty constant is too large then it worsens substantially the properties of the objective function in the transformed problem, in particular, the Lipschitz constant can increase essentially.

The second approach is based on building the so called index function [36] that contains no tuning parameters but generates, in general case, a discontinuous objective function in the transformed problem and requires, as a consequence, application of special global optimization techniques oriented at this class of functions.

When solving the transformed problem in the framework of both the approaches (penalty and index methods), the optimization algorithm places trial points not only in the feasible domain of the constrained problem but out of it as well.

In this paper we consider the approach which allows one to avoid performing trials at non-feasible points and does not include any tuning parameters. The core of this approach is the nested optimization scheme applied to multiextremal optimization in domains with special type of constraints, namely, in domains with computable boundaries. These domains can be very complicated, in particular, non-convex, non-simply connected, and even disconnected domains. An algorithm for Lipschitzian optimization on the base of classical nested scheme for domains with computable boundaries has been described in the paper [16]. In the present paper we propose its generalization that applies the more efficient recursive technique of global search in the framework of the adaptive nested optimization [9]. To demonstrate the advantages of the proposed constraint satisfaction approach the results of comparison with the penalty function method are given on two known test classes that are classical for estimating the efficiency of global optimization algorithms.

The rest of the paper is organized as follows. Section 2 contains statement of the multiextremal constrained problem to be studied and description of a generalization of the adaptive nested scheme for the case of computable boundaries. Section 3 is devoted to computational testing the proposed technique in comparison with the classical nested scheme and the method of penalty functions. Section 4 concludes the paper.

2   Nested Optimization and Computable Boundaries

The optimization problem under consideration is formulated in the following way. It is required to find the least value (global minimum) and its coordinates (global minimizer) of an objective function $f(x)$ in a domain D of the N-dimensional Euclidean space $\mathbb{R}^N$. This problem will be denoted as

$$f(x) \to \min, \ x = (x_1, \ldots, x_N) \in D \subseteq \mathbb{R}^N. \tag{1}$$

The feasible domain D is supposed to be given by constraints-inequalities

$$D = \{x \in X : h_s(x) \leq 0, \, 1 \leq s \leq q\}, \tag{2}$$

where the region X is determined by simple coordinate constraints as

$$X = \{x \in \mathbb{R}^N : a_j \leq x \leq b_j, \, 1 \leq j \leq N\}. \tag{3}$$

The objective function $f(x)$ and constraints $h_s(x)$, $1 \leq s \leq q$, are supposed to satisfy in the domain X the Lipschitz condition

$$|h_s(x') - h_s(x'')| \leq L_s \|x' - x''\|, \, x', x'' \in X, \, 1 \leq s \leq q+1, \tag{4}$$

where the function $h_{q+1}(x) = f(x)$, $L_s > 0$ is a finite value called the Lipschitz constant of the function $h_s(x)$, $1 \leq s \leq q+1$, and $\|\cdot\|$ denotes the Euclidean norm in $\mathbb{R}^N$. In general case, the objective function and constraints of the problem (1)–(2) are multiextremal and non-smooth.

If the problem (1) does not contain constraints (2) ($q = 0$), i.e., $D = X$, for solving such the problem the known nested scheme of dimensionality reduction [4,36] can be applied. For example, it can be done if the constrained problem (1) has been transformed to the unconstrained one in the framework of the penalty function method [7,20,37]. According to this method, instead of the problem (1)–(2), the problem

$$F(x) \to \min, \, x \in X \subseteq \mathbb{R}^N, \tag{5}$$

is considered with the "penalized" objective function

$$F(x) = f(x) + PH(x), \tag{6}$$

where $P > 0$ is the penalty constant and $H(x)$ is the penalty function such that $H(x) = 0$, if $x \in D$, and $H(x) > 0$, if $x \notin D$. If to choose the penalty function as

$$H(x) = \max\{0, h_1(x), \ldots, h_q(x)\}, \tag{7}$$

then $F(x)$ meets the Lipschitz condition under requirements (4).

In its original classical form the nested optimization scheme was oriented at unconstrained optimization, or more detailed, at solving problems (1) when constraints of the type (2) are absent, i.e., $D = X$. In this situation there takes place [4] the relation

$$\min_{x \in X} f(x) = \min_{x_1 \in X_1} \min_{x_2 \in X_2} \cdots \min_{x_N \in X_N} f(x_1, \ldots, x_N). \tag{8}$$

where X_i is a line segment $[a_i, b_i]$, $1 \leq i \leq N$.

This approach can be generalized (see, for example, [16]) to the case with continuous constraints (2) that allows one to present (8) for the domain D in the form

$$\min_{x \in D} f(x) = \min_{x_1 \in \Lambda_1} \min_{x_2 \in \Lambda_2(\xi_1)} \cdots \min_{x_N \in \Lambda_N(\xi_{N-1})} f(x_1, \ldots, x_N), \tag{9}$$

where $\xi_s = (x_1, \ldots, x_s)$, $1 \leq s \leq N$, and the region $\Lambda_s(\xi_{s-1})$ is the projection of the set

$$\Omega_s(\xi_s) = \{\xi_s \in \mathbb{R}^s : (\xi_s, x_{s+1}, \ldots, x_N) \in D\}, \tag{10}$$

onto the coordinate axis x_s.

Now the nested optimization scheme applied for the case (9) can be described as follows.

Let us introduce a family of reduced function $f^s(\xi_s)$, $1 \leq s \leq N$, in the following manner:

$$f^{s-1}(\xi_{s-1}) = \min_{x_s \in \Lambda_s(\xi_{s-1})} f^s(\xi), \ 2 \leq s \leq N, \tag{11}$$

$$f^N(x) \equiv f(x). \tag{12}$$

Then, the solving the multidimensional problem (1) can be substituted with searching for the global minimum of the univariate function $f^1(x_1)$ in the domain Λ_1, as according to (9)–(11)

$$\min_{x \in D} f(x) = \min_{x_1 \in \Lambda_1} f^1(x_1). \tag{13}$$

But any evaluation of the function $f^1(x_1)$ at a chosen point x_1 requires solving the problem

$$f^2(x_1, x_2) \rightarrow \min, \ x_2 \in \Lambda_2(x_1), \tag{14}$$

which is one-dimensional because the coordinate x_1 is fixed.

The necessity of evaluation of the function $f^2(x_1, x_2)$ generates solving the problem of minimization of the function $f^3(\xi_2, x_3)$ in the domain $\Lambda_3(\xi_2)$, and this problem is univariate as well, because the vector ξ_2 is fixed.

This recursive procedure is in progress until we reach the level N where it is required to solve problem

$$f^N(\xi_{N-1}, x_N) \rightarrow \min, \ x_N \in \Lambda_N(\xi_{N-1}) \tag{15}$$

This problem is univariate too because the vector ξ_{N-1} has been given at previous levels and is fixed for the problem (15). Moreover, in this problem an evaluation of the objective function consists in computation of the value $f(\xi_{N-1}, x_N)$ of the function $f(x)$ from the original problem (1).

The approach of reducing the multidimensional problem (1) to solving the family of one-dimensional subproblems

$$f^s(\xi_{s-1}, x_s) \rightarrow \min, \ x_s \in \Lambda_s(\xi_{s-1}), \ 1 \leq s \leq N, \tag{16}$$

in accordance of the above procedure is called *the nested scheme of dimensionality reduction or the nested scheme of optimization*.

The structure of domains $\Lambda_s(\xi_{s-1})$ depends on the properties and complexity of the constraints $h_s(x)$ from (2). For example, if all the functions $h_s(x)$ are convex, then the domain is a convex set, and any projection $\Lambda_s(\xi_{s-1})$ is a single interval of the axis x_s. In general case, when the constraints $h_s(x)$ are continuous, the domain $\Lambda_s(\xi_{s-1})$ is a union of closed intervals, i.e.,

$$\Lambda_s(\xi_{s-1}) = \cup_{m=1}^{M}[a_s^m, b_s^m], \ 1 \le s \le N, \tag{17}$$

where the end points a_s^m, b_s^m of intervals and even the number of interval M can depend on the vector ξ_{s-1}.

If all the end points a_s^m, b_s^m and all numbers M can be given explicitly (for example, as analytical expressions or by means of a computational procedure) in all the subtasks (16) then the domain D is called as the domain with computable boundaries. These domains can have very complicated structure, in particular, can be non-convex and even disconnected.

As an example, let us consider a 2-dimensional domain (2) determined by the following constraints:

$$h_1(x_1, x_2) = 1 - \frac{(x_2 - 0.5(u_1(x_1) + u_2(x_1)))^2}{0.25(u_1(x_1) - u_2(x_1))^2}, \tag{18}$$

$$h_2(x_1, x_2) = 0.04 - (x_1 - 0.6)^2 - (x_2 - 0.59)^2, \tag{19}$$

$$h_3(x_1, x_2) = x_2 - u_3(x_1), \tag{20}$$

where

$$u_1(x_1) = -0.05 \cos(40x_1) - 0.1x_1 + 0.15,$$
$$u_2(x_1) = -0.05 \cos(45x_1) + 0.1x_1 - 0.22,$$
$$u_3(x_1) = 0.1 \sin(50x_1) + 0.5x_1 + 0.6,$$

and coordinate constraints (3), $0 \le x_1, x_2 \le 1$. For this domain

$$\Lambda_1 = [0, 1], \tag{21}$$

$$\Lambda_2(x_1) = \begin{cases} \cup_{m=1}^{3} [a_2^m, b_2^m], & |x_1 - 0.6| \le 0.2, \\ \cup_{m=1}^{2} [\alpha_2^m, \beta_2^m], & \text{otherwise}, \end{cases} \tag{22}$$

where

$$a_2^1 = \alpha_2^1 = 0, \qquad\qquad\qquad b_2^1 = \beta_2^1 = u_1(x_1),$$
$$a_2^2 = \alpha_2^2 = u_2(x_1), \qquad\qquad b_2^2 = 0.59 - \sqrt{0.04 - (x_1 - 0.6)^2},$$
$$a_2^3 = 0.59 + \sqrt{0.04 - (x_1 - 0.6)^2}, \qquad b_2^3 = \beta_2^2 = \min\{u_3(x_1), 1\}.$$

The domain D corresponding to these constraints is shown in Fig. 1, where inaccessible part is dark. The domain consists of two disconnected parts and inside the upper part there is a removed circle. Moreover, the boundaries have complicated "oscillating" structure.

The nested optimization scheme in combination with univariate global search methods providing optimization on several intervals like characteristical algorithms [19] allows one to execute trials in the feasible domain only and not to spend resources to evaluate the objective function at inaccessible points as opposed to penalty function or index methods.

In present paper we propose to apply the adaptive nested scheme to solving problems with computable boundaries in combination with information-statistical univariate algorithm of global search [36] adapted to optimization in the domain of type (17). In the adaptive scheme all one-dimensional sub-problems (16) are considered in dynamics simultaneously and to each of them a numerical value called the characteristic of the subproblem is assigned. The characteristic depends on the domain (17) and values of the subproblem objective function. The iteration of the multidimensional search consists in the choice of the subproblem with the best characteristic and executing a new trial in it. Such organization allows one to take into account the full information about the multidimensional problem obtained in the course of optimization and to focus on the most perspective subproblems. The effectiveness of the new proposed adaptive nested technique is demonstrated in the next section on the base of representative experiment on test classes of multiextremal problems in domains with computable boundaries of complicated structure.

3   Numerical Experiments

The efficiency estimation of different approaches to solving constrained global optimization problems was executed experimentally on two test classes of multiextremal functions which are often used for testing the global search algorithms [9,10,18,32,36]. The first class GLOB2 included 2-dimensional functions

$$f(x_1, x_2) = -\left\{ \left(\sum_{i=1}^{7} \sum_{j=1}^{7} u_{ij}(x_1, x_2) \right)^2 + \left(\sum_{i=1}^{7} \sum_{j=1}^{7} v_{ij}(x_1, x_2) \right)^2 \right\}^{\frac{1}{2}} \quad (23)$$

where

$$u_{ij}(x_1, x_2) = \alpha_{ij} \sin(\pi i x_1) \sin(\pi j x_2) + \beta_{ij} \cos(\pi i x_2) \cos(\pi j x_2),$$
$$v_{ij}(x_1, x_2) = \gamma_{ij} \sin(\pi i x_1) \sin(\pi j x_2) - \delta_{ij} \cos(\pi i x_2) \cos(\pi j x_2),$$

and the parameters a_{ij}, β_{ij}, γ_{ij}, δ_{ij}, $1 \leq i, j \leq 7$, are the independent random numbers, distributed uniformly over the interval $[-1, 1]$. The functions (23) were considered in the box $X = \{x \in \mathbb{R}^2 : 0 \leq x_1, x_2 \leq 1\}$.

The multiextremal class GKLS [8] was chosen as the second class of objective functions in the problem (1). The functions were taken from the hard GKLS subclass of the dimension 3 and for them $X = \{x \in \mathbb{R}^3 : -1 \leq x_1, x_2, x_3 \leq 1\}$.

For building constraints (2) for both GLOB2 and GKLS the idea close to making constraints in the EMMENTAL GKLS [31] was used. Namely, in the domain X several random points were generated which are considered as centers of spheres with random radii. The hyperparallelepiped X without internal parts of the generated spheres was considered as the feasible domain D. Such way allows one to form complicated domains with computable boundaries because the information about centers and radii of spheres enables to build explicitly the regions (17) in univariate subproblems of the nested scheme.

Three methods were compared in experiments:

- **CNS-CB** Classical nested scheme with computable boundaries;
- **ANS-CB** Adaptive nested scheme with computable boundaries;
- **ANS-PF** Adaptive nested scheme combined with penalty function method.

In all three methods for solving univariate problems (16) the information-statistical Global Search Algorithm (GSA) was used.

An example of comparative behavior of the methods taking computable boundaries into account (ANS-CB) and applying penalty function approach (5), (6) (ANS-PF) is presented in Fig. 1 for a function from class GLOB2 and constraints (18)–(20). The pictures contain level curves of the function, points of trials, and the infeasible part $X \setminus D$ is dark.

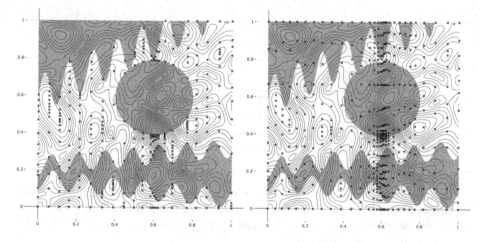

Fig. 1. Distribution of trials by ANS-CB (the left panel) and by ANS-PF (the right panel).

Comparison of the algorithms on the test classes was carried out according to the method of operational characteristics introduced in [15]. In the framework of this method a set of test problems is taken, the problems of the set are solved by an optimization algorithm with different parameters and two criteria are used for evaluating the algorithm's quality: average number K of trials executed (search spending) and number Δ of problems solved successfully (search reliability). For launches of the algorithm with different parameters we obtain several pairs $(\tilde{K}, \tilde{\Delta})$. The set of these pairs on the plane (K, Δ) is called the operational characteristic of the algorithm.

Fig. 2. Operational characteristics on 2-dimensional class GLOB2

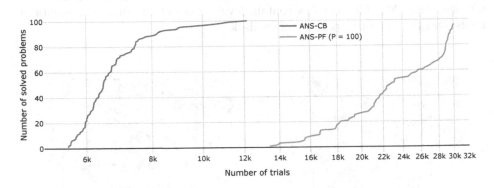

Fig. 3. Operational characteristics on 3-dimensional class GKLS

Figure 2 shows the operational characteristics (from left to right) of ANS-CB, CNS-CB and 3 operational characteristics of ANS-PF for different values of penalty factor P from (6) on the class (23) with 100 test problems. The axis K is presented in the logarithmic scale.

As it follows from the results presented in Fig. 2 the adaptive and classical schemes using the computable boundaries approach excel significantly the version with the penalty function method. With the value of penalty constant $P = 100$ the algorithm with transformation to the penalized function (6) did not provide solving all test problems and spent considerably more trials.

As the functions of the test class are very complicated, attempts to enlarge the penalty constant have demonstrated one of the drawbacks of the penalty function method for Lipschitzian optimization problems, namely, such the enlargement leads to increasing the Lipschitz constant for the function (6) and, as a consequence, to the growth of the trial number. Moreover, the adaptive nested scheme is better than its classical prototype CNS-CB.

The experiment with 100 3-dimensional functions from the class GKLS has shown even more advantage of the computable boundaries approach over the penalty function technique. Figure 3 presents the operational characteristic for ANS-CB (the left plot) and the operational characteristic for ANS-PF (the right plot) with the penalty constant $P = 100$.

The algorithm ANS-CB has solved all the test problems for about 12000 objective function evaluations and its rival ANS-PF having spent 30000 trials could not find all the global minima.

4 Conclusion

In the paper the multidimensional global optimization problems with non-linear and multiextremal objective functions and constraints generating domains with computable boundaries have been considered. The domains of this type can have a complicated structure, in particular, can be non-convex and disconnected. For solving the problems under consideration a new global optimization algorithm based on the adaptive nested scheme has been proposed. The algorithm reduces the initial multidimensional problem to a family of univariate subproblems in which the domains of one-dimensional optimization can be presented as systems of closed intervals with explicitly given boundary points. For solving univariate subproblems a modification of the information-statistic algorithm of global search is used which execute iteration within the feasible intervals only. It provides evaluation of multidimensional objective function in the accessible domain only and distinguishes the proposed method from known approaches to solving global constrained optimization such as penalty function and index methods which can carry out iterations at infeasible points.

The more economical behavior of the new method has been confirmed in the experiment where the proposed adaptive nested algorithm was compared with the classical nested scheme and adaptive scheme combined with penalty function method. The results of the experiment have demonstrated the significant advantage of the suggested adaptive scheme over its opponents.

As continuation of the research it is interesting to evaluate the efficiency of the new adaptive scheme via comparison with the global optimization methods of different nature, for example, with some component methods of DIRECT-type. Moreover, it would be perspective to develop a parallel version of the algorithm and to study its effectiveness of parallelizing on various computational architectures.

Acknowledgements. The research of the first author has been supported by the Russian Science Foundation, project No 16-11-10150 "Novel efficient methods and software tools for timeconsuming decision make problems using superior-performance supercomputers."

References

1. Bartholomew-Biggs, M.C., Parkhurst, S.C., Wilson, S.P.: Using DIRECT to solve an aircraft routing problem. Comput. Optim. Appl. **21**(3), 311–323 (2002)

2. Boender, C.G.E., Rinnooy Kan, A.H.G.: Bayesian stopping rules for multistart global optimization methods. Math. Program. **37**(1), 59–80 (1987)
3. Butz, A.R.: Space-filling curves and mathematical programming. Inf. Control **12**, 314–330 (1968)
4. Carr, C.R., Howe, C.W.: Quantitative Decision Procedures in Management and Economic: Deterministic Theory and Applications. McGraw-Hill, New York (1964)
5. Dam, E.R., Husslage, B., Hertog, D.: One-dimensional nested maximin designs. J. Global Optim. **46**, 287–306 (2010)
6. Famularo, D., Pugliese, P., Sergeyev, Y.D.: A global optimization technique for checking parametric robustness. Automatica **35**, 1605–1611 (1999)
7. Fiacco, A.V., McCormick, G.P.: Nonlinear Programming: Sequential Unconstrained Minimization Techniques. Wiley, New York (1968)
8. Gaviano, M., Kvasov, D.E., Lera, D., Sergeyev, Y.D.: Software for generation of classes of test functions with known local and global minima for global optimization. ACM Trans. Math. Softw. **29**(4), 469–480 (2003)
9. Gergel, V.P., Grishagin, V.A., Gergel, A.V.: Adaptive nested optimization scheme for multidimensional global search. J. Global Optim. **66**, 35–51 (2016)
10. Gergel, V.P., Grishagin, V.A., Israfilov, R.A.: Local tuning in nested scheme of global optimization. Procedia Comput. Sci. **51**, 865–874 (2015)
11. Gergel, V.P., Grishagin, V.A., Israfilov, R.A.: Adaptive dimensionality reduction in multiobjective optimization with multiextremal criteria. In: Nicosia, G., Pardalos, P., Giuffrida, G., Umeton, R., Sciacca, V. (eds.) Machine Learning, Optimization, and Data Science. LNCS, vol. 11331, pp. 129–140. Springer, Cham (2019). https://doi.org/10.1007/978-3-030-13709-0_11
12. Gergel, V.P., Grishagin, V.A., Israfilov, R.A.: Parallel dimensionality reduction for multiextremal optimization problems. In: Malyshkin, V. (ed.) PaCT 2019. LNCS, vol. 11657, pp. 166–178. Springer, Cham (2019). https://doi.org/10.1007/978-3-030-25636-4_13
13. Gergel, V.P., Kuzmin, M.I., Solovyov, N.A., Grishagin, V.A.: Recognition of surface defects of cold-rolling sheets based on method of localities. Int. Rev. Autom. Control **8**, 51–55 (2015)
14. Goertzel, B.: Global optimization with space-filling curves. Appl. Math. Lett. **12**, 133–135 (1999)
15. Grishagin, V.A.: Operating characteristics of some global search algorithms. In: Problems of Stochastic Search, vol. 7, pp. 198–206. Zinatne, Riga (1978). (in Russian)
16. Grishagin, V.A., Israfilov, R.A.: Multidimensional constrained global optimization in domains with computable boundaries. In: CEUR Workshop Proceedings, vol. 1513, pp. 75–84 (2015)
17. Grishagin, V.A., Israfilov, R.A.: Global search acceleration in the nested optimization scheme. In: AIP Conference Proceedings, vol. 1738, p. 400010 (2016)
18. Grishagin, V.A., Israfilov, R.A., Sergeyev, Y.D.: Convergence conditions and numerical comparison of global optimization methods based on dimensionality reduction schemes. Appl. Math. Comput. **318**, 270–280 (2018)
19. Grishagin, V.A., Sergeyev, Y.D., Strongin, R.G.: Parallel characteristic algorithms for solving problems of global optimization. J. Global Optim. **10**, 185–206 (1997)
20. Han, S.P., Mangasarian, O.L.: Exact penalty functions in nonlinear programming. Math. Program. **17**(1), 251–269 (1979)
21. Jones, D.R.: The DIRECT global optimization algorithm. In: Floudas, C.A., Pardalos, P.M. (eds.) Encyclopedia of Optimization, pp. 431–440. Kluwer Academic Publishers, Dordrecht (2001)

22. Kvasov, D.E., Menniti, D., Pinnarelli, A., Sergeyev, Y.D., Sorrentino, N.: Tuning fuzzy power-system stabilizers in multi-machine systems by global optimization algorithms based on efficient domain partitions. Electric. Power Syst. Res. **78**, 1217–1229 (2008)
23. Kvasov, D.E., Pizzuti, C., Sergeyev, Y.D.: Local tuning and partition strategies for diagonal GO methods. Numer. Math. **94**, 93–106 (2003)
24. Lera, D., Sergeyev, Y.D.: Lipschitz and Hölder global optimization using space-filling curves. Appl. Numer. Math. **60**, 115–129 (2010)
25. Lera, D., Sergeyev, Y.D.: Deterministic global optimization using space-filling curves and multiple estimates of Lipschitz and Holder constants. Commun. Nonlinear Sci. Numer. Simul. **23**, 328–342 (2015)
26. Oliveira Jr., H.A., Petraglia, A.: Global optimization using space-filling curves and measure-preserving transformations. In: Gaspar-Cunha, A., Takahashi, R., Schaefer, G., Costa, L. (eds.) Soft Computing in Industrial Applications. AINSC, vol. 96, pp. 121–130. Springer, Heidelberg (2011). https://doi.org/10.1007/978-3-642-20505-7_10
27. Paulavičius, R., Žilinskas, J.: Simplicial Global Optimization. Springer, New York (2014). https://doi.org/10.1007/978-1-4614-9093-7
28. Pintér, J.D.: Global Optimization in Action. Kluwer, Dordrecht (1996)
29. Sergeyev, Y.D., Grishagin, V.A.: Parallel asynchronous global search and the nested optimization scheme. J. Comput. Anal. Appl. **3**, 123–145 (2001)
30. Sergeyev, Y.D., Kvasov, D.E.: Deterministic Global Optimization: An Introduction to the Diagonal Approach. Springer, New York (2017). https://doi.org/10.1007/978-1-4939-7199-2
31. Sergeyev, Y.D., Kvasov, D.E., Mukhametzhanov, M.S.: Emmental-type GKLS-based multiextremal smooth test problems with non-linear constraints. In: Battiti, R., Kvasov, D.E., Sergeyev, Y.D. (eds.) LION 2017. LNCS, vol. 10556, pp. 383–388. Springer, Cham (2017). https://doi.org/10.1007/978-3-319-69404-7_35
32. Sergeyev, Y.D., Strongin, R.G., Lera, D.: Introduction to Global Optimization Exploiting Space-Filling Curves. Springer, New York (2013). https://doi.org/10.1007/978-1-4614-8042-6
33. Shevtsov, I.Y., Markine, V.L., Esveld, C.: Optimal design of wheel profile for railway vehicles. In: Proceedings 6th International Conference on Contact Mechanics and Wear of Rail/Wheel Systems, Gothenburg, Sweden, pp. 231–236 (2003)
34. Shi, L., Ólafsson, S.: Nested partitions method for global optimization. Oper. Res. **48**, 390–407 (2000)
35. Snyman, J.A., Fatti, L.P.: A multi-start global minimization algorithm with dynamic search trajectories. J. Optimi. Theory Appl. **54**(1), 121–141 (1987)
36. Strongin, R.G., Sergeyev, Y.D.: Global Optimization with Non-convex Constraints: Sequential and Parallel Algorithms, 3rd edn. Springer, New York (2014). https://doi.org/10.1007/978-1-4615-4677-1
37. Zangwill, W.I.: Non-linear programming via penalty functions. Manag. Sci. **13**(5), 344–358 (1967)
38. Zhao, Z., Meza, J.C., Hove, V.: Using pattern search methods for surface structure determination of nanomaterials. J. Phys. Condens. Matter **18**(39), 8693–8706 (2006)
39. Zhigljavsky, A.A., Žilinskas, A.: Stochastic Global Optimization. Springer, New York (2008). https://doi.org/10.1007/978-0-387-74740-8

On Polyhedral Estimates of Reachable Sets of Discrete-Time Systems with Uncertain Matrices and Integral Bounds on Additive Terms

Elena K. Kostousova[(⊠)] [iD]

Krasovskii Institute of Mathematics and Mechanics,
Ural Branch of the Russian Academy of Sciences,
S.Kovalevskaja Street 16, 620108 Ekaterinburg, Russia
kek@imm.uran.ru

Abstract. We consider discrete-time systems of bilinear type for the case when interval bounds on the coefficients of the system are imposed, additive input terms are restricted by integral non-quadratic constraints, and initial states belong to given sets, which are assumed to be parallelepipeds. An approach for estimating the reachable sets is presented. It is based on considering reachable sets in the "extended" space and constructing external and internal estimates of them in the form of polytopes of some special shape. The specific cross-sections of these polytopes provide the parallelepiped-valued or parallelotope-valued estimates of the reachable sets in the "initial" space. Evolution of the estimates in the "extended" space is determined by recurrence relations. All the estimates can be calculated by explicit formulas. The main attention is paid to internal estimates. Illustrative examples are presented.

Keywords: Discrete-time systems · Reachable sets · Integral constraints · Uncertain matrices · Polyhedral estimates · Parallelepipeds · Parallelotopes

1 Introduction

The reachability problem may be considered as one of the fundamental problems of the mathematical control theory [19–21]. Exact calculation of reachable sets is as a rule a very complicated problem, therefore different numerical methods were developed for their approximations, in particular using polytopes with a large number of vertices or unions of a large number of points (see, for example, [1,2,7,26,27]; here and below we cite for instances only some of numerous publications; see references therein too). But the methods meant for constructing approximations as accurate as possible can require much calculations, especially for large dimensional systems. Another approach is based on estimates of sets by domains of some fixed shape such as ellipsoids, parallelepipeds, and some others [3,5,6,9–16,19–21,24]. Its main advantage is that it enables to calculate

© Springer Nature Switzerland AG 2020
Y. D. Sergeyev and D. E. Kvasov (Eds.): NUMTA 2019, LNCS 11974, pp. 124–138, 2020.
https://doi.org/10.1007/978-3-030-40616-5_10

approximate solutions using relatively simple tools in opposite to the mentioned above methods for obtaining the most accurate approximations. More accurate approximations can be obtained by using parametric families of fixed shape estimates similarly to [20,21]. The interval analysis methods based on subpavings of interval vectors [9] serve the same aim, but can require much computations and memory for high-dimensional systems.

Fair techniques for constructing effective fixed shape estimates were developed for linear systems with hard bounds on controls. It is also important to study linear systems under integral constraints and moreover to study the systems with uncertain coefficients (matrices). This leads to bilinearity and additional difficulties caused by nonlinearity (specifically reachable sets can be non-convex). Some approaches to investigation and approximation of reachable sets for systems with integral constraints and different impulsive systems, for bilinear systems, and for some combinations of such types can be found, for instance, in [1,2,6–8,11,12,18,19,21,26], in [3,10,13,14,21,22,24,25], and in [5,15,16] respectively.

The paper develops research [11] to the more complicated case of systems with uncertain matrices. The first such extensions are given in [15,16]. There, in [15], a family of external estimates for reachable sets of the systems under consideration is proposed, and, in [16], another family of external estimates is constructed, which can provide more accurate estimates. The last mentioned estimates [16] are obtained by two ways: using considerations in the initial space just like [15] and using considerations in an "extended" space. This paper presents an approach for two-sided estimation of the reachable sets based on considering reachable sets in the "extended" space and constructing external and internal estimates for them in the form of polytopes of some special shape. The specific cross-sections of these polytopes provide the parallelepiped-valued or parallelotope-valued estimates of the reachable sets in the initial space. The main attention is paid to the internal estimates. Note that the task of constructing internal estimates is usually more difficult than the task for external ones. We construct new (in comparison with [13]) primary internal estimates in $\mathbb{R}^n$ for the result of multiplying a parallelotope by an interval matrix, then construct primary internal estimates for results of two operations with sets in $\mathbb{R}^{n+1}$, and then derive systems of recurrence relations for calculating parametric families of internal estimates of the reachable sets in the "extended" space and in the initial one. For completeness of description of the unified technique for both-sided estimation of the reachable sets we also briefly recall the way of construction of the external estimates from [16]. Calculation of both external and internal estimates, first, provide more information about the exact reachable sets and, second, can provide some insight into a quality of the estimates by comparing them. Illustrative examples are presented.

The following notation is used below: $\mathbb{R}^n$ is the n-dimensional vector space; $\top$ is the transposition symbol; $\|x\|_2 = (x^\top x)^{1/2}$, $\|x\|_\infty = \max_{1 \le i \le n} |x_i|$ are vector norms for $x = (x_1, \ldots, x_n)^\top \in \mathbb{R}^n$; $e^i = (0, \ldots, 0, 1, 0, \ldots, 0)^\top$ is the unit vector oriented along the axis x_i (the unit stands at position i); $c = (1, 1, \ldots, 1)^\top$; $\mathbb{R}^{n \times m}$ is the space of real $n \times m$-matrices $A = \{a_i^j\} = \{a^j\}$ with elements a_i^j and columns a^j; 0 is the zero matrix (vector); I is the identity matrix; $\mathrm{Abs}\, A = \{|a_i^j|\}$

for $A = \{a_i^j\} \in \mathbb{R}^{n \times m}$; diag π, diag $\{\pi_i\}$ are the diagonal matrix A with $a_i^i = \pi_i$ (π_i are the components of the vector π); det A is the determinant of $A \in \mathbb{R}^{n \times n}$; $\|A\| = \max_{1 \le i \le n} \sum_{j=1}^m |a_i^j|$ is the matrix norm for $A \in \mathbb{R}^{n \times m}$ induced by the vector norm $\|x\|_\infty$; co Q is the convex hull of a set $Q \subset \mathbb{R}^n$; $\rho(l|Q) = \sup\{l^\top x \mid x \in Q\}$ is the support function of $Q \subset \mathbb{R}^n$, vol Q is the volume of $Q \subset \mathbb{R}^n$; and the notation $k = 1, \ldots, N$ is used instead of $k = 1, 2, \ldots, N$ for brevity.

2   Problem Formulation

We consider a system (with states $x \in \mathbb{R}^n$)

$$x[j] = A[j]x[j-1] + B[j]u[j] + v[j], \quad j = 1, \ldots, N; \tag{1}$$

$$x[0] \in \mathcal{X}_0 \subset \mathbb{R}^n; \quad \sum_{j=1}^N \|u[j]\|_\infty \le \mu_0; \tag{2}$$

$$u[j] \in \mathcal{K}[j] \subseteq \mathbb{R}^r, \quad j = 1, \ldots, N, \tag{3}$$

where terms $v[j] \in \mathbb{R}^n$ and matrices $B[j] \in \mathbb{R}^{n \times r}$ ($r \le n$) are given. The initial state $x[0] = x_0 \in \mathbb{R}^n$ and the inputs (controls/disturbances) $u[j] \in \mathbb{R}^r$ are unknown but satisfy constraints (2)–(3). Here $\mathcal{X}_0$ is a given convex compact set, $\mu_0 > 0$, $\mathcal{K}[j] \subseteq \mathbb{R}^r$ are convex closed cones in $\mathbb{R}^r$. Matrices $A[j] \in \mathbb{R}^{n \times n}$ are also unknown but subjected to constraints of an interval type

$$
\begin{aligned}
A[j] \in \mathcal{A}[j] &= \{A \in \mathbb{R}^{n \times n} \mid \underline{A}[j] \le A \le \overline{A}[j]\} \\
&= \{A \mid \mathrm{Abs}\,(A - \tilde{A}[j]) \le \hat{A}[j]\}, \quad j = 1, \ldots, N,
\end{aligned}
\tag{4}
$$

where $\tilde{A}[j] = (\underline{A}[j] + \overline{A}[j])/2$, $\hat{A}[j] = (\overline{A}[j] - \underline{A}[j])/2$. Here and below, matrix and vector inequalities and also the operations of maximum and minimum are understood elementwise.

Let us start with some definitions.

The *reachable set* $\mathcal{X}[k]$ for the system (1)–(4) at time $k \in \{1, \ldots, N\}$ is a set of all points $x \in \mathbb{R}^n$ for each of which there exists a triple $\{x[0], u[\cdot], A[\cdot]\}$ that satisfies (2)–(4) and generates a solution $x[\cdot]$ of (1) that satisfies $x[k] = x$. Set-valued map $\mathcal{X}[k]$, as a function of k, defines a so-called *trajectory tube* $\mathcal{X}[\cdot]$.

By a *parallelepiped* $\mathcal{P}(p, P, \pi) \subset \mathbb{R}^n$ we mean a set such that $\mathcal{P} = \mathcal{P}(p, P, \pi) = \{x \in \mathbb{R}^n \mid x = p + P\mathrm{diag}\,\pi\,\xi, \|\xi\|_\infty \le 1\}$, where $p \in \mathbb{R}^n$; $P = \{p^i\} \in \mathbb{R}^{n \times n}$ is such that det $P \ne 0$, $\|p^i\|_2 = 1$ (the normality condition $\|p^i\|_2 = 1$ may be omitted to simplify formulas); $\pi \in \mathbb{R}^n$, $\pi \ge 0$. It may be said that p determines the center of the parallelepiped, P is the orientation matrix, p^i are the "directions", and π_i are the values of its "semi-axes" Fig. 1(a).

By a *parallelotope* $\mathcal{P}[p, \bar{P}] \subset \mathbb{R}^n$ we mean a set $\mathcal{P} = \mathcal{P}[p, \bar{P}] = \{x \mid x = p + \bar{P}\xi, \|\xi\|_\infty \le 1\}$, where $p \in \mathbb{R}^n$ and $\bar{P} = \{\bar{p}^i\} \in \mathbb{R}^{n \times m}$, $m \le n$. We call a parallelotope $\mathcal{P}$ *nondegenerate* if $m = n$ and det $\bar{P} \ne 0$.

Each parallelepiped $\mathcal{P}(p, P, \pi)$ is a parallelotope $\mathcal{P}[p, \bar{P}]$ with $\bar{P} = P \,\mathrm{diag}\,\pi$. Each nondegenerate parallelotope is a parallelepiped with $P = \bar{P}$, $\pi = \mathrm{e}$.

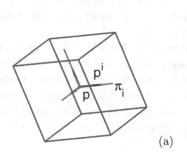

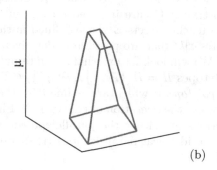

(a)   (b)

Fig. 1. Forms of polyhedral estimates: (a) parallelepiped in $\mathbb{R}^3$, (b) Π-polytope in $\mathbb{R}^3$

Let us introduce the following sets, which will be used below:

$$\mathcal{R}[j] = \mathcal{C} \cap \mathcal{K}[j], \quad \mathcal{C} = \mathcal{P}(0, I, \mathrm{e}) \subset \mathbb{R}^r, \tag{5}$$

where $\mathcal{C}$ is the unit cube with the center at origin.

We consider the system under the following assumption.

Assumption 1. *The initial set $\mathcal{X}_0$ is a parallelepiped $\mathcal{X}_0 = \mathcal{P}_0 = \mathcal{P}(p_0, P_0, \pi_0) \subset \mathbb{R}^n$, and all the sets $\mathcal{R}[j]$ defined in (5) are parallelepipeds in $\mathbb{R}^r$.*

Exact computing the reachable sets $\mathcal{X}[k]$ can be rather cumbersome especially for systems with uncertain matrices because $\mathcal{X}[k]$ may be nonconvex in this case. We will look for external parallelepiped-valued and internal parallelotope-valued estimates for $\mathcal{X}[k]$.

We call $\mathcal{P}$ *external (internal) estimate for* $\mathcal{Q} \subset \mathbb{R}^n$ if $\mathcal{Q} \subseteq \mathcal{P}$ ($\mathcal{P} \subseteq \mathcal{Q}$). The estimate $\mathcal{P}$ for $\mathcal{Q} \subset \mathbb{R}^n$ is called *tight (in direction l)* [21] if $\mathcal{Q} \subseteq \mathcal{P}$ ($\mathcal{P} \subseteq \mathcal{Q}$) and there exists $l \in \mathbb{R}^n$ such that $\rho(\pm l|\mathcal{P}) = \rho(\pm l|\mathcal{Q})$. We call a parallelepiped $P_V^+(\mathcal{Q}) = \mathcal{P}(p^+, V, \pi^+)$ *touching external estimate for* $\mathcal{Q}$, denoting it by $P_V^+(\mathcal{Q})$, if it is tight estimate in n specified directions $l^i = (V^{-1})^\top e^i$, $i = 1, \ldots, n$.

Thus we consider the following problem.

Problem 1. Find some external parallelepiped-valued estimates $\mathcal{P}^+[k] = \mathcal{P}(p^+[k], P^+[k], \pi^+[k])$ and internal parallelotope-valued ones $\mathcal{P}^-[k] = \mathcal{P}[p^-[k], \bar{P}^-[k]]$ for the reachable sets $\mathcal{X}[k]$: $\mathcal{P}^-[k] \subseteq \mathcal{X}[k] \subseteq \mathcal{P}^+[k]$, $k = 1, \ldots, N$. Moreover introduce some families of such estimates $\mathcal{P}^\pm[k]$.

To investigate the reachable sets $\mathcal{X}[k]$ it is useful to introduce reachable sets $\mathcal{Z}[k]$ of states $z = \{x, \mu\} = (x^\top, \mu)^\top \in \mathbb{R}^{n+1}$ for system (1), (3), (4), (6)–(8):

$$\mu[j] = \mu[j-1] - \|u[j]\|_\infty, \quad j = 1, \ldots, N; \tag{6}$$

$$\mu[j] \geq 0, \quad j = 1, \ldots, N; \tag{7}$$

$$z[0] = \{x[0], \mu[0]\} \in \mathcal{Z}_0 = \mathcal{X}_0 \times [0, \mu_0] \tag{8}$$

in the "extended" space, where μ corresponds to a current stock of u and state constraints (7) are imposed on $\mu[j]$ instead of integral constraints (4) on $u[\cdot]$. The reachable sets $\mathcal{Z}[k]$ are defined in the standard way (in short, the sets of all states $z[k]$ that are possible under given constraints).

We will look for external and internal estimates $\Pi^\pm[k]$ for $\mathcal{Z}[k]$ in the form of polytopes $\Pi = \Pi(\{\mathcal{P}^b, 0\}, \{\mathcal{P}^t, \mu^t\}) \subset \mathbb{R}^{n+1}$ of some specific form (we call them Π-*polytopes*), which are defined by their "lower" and "upper" cross-sections through the operation of convex hull, Fig. 1(b) where the both mentioned cross-sections are either the parallelepipeds with the identical orientation matrices or the identical parallelotopes (then we also say that the Π-polytope is a Π-*cylinder*):

$$\Pi = \Pi(\{\mathcal{P}^b, 0\}, \{\mathcal{P}^t, \mu^t\}) = \operatorname{co}(\{\mathcal{P}^b, 0\} \cup \{\mathcal{P}^t, \mu^t\}), \quad \mu^t \geq 0,$$
$$\mathcal{P}^b = \mathcal{P}(p^b, P, \pi^b), \quad \mathcal{P}^t = \mathcal{P}(p^t, P, \pi^t), \tag{9}$$
$$\text{or} \quad \mathcal{P}^b = \mathcal{P}^t = \mathcal{P}[p^t, \bar{P}^t].$$

Problem 2. Find some external and internal estimates $\Pi^\pm[k]$ for $\mathcal{Z}[k]$: $\Pi^-[k] \subseteq \mathcal{Z}[k] \subseteq \Pi^+[k]$, $k = 1, \ldots, N$. Introduce some families of such estimates $\Pi^\pm[k]$.

We will find estimates $\mathcal{P}^\pm[k]$ for $\mathcal{X}[k]$ using estimates $\Pi^\pm[k]$ for $\mathcal{Z}[k]$.
We call both estimates $\mathcal{P}^\pm[k]$ and $\Pi^\pm[k]$ *polyhedral estimates* for brevity.

3   Auxiliary Results

It is convenient to represent the reachable sets $\mathcal{Z}[k]$ of the system (1), (3), (4), (6)–(8) in the form of the union of their μ-cross-sections $\mathcal{X}(\mu, k)$: $\mathcal{Z}[k] = \bigcup_{0 \leq \mu \leq \mu^t[k]} \{\mathcal{X}(\mu, k), \mu\}$.

The sets $\mathcal{Z}[k]$ (unlike $\mathcal{X}[k]$) satisfy the semigroup property (for the definitions see, for example, [10,20,21]) and therefore satisfy some recurrence relations.

Theorem 1 (*See* [16]). *Let $\mathcal{Z}[k]$ be the reachable sets for the system (1), (3), (4), (6)–(8) with the initial set $\mathcal{Z}_0 = \mathcal{X}_0 \times [0, \mu_0]$. Then we have $\mu^t[k] = \mu_0$, $k = 1, \ldots, N$, and $\mathcal{Z}[k]$ satisfy the following recurrence relations:*

$$\mathcal{Z}[k] = (\mathcal{A}[k] \otimes \mathcal{Z}[k-1] \oplus v[k]) \uplus B[k]\mathcal{R}[k], \quad k = 1, \ldots, N; \quad \mathcal{Z}[0] = \mathcal{X}_0 \times [0, \mu_0],$$

and we have $\mathcal{X}[k] = \bigcup\{\mathcal{X}(\mu, k) \mid 0 \leq \mu \leq \mu_0\} = \mathcal{X}(0, k)$, $k = 1, \ldots, N$.

Here are involved the following operations with the sets of the form $\mathcal{Z} = \bigcup_{0 \leq \mu \leq \mu^t}\{\mathcal{X}(\mu), \mu\} \subseteq \mathbb{R}^{n+1}$:

$$\mathcal{Z} \oplus v = \bigcup_{0 \leq \mu \leq \mu^t} \{\mathcal{X}(\mu) + v, \mu\}, \quad \forall v \in \mathbb{R}^n;$$

$$\mathcal{A} \otimes \mathcal{Z} = \bigcup_{0 \leq \mu \leq \mu^t} \{\mathcal{A} \circ \mathcal{X}(\mu), \mu\};$$

$$\mathcal{Z} \uplus \mathcal{R} = \tilde{\mathcal{Z}} = \bigcup_{0 \leq \mu \leq \mu^t} \{\tilde{\mathcal{X}}(\mu), \mu\}, \quad \tilde{\mathcal{X}}(\mu) = \bigcup_{\mu \leq \zeta \leq \mu^t} (\mathcal{X}(\zeta) + (\zeta - \mu)\mathcal{R}), \forall \mathcal{R} \subset \mathbb{R}^n,$$

$$\tag{10}$$

which in their turn involve the results of set operations in $\mathbb{R}^n$ such as the *Minkowski sum* $\mathcal{X}^1 + \mathcal{X}^2 = \{y \mid y = x^1 + x^2, x^k \in \mathcal{X}^k\}$ and *multiplying a set $\mathcal{X}$ by an interval matrix* $\mathcal{A} = \{A \mid \underline{A} \le A \le \overline{A}\}$: $\mathcal{A} \circ \mathcal{X} = \{y \in \mathbb{R}^n \mid y = Ax, A \in \mathcal{A}, x \in \mathcal{X}\}$. In (10), the first two operations act on the each cross-section independently, the last one combines operations of Minkowski sum and union over cross-sections.

To construct polyhedral estimates for $\mathcal{Z}[k]$ we use properties of operations with sets and primary polyhedral estimates for the results of set operations.

The touching external parallelepiped-valued estimates $\boldsymbol{P}_V^+(\mathcal{Q})$ with a given orientation matrix V can be calculated by known *explicit* formulas for the cases when $\mathcal{Q}$ is a sum or union of two parallelotopes ($\mathcal{Q} = \mathcal{P}^1 + \mathcal{P}^2$ or $\mathcal{Q} = \mathcal{P}^1 \cup \mathcal{P}^2$), and also for $\mathcal{Q} = \mathcal{A} \circ \mathcal{P}$, where $\mathcal{A}$ is an interval matrix and $\mathcal{P}$ is a parallelepiped or a parallelotope (see [10,13,16]). For the reader's convenience, let us recall these primary estimates.

The affine transformation of a parallelepiped and of a parallelotope is a parallelotope: $A\mathcal{P}[p, \bar{P}] + a = \mathcal{P}[Ap + a, A\bar{P}]$ for $A \in \mathbb{R}^{n \times r}$, $p \in \mathbb{R}^r$, $\bar{P} \in \mathbb{R}^{r \times r}$, $a \in \mathbb{R}^n$; $A\mathcal{P}(p, P, \pi) = \mathcal{P}(Ap, AP, \pi)$ if $\det A \ne 0$.

For further it is useful to bear in mind the equivalent representation of the parallelepiped $\mathcal{P} = \mathcal{P}(p, P, \pi)$: $\mathcal{P} = \mathcal{P}(P, \gamma^{(-)}, \gamma^{(+)}) = \{x \mid \gamma^{(-)} \le P^{-1}x \le \gamma^{(+)}\}$, where $\gamma_i^{(\pm)} = \pm\rho(\pm(P^{-1})^\top e^i | \mathcal{P})$, $i = 1, \ldots, n$, and we have the following interconnections: $\gamma^{(\pm)} = P^{-1}p \pm \pi$; $p = P(\gamma^{(-)} + \gamma^{(+)})/2$, $\pi = (\gamma^{(+)} - \gamma^{(-)})/2$.

The touching external estimate for a bounded set $\mathcal{Q} \subset \mathbb{R}^n$ with a given orientation matrix V is determined by the formula $\boldsymbol{P}_V^+(\mathcal{Q}) = \mathcal{P}(V, \gamma^{(-)}, \gamma^{(+)})$, $\gamma_i^{(\pm)} = \pm\rho(\pm V^{-1}^\top e^i | \mathcal{Q})$.

The support functions of a parallelepiped and a parallelotope are determined by formulas $\rho(l | \mathcal{P}(p, P, \pi)) = l^\top p + \mathrm{Abs}\,(l^\top P)\pi$, $\rho(l | \mathcal{P}[p, \bar{P}]) = l^\top p + \mathrm{Abs}\,(l^\top \bar{P})e$.

The touching estimates for the sum of two parallelepipeds can be found by the explicit formula $\boldsymbol{P}_V^+(\sum_{k=1}^2 \mathcal{P}(p^k, P^k, \pi^k)) = \mathcal{P}(\sum_{k=1}^2 p^k, V, \sum_{k=1}^2 (\mathrm{Abs}\,(V^{-1}P^k))\pi^k)$. For the sum of two parallelotopes we have $\boldsymbol{P}_V^+(\sum_{k=1}^2 \mathcal{P}[p^k, \bar{P}^k]) = \mathcal{P}[\sum_{k=1}^2 p^k, V \operatorname{diag}(\sum_{k=1}^2 \mathrm{Abs}\,(V^{-1}\bar{P}^k)e)]$.

The estimates $\boldsymbol{P}_V^+(\mathcal{P}^1 \cup \mathcal{P}^2)$ for the union of two parallelotopes $\mathcal{P}^k = \mathcal{P}[p^k, \bar{P}^k]$ are determined by the formula

$$\boldsymbol{P}_V^+(\mathcal{P}^1 \cup \mathcal{P}^2) = \mathcal{P}(V, \gamma^{+(-)}, \gamma^{+(+)}), \gamma^{+(\pm)} = \pm \max_{1 \le k \le 2} \{\pm V^{-1}p^k + \mathrm{Abs}\,(V^{-1}\bar{P}^k)e\}.$$

For calculating touching external estimate $\boldsymbol{P}_{P+}^+(\mathcal{Q})$ for $\mathcal{Q} = \mathcal{A} \circ \mathcal{P}$, where $\mathcal{A}$ is an interval matrix and $\mathcal{P}$ is a parallelepiped, we can use each of two expressions

$$\rho(l | \mathcal{Q}) = \max_{x \in \mathbb{E}(\mathcal{P})} \{l^\top \tilde{A}x + (\mathrm{Abs}\, l)^\top \hat{A}(\mathrm{Abs}\, x)\},$$
$$\rho(l | \mathcal{Q}) = \max_{A \in \mathbb{E}(\mathcal{A})} \{l^\top Ap + (\mathrm{Abs}\,(l^\top AP))\pi\},$$

where $\mathbb{E}(\mathcal{P})$ and $\mathbb{E}(\mathcal{A})$ denote sets of all vertices of $\mathcal{P}$ and $\mathcal{A}$ (i.e., the set of points $p + \sum_{i=1}^m p^i \pi_i \xi_i$ with $\xi_i \in \{-1, 1\}$ and the set of matrices with elements $a_i^j \in \{\underline{a}_i^j, \overline{a}_i^j\}$).

To construct *external* estimates $\Pi^+[k]$ for the reachable sets $\mathcal{Z}[k]$ we use primary polyhedral estimates for the results of set operations $\Pi \uplus \mathcal{P}$ and $\mathcal{A} \otimes \Pi$ from (10), where $\Pi = \Pi(\{\mathcal{P}^b, 0\}, \{\mathcal{P}^t, \mu^t\})$ is a Π-polytope defined in (9).

External estimates $\mathbf{\Pi}_V^+(\Pi \uplus \mathcal{P}) = \Pi(\{\mathcal{P}^{+b}, 0\}, \{\mathcal{P}^{+t}, \mu^{+t}\})$, where $\mathcal{P}$ is a paralelotope, can be found [11, 16] by formulas

$$\mathcal{P}^{+t} = \mathbf{P}_V^+(\mathcal{P}^t), \quad \mu^{+t} = \mu^t, \quad \mathcal{P}^{+b} = \mathbf{P}_V^+(\mathcal{P}^b \cup \mathbf{P}_V^+(\mathcal{P}^t + \mu^t \mathcal{P})), \qquad (11)$$

and all its μ-cross-sections turn out to be touching estimates for μ-cross-sections of the set $\mathcal{Z} = \Pi \uplus \mathcal{P}$.

External estimates $\mathbf{\Pi}_V^+(\mathcal{A} \otimes \Pi) = \Pi(\{\mathcal{P}^{+b}, 0\}, \{\mathcal{P}^{+t}, \mu^{+t}\})$, where $\mathcal{A}$ is an interval matrix, can be found [16] by formulas

$$\mathcal{P}^{+t} = \mathbf{P}_V^+(\mathcal{A} \circ \mathcal{P}^t), \quad \mu^{+t} = \mu^t, \quad \mathcal{P}^{+b} = \mathbf{P}_V^+(\mathcal{A} \circ \mathcal{P}^b). \qquad (12)$$

Here the orientation matrix V appears as a parameter, which determines the parametric families of the estimates.

Now let us consider ways for constructing primary *internal* polyhedral estimates.

Introduce the following set of matrices (where $\|\Gamma\| = \max_{1 \leq \alpha \leq r} \sum_{\beta=1}^n |\gamma_\alpha^\beta|$):

$$\mathcal{G}^{r \times n} = \{\Gamma = \{\gamma_\alpha^\beta\} \in \mathbb{R}^{r \times n} \mid \|\Gamma\| \leq 1\}.$$

Let $\mathcal{P}^k = \mathcal{P}[p^k, \bar{P}^k]$, $k = 1, 2$, $\bar{P}^1 \in \mathbb{R}^{n \times n}$, $\bar{P}^2 \in \mathbb{R}^{n \times r}$. Internal estimates for the sum of parallelotopes $\mathcal{Q} = \mathcal{P}^1 + \mathcal{P}^2$ can be found similarly to [13] in the form of a parallelotope

$$\mathcal{P}^- = \mathbf{P}_{\Gamma^1, \Gamma^2}^-(\mathcal{P}^1 + \mathcal{P}^2) = \mathcal{P}[p^1 + p^2, \bar{P}^1 \Gamma^1 + \bar{P}^2 \Gamma^2], \qquad (13)$$

where $\Gamma^1 \in \mathcal{G}^{n \times n}$, $\Gamma^2 \in \mathcal{G}^{r \times n}$. Matrices Γ^1, Γ^2 serve as admissible parameters.

Let's pass to the new results concerning ways of constructing internal estimates for $\mathcal{A} \circ \mathcal{P}$ and $\Pi \uplus \mathcal{P}$, $\mathcal{A} \otimes \Pi$.

First we present a parametric family of internal estimates for $\mathcal{Q} = \mathcal{A} \circ \mathcal{P}$.

Proposition 1. *Let* $\mathcal{A} = \{A \mid \text{Abs}\,(A - \tilde{A}) \leq \hat{A}\}$ *and* $\mathcal{P} = \mathcal{P}[p, \bar{P}]$ *with* $\bar{P} \in \mathbb{R}^{n \times n}$. *Let* $J = \{j_1, \ldots, j_n\}$ *be an arbitrary permutation for* $\{1, \ldots, n\}$ *and* Γ^1, Γ^2 *be arbitrary matrices such that* $\Gamma^1, \Gamma^2 \in \mathcal{G}^{n \times n}$. *Then the parallelotope*

$$\begin{aligned} \mathcal{P}^- &= \mathbf{P}_{J, \Gamma^1, \Gamma^2}^-(\mathcal{A} \circ \mathcal{P}) = \mathcal{P}[\tilde{A}p, \tilde{A}\bar{P}\Gamma^1 + (\text{diag}\,\nu)\Gamma^2], \\ \nu_i &= \hat{a}_i^{j_i} \eta_{j_i}, \ i = 1, \ldots, n, \quad \eta = \max\{0, \text{Abs}\,p - \text{Abs}\,(\bar{P}\Gamma^1)\}e\}, \end{aligned} \qquad (14)$$

is an internal estimate for $\mathcal{A} \circ \mathcal{P}$*, i.e.,* $\mathcal{P}^- \subseteq \mathcal{A} \circ \mathcal{P}$.

Proof. To prove the inclusion $\mathcal{P}^- \subseteq \mathcal{A} \circ \mathcal{P}$, let us make sure that for any $y \in \mathcal{P}^-$ (i.e., $y = p^- + \bar{P}^-\xi$, $\|\xi\|_\infty \leq 1$) we can find $A \in \mathcal{A}$ (i.e., $A = \tilde{A} + \Delta A$, $\text{Abs}\,(\Delta A) \leq \hat{A}$) and $x \in \mathcal{P}$ (i.e., $x = p + \bar{P}\zeta$, $\|\zeta\|_\infty \leq 1$), such that $y = Ax$, i.e.,

$$\tilde{A}p + (\tilde{A}\bar{P}\Gamma^1 + (\text{diag}\,\nu)\Gamma^2)\xi = \tilde{A}p + \tilde{A}\bar{P}\zeta + \Delta A(p + \bar{P}\zeta). \qquad (15)$$

Set $\zeta = \Gamma^1 \xi$. Then we have $\|\zeta\|_\infty \le \|\Gamma^1\| \, \|\xi\|_\infty \le 1$. Set $\Delta A = (\mathrm{diag}\,\delta) D$, where $D = \{e^{j_1} \cdots e^{j_n}\}^\top$ is the matrix obtained by permuting rows of the identity matrix according to the permutation J, and components of the vector δ are calculated as follows. If $\nu_i = 0$ for some i, then set $\delta_i = 0$. If $\nu_i > 0$, then $\eta_{j_i} > 0$ due to (14), and $|p_{j_i}| > {e^{j_i}}^\top \mathrm{Abs}\,(\bar P \Gamma^1) e$. Then ${e^{j_i}}^\top (p + \bar P \Gamma^1 \xi) \ne 0$ and we set $\delta_i = \nu_i {e^{i}}^\top \Gamma^2 \xi / ({e^{j_i}}^\top (p + \bar P \Gamma^1 \xi))$. The equality (15) is provided. It remains to check the inequalities $|\delta_i| \le \hat a_i^{j_i}$, which ensure $\mathrm{Abs}\,(\Delta A) \le \hat A$. They are evident for indices i such that $\nu_i = 0$. For the rest ones (i.e., for i such that $\nu_i > 0$), we have $|\delta_i| \le \hat a_i^{j_i} \eta_{j_i} {e^{i}}^\top (\mathrm{Abs}\, \Gamma^2) e / (|p_{j_i}| - {e^{j_i}}^\top (\mathrm{Abs}\,(\bar P \Gamma^1)) e) \le \hat a_i^{j_i} {e^{i}}^\top (\mathrm{Abs}\, \Gamma^2) e \le \hat a_i^{j_i}$ because $\|\Gamma^2\| \le 1$. $\square$

Remark 1. Under conditions of Proposition 1, we can choose $\Gamma^2 \in \mathcal{G}^{n\times n}$ such that

$$ {e^{i}}^\top \Gamma^2 = \beta_i {e^{i}}^\top \tilde A \bar P \Gamma^1, \quad \beta_i = ({e^{i}}^\top \mathrm{Abs}\,(\tilde A \bar P \Gamma^1) e)^{-1}, \quad i \in I_*, \qquad (16) $$

where $I_* = \{i \in \{1,\ldots,n\} \mid {e^{i}}^\top \mathrm{Abs}\,(\tilde A \bar P \Gamma^1) e \ne 0\}$. If we also put

$$ {e^{i}}^\top \Gamma^2 = 0 = \beta_i {e^{i}}^\top \tilde A \bar P \Gamma^1, \quad \beta_i = 0, \quad i \in I_{**} = \{1,\ldots,n\} \backslash I_*, \qquad (17) $$

then we obtain $\mathcal{P}^- = \mathcal{P}[\tilde A p, (I + \mathrm{diag}\,\nu \, \mathrm{diag}\,\beta) \tilde A \bar P \Gamma^1]$, which is the same estimate as in [13, Theorem 3.3]. If in addition the parallelotope $\mathcal{P}$ is nondegenerate, $\det \tilde A \ne 0$, and $\det \Gamma^1 \ne 0$, then the above parallelotope $\mathcal{P}^-$ turns out to be nondegenerate.

Remark 2. It is not difficult to provide examples, where $\mathcal{P}$ is a degenerate parallelotope and, under some conditions, matrices $\Gamma^2 \in \mathcal{G}^{n\times n}$ with (16) but without zero row vectors from (17) can give nondegenerate estimates $\mathcal{P}^- \subseteq \mathcal{A} \circ \mathcal{P}$.

Example 1. Let us consider a simplest example, where $n = 2$, $\tilde A = I$, $\bar P = \begin{bmatrix} 1 & 0 \\ 0 & 0 \end{bmatrix}$, $\Gamma^1 = I$, $\Gamma^2 = \begin{bmatrix} 1 & 0 \\ \gamma_2^1 & \gamma_2^2 \end{bmatrix} \in \mathcal{G}^{n\times n}$. In this case, $\tilde A \bar P \Gamma^1 = \bar P$, $(\mathrm{Abs}\,(\tilde A \bar P \Gamma^1)) e = (\mathrm{Abs}\,(\bar P \Gamma^1)) e = e^1$, $\eta = \max\{0, \mathrm{Abs}\, p - e^1\} = (\max\{0, |p_1| - 1\}, |p_2|)^\top$. Let $\bar P^{(1)-}$ and $\bar P^{(2)-}$ correspond to $J^{(1)} = \{1, 2\}$ and $J^{(2)} = \{2, 1\}$ according to (14), where we have $\nu = \nu^{(1)} = (\hat a_1^1 \cdot \max\{0, |p_1| - 1\}, \hat a_2^2 \cdot |p_2|)^\top$ and $\nu = \nu^{(2)} = (\hat a_1^2 \cdot |p_2|, \hat a_2^1 \cdot \max\{0, |p_1| - 1\})^\top$ respectively. We obtain $\det \bar P^{(1)-} = (1 + \hat a_1^1 \max\{0, |p_1| - 1\}) \cdot \hat a_2^2 |p_2| \gamma_2^2$, $\det \bar P^{(2)-} = (1 + \hat a_1^2 |p_2|) \cdot \hat a_2^1 \max\{0, |p_1| - 1\} \gamma_2^2$. The estimates $\mathcal{P}^- = \mathcal{P}[p^-, \bar P^-]$ with maximal volumes correspond to the ones with maximal absolute value of $\det \bar P^-$. Maximal values of both $|\det \bar P^{(1)-}|$ and $|\det \bar P^{(2)-}|$ under the above $\Gamma^2 \in \mathcal{G}^{n\times n}$ are obtained at Γ^2 with $\gamma_2^2 \in \{1, -1\}$, $\gamma_2^1 = 0$. Thus such choice of Γ^2 together with $J^{(1)}$ under conditions $\hat a_2^2 > 0$, $|p_2| > 0$ and together with $J^{(2)}$ under the conditions $\hat a_2^1 > 0$, $|p_1| > 1$ generates nondegenerate estimates, while the estimates from Remark 1 are degenerate because $\det \bar P = 0$.

Remark 3. Let $\det \bar P \ne 0$, $\det \tilde A \ne 0$, and $\hat A$ has the unique nonzero element $a_{i_*}^{j_*}$. Let Γ^1 satisfies $\Gamma^1 \in \mathcal{G}^{n\times n}$, $\det \Gamma^1 \ne 0$ (in particular, $\Gamma^1 = I$) and Γ^2 be determined by (16), where $I_* = \{1,\ldots,n\}$. Then the parallelotope $\mathcal{P}[\tilde A p, \mathrm{diag}\,(e + a_{i_*}^{j_*} \eta_{j_*} \beta_{i_*} e^{i_*}) \tilde A \bar P \Gamma^1]$ is a solution of the optimization problem $\mathrm{vol}\, \boldsymbol{P}^-_{J, \Gamma^1, \Gamma^2} (\mathcal{A} \circ \mathcal{P}) \to \max_J$.

Remark 4. The estimates from Proposition 1 possess the peculiarity that we have $\nu = \eta = 0$ in (14) if $\mathcal{P}$ contains 0. In this case it may be useful to apply the estimates $\boldsymbol{P}_A^-(\mathcal{A} \circ \mathcal{P})$ of the form $\boldsymbol{P}_A^-(\mathcal{A} \circ \mathcal{P}) = A\mathcal{P} = \mathcal{P}[Ap, A\bar{P}]$ with $A \in \mathcal{A}$, which were called simple in [13]. Recall that volume-maximal simple estimates can be found among those that correspond to vertices of $\mathcal{A}$: $\max_{A \in \mathcal{A}} \mathrm{vol}\,(A\mathcal{P}) = \max_{A \in \mathbb{E}(\mathcal{A})} \mathrm{vol}\,(A\mathcal{P}) = \max_{A \in \mathbb{E}(\mathcal{A})} |\det A| \cdot \mathrm{vol}\,\mathcal{P}$ [13].

To construct estimates $\Pi^-[k] \subseteq \mathcal{Z}[k]$ we use primary Π-cylinder-valued estimates for the results of set operations $\Pi \uplus \mathcal{P}$ and $\mathcal{A} \otimes \Pi$, where Π is a Π-cylinder.

Proposition 2. *Let* $\Pi = \Pi(\{\mathcal{P}^b, 0\}, \{\mathcal{P}^t, \mu^t\})$ *be a* Π-*cylinder with* $\mathcal{P}^b = \mathcal{P}^t$ *and* $\mathcal{P} = \mathcal{P}[p, \bar{P}]$ *be a parallelotope with* $\bar{P} \in \mathbb{R}^{n \times r}$ *and* $0 \in \mathcal{P}$. *Then an arbitrary* Π-*cylinder* $\Pi^- = \Pi(\{\mathcal{P}^{-b}, 0\}, \{\mathcal{P}^{-t}, \mu^{-t}\})$ *with* $\mu^{-t} = \mu^t - h$, $0 \leq h \leq \mu^t$, $\mathcal{P}^{-b} = \mathcal{P}^{-t} \subseteq \mathcal{P}^t + h\mathcal{P}$ *serves as an internal estimate for* $\Pi \uplus \mathcal{P}$. *In particular,* Π-*cylinders*

$$\Pi^- = \mathbf{\Pi}_{h, \Gamma^1, \Gamma^2}^-(\Pi \uplus \mathcal{P}) = \Pi(\{\mathcal{P}^{-b}, 0\}, \{\mathcal{P}^{-t}, \mu^{-t}\}),$$
$$\mu^{-t} = \mu^t - h, \quad \mathcal{P}^{-b} = \mathcal{P}^{-t} = \boldsymbol{P}_{\Gamma^1, \Gamma^2}^-(\mathcal{P}^t + h\mathcal{P}) = \mathcal{P}[p^t + h\,p, \bar{P}^t \Gamma^1 + h\bar{P}\Gamma^2]$$

(18)

are internal estimates for $\Pi \uplus \mathcal{P}$ *whatever are admissible parameters* $0 \leq h \leq \mu^t$, $\Gamma^1 \in \mathcal{G}^{n \times n}$, *and* $\Gamma^2 \in \mathcal{G}^{r \times n}$.

Proposition 3. *Let* $\Pi = \Pi(\{\mathcal{P}^b, 0\}, \{\mathcal{P}^t, \mu^t\})$ *be a* Π-*cylinder and* $\mathcal{A}$ *be an iterval matrix. Then an arbitrary* Π-*cylinder* $\Pi^- = \Pi(\{\mathcal{P}^{-b}, 0\}, \{\mathcal{P}^{-t}, \mu^{-t}\})$ *with* $\mu^{-t} = \mu^t$, $\mathcal{P}^{-b} = \mathcal{P}^{-t} \subseteq \mathcal{A} \circ \mathcal{P}^t$ *serves as an internal estimate for* $\mathcal{A} \otimes \Pi$. *In particular,* Π-*cylinders*

$$\Pi^- = \mathbf{\Pi}_{J, \Gamma^1, \Gamma^2}^-(\mathcal{A} \otimes \Pi) = \Pi(\{\mathcal{P}^{-b}, 0\}, \{\mathcal{P}^{-t}, \mu^{-t}\}),$$
$$\mu^{-t} = \mu^t, \quad \mathcal{P}^{-b} = \mathcal{P}^{-t} = \boldsymbol{P}_{J, \Gamma^1, \Gamma^2}^-(\mathcal{A} \circ \mathcal{P}^t)$$

(19)

are internal estimates for $\mathcal{A} \otimes \Pi$ *whatever are admissible parameters* J *(which is an arbitrary permutation of* $\{1, \ldots, n\}$*),* $\Gamma^1, \Gamma^2 \in \mathcal{G}^{n \times n}$.

Proof. Both propositions follow from the definition of Π-cylinder and properties of the used primary estimates in $\mathbb{R}^n$ that were described above. □

4   Polyhedral Estimates for Reachable Sets

First we recall the way of constructing external estimates for the reachable sets $\mathcal{Z}[k]$ and $\mathcal{X}[k]$.

Theorem 2 *(See [16]). Let* $\mathcal{Z}[k]$ *be the reachable sets of the system (1), (3), (4), (6)–(8) under Assumption 1. Let* Π-*polytopes* $\Pi^+[k]$ *satisfy the relations*

$$\Pi^{1+}[k] = \mathbf{\Pi}_{P+[k]}^+(A[k] \otimes \Pi^+[k-1]) \oplus v[k], \quad k = 1, \ldots, N; \quad \Pi^+[0] = \mathcal{P}_0 \times [0, \mu_0];$$
$$\Pi^+[k] = \mathbf{\Pi}_{P+[k]}^+(\Pi^{1+}[k] \uplus B[k]\mathcal{R}[k]), \quad k = 1, \ldots, N,$$

(20)

where formulas (11) and (12) are applied. Then $\mathcal{Z}[k] \subseteq \Pi^+[k]$, $k = 1, \ldots, N$, whatever are nonsingular orientation matrices $P^+[k] \in \mathbb{R}^{n \times n}$, $k = 1, \ldots, N$, and parallelepipeds $\mathcal{P}^+[k] = \mathcal{P}^{+b}[k]$ that coincide with the "lower" cross-sections of $\Pi^+[k]$ are external estimates for the reachable sets $\mathcal{X}[k]$ of the system (1)–(4): $\mathcal{X}[k] \subseteq \mathcal{P}^+[k]$, $k = 1, \ldots, N$.

Remark 5. Theorem 2 describes the parametric family of estimates, where the function $P^+[\cdot]$, which determines a dynamics of orientation matrices of cross-sections, serves as a parameter of the family. Under the condition $\det \tilde{A}[j] \neq 0$, $j = 1, \ldots, N$, we can construct the tubes for which the orientation matrices satisfy

$$P^+[k] = \tilde{A}[k]P^+[k-1], \quad k = 1, \ldots, N; \quad P^+[0] = P, \qquad (21)$$

where P is an arbitrary nonsingular matrix (recall that for the case $\hat{A}[k] \equiv 0$ the corresponding estimates $\mathcal{P}^+[k]$ turn out to be touching [16, Remark 7]). The choice of constant orientation matrices $P^+[k] \equiv P$ can lead to much more conservative estimates due to the well-known in interval analysis "wrapping effect".

Now we present the technique for constructing internal estimates for the reachable sets $\mathcal{Z}[k]$ and $\mathcal{X}[k]$.

Let us introduce the following family of tubes $\Pi^-[\cdot]$ that satisfy the relations

$$\Pi^{1-}[k] = \mathbf{\Pi}^-_{J[k], \Gamma^1[k], \Gamma^2[k]}(A[k] \otimes \Pi^-[k-1]) \oplus v[k], \quad k = 1, \ldots, N;$$
$$\Pi^-[k] = \mathbf{\Pi}^-_{h[k], \Gamma^3[k], \Gamma^4[k]}(\Pi^{1-}[k] \uplus B[k]\mathcal{R}[k]), \quad k = 1, \ldots, N; \qquad (22)$$
$$\Pi^-[0] = \mathcal{P}_0 \times [0, \mu_0] = \Pi(\{\mathcal{P}_0, 0\}, \{\mathcal{P}_0, \mu_0\}),$$

where the admissible parameters satisfy the conditions

$$h[j] \geq 0, \quad \sum_{j=1}^{N} h[j] \leq \mu_0, \quad \Gamma^1[j], \Gamma^2[j], \Gamma^3[j] \in \mathcal{G}^{n \times n}, \quad \Gamma^4[j] \in \mathcal{G}^{r \times n}, \qquad (23)$$

$J[j]$ are arbitrary permutations of $\{1, \ldots, n\}$, and formulas (18), (19) are used.

Theorem 3. *Let $\mathcal{Z}[k]$ be the reachable sets of the system (1), (3), (4), (6)–(8) under Assumption 1. Then Π-cylinders $\Pi^-[k]$ that satisfy (22) are internal estimates for $\mathcal{Z}[k]$: $\Pi^-[k] \subseteq \mathcal{Z}[k]$, $k=1, \ldots, N$, whatever are the above-mentioned admissible parameters satisfying (23), and parallellotopes $\mathcal{P}^-[k] = \mathcal{P}^{-b}[k]$ that coincide with the cross-sections of Π-cylinders $\Pi^-[k]$ are internal estimates for the reachable sets $\mathcal{X}[k]$ of the system (1)–(4): $\mathcal{P}^-[k] \subseteq \mathcal{X}[k]$, $k = 1, \ldots, N$.*

Proof. The proof is obtained by using Theorem 1 and Propositions 2 and 3. $\square$

As a result, we obtain the following explicit formulas for constructing the parametric family of internal estimates the reachable sets $\mathcal{X}[k]$.

134   E. K. Kostousova

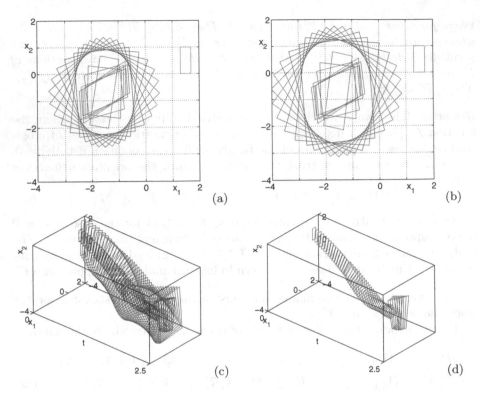

Fig. 2. External and internal estimates for $\mathcal{X}[N]$ in Example 2 with $\mathcal{K}[j] \equiv (-\infty, \infty)$: (a) case (i), (b) case (ii); external and internal estimates for $\mathcal{X}[\cdot]$: (c),(d) case (ii)

Corollary 1. *Let* $\mathcal{X}[k]$ *be the reachable sets of the system (1)–(4) under Assumption 1. Let* $h[\cdot]$ *and* $\Gamma^1[\cdot]$, $\Gamma^2[\cdot]$, $\Gamma^3[\cdot]$, $\Gamma^4[\cdot]$ *be parameters satisfying conditions (23) and parallelotopes* $\mathcal{P}^-[k]$ *be constructed by the following formulas:*

$$\mathcal{P}^{1-}[k] = \boldsymbol{P}^-_{J[k],\Gamma^1[k],\Gamma^2[k]}(\mathcal{A}[k] \circ \mathcal{P}^-[k-1]) + v[k], \ k = 1,\ldots,N; \ \mathcal{P}^-[0] = \mathcal{P}_0;$$
$$\mathcal{P}^-[k] = \boldsymbol{P}^-_{\Gamma^3[k],\Gamma^4[k]}(\mathcal{P}^{1-}[k] + h[k]B[k]\mathcal{R}[k]), \quad k = 1,\ldots,N,$$

$$(24)$$

where primary estimates (13) and (14) are used. Then $\mathcal{P}^-[k]$ *are internal estimates for* $\mathcal{X}[k]$: $\mathcal{P}^-[k] \subseteq \mathcal{X}[k]$, $k = 1,\ldots,N$.

Introducing the families of estimates instead of single ones, we can estimate reachable sets more accurately in the form of intersections of several external estimates and unions of several internal ones.

Example 2. For illustration, we present simulation results for discrete-time systems which can be obtained by discretization of impulsive differential ones considered on a time interval $[0,\theta]$. Let $\tilde{A}[j] \equiv I + h_N \cdot \begin{bmatrix} 0 & 1 \\ -1.5 & 0 \end{bmatrix}$, $\hat{A}[j] \equiv 0$

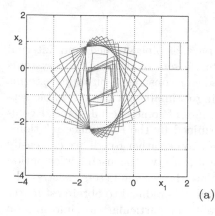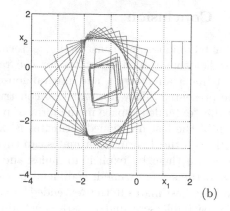

(a)   (b)

Fig. 3. External and internal estimates for $\mathcal{X}[N]$ in Example 2 with $\mathcal{K}[j] \equiv [0, \infty)$: (a) case (i), (b) case (ii)

(case (i), the system is linear) or $\hat{A}[j] \equiv h_N \cdot \begin{bmatrix} 0 & 0 \\ 0.1 & 0 \end{bmatrix}$ (case (ii), the system turns out to be bilinear), $h_N = \theta N^{-1}$, $\theta = 2.5$, $N = 100$, $B[j] \equiv (0, 1)^\top$, $v[j] \equiv 0$, $\mathcal{P}_0 = \mathcal{P}((1.5, 0.5)^\top, I, (0.2, 0.5)^\top)$, $\mu_0 = 1$, $\mathcal{K}[j] \equiv (-\infty, \infty)$ or $\mathcal{K}[j] \equiv [0, \infty)$.

First we consider the case with $\mathcal{K}[j] \equiv (-\infty, \infty)$. Figure 2(a) and 2(b) show $\mathcal{X}_0$ ("small parallelepiped") and several external and internal estimates $\mathcal{P}^+[N]$ and $\mathcal{P}^-[N]$ for the reachable set $\mathcal{X}[N]$ for cases (i) and (ii) respectively. External estimates are calculated by (20), (21), where the matrices P are taken as orthogonal ones of the form $P = \begin{bmatrix} \cos\varphi & -\sin\varphi \\ \sin\varphi & \cos\varphi \end{bmatrix}$ with $\varphi_i = 0.5(i-1)\pi/n_\varphi$, $i = 1, \ldots, n_\varphi$, $n_\varphi = 9$. Internal estimates are calculated due to (24), where $\Gamma^1[k] \equiv \Gamma^3[k] \equiv I$; $\Gamma^2[k]$ and $\Gamma^4[k]$ are calculated according to (16) and similarly to [13, Sec. 4] respectively; $J[k]$ are found by maximization of $\mathrm{vol}\, \boldsymbol{P}^-_{J[k], I, \Gamma^2[k]}(\mathcal{A}[k] \circ \mathcal{P}^-[k-1])$ under fixed $\Gamma^2[k]$; $h[\cdot]$ are taken as $h[j] \equiv \mu_0/N$ and as several random realizations of $h[\cdot]$ satisfying (23). The external estimates for the case (i) in Fig. 2(a) are touching; in aggregate, they "outline" $\mathcal{X}[N]$. Both external and internal estimates for the case (ii) in Fig. 2(b) (when the system is bilinear) as expected turned out to be larger than for the case (i), this is consistent with the fact that reachable sets for systems with uncertain matrices should be larger. Figure 2(d) presents one of the corresponding tubes $\mathcal{P}^-[\cdot]$ for the case (ii) (drawing is carried out one time per each 2 stages k). Figure 2(c) shows both some external and internal polyhedral tubes, namely $\mathcal{P}^+[\cdot]$ that correspond to $P = P_0$ and $\mathcal{P}^-[\cdot]$ presented in Fig. 2(d).

Figure 3(a) and 3(b) are obtained for the case with the cone constraint $\mathcal{K}[j] \equiv [0, \infty)$ (i.e., only nonnegative values of $u[j]$ are allowed) and are similar to Fig. 2(a) and 2(b). Here both external and internal estimates for $\mathcal{X}[N]$ are smaller than for the case $\mathcal{K}[j] \equiv (-\infty, \infty)$.

5   Conclusion

The techniques for constructing external parallelepiped-valued and internal parallelotope-valued estimates for the reachable sets of discrete-time systems with uncertain matrices and integral non-quadratic constraints on additive terms are presented. They are based on constructing estimates for the reachable sets in the "extended" space in the form of polytopes of some special shape. Evolution of the last mentioned estimates is determined by the recurrence relations. Although the described estimates can turn out to be rather rough, we can easily calculate them by explicit formulas and they can give the useful information about the system while it is difficult to calculate the reachable sets exactly. The proposed estimates in the "extended" space can be modified to obtain estimates for reachable sets under state constraints and, in particular, for information sets similarly to [11], where external estimates were constructed for the case of linear systems. The proposed estimates can be used for constructing estimates of reachable sets for impulsive differential systems similarly to [12]. Models of linear impulsive differential systems, for which considerations under uncertain matrices is also of importance, arise (including linearization) in many applied areas, for example, space navigation, automation, biomedical issues, problems in economics, investment problems, and others (see, for example, [17, Sec. 3], [4, Sec. 4.4, Ch. 6], [21, p. 253], [19, Ch. 1], [23] and references therein).

Acknowledgments. The research was supported by the Russian Foundation for Basic Research (RFBR) under Project 18-01-00544a.

References

1. Baier, R., Donchev, T.: Discrete approximation of impulsive differential inclusions. Numer. Funct. Anal. Optim. **31**(6), 653–678 (2010)
2. Baturin, V.A., Goncharova, E.V., Pereira, F.L., Sousa, J.B.: Measure-controlled dynamic systems: polyhedral approximation of their reachable set boundary. Autom. Remote Control **67**(3), 350–360 (2006)
3. Chernousko, F.L., Rokityanskii, D.Y.: Ellipsoidal bounds on reachable sets of dynamical systems with matrices subjected to uncertain perturbations. J. Optim. Theory Appl. **104**, 1–19 (2000)
4. Dykhta, V.A., Sumsonuk, O.N.: Optimal Impulse Control with Applications. Fizmatlit, Moscow (2000). (Russian)
5. Filippova, T.F.: Estimates of reachable sets of impulsive control problems with special nonlinearity. In: Todorov, M.D. (ed.) Application of Mathematics in Technical and Natural Sciences – AMiTaNS 2016. AIP Conference Proceedings, vol. 1773, pp. 100004–1–100004–10. Melville, New York (2016). https://doi.org/10.1063/1.4964998

6. Filippova, T.F., Matviychuk, O.G.: Reachable sets of impulsive control system with cone constraint on the control and their estimates. In: Lirkov, I., Margenov, S., Waśniewski, J. (eds.) LSSC 2011. LNCS, vol. 7116, pp. 123–130. Springer, Heidelberg (2012). https://doi.org/10.1007/978-3-642-29843-1_13

7. Guseinov, K.G., Ozer, O., Akyar, E., Ushakov, V.N.: The approximation of reachable sets of control systems with integral constraint on controls. Nonlinear Differ. Equ. Appl. **14**(1–2), 57–73 (2007)

8. Gusev, M.I.: On convexity of reachable sets of a nonlinear system under integral constraints. IFAC-PapersOnLine **51**(32), 207–212 (2018)

9. Jaulin, L., Kieffer, M., Didrit, O., Walter, É.: Applied Interval Analysis: With Examples in Parameter and State Estimation, Robust Control and Robotics. Springer, London (2001). https://doi.org/10.1007/978-1-4471-0249-6

10. Kostousova, E.K.: Outer polyhedral estimates for attainability sets of systems with bilinear uncertainty. J. Appl. Math. Mech. **66**(4), 547–558 (2002)

11. Kostousova, E.K.: External polyhedral estimates for reachable sets of linear discrete-time systems with integral bounds on controls. Int. J. Pure Appl. Math. **50**(2), 187–194 (2009)

12. Kostousova, E.K.: State estimation for linear impulsive differential systems through polyhedral techniques. Discrete Continuous Dyn. Syst. (Issue Suppl.) 466–475 (2009)

13. Kostousova, E.K.: On polyhedral estimates for trajectory tubes of dynamical discrete-time systems with multiplicative uncertainty. Discrete Continuous Dyn. Syst. (Issue Suppl.) 864–873 (2011)

14. Kostousova, E.K.: State estimation for control systems with a multiplicative uncertainty through polyhedral techniques. In: Hömberg, D., Tröltzsch, F. (eds.) CSMO 2011. IFIPAICT, vol. 391, pp. 165–176. Springer, Heidelberg (2013). https://doi.org/10.1007/978-3-642-36062-6_17

15. Kostousova, E.K.: External polyhedral estimates of reachable sets of linear and bilinear discrete-time systems with integral bounds on additive terms. In: Proceedings of 2018 14th International Conference Stability and Oscillations of Nonlinear Control Systems (Pyatnitskiys Conference), STAB 2018, pp. 1–4. IEEE Xplore Digital Library (2018). https://doi.org/10.1109/STAB.2018.8408370

16. Kostousova, E.K.: State estimates of bilinear discrete-time systems with integral constraints through polyhedral techniques. IFAC-PapersOnLine **51**(32), 245–250 (2018)

17. Krasovskii, N.N.: Theory of Control of Motion: Linear Systems. Nauka, Moskow (1968). (Russian)

18. Kurzhanski, A.B., Dar'in, A.N.: Dynamic programming for impulse controls. Annu. Rev. Control **32**, 213–227 (2008)

19. Kurzhanski, A.B., Daryin, A.N.: Dynamic Programming for Impulse Feedback and Fast Controls. LNCIS, vol. 468. Springer, London (2020). https://doi.org/10.1007/978-1-4471-7437-0

20. Kurzhanski, A.B., Vályi, I.: Ellipsoidal Calculus for Estimation and Control. Birkhäuser, Boston (1997)

21. Kurzhanski, A.B., Varaiya, P.: Dynamics and Control of Trajectory Tubes: Theory and Computation. Birkhäuser, Basel (2014)

22. Mazurenko, S.S.: Partial differential equation for evolution of star-shaped reachability domains of differential inclusions. Set Valued Var. Anal. **24**, 333–354 (2016)

23. Pierce, J.G., Schumitzky, A.: Optimal impulsive control of compartment models, I: qualitative aspects. J. Optim. Theory Appl. **18**(4), 537–554 (1976)
24. Polyak, B.T., Nazin, S.A., Durieu, C., Walter, E.: Ellipsoidal parameter or state estimation under model uncertainty. Automatica **40**(7), 1171–1179 (2004)
25. Sinyakov, V.V.: Method for computing exterior and interior approximations to the reachability sets of bilinear differential systems. Differ. Equ. **51**(8), 1097–1111 (2015)
26. Vdovina, O.I., Sesekin, A.N.: Numerical construction of attainability domains for systems with impulse control. In: Proc. Steklov Inst. Math. (Suppl. 1), S246–S255 (2005)
27. Veliov, V.M.: On the relationship between continuous- and discrete-time control systems. Cent. Eur. J. Oper. Res. **18**(4), 511–523 (2010)

Numerical Simulation of Hyperbolic Conservation Laws Using High Resolution Schemes with the Indulgence of Fuzzy Logic

Ruchika Lochab[ID] and Vivek Kumar[(✉)][ID]

Department of Applied Mathematics, Delhi Technological University,
Delhi 110042, India
rchklchb@gmail.com, vivekkumar.ag@gmail.com

Abstract. The aim of this paper is to solve numerically a class of problems on conservation laws, modelled by hyperbolic partial differential equations. In this paper, primary focus is over the idea of fuzzy logic-based operators for the simulation of problems related to hyperbolic conservation laws. Present approach considers a novel computational procedure which relies on using some operators from fuzzy logic to reconstruct several higher-order numerical methods known as the flux-limited methods. Further optimization of the flux limiters is discussed. The approach ensures better convergence of the approximation and preserves the basic properties of the solution of the problem under consideration. The new limiters are further applied to several real-life problems like the advection problem to demonstrate that the optimized schemes ensure better results. Simulation results are included wherever required.

Keywords: Conservation laws · Flux limiters · Fuzzy logic

1 Introduction

Conservation laws can also be stated as the fundamental laws of nature, they have various applications in real life and they are an interesting topic of research in multiple fields like Biology, Physics, Geology, Chemistry, and many engineering sciences like astronomy, civil, electrical to name a few. The simulation of the partial differential equations associated with conservation laws has been a popular branch of computational mathematics. It is well known that numerically solving hyperbolic system of conservation laws is a difficult task due to the possible interaction between the shock and rarefaction waves, the undesirable propagation of discontinuities and the main difficulty is the evolution of discontinuities after some time no matter how smooth our initial condition is. For any numerical method, the essential requirements for the convergence of the approximate solution to the real solution are the preservation of basic properties and the efficiency of the procedure in reducing the rounding and systematic errors.

© Springer Nature Switzerland AG 2020
Y. D. Sergeyev and D. E. Kvasov (Eds.): NUMTA 2019, LNCS 11974, pp. 139–153, 2020.
https://doi.org/10.1007/978-3-030-40616-5_11

There are different methods in the literature [1–3] to control these undesirable obstacles, which also ensure the essential requirements stated above.

The central theme of this work is to resolve the two contradictory but necessary needs for numerical methods, one is to attain high order accuracy and the other is to reduce non-physical oscillations near the discontinuities. High-resolution methods are significant for hyperbolic conservation laws because they provide better results as compared to the first generation methods [3] which are in general, least concerned with the type of initial solution. Another category of methods involves TVD (Total variation diminishing) methods which are one of the most important tools in the development of numerical methods for problems in computational fluid dynamics. A lot of effective methods have been approached from classical methods like the first generation methods to the advanced methods like the essentially non oscillatory/weighted essentially non-oscillatory (ENO/WENO) methods [4,5], the flux limited methods [5]. These methods have been regarded as successful in simulating the conservation laws. But, the design of some high order methods requires expert knowledge and the coding of such methods is also a tedious task. To ease this coding process one can somehow take the help of operators from the fuzzy logic branch. This work presents a unique combination of two entirely different subjects namely Fuzzy Mathematics and the Computational Fluid Dynamics. The main concern among various robust methods is modern HR (High resolution) methods which blend two or more first-generation methods to produce some new schemes which give a more stable and high order accurate output. Flux limited methods are the ones belonging in this category. Flux limiters play an important role in switching up efficiently from smooth regions to the region having discontinuities. Thus, limiters help in providing a high-resolution approximation to the system. This work provides a clear observation of the effect of these limiters in the light of fuzzy expert rules, applied to a Fuzzy inference system to the same setup of flux limiters based upon their behaviour with respect to different solution parts and regenerated them in fuzzy logic. Apart from that, hedges provide better optimization to the existing limiters, and the new limiters obtained after optimizing the limiters with the aid of modifiers are able to provide even better results for the problems in conservation laws.

The construction of new and more efficient numerical methods for Hyperbolic conservation laws using a few tools from fuzzy is explored, these techniques can be considered as an easier analogue of the flux-limited methods and also this sort of application has not been used much in the literature, therefore there is not much theory developed in the context of this application. The main objective of this paper is to design new computational methods in an autonomous way. In upcoming sections, a brief review of the concepts used in this paper from fuzzy logic, namely: fuzzy sets, fuzzy inference system and fuzzy modifiers is given. Later in the same section, a summarized mathematical background on hyperbolic conservation laws is written, specifically flux limiters. Then in the third section, the experimental part is disclosed where reconstruction of the limiters in fuzzy logic using a built-in Matlab toolbox can be seen, followed by the optimization of limiters. After that, approximation of some popular test cases on advection equation based on square pulse test and a combination of sine wave and square pulse using the new optimized limiters are presented. The paper ends with a summary where the final conclusions and the benefits of this new approach are discussed.

2   Preliminaries

2.1   Fuzzy Logic Concepts Required for This Work

Fuzzy logic serves as an important concept in various real-life applications. It permits to control various complex processes based upon a few rules where a knowledge base is created which give the idea about the behavior of the system so considered. Fuzzy logic aims to generalize the concept of classical logic for reasoning under certainty.

Fuzzy Sets. Just like the classical set theory, fuzzy sets are studied in fuzzy logic. Fuzzy sets can be considered as the extension of classical sets. It can be best understood in the context of set membership. Unlike Classical sets known as crisp sets in fuzzy logic, fuzzy sets talk about the degree of compatibility of each member of the set with the set itself. The main idea is, in fuzzy one takes membership values which may lie in the interval $[0, 1]$ but in case of classical sets, it was either 0 or 1 based upon whether the element belongs to the set or not. A fuzzy set is defined as follows:

Definition 1. *Fuzzy Set: Let U be the universe of discourse and $K \subseteq U$ then the fuzzy set K is the collection of ordered pairs $(x, \mu(x))$, where $\mu(x)$ is the degree of compatibility of the element x.*

$$K = \{(x, \mu(x))|x \in U\}$$

In crisp sets, the total number of elements in the set gives the cardinality of the set, but in case of fuzzy logic, we have a different approach. The cardinality of a fuzzy set K denoted by card(K) is:

$$card(K) = \sum \mu(x_i)$$

Just like the classical sets, the mathematical and logical operations can be carried out in fuzzy theory also. These operations enable us to put these sets into practical use. For detailed theory refer [6]. Although, there are many fuzzy sets in the fuzzy logic theory the ones we need in our work are: Singleton fuzzy sets, triangular fuzzy sets, trapezoidal fuzzy sets, as shown in the Fig. 1. Here, we need to mark that the interval $[0, 1]$ is the main reason for making us capable of building the foundation of approximate reasoning and fuzzy control.

Some standard fuzzy sets to be used in this work are the triangular fuzzy set, the trapezoidal fuzzy set, and the singleton fuzzy set (see Fig. 1). Additionally,

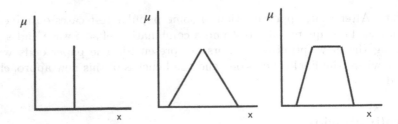

Fig. 1. Some standard fuzzy sets: Singleton(leftmost), triangular(middle), trapezoidal(right)

in order to adapt the fuzzy sets, we can use fuzzy modifiers, also known as fuzzy hedges, which are indeed a powerful adaption tool resulting in a change in the shapes of original fuzzy sets. These operators modify the membership values related to the fuzzy set, due to which its geometry gets altered.

Hedges/Modifiers: Another important concept which is to be considered from fuzzy systems are the "hedges", or modifier of fuzzy values. These operations are used in an effort to get closer to the natural human language, and they help in the generation of fuzzy statements with some mathematical calculations. As such, the initial definition of modifiers and corresponding algebra upon them will be quite a subjective process and may vary from project to project. Nonetheless, the system ultimately derived operates with the same formality as the classic theory of logic. These are special terms aimed to make modifications in fuzzy theory. Hedges modify the meaning of existing data by changing the membership values corresponding to the relevant fuzzy sets.

Consider a fuzzy set $A = \{(x, \mu(x)) \mid x \in K\}$ corresponding to a crisp set K lying in some universe U, for such case some popular modifiers are defined as follows: (see figure for their geometric representation)

– **Concentration operator:**

$$CON(A) = \{(x, (\mu(x))^n)) \mid x \in U\}$$

– **Dilation operator:**

$$DILT(A) = \{(x, \sqrt[n]{(\mu(x))}) \mid x \in U\}$$

– **Contrast operator:** $CONT(A) = \begin{cases} 2(\mu(x))^n & \mu(x) \leq 0.5 \\ 1 - 2(1 - \mu(x))^n & else \end{cases}$

In this work, some standard modifiers are used, namely: the contrast operator, dilation operator, and the concentration operator. This work modifies the existing limiters [7] using these tools and molds them into more efficient limiters. The results obtained by doing such a thing is shown in the third section of the paper.

Fuzzy Inference System (FIS): It is a system based upon the popular "Input-processing-Output" theory. One has to provide some crisp data as input to this inference system, which is then fuzzified i.e, changed into the equivalent fuzzy set using the fuzzifiers. This fuzzified data is then evaluated based upon some fuzzy rules, which ultimately infer something in terms of fuzzy. This fuzzy output is then translated to the crisp set by the aid of defuzzifier. This is how a fuzzy inference system works. There are various fuzzy inference systems available in fuzzy systems and the most popular ones are the Mamdani and the Sugeno FIS [8]. In this work, Mamdani FIS is used (see Fig. 2 for fuzzy inference system). It is an expert system based on fuzzy logic. The fuzzy rules to be used in this system are decided on the basis of the behavior of the limiter to be reconstructed with respect to various solution areas. In short, fuzzy rules are extracted from a given data which is considered at the very initial stage. The good thing about using fuzzy inference system is that one does not need to seriously code up the things to produce results because the fuzzy rule base consisting of a few three to four fuzzy rules is serving our goal. Therefore, a better approximation is obtained without messy coding work which saves both the time and money and hence this new approach to reconstruct the limiters via inference system seems easier and doing optimization is also quite handy in this place. In a way, the fuzzy theory provides yet another way to carry out the study based upon interpolation in a nice way. Specifically, it provides the optimization tools for the hyperbolic partial differential equations. Fuzzy systems, mainly fuzzy logic and fuzzy set theory, gives a rich and clear version addition to standard logic. The mathematics generated by these theories is consistent, and fuzzy logic may be considered as a generalization of classical logic. The applications which may be generated from or adapted to fuzzy logic are quite wide-ranging, and they provide the opportunity for modeling of situations which are inherently imprecisely defined, despite the concerns of classical logicians. The important and nice thing is any systems may be re-modeled, and even replicated with the help of fuzzy systems, not the least of which is human reasoning itself.

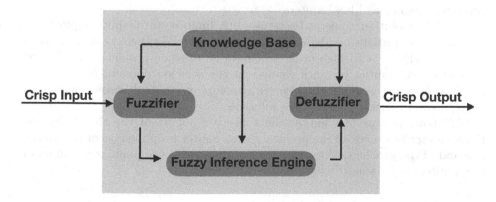

Fig. 2. Fuzzy inference system

2.2   Concepts Required from Hyperbolic Conservation Laws

High Resolution Methods: High-resolution methods are basically the non-linear methods which are more solution sensitive as they take into account the flux/solution gradients. Such methods are also known commonly as high resolution or TVD (Total Variation Diminishing) methods. In such methods, nonlinear stability conditions are enforced which help in reducing the spurious oscillations but they sometimes lead to clipping errors (a form of distortion that limits a wavefront once it exceeds a threshold) at the solution extrema.

Here main concern is flux-limited methods for scalar conservation laws. Flux limited methods are based upon first-generation methods, they are basically the adaptive linear combination of two first-generation methods. For scalar conservation laws such methods are defined as follows:

Consider the computational domain $\mathbb{R} \times \mathbb{R}_0^+$ covered uniformly by the cells $[x_{i-1/2}, x_{i+1/2}] \times [t^n, t^{n+1}]$, where $x_j = j \triangle x$ and $t^n = n \triangle t$ being the space size and the time step respectively. Over these cells, the unknown u(x, t) is given by the cell averages

$$U_j^n = \frac{1}{\triangle x} \int_{x_{i-1/2}}^{x_{i+1/2}} u(x, t^n) dx$$

Then, the general form of numerical scheme by taking forward difference in time and central difference in space reads as

$$u_j^{n+1} = u_j^n - \lambda(F_{j+\frac{1}{2}}^n - F_{j-\frac{1}{2}}^n),$$

where

$$F_{j+\frac{1}{2}}^n = F_{j+\frac{1}{2}}^{(l)} + \phi_{j+\frac{1}{2}}^n (F_{j+\frac{1}{2}}^{(h)} - F_{j+\frac{1}{2}}^{(l)})$$

here $F_{j+\frac{1}{2}}^{(l)}$ and $F_{j+\frac{1}{2}}^{(h)}$ are the conservative flux terms (edge fluxes for the jth cell) which are selected from non adaptive (first generation schemes) having complimentary properties (l means low precision and h means high precision) and $\phi_{j+\frac{1}{2}}^n$ is the main adaptive parameter for such adaptive schemes which is commonly known as Flux Limiter.

In high-resolution numerical schemes, flux limiters are mainly employed to deal with the spurious oscillations (wiggles) that would otherwise arise in non-adaptive methods with high order schemes due to some problems like shocks, contact discontinuities or quick changes in the solution domain. More importantly, the proper use of flux limiters with an appropriate high-resolution scheme leads to total variation diminishing solutions.

Till now, various flux limiters have become a part of the theory but no single limiter can serve all the problems, each limiter is applied according to the demand of the problem we want to work with some of the popular examples of flux limiters are discussed here (see Fig. 3):

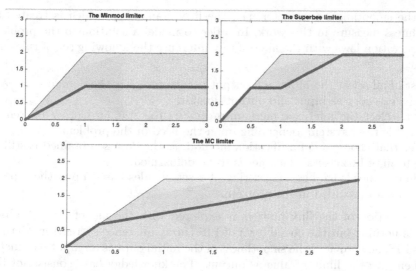

Fig. 3. Red line indicates classical limiters (Color figure online)

The minmod limiter selects the values with smaller modulus, else returns 0 and in MC limiter, we compare the second-order central difference with twice the forward and twice the backward differences. Minmod scheme gives quite dissipative output at the discontinuous parts, so this superbee limiter takes higher modulus near the discontinuities. The expressions for Minmod, superbee and the MC limiter are given by:

- **Minmod limiter:** $\phi_{mm}(r) = max[0, min(1, r)]$
- **Monotonized Central(MC) limiter:**

$$\phi_{mc}(r) = max[0, min(2r, 0.5(1 + r), 2)]$$

- **Superbee Limiter:** $\phi_{sb}(r) = max[0, max(min(2r, 1)), min(r, 2)]$
  where 'r' is the smoothness measure, also called the slope gradient and mm, mc and sb are just the notations for respective limiters.

3 Experiments: New Approach to Flux Limiters

High-resolution schemes can be approached via fuzzy logic as well, in fact, the study of flux limiters becomes easier in this manner, especially for the coding part as already mentioned earlier in the aim of this paper, the focus is on constructing the flux limiters by using fuzzy logic.

3.1 Reconstruction of Flux Limiters in Fuzzy Logic

In this fuzzy limiter reconstruction, as an abstract view, the purpose here is to determine the value of flux limiter ϕ in the range $[0, 2]$, this limiter depends upon the smoothness parameter, i.e., the flux gradient 'r' which is treated as smoothness measure in this work. In order to model a solution to the problem in conservation laws with the aid of fuzzy inference the following set of rules are obeyed:

1. First of all select the input and output parameters. Fix the domains for each of the category as input and output domain.
2. Next, select the inference mechanism among the standard fuzzy inferences available to in system, depending upon the need of the problem.
3. After that choose a defuzzification method, as the data so obtained is still in the form of fuzzy sets which needs to be defuzzified.
4. Define a knowledge base, consisting of a set of rules based upon the type of initial data provided in the problem.

Here recreation of the flux limiters is explored with the aid of fuzzy toolbox which is nothing but the coded form of FIS (fuzzy inference system) in Matlab. In this FIS, one provides the smoothness of the initial data as an input parameter and then give the limiter value as output. The knowledge base consists of the set of if-then rules, which are based upon the classic limiter setup as available in the literature. This theory can be concluded in the following points:

- A flux limiter takes initial data type features as input unit and returns a suitable limiter value as output unit. So, here look for an ideal output limiter for corresponding characteristic input situations.
- For fuzzy flux limiter, as a first step, one has to choose the input and output parameters (known as linguistic variables) and determine their respective domains.
- Then partition the variables by defining some terms with their membership functions.
- After that final and important step is to specify the knowledge base (if-then type of rules).

Demonstrating the Reconstruction of the Minmod Limiter: Minmod limiter is supposed to be the simplest limiter among the family of flux limiters in computational fluid dynamics. The functioning of minmod function can be explained in three parts, the two regions where the limiters are constant with functional values 0 and 1, and the intermediate part where it seems like a straight line passing through the origin. The main rule is to use the high order scheme at the smooth regions and the lower first-order monotone scheme at the problematic parts pertaining to discontinuous solution features. So, as far as the smoothness measure is concerned, lookup for the subset $[-1, 2]$ of the real line as the working domain. Then the rule base to be used in order to reconstruct the minmod scheme is selected.

Also, 'min' operator is taken as the implication operator, for aggregation the 'max' operator is used and 'centroid method' is selected for defuzzification of the fuzzified output (Fuzzy toolbox is a graphical user interface, there one can select from the available options according to the demand of the problem, see Fig. 4).

- Classify the input variable "datatype" as "type1" and "type2" for the extremum and smooth initial profiles respectively, similarly classify the output parameter "limiter" as "upwind" and "laxwend" for the upwind and lax wendroff schemes respectively (see Fig. 5)
  Taking 'datatype' as the input variable with linguistic terms 'type1' and 'type2' as the trapezoidal functions.
- Taking 'limiter' as the output variable with terms 'upwind' and 'laxwend' as singleton membership functions (see Fig. 6).
- Then the **knowledge base** consists of the rules which says the input variable is 'type1' then the output value will be 'upwind' and if the input is 'type2' then the output will be 'laxwend' (see Fig. 7).
  - If smoothness is **type1** then flux is **upwind**.
  - If smoothness is **type2** then flux is **laxwend**.
- Then the FIS so obtained is analogous to **Minmod scheme** from computational fluid dynamics to that in fuzzy logic (see Fig. 8).

This reconstruction of minmod limiter only requires two rules in the rule base, but in general, there may be more than two rules in the rule base for other limiters. Like while reconstructing the superbee limiter one has to use three rules, for superbee the range is $[0, 2]$ for the output variable and select $[-1, 3]$ as the domain of the input variable and the three key features are: superbee limiter takes either three functional values namely 0, 1, 2 (remaining constant at these values) and in the connecting parts it is a straight line, so one can use the rule base which allows using three linguistic terms for the output variable and three linguistic terms namely extremum, smooth and excursive for the input variable.

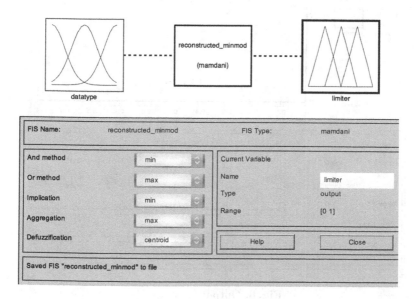

Fig. 4. Creating environment for FIS

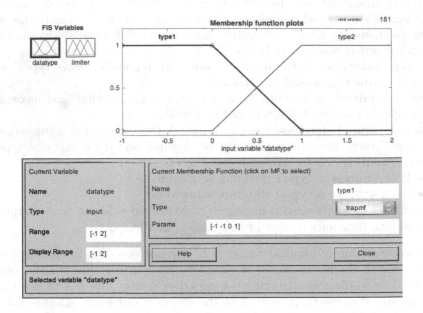

Fig. 5. Input variables

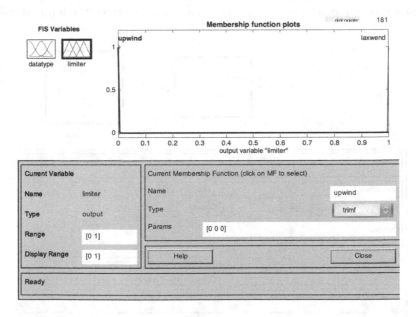

Fig. 6. Output variables

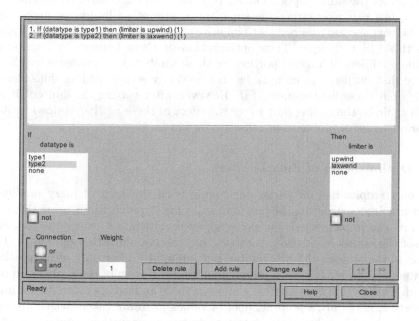

Fig. 7. Rules insertion

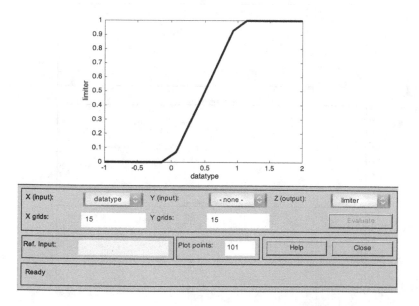

Fig. 8. The analogous minmod limiter

Hence, under the same implication, aggregation, and the centroid defuzzification method, one can reconstruct the superbee or any other limiter in fuzzy logic. Now the next section uses hedges to improve these analogous limiters so obtained using this FIS technique. Then optimization of these limiters by considering suitable modifiers at various portions of these limiters is discussed. One should observe that until now, this work has not used any serious coding skills as such to carry out these limiters using FIS. However, after getting the limiters, it will switch again to the coding part to see the effect of these newly obtained limiters on various class of problems in conservation laws.

3.2   Optimization of Flux Limiters

This part emphasizes on parameter tuning with the help of fuzzy modifiers. Apart from reconstructing the flux limiters, some improvisations using the modifiers are also implemented. For doing so, systematically consider the combinations of the standard dilation, contrast, and the concentration operators. This section is focusing on the MC limiter and minmod limiter only although these operators can be implemented on any flux limiter if it suits the problem so considered. This section systematically imposes the standard operators to the input parameters, here only a finite number of values for n are taken, which are used in the membership value of the operators, but one can further extend this set to obtain, even more, results (see Fig. 9) for the following optimized limiters:

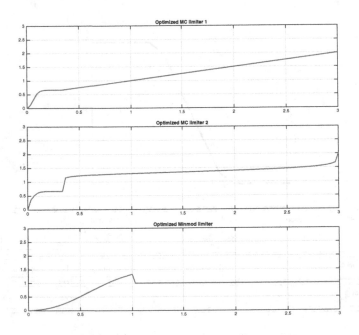

Fig. 9. Newly obtained optimized limiters

1. Optimized MC limiter 1 obtained by applying concentration operator with $n = 8$ to the extremum part.
2. Optimized MC limiter 2 by applying concentration operator with $n = 6$ to the extremum part and the dilation operator with $n = 8$ to sharp regions.
3. Optimized Minmod limiter by applying $n = 2$ to the extremum and the dilation operator with $n = 8$ to smooth regions.

Further, in order to examine the new limiters, the linear advection equation is considered, which is the benchmark problem to check some new scheme or something, it is the first test problem which flashes into mind. Here some experiments based on these optimized limiters are applied to the standard model problem in computational fluid dynamics, the linear advection problem.

3.3   Application of Optimized Limiters to Real Life Problem

The Advection Problem: The advection equation in one space dimension is of the form:

$$\frac{\partial u}{\partial t} + a\frac{\partial u}{\partial x} = 0 \tag{1}$$

here, $u := (x, t)$ is some scalar field, $x \in \mathbb{R}$ and $t \in \mathbb{R}^+$ are the space and time components respectively. Here, 'a' is a nonzero constant (in most cases we refer 'a' as some velocity vector field.)

We will work with the Eq. (1) with the conditions

$$u(x, 0) = f(x) \tag{2}$$

where f is some conserved quantity [9, 10].

Note: Here, the numerical scheme is obtained by taking forward difference in time and a central difference in space which is written as:

$$u_j^{n+1} = u_j^n - \lambda(F_{j+\frac{1}{2}}^n - F_{j-\frac{1}{2}}^n),$$

where

$$F_{j+\frac{1}{2}}^n = F_{j+\frac{1}{2}}^{(l)} + \phi_{j+\frac{1}{2}}^n (F_{j+\frac{1}{2}}^{(h)} - F_{j+\frac{1}{2}}^{(l)})$$

here

$$\phi = max(0, min(min(0.5(1 + r), 2), 2r)/((1 - 3r)^8 + 3r)).$$

Also, $F^{(l)}$ and $F^{(h)}$ are the Upwind scheme and the Lax Wendroff schemes respectively. We are taking time step $\triangle t = 0.0025$ and space size $\triangle x = 0.01$ and checking the results over varying points (200, 400, 600, 800), taking the speed of advection $a = 1.0$. Here, two test cases have been considered, the square pulse test, and the combination of a sine wave and square pulse (see Fig. 10, 11). It is interesting to note that we get better result as compared to the existing minmod limiter for both tests. On same lines, many other robust limiters can be used to approximate the numerical solutions for the problems in conservation laws.

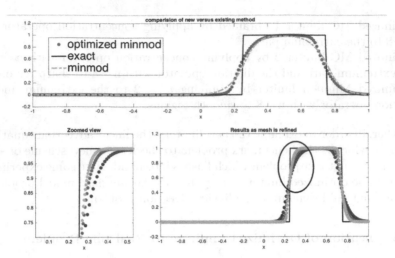

Fig. 10. Results for the optimized Minmod limiter for square pulse initial conditions, leftmost figure in bottom row shows the zoomed view of circled part in the corresponding right image

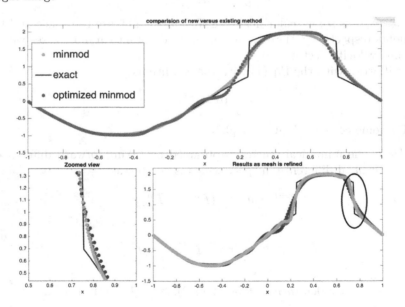

Fig. 11. Results for the optimized Minmod limiter for combined initial conditions, leftmost figure in bottom row shows the zoomed view of circled part in the corresponding right image

4   Summary

In this work, a novel approach to study flux limiters methods using fuzzy logic theory is described. In the test cases like the two discussed in section four namely the square pulse test and the mixed case, the modified limiters are able to approximate the solution in a nice way resulting in even much better results. The main advantage of doing so can be the easy interpretation of flux limiters. Using fuzzy toolbox the modification of flux limiters becomes easier. This approach can be beneficial in providing more efficient numerical methods for solving various problems arising in the computational fluid dynamics.

Acknowledgement. RL thanks the Delhi Technological University for the partial financial support to attend NUMTA 2019 and UGC for PhD fellowship.

References

1. Strikwerda, J.: Finite Difference Schemes and Partial Differential Equations, 2nd edn. Society for Industrial and Applied Mathematics, Philadelphia (2004)
2. LeVeque, R.: Finite-Volume Methods for Hyperbolic Problems. Cambridge University Press, Cambridge (2002)
3. Laney, C.: Computational Gasdynamics, 1st edn. Cambridge University Press, New York (1998)
4. Hirsch, C.: Numerical Computation of Internal and External Flows. Elsevier (2007)
5. Toro, E.: Riemann Solvers and Numerical methods for Fluid Dynamics. Springer, Heidelberg (1999). https://doi.org/10.1007/978-3-662-03915-1
6. Klir, G., Yuan, B.: Fuzzy Sets and Fuzzy Logic, Theory and Applications (1995)
7. Breuss, M., Dietrich, D.: On the optimization of flux limiters for hyperbolic conservation laws. Numer. Methods Part. Differ. Equ. **29**, 884–896 (2013)
8. Chin, T., Qi, X.: Genetic algorithms for learning the rule base of fuzzy logic controller. Fuzzy Sets Syst. **97**, 1–7 (1998)
9. Kumar, V., Srinivasan, B.: An adaptive mesh strategy for singularly perturbed convection diffusion problem. Appl. Math. Model. **39**, 2081–2091 (2015)
10. Kumar, V., Rao, R.: Composite scheme using localized relaxation non-standard finite difference method for hyperbolic conservation laws. J. Sound Vib. **311**, 786–801 (2008)

Methodology for Interval-Valued Matrix Games with 2-Tuple Fuzzy Linguistic Information

Tanya Malhotra[1] ⓘ, Anjana Gupta[1]([✉]) ⓘ, and Anjali Singh[2] ⓘ

[1] Delhi Technological University, Shahbad Daulatpur, Main Bawana Road, Delhi 110042, India
`malhotra.tanya92@gmail.com`, `anjanagupta@dce.ac.in`
[2] Mahatma Gandhi Institute of Technology, JNTUH, Hyderabad 500085, India
`anjalisingh_maths@mgit.ac.in`

Abstract. In this paper, we consider a non-cooperative 2-player zero-sum interval-valued 2-tuple fuzzy linguistic (IVTFL) matrix game and develop a methodology to evaluate its saddle point and optimal interval-valued linguistic value of the game. In this direction, we have constructed an auxiliary pair of interval-valued linguistic linear programming (IVLLP) problem that is further transformed into conventional interval linear programming (ILP) problem to obtain optimal strategy sets of both players as the region that is not only completely feasible but also totally optimal. The proposed method is illustrated via a hypothetical example to show its applicability in the real world. To validate the suggested solution scheme, the transformed ILP problems are solved using best-worst case (BWC) approach, enhanced-interval linear programming (EILP) method and linguistic linear programming (LLP) technique of solving interval linguistic matrix game problems and lastly the obtained results are compared.

Keywords: 2-tuple fuzzy linguistic model · Interval-valued 2-tuple fuzzy linguistic model · Interval linear programming · Interval-valued linguistic linear programming · Matrix game problem

1 Introduction

Non-cooperative game theory in its classical set up was introduced by Von-Neumann and Morgenstern [6] in 1944. It asserts that every player is exposed to the game's precisely known information. The prevailing knowledge of the game permits each player to furnish appropriate evaluations to their utility functions corresponding to different pair of strategies. The postulations made for the exact payoffs can be considered as the stringent ideology in the real world scenario which involves uncertain and ambiguous information. Imprecision and uncertainty have been incorporated in game theory by using various frameworks like fuzzy, stochastic etc. Several researchers have contributed significantly in enhancing the literature of fuzzy games [17–19] and stochastic games [21,22]. However, in the world of uncertainties, it is also challenging for players to express payoffs in terms of membership functions in fuzzy environment or probability distribution functions in stochastic environment. To facilitate the players with

Supported by Delhi Technological University (Ref. No. DTU/IRD/619/2019/2107).

ⓒ Springer Nature Switzerland AG 2020
Y. D. Sergeyev and D. E. Kvasov (Eds.): NUMTA 2019, LNCS 11974, pp. 154–168, 2020.
https://doi.org/10.1007/978-3-030-40616-5_12

effortless choice of payoffs, a new version of matrix games under uncertainty is proposed by Arfi [23,24] based on linguistic fuzzy logic. To annex a new dimension to the matrix game problems under linguistic environment, Singh et al. [2] defined matrix games with linguistic information and proposed a linguistic linear programming (LLP) approach to solve such class of games. Singh et al. [3] further extended the matrix games to interval-valued 2-tuple fuzzy linguistic framework to increase the level of uncertainty in game problems and adopted LLP approach to solve it. The authors formulated a pair of auxiliary LLP problems to obtain the linguistic lower and upper bounds of interval linguistic value of the game.

In this study, we extend the work of solving interval linguistic matrix game (ILG) problems one step forward. Here, we propose a mechanism to compare IVTFL variables using the bounds of the intervals and subsequently, define interval linguistic lower value (ILLV) and interval linguistic upper value (ILUV) of the matrix game by introducing the concept of max-min and min-max principle. In the absence of pure strategies, we suggest IVLLP formulation to obtain the interval linguistic value of game with the optimal strategies of both players by transforming it to conventional ILP problem. To validate the proposed methodology, Best Worst Case (BWC) method [20], Enhanced Interval Linear Programming (EILP) method [25] and Linguistic Linear Programming (LLP) method [3] are adopted to solve the transformed ILP problems and provide a comparative analysis. The duality principle of IVLLP is also taken into consideration in order to prove the equality of ILLV and ILUV of the game for player I and II, respectively.

The remaining paper unfolds as follows. In subsequent section, the fundamentals of subscript symmetric linguistic variables are elucidated with matrix games under linguistic framework. Section 3 explains a new approach to compare two IVTFL variables based on the end point approach. In Sect. 4, a zero sum interval-valued 2-tuple fuzzy linguistic matrix game is defined with its interval linguistic lower and upper values using max-min principle. Section 5 discusses interval-valued linguistic linear programming approach to solve the game in absence of pure strategies with a hypothetical illustration. The paper concludes in Sect. 6.

2 Preliminaries

In this section, we review the fundamentals of subscript symmetric linguistic variables followed by the foundations of matrix games with linguistic information.

2.1 Subscript-Symmetric Linguistic Computational Model

Definition 1 [5]. Let $LT = \{\ell_{-g}, \ldots, \ell_0, \ldots, \ell_g\}$ be a finite and totally ordered predefined linguistic term set with the following properties.

(i) The set LT is ordered i.e. $\ell_i > \ell_j$ if and only if $i > j$,
(ii) Negation of any linguistic variable $\ell_i \in LT$ is given as ℓ_{-i}.

The 2-tuple linguistic computational model, defined by Herrera and Martinez [1] can be easily enhanced to the above defined subscript symmetric linguistic term set LT. Extending the notion of operators Δ and Δ^{-1}, formalized by Herrera and Martinez [1] to set LT, the translation function converting a numerical value to a 2-tuple linguistic variables can be stipulated as follows.

Definition 2. Let $LT = \{\ell_i \mid i = -g,\ldots,0,\ldots,g\}$ be a finite set of linguistic terms having cardinality $2g + 1$ and let $\beta \in [-g,g]$ be a value that represents the outcome of a symbolic aggregation operation. The 2-tuple linguistic variable that depicts the identical information to β is defined in the following aspect.

$$\Delta : [-g,g] \to LT \times [-0.5,0.5]$$

$$\Delta(\beta) = (\ell_i, \alpha) \text{ with } \begin{cases} \ell_i, & i = \text{round}(\beta), \\ \alpha = \beta - i, & \alpha \in [-0.5,0.5]. \end{cases}$$

where round(.) is the usual round operation, ℓ_i being the linguistic term closest to β, and α is the symbolic translation.

Clearly, it has been observed that the aforementioned function Δ is a bijection [1] and hence, its inverse is given by,

$$\Delta^{-1} : LT \times [-0.5,0.5] \to [-g,g] \quad \text{as} \quad \Delta^{-1}(\ell_i,\alpha) = i + \alpha = \beta.$$

Furthermore, Herrera and Martinez [1] have expressed the comparison of 2-tuple linguistic information by using conventional lexicographic ordering. On the similar grounds we can propose the following definition stating the ranking order.

Definition 3. Let (ℓ_i, α_i) and (ℓ_j, α_j) be 2-tuple linguistic variables using the term set LT. Then,

(i) If $i < j$ then $(\ell_i, \alpha_i) < (\ell_j, \alpha_j)$.
(ii) If $i = j$, i.e., $\ell_i = \ell_j$, then
    (a) if $\alpha_i = \alpha_j$, then $(\ell_i, \alpha_i) = (\ell_j, \alpha_j)$, that is, (ℓ_i, α_i) and (ℓ_j, α_j) express the identical information;
    (b) if $\alpha_i > \alpha_j$, then $(\ell_i, \alpha_i) > (\ell_j, \alpha_j)$;
    (c) if $\alpha_i < \alpha_j$, then $(\ell_i, \alpha_i) < (\ell_j, \alpha_j)$.

The literature concerning operators for the set of 2-tuple linguistic variables is vast and extensive. Here, we recall the weighted average operator defined in [4] after extending it to subscript symmetric linguistic term set LT.

Definition 4. Let $\{(\ell_{r_i}, \alpha_{r_i}), r_i \in \{-g,\ldots,0,\ldots,g\}, i = 1,\ldots,q\}$ be a set of 2-tuple linguistic variables and $w = (w_1,\ldots,w_q)^T$ be the weight vector satisfying $0 \le w_i \le 1, i = 1,\ldots,q, \sum_{i=1}^{q} w_i = 1$. Then, the weighted average operator is defined as

$$LWA[(\ell_{r_i},\alpha_{r_i}) : i = 1,\ldots,q] = (\ell_{r_1},\alpha_{r_1})w_1 \oplus (\ell_{r_2},\alpha_{r_2})w_2 \oplus \ldots \oplus (\ell_{r_q},\alpha_{r_q})w_q$$

$$= \Delta\left(\sum_{i=1}^{q} w_i \Delta^{-1}(\ell_{r_i},\alpha_{r_i})\right).$$

Consequently,

$$\Delta^{-1}(\bigoplus_{i=1}^{q}(\ell_{r_i},\alpha_{r_i})w_i) = \sum_{i=1}^{q} w_i \Delta^{-1}(\ell_{r_i},\alpha_{r_i}).$$

To extend the degree of uncertainty, an interval-valued 2-tuple linguistic variable [7] can also be defined as follows in frame of aforementioned linguistic term set LT.

Definition 5. Let $LT = \{\ell_{-g},\ldots,0,\ldots,\ell_g\}$ be a predefined linguistic term set. Then, an interval-valued 2-tuple linguistic variable is defined as $[(\ell_i^{(L)},\alpha^{(L)}),(\ell_i^{(U)},\alpha^{(U)})]$ where $\ell_i^{(L)},\ell_i^{(U)} \in LT$ with $\ell_i^{(L)} \leq \ell_i^{(U)}$ and $\alpha^{(L)},\alpha^{(U)}$ are the symbolic translations.

In literature, Singh et al. [2,3] have introduced matrix games with 2-tuple linguistic and interval-valued 2-tuple linguistic information based on the set of predefined linguistic term, $\ell_i, i = 0,1,\ldots,g$. In that paper, authors have defined a methodology for solving two players constant-sum linguistic matrix games primarily based on 2-tuple linguistic and interval-valued 2-tuple linguistic information. Authors also proposed an LLP approach to solve such class of games to evaluate optimal mixed strategies with linguistic value of the game. For thorough study of the methodology, one may refer to the paper [2,3].

2.2 Zero-Sum Linguistic Matrix Game

The matrix game problem where the sum of the payoffs corresponding to any given set of strategies is zero is termed as two players zero-sum game [26]. A zero-sum game is a particular case of the constant sum game and has subjected to several findings both in the fuzzy as well as conventional set up. However, the game problems with linguistic payoff matrices are pristine and required to be explored.

In the present subsection, we review the basic terminologies and notations related to the zero-sum matrix games within a 2-tuple linguistic framework. The following definitions are taken from [2] and can be easily extended to the subscript symmetric linguistic term set, LT as mentioned above.

Definition 6. A two-player zero-sum linguistic game $\widetilde{G}$ is defined by a quadruplet $(S^n, S^m, LT, \widetilde{A})$, where S^n and S^m are the strategy sets of player I and II, $LT = \{\ell_{-g},\ldots,\ell_0,\ell_1,\ldots,\ell_g\}$, with cardinality $2g+1$, is a subscript symmetric linguistic term set for both the players, $\widetilde{A}$ is the linguistic payoff matrix of player I against player II, and $neg(\widetilde{A})$ is the payoff matrix for player II.

Since the lexicographic ordering is available in the 2-tuple linguistic variables, one can easily extend the notion of the value of the game to the linguistic matrix game $\widetilde{G}$.

Definition 7. A matrix game $\widetilde{G}$ with payoff matrix $\widetilde{A} = [\widetilde{a}_{ij}]_{n \times m}$ has the linguistic lower value and the linguistic upper value defined as,

$$\widetilde{v}^- = \max_{i=1,\ldots,n} \min_{j=1,\ldots,m} \widetilde{a}_{ij}, \qquad \widetilde{v}^+ = \min_{j=1,\ldots,m} \max_{i=1,\ldots,n} \widetilde{a}_{ij}.$$

Here, it is considered that $\widetilde{v}^-$ (player I gain floor) is the minimum linguistic payoff that player I is assured to receive while $\widetilde{v}^+$ (player II loss ceiling) is the maximum linguistic loss of player II. The value of the game $\widetilde{G}$ exists if and only if $\widetilde{v}^- = \widetilde{v}^+$. The strategies i^* and j^*, yielding the payoff $\widetilde{a}_{i^* j^*} = \widetilde{v}^- = \widetilde{v}^+$, are optimal for player I and player II, respectively. The pair (i^*, j^*) is also known as the saddle point of the game $\widetilde{G}$.

In the case, where solution set of the game $\widetilde{G}$ does not possess pure strategies. We define the solution set as mixed strategies.

Definition 8. A mixed strategy is an ordered pair of vectors $(x, y) \in S^n \times S^m$, where

$$S^n = \{(x_1, \ldots, x_n) \; : \; x_i \geq 0, \; i = 1 \ldots, n, \; \Sigma_{i=1}^n x_i = 1\};$$
$$S^m = \{(y_1, \ldots, y_m) \; : \; y_j \geq 0, \; j = 1, \ldots, m, \; \Sigma_{j=1}^m y_j = 1\}.$$

Here, x_i is the probability of choosing strategy i by player I and y_j is the probability of selecting strategy j by player II.

In the subsequent section, we define the comparison of linguistic intervals to propose the interval linguistic lower and upper values using max-min principle.

3   Comparison of Interval-Valued 2-Tuple Fuzzy Linguistic Variables

In literature, Zhang [7] defined the comparison of interval-valued 2-tuple linguistic variables using score and accuracy values. It gives a total ordering of the linguistic intervals that does not show analogy with classical numeric intervals [8–16].

So, in this study, we present a new comparison scheme of interval-valued 2-tuple linguistic variables. The approach involves the bounds of the intervals that allows to define a partial ordering of the linguistic intervals. Here, we consider the following cases to encompass all possible pair of intervals.

(I) **Case of Disjoint Intervals:** Let $\widetilde{\mu} = [(\ell_{i(L)}, \alpha_{i(L)}), (\ell_{i(U)}, \alpha_{i(U)})]$, $\widetilde{v} = [(\ell_{j(L)}, \alpha_{j(L)}), (\ell_{j(U)}, \alpha_{j(U)})]$ be two disjoint IVTFL variables. Then

$$\widetilde{\mu} < \widetilde{v} \; \text{iff} \; (\ell_{i(U)}, \alpha_{i(U)}) < (\ell_{j(L)}, \alpha_{j(L)}).$$

(II) **Case of Nested Intervals :** Let $\widetilde{\mu} = [(\ell_{i(L)}, \alpha_{i(L)}), (\ell_{i(U)}, \alpha_{i(U)})]$, $\widetilde{v} = [(\ell_{j(L)}, \alpha_{j(L)}), (\ell_{j(U)}, \alpha_{j(U)})]$ be the two IVTFL variables such that one of the following cases occur :

(i) If $i^{(L)} \leq j^{(L)} < j^{(U)} \leq i^{(U)} \Rightarrow (\ell_{i(L)}, \alpha_{i(L)}) \leq (\ell_{j(L)}, \alpha_{j(L)}) < (\ell_{j(U)}, \alpha_{j(U)}) \leq (\ell_{i(U)}, \alpha_{i(U)})$.

(ii) If $i^{(L)} = j^{(L)} = j^{(U)} = i^{(U)} \Rightarrow \alpha_{i(L)} \leq \alpha_{j(L)} < \alpha_{j(U)} \leq \alpha_{i(U)}$ such that $(\ell_{i(L)}, \alpha_{i(L)}) \leq (\ell_{j(L)}, \alpha_{j(L)}) < (\ell_{j(U)}, \alpha_{j(U)}) \leq (\ell_{i(U)}, \alpha_{i(U)})$.

(iii) If $i^{(L)} = j^{(L)} < j^{(U)} < i^{(U)} \Rightarrow \alpha_{i(L)} \leq \alpha_{j(L)} < \alpha_{j(U)} \leq \alpha_{i(U)}$ such that $(\ell_{i(L)}, \alpha_{i(L)}) \leq (\ell_{j(L)}, \alpha_{j(L)}) < (\ell_{j(U)}, \alpha_{j(U)}) \leq (\ell_{i(U)}, \alpha_{i(U)})$.

(iv) If $i^{(L)} = j^{(L)} = j^{(U)} < i^{(U)} \Rightarrow \alpha_{i(L)} \leq \alpha_{j(L)} < \alpha_{j(U)} \leq \alpha_{i(U)}$ such that $(\ell_{i(L)}, \alpha_{i(L)}) \leq$
$(\ell_{j(L)}, \alpha_{j(L)}) < (\ell_{j(U)}, \alpha_{j(U)}) \leq (\ell_{i(U)}, \alpha_{i(U)})$.

(v) If $i^{(L)} \leq j^{(L)} = j^{(U)} \leq i^{(U)} \Rightarrow \alpha_{i(L)} \leq \alpha_{j(L)} < \alpha_{j(U)} \leq \alpha_{i(U)}$ such that $(\ell_{i(L)}, \alpha_{i(L)}) \leq$
$(\ell_{j(L)}, \alpha_{j(L)}) < (\ell_{j(U)}, \alpha_{j(U)}) \leq (\ell_{i(U)}, \alpha_{i(U)})$.

(vi) If $i^{(L)} \leq j^{(L)} < j^{(U)} = i^{(U)} \Rightarrow \alpha_{i(L)} \leq \alpha_{j(L)} < \alpha_{j(U)} \leq \alpha_{i(U)}$ such that $(\ell_{i(L)}, \alpha_{i(L)}) \leq$
$(\ell_{j(L)}, \alpha_{j(L)}) < (\ell_{j(U)}, \alpha_{j(U)}) \leq (\ell_{i(U)}, \alpha_{i(U)})$.

All above cases infer that the linguistic interval $\widetilde{v}$ is contained in $\widetilde{\mu}$, denoted as $\widetilde{v} \subset \widetilde{\mu}$. It demonstrates the inclusion property of linguistic intervals i.e. the interval $\widetilde{v}$ is nested within $\widetilde{\mu}$ and cannot be ordered in respect of values.

(III) **Case of Overlapped Intervals:** Let $\widetilde{\mu} = [(\ell_{i(L)}, \alpha_{i(L)}), (\ell_{i(U)}, \alpha_{i(U)})]$, $\widetilde{v} = [(\ell_{j(L)}, \alpha_{j(L)}), (\ell_{j(U)}, \alpha_{j(U)})]$ be two overlapping IVTFL variables such that

$$(\ell_{i(L)}, \alpha_{i(L)}) \leq (\ell_{j(L)}, \alpha_{j(L)}) < (\ell_{i(U)}, \alpha_{i(U)}) \leq (\ell_{j(U)}, \alpha_{j(U)}),$$

then $\widetilde{\mu} < \widetilde{v}$.

For instance, consider the predefined linguistic term set LT = $\{\ell_{-2} :$ Very Bad (VB), $\ell_{-1} :$ Bad (B), $\ell_0 :$ Medium (M), $\ell_1 :$ Good (G), $\ell_2 :$ Very Good (VG)$\}$. Suppose $S = \{\widetilde{\mu}_1 = [(\ell_{-2}, 0), (\ell_0, 0)], \widetilde{\mu}_2 = [(\ell_{-2}, 0.8), (\ell_{-1}, 0.23)], \widetilde{\mu}_3 = [(\ell_0, 0.05), (\ell_2, -0.5)], \widetilde{\mu}_4 = [(\ell_{-1}, 0), (\ell_1, 0)]\}$ be a set of IVTFL variables using the predefined linguistic term set LT. Here, μ_1 and μ_2 are nested linguistic intervals whereas interval μ_1 is disjoint with μ_3 and overlapping with μ_4, comparing which we obtain that $\mu_1 < \mu_3$, $\mu_1 < \mu_4$ but μ_1 and μ_2 can not be compared. Only the inclusion property can be discussed i.e. $\mu_2 \subset \mu_1$. On the similar grounds, the other pair of intervals can be compared.

In literature, Singh et al. [3] adopted the matrices formulated using lower bounds and upper bounds of the payoff intervals to define interval-valued linguistic (IVL) value of the game. However, the authors suggested linguistic linear programming approach to solve interval-valued linguistic matrix game in case of mixed strategies. Unlike the existing solution scheme, here in this work, using the comparison of linguistic intervals defined in the preceding section, the value of the interval fuzzy linguistic game is defined in the light of min-max principle and subsequently, interval linguistic linear programming problem approach is proposed to solve such games.

4   A Zero-Sum Interval-Valued Linguistic Matrix Game

Definition 9. A two-player zero-sum interval-valued linguistic matrix game $\widetilde{G}_{Int}$ is characterized by a quadruplet $(S^n, S^m, LT, \widetilde{A}_{Int})$, where S^n, S^m are strategy sets for player I and II respectively and LT = $\{\ell_{-g}, \ldots, \ell_0, \ldots, \ell_g\}$ is the predefined subscript-symmetric linguistic term set. The matrix $\widetilde{A}_{Int} = \left([\widetilde{a}_{ij}^{(L)}, \widetilde{a}_{ij}^{(U)}]\right)_{n \times m}$; $i = 1, \ldots n$, $j = 1, \ldots, m$ is the interval-valued linguistic payoff matrix of player I in defiance of player II whereas $neg\widetilde{A}_{Int} = \left([neg(\widetilde{a}_{ij}^{(U)}), neg(\widetilde{a}_{ij}^{(L)})]\right)_{n \times m}$ depicts the payoff matrix of player II such that the payoffs of two players sum up to $(\ell_0, 0)$.

Since the comparison of IVTFL variables are proposed in the preceding section, the IVL value of the game can be defined as follows.

Definition 10. For a given IVL matrix game $\widetilde{G}_{Int}$ with payoff matrix $\widetilde{A}_{Int}$, the interval-valued linguistic lower value, $\widetilde{v}_{Int}^{-}$ and interval-valued linguistic upper value, $\widetilde{v}_{Int}^{+}$ of the game is defined as,

$$\widetilde{v}_{Int}^{-} = \max_{i=1,\ldots,n} \min_{j=1,\ldots,m} [\widetilde{a}_{ij}^{(L)}, \widetilde{a}_{ij}^{(U)}],$$

$$\widetilde{v}_{Int}^{+} = \min_{j=1,\ldots,m} \max_{i=1,\ldots,n} [\widetilde{a}_{ij}^{(L)}, \widetilde{a}_{ij}^{(U)}].$$

The IVL value, $\widetilde{v}_{Int}$ of the game exists when $\widetilde{v}_{Int}^{-} = \widetilde{v}_{Int}^{+} = \widetilde{v}_{Int}$.

The strategy set (i^*, j^*) for which these values are equal is called the saddle point of the game and i^*, j^* are optimal strategies of players I and II respectively.

For any IVL matrix game, the following inequality holds.

Theorem 1. Suppose $\widetilde{v}_{Int}^{-} = [\widetilde{v}^{-(L)}, \widetilde{v}^{-(U)}]$ and $\widetilde{v}_{Int}^{+} = [\widetilde{v}^{+(L)}, \widetilde{v}^{+(U)}]$ be the interval-valued linguistic lower and upper values of an interval linguistic matrix game $\widetilde{G}_{Int}$ such that both values exist. Then, $\widetilde{v}_{Int}^{-} \leq \widetilde{v}_{Int}^{+}$.

Proof. We are given that $\widetilde{v}_{Int}^{-}$ and $\widetilde{v}_{Int}^{+}$ both exist, so for some column j and fixed row i, we have,

$$\min_{j=1,\ldots,m} [\widetilde{a}_{ij}^{(L)}, \widetilde{a}_{ij}^{(U)}] \leq [\widetilde{a}_{ij}^{(L)}, \widetilde{a}_{ij}^{(U)}],$$

By taking max over $i = 1, \ldots, n$ on both sides, we obtain,

$$\widetilde{v}_{Int}^{-} \equiv \max_{i=1,\ldots,n} \min_{j=1,\ldots,m} [\widetilde{a}_{ij}^{(L)}, \widetilde{a}_{ij}^{(U)}] \leq \max_{i=1,\ldots,n} [\widetilde{a}_{ij}^{(L)}, \widetilde{a}_{ij}^{(U)}]$$

$$\Rightarrow \widetilde{v}_{Int}^{-} \leq \max_{i=1,\ldots,n} [\widetilde{a}_{ij}^{(L)}, \widetilde{a}_{ij}^{(U)}].$$

Since the above inequality holds for any j. Hence, we obtain the following inequality.

$$\widetilde{v}_{Int}^{-} \leq \min_{j=1,\ldots,m} \max_{i=1,\ldots,n} [\widetilde{a}_{ij}^{(L)}, \widetilde{a}_{ij}^{(U)}]$$

Hence, $\widetilde{v}_{Int}^{-} \leq \widetilde{v}_{Int}^{+}$.

Now, we exemplify the above theory using an illustration.

Example 1. Two firms need to introduce a number of essentially equivalent new products. In the next two months, the companies are planning to launch the products. The payoffs are the companies' share, which it will acquire taking into account the months during which production takes place. The payoffs of the companies appear in the form of IVTFL variables from the set of predefined linguistic terms, LT = $\{\ell_{-2} :$

Very Low (VL), ℓ_{-1} : Low (L), ℓ_0 : Fair (F), ℓ_1 : Good (G), ℓ_2 : Very Good (VG).}. The interval linguistic payoff matrix for player I is given as.

$$\tilde{A} = \begin{bmatrix} VL & [(VL,0.2),(L,0.4561)] \\ [(G,0.4),VG] & G \end{bmatrix}$$

Here,

$$\widetilde{v_{Int}^-} = \max_{i=1,2} \min_{j=1,2} \left\{ [\tilde{a}_{ij}^{(L)}, \tilde{a}_{ij}^{(U)}] \right\}$$

$$= \max\{\min\{VL, [(VL,0.2),(L,0.4561)]\}, \min\{[(G,0.4),VG],G\}\}$$

$$= \max\{VL, G\} = G.$$

Also, $\widetilde{v_{Int}^+} = \min_{j=1,2} \max_{i=1,2} \left\{ [\tilde{a}_{ij}^{(L)}, \tilde{a}_{ij}^{(U)}] \right\}$

$$= \min\{\max\{VL, [(G,0.4),VG)]\}, \max\{[(VL,0.2),(L,0.4561)],G\}\}$$

$$= \min\{[(G,0.4),VG],G\} = G.$$

Here, $(2,2)$ is the saddle point and $\widetilde{v_{Int}} = G$ is the interval-valued linguistic value of the matrix games. This shows that in order to maximize the profit both the firms should launch their products in the second month simultaneously.

In the above example, if we replace the entry $[\tilde{a}_{11}^{(L)}, \tilde{a}_{11}^{(U)}]$ as $[F,G]$ and $[\tilde{a}_{21}^{(L)}, \tilde{a}_{21}^{(U)}]$ as VL, then $\widetilde{v_{Int}^-} = [(VL,0.2),(L,0.4561)]$, and $\widetilde{v_{Int}^+} = [F,G]$, it depicts the absence of pure strategies. The validity of Theorem 1 can also be deduced from the example as in case of pure strategy, the equality holds whereas in another case, $\widetilde{v_{Int}^-} < \widetilde{v_{Int}^+}$.

To evaluate the strategy sets and optimal value of the game in absence of pure strategy, here we define the interval-valued linguistic linear programming approach to solve such games.

5   Interval-Valued Linguistic Linear Programming Approach to Solve Interval Linguistic Matrix Games

Suppose, we have the interval linguistic payoff matrix $\tilde{A}_{Int}$ using the predefined linguistic term set $LT = \{\ell_{-g}, \ldots, \ell_0, \ldots, \ell_g\}$ as follows.

$$\tilde{A}_{Int} = \begin{pmatrix} [\tilde{a}_{11}^{(L)}, \tilde{a}_{11}^{(U)}] & [\tilde{a}_{12}^{(L)}, \tilde{a}_{12}^{(U)}] & \cdots & [\tilde{a}_{1m}^{(L)}, \tilde{a}_{1m}^{(U)}] \\ [\tilde{a}_{21}^{(L)}, \tilde{a}_{21}^{(U)}] & [\tilde{a}_{22}^{(L)}, \tilde{a}_{22}^{(U)}] & \cdots & [\tilde{a}_{2m}^{(L)}, \tilde{a}_{2m}^{(U)}] \\ \vdots & \ddots & \vdots & \vdots \\ [\tilde{a}_{n1}^{(L)}, \tilde{a}_{n1}^{(U)}] & [\tilde{a}_{n2}^{(L)}, \tilde{a}_{n2}^{(U)}] & \cdots & [\tilde{a}_{nm}^{(L)}, \tilde{a}_{nm}^{(U)}] \end{pmatrix}$$

where $[\tilde{a}_{ij}^{(L)}, \tilde{a}_{ij}^{(U)}]$ is the payoff of player I on selecting i^{th} strategy when player II selects the jth strategy.

Here, we may assume that each entries of the interval linguistic payoff matrix is either $[\tilde{a}_{ij}^{(L)}, \tilde{a}_{ij}^{(U)}] < 0$ or $[\tilde{a}_{ij}^{(L)}, \tilde{a}_{ij}^{(U)}] > 0$. Let $S^n = \{\mathbf{X} = (x_1, x_2, \ldots, x_n) \mid x_i \geq 0, \sum_{i=1}^{n} x_i = 1\}$ and $S^m = \{\mathbf{Y} = (y_1, y_2, \ldots, y_m) \mid y_j \geq 0, \sum_{j=1}^{m} y_j = 1\}$ be the mixed strategy set for player I and II, respectively. Then, the expected payoff of player I when player II selects j^{th} strategy, is taken as the weighted average of the interval-valued linguistic variables in the jth column i.e. $[\tilde{a}_{1j}^{(L)}, \tilde{a}_{1j}^{(U)}]x_1 \oplus \ldots \oplus [\tilde{a}_{nj}^{(L)}, \tilde{a}_{nj}^{(U)}]x_n$.

Hence, the required IVLLP problem for player I is given as.

$$\max \quad \widetilde{v_{Int}} \qquad\qquad\qquad \text{(IVLLP1)}$$

subject to

$$[\tilde{a}_{11}^{(L)}, \tilde{a}_{11}^{(U)}]x_1 \oplus \ldots \oplus [\tilde{a}_{n1}^{(L)}, \tilde{a}_{n1}^{(U)}]x_n \geq \widetilde{v_{Int}}$$

$$[\tilde{a}_{12}^{(L)}, \tilde{a}_{12}^{(U)}]x_1 \oplus \ldots \oplus [\tilde{a}_{n2}^{(L)}, \tilde{a}_{n2}^{(U)}]x_n \geq \widetilde{v_{Int}}$$

$$\vdots$$

$$[\tilde{a}_{1m}^{(L)}, \tilde{a}_{1m}^{(U)}]x_1 \oplus \ldots \oplus [\tilde{a}_{nm}^{(L)}, \tilde{a}_{nm}^{(U)}]x_n \geq \widetilde{v_{Int}}$$

$$x_1 + x_2 + \ldots + x_n = 1$$

$$x_1, x_2, \ldots, x_n \geq 0.$$

Using the monotonicity of Δ^{-1} operator, the inequality constraints of above IVLLP model can be rewritten as follows,

$$\Delta^{-1}([\tilde{a}_{11}^{(L)}, \tilde{a}_{11}^{(U)}])x_1 \oplus \ldots \oplus \Delta^{-1}([\tilde{a}_{n1}^{(L)}, \tilde{a}_{n1}^{(U)}])x_n \geq \Delta^{-1}(\widetilde{v_{Int}})$$

$$\Delta^{-1}([\tilde{a}_{12}^{(L)}, \tilde{a}_{12}^{(U)}])x_1 \oplus \ldots \oplus \Delta^{-1}([\tilde{a}_{n2}^{(L)}, \tilde{a}_{n2}^{(U)}])x_n \geq \Delta^{-1}(\widetilde{v_{Int}})$$

$$\vdots$$

$$\Delta^{-1}([\tilde{a}_{1m}^{(L)}, \tilde{a}_{1m}^{(U)}])x_1 \oplus \ldots \oplus \Delta^{-1}([\tilde{a}_{nm}^{(L)}, \tilde{a}_{nm}^{(U)}])x_n \geq \Delta^{-1}(\widetilde{v_{Int}})$$

and the objective function $\max \widetilde{v_{Int}} \equiv \max \Delta^{-1}(\widetilde{v_{Int}})$.

By taking $\Delta^{-1}([\tilde{a}_{ij}^{(L)}, \tilde{a}_{ij}^{(U)}]) = [a_{ij}^{(L)}, a_{ij}^{(U)}]$, $i = 1, \ldots, n$, $j = 1, \ldots, m$, and $\Delta^{-1}(\widetilde{v_{Int}}) = v_{Int}$, the constraints of model IVLLP1 is given as.

$$[a_{11}^{(L)}, a_{11}^{(U)}]x_1 + \ldots + [a_{n1}^{(L)}, a_{n1}^{(U)}]x_n \geq v_{Int}$$

$$[a_{12}^{(L)}, a_{12}^{(U)}]x_1 + \ldots + [a_{n2}^{(L)}, a_{n2}^{(U)}]x_n \geq v_{Int}$$

$$\vdots$$

$$[a_{1m}^{(L)}, a_{1m}^{(U)}]x_1 + \ldots + [a_{nm}^{(L)}, a_{nm}^{(U)}]x_n \geq v_{Int}$$

$$x_1 + x_2 + \ldots + x_n = 1$$

$$x_1, x_2, \ldots, x_n \geq 0.$$

Here, we assume that $0 \notin \Delta^{-1}(\widetilde{v}_{Int}^-) = v_{Int}$.

We set, $X_i = \dfrac{x_i}{v_{Int}}$, $i = 1,\ldots n$, and $V_{Int} = \dfrac{1}{v_{Int}} = \left[\dfrac{1}{v^{(U)}}, \dfrac{1}{v^{(L)}}\right]$. Hence, by making the above substitutions, the model IVLLP1 is transformed into standard ILP problem for player I, given below.

$$\min V_{Int} = \quad X_1 + X_2 + \ldots + X_n \tag{ILP1}$$

subject to

$$[a_{11}^{(L)}, a_{11}^{(U)}]X_1 + \ldots + [a_{n1}^{(L)}, a_{n1}^{(U)}]X_n \geq [1,1]$$

$$[a_{12}^{(L)}, a_{12}^{(U)}]X_1 + \ldots + [a_{n2}^{(L)}, a_{n2}^{(U)}]X_n \geq [1,1]$$

$$\vdots$$

$$[a_{1m}^{(L)}, a_{1m}^{(U)}]X_1 + \ldots + [a_{nm}^{(L)}, a_{nm}^{(U)}]X_n \geq [1,1]$$

$$X_1, X_2, \ldots, X_n \geq 0.$$

Analogously, we can formulate an IVLLP problem for player II.

$$\min \widetilde{v}_{Int}^+ \tag{IVLLP2}$$

subject to

$$[\widetilde{a}_{11}^{(L)}, \widetilde{a}_{11}^{(U)}]y_1 \oplus \ldots \oplus [\widetilde{a}_{1m}^{(L)}, \widetilde{a}_{1m}^{(U)}]y_m \leq \widetilde{v}_{Int}^+$$

$$[\widetilde{a}_{21}^{(L)}, \widetilde{a}_{21}^{(U)}]y_1 \oplus \oplus \ldots \oplus [\widetilde{a}_{2m}^{(L)}, \widetilde{a}_{2m}^{(U)}]y_m \leq \widetilde{v}_{Int}^+$$

$$\vdots$$

$$[\widetilde{a}_{n1}^{(L)}, \widetilde{a}_{n1}^{(U)}]y_1 \oplus \ldots \oplus [\widetilde{a}_{nm}^{(L)}, \widetilde{a}_{nm}^{(U)}]y_m \leq \widetilde{v}_{Int}^+$$

$$y_1 + y_2 + \ldots + y_m = 1$$

$$y_1, y_2, \ldots, y_m \geq 0.$$

Recall $0 \notin v_{Int} = \Delta^{-1}(\widetilde{v}_{Int}^-)$. If v_{Int} is the value of the interval linguistic game then $v_{Int} = \Delta^{-1}(\widetilde{v}_{Int}^+)$.

By taking $Y_j = \dfrac{y_j}{v_{Int}}$, $j = 1,\ldots,m$, and as earlier we discussed that $\Delta^{-1}([\widetilde{a}_{ij}^{(L)}, \widetilde{a}_{ij}^{(U)}]) = [a_{ij}^{(L)}, a_{ij}^{(U)}]$, the corresponding model IVLLP2 reduces to the following standard ILP problem for player II.

$$\max V_{Int} = \quad Y_1 + Y_2 + \ldots + Y_m \tag{ILP2}$$

subject to

$$[a_{11}^{(L)}, a_{11}^{(U)}]Y_1 \ldots + [a_{1m}^{(L)}, a_{1m}^{(U)}]Y_m \leq [1,1]$$

$$[a_{21}^{(L)}, a_{21}^{(U)}]Y_1 + \ldots + [a_{2m}^{(L)}, a_{2m}^{(U)}]Y_m \leq [1,1]$$

$$\vdots$$

$$[a_{n1}^{(L)}, a_{n1}^{(U)}]Y_1 + \ldots + [a_{nm}^{(L)}, a_{nm}^{(U)}]Y_m \leq [1,1]$$

$$Y_1, Y_2, \ldots, Y_m \geq 0.$$

Here, models ILP1 and ILP2 can be solved using any existing methods for solving interval linear programming problems to obtain the optimal mixed strategies $\mathbf{X}_{Int}^* \in S^n$ and $\mathbf{Y}_{Int}^* \in S^m$ along with the interval linguistic value of the game V_{Int}^*. It is also noteworthy that both models ILP1 and ILP2 form a primal-dual interval linear programs in the crisp set-up.

Example 2. Consider the zero-sum IVTFL matrix game with interval payoffs defined from the predefined subscript symmetric linguistic term set $LT = \{\ell_{-3} : \text{Very Low}(VL),$ $\ell_{-2} : \text{Low}(L), \ell_{-1} : \text{Moderately Low}(ML), \ell_0 : \text{Average}(Avg), \ell_1 : \text{Moderately High}$ $(MH), \ell_2 : \text{High}(H), \ell_3 : \text{Very High}(VH)\}$ with payoff matrix,

$$\tilde{A}_{Int} = \begin{pmatrix} [(VL,0.2),(L,0)] & [(H,-0.2),(VH,-0.3)] & [(MH,0.3),(H,0.2)] & [(ML,-0.13),(Avg,0)] \\ [(VH,0),(VH,0)] & [(H,0),(VH,-0.2)] & [(L,-0.4),(ML,0)] & [(MH,0),(II,-0.2)] \\ [(MH,-0.28),(MH,-0.28)] & [(L,0),(ML,0)] & [(VH,0),(VH,0)] & [(VH,0),(VH,0)] \end{pmatrix}$$

Let player I's mixed strategies be given as $\mathbf{x}^{(L)} = (x_1^{(L)}, x_2^{(L)}, x_3^{(L)})$, $x_i^{(L)} \geq 0$, $i = 1,\ldots,3$, $\sum_{i=1}^3 x_i^{(L)} = 1$, and $\mathbf{x}^{(U)} = (x_1^{(U)}, x_2^{(U)}, x_3^{(U)})$, $x_i^{(U)} \geq 0$, $i = 1,\ldots,3$, $\sum_{i=1}^3 x_i^{(U)} = 1$ for the given interval payoff matrix, $\tilde{A}_{Int}$. Additionally, player II's mixed strategies are defined as $\mathbf{y}^{(L)} = (y_1^{(L)}, y_2^{(L)}, y_3^{(L)}, y_4^{(L)})$, $y_j^{(L)} \geq 0$, $j = 1,\ldots,4$, $\sum_{j=1}^4 y_j^{(L)} = 1$, and $\mathbf{y}^{(U)} = (y_1^{(U)}, y_2^{(U)}, y_3^{(U)}, y_4^{(U)}, y_5^{(U)})$, $y_j^{(U)} \geq 0$, $j = 1,\ldots,5$, $\sum_{j=1}^5 y_j^{(U)} = 1$.

In view of the proposed methodology, we formulate models (IVLLP1 and IVLLP2) that is further converted into standard ILP problems to obtain optimal strategy set for player I and II, respectively.

Method 1: Best-Worst Method

For Player I:

Best-sub model	Worst-sub model
min $V^{-(U)} = X_1 + X_2 + X_3$ subject to $-2.8X_1 + 3X_2 + 2.4X_3 \geq 1,$ $1.8X_1 + 2X_2 - 2X_3 \geq 1,$ $1.3X_1 - 2.4X_2 + 3X_3 \geq 1,$ $-1.13X_1 + X_2 + 3X_3 \geq 1,$ $X_1, X_2, X_3 \geq 0.$	min $V^{-(L)} = X_1 + X_2 + X_3$ subject to $-2X_1 + 3X_2 + 3X_3 \geq 1,$ $2.7X_1 + 2.8X_2 - X_3 \geq 1,$ $2.2X_1 - X_2 + 3X_3 \geq 1,$ $0X_1 + 1.8X_2 + 3X_3 \geq 1,$ $X_1, X_2, X_3 \geq 0.$

Solving these two problems, we obtain the optimal strategy of player I as $x_1 = [0.3146, 0.3685]$, $x_2 = [0.3149, 0.3289]$, $x_3 = [0.3149, 0.3575]$ with interval-valued linguistic lower value of the game given as, $v_{Int}^- = [(\ell_1, -0.33), (\ell_1, 0.43)]$.

For Player II:

Best-sub model	Worst-sub model
max $\quad V^{+(U)} = Y_1 + Y_2 + Y_3 + Y_4$ subject to $-2.8Y_1 + 1.8Y_2 + 1.3Y_3 - 1.13Y_4 \leq 1,$ $3Y_1 + 2Y_2 - 2.4Y_3 + Y_4 \leq 1,$ $2.4Y_1 - 2Y_2 + 3Y_3 + 3Y_4 \leq 1,$ $Y_1, Y_2, Y_3, Y_4 \geq 0.$	max $\quad V^{+(L)} = Y_1 + Y_2 + Y_3 + Y_4$ subject to $-2Y_1 + 2.7Y_2 + 2.2Y_3 + 0Y_4 \leq 1,$ $3Y_1 + 2.8Y_2 - 1Y_3 + 1.8Y_4 \leq 1,$ $3Y_1 - 1Y_2 + 3Y_3 + 3Y_4 \leq 1,$ $Y_1, Y_2, Y_3, Y_4 \geq 0.$

For player II, the optimal strategy set is $y_1 = [0.2077, 0.2288]$, $y_2 = [0.4004, 0.4422]$, $y_3 = [0.3484, 0.3718]$, $y_4 = 0$ with interval-valued linguistic upper value of the game, $v_{Int}^+ = [(\ell_1, -0.33), (\ell_1, 0.43)]$.

Method 2: Enhanced Interval-Valued Linear Programming Method

For Player I:

Sub-problem I	Sub-problem II
min $\quad V^{-(U)} = X_1^U + X_2^U + X_3^U$ subject to $-2X_1^U + 3X_2^U + 2.4X_3^U \geq 1,$ $1.8X_1^U + 2X_2^U - X_3^U \geq 1,$ $1.3X_1^U - X_2^U + 3X_3^U \geq 1,$ $0X_1^U + X_2^U + 3X_3^U \geq 1,$ $X_1^U, X_2^U, X_3^U \geq 0.$	min $\quad V^{-(L)} = X_1^L + X_2^L + X_3^L$ subject to $-2.8X_1^L + 3X_2^L + 3X_3^L \geq 1,$ $2.7X_1^L + 2.8X_2^L - 2X_3^L \geq 1,$ $2.2X_1^L - 2.4X_2^L + 3X_3^L \geq 1,$ $-1.13X_1^L + 1.8X_2^L + 3X_3^L \geq 1,$ $X_1^L, X_2^L, X_3^L \geq 0.$

For Player II:

Sub-problem I	Sub-problem II
max $\quad V^{+(U)} = Y_1^U + Y_2^U + Y_3^U + Y_4^U$ subject to $-2Y_1^U + 1.8Y_2^U + 1.3Y_3^U \leq 1,$ $3Y_1^U + 2Y_2^U - Y_3^U + Y_4^U \leq 1,$ $2.4Y_1^U - Y_2^U + 3Y_3^L + 3Y_4^L \leq 1,$ $Y_1^U, Y_2^U, Y_3^U, Y_4^U \geq 0.$	max $\quad V^{+(L)} = Y_1^L + Y_2^L + Y_3^L + Y_4^L$ subject to $-2.8Y_1^L + 2.7Y_2^L + 2.2Y_3^L - 1.13Y_4^L \leq 1,$ $3Y_1^L + 2.8Y_2^L - 2.4Y_3^L - 1.8Y_4^L \leq 1,$ $3Y_1^L - 2Y_2^L + 3Y_3^L + 3Y_4^U \leq 1,$ $Y_1^L, Y_2^L, Y_3^L, Y_4^L \geq 0.$

Solving the above models, the optimal strategies of player I and II are evaluated as $x_1 = [0.3008, 0.3395]$, $x_2 = [0.2726, 0.3201]$, $x_3 = [0.2813, 0.3196]$ and

$y_1 = [0.1455, 0.235]$, $y_2 = [0.3478, 0.4462]$, $y_3 = [0.3102, 0.3589]$, $y_4 = 0$ respectively with $v_{Int} = [(\ell_1, -0.06), (\ell_1, -0.03)]$.

Method 3: Linguistic Linear Programming (LLP) Method

We split our matrix $\widetilde{A}_{Int}$ into linguistic lower matrix and linguistic upper matrix to obtain interval linguistic lower and upper values of the interval linguistic matrix game. The mathematical formulation for this problem is similar to that of BWC.

The optimal strategies of both players and value of the game using various existing methodologies to solve ILP problems, are tabulated below.

For Player I:

Method	$\tilde{x}_1$	$\tilde{x}_2$	$\tilde{x}_3$	Optimal value
BWC	$[0.3146, 0.3685]$	$[0.3149, 0.3289]$	$[0.3149, 0.3575]$	$[(\ell_1, -0.33), (\ell_1, 0.43)]$
EILP	$[0.3008, 0.3395]$	$[0.2726, 0.3201]$	$[0.2813, 0.3196]$	$[(\ell_1, -0.06), (\ell_1, -0.03)]$
LLP	$[0.3146, 0.3685]$	$[0.3149, 0.3289]$	$[0.3149, 0.3575]$	$[(\ell_1, -0.33), (\ell_1, 0.43)]$

For Player II:

Method	$\tilde{y}_1$	$\tilde{y}_2$	$\tilde{y}_3$	$\tilde{y}_4$	Optimal value
BWC	$[0.2077, 0.2288]$	$[0.4004, 0.4422]$	$[0.3484, 0.3718]$	0	$[(\ell_1, -0.33), (\ell_1, 0.43)]$
EILP	$[0.1455, 0.235]$	$[0.3478, 0.4462]$	$[0.3102, 0.3589]$	0	$[(\ell_1, -0.06), (\ell_1, -0.03)]$
LLP	$[0.2077, 0.2288]$	$[0.4004, 0.4422]$	$[0.3484, 0.3718]$	0	$[(\ell_1, -0.33), (\ell_1, 0.43)]$

Here, the solution region obtained using EILP method is completely optimal and feasible. However, BWC and LLP approach provides a solution region which is completely optimal but may not be feasible. This is because it incorporates some infeasible points within the solution set.

6   Conclusion

In this paper, we have studied the 2-player zero sum interval-valued linguistic matrix game problems. We proposed a new methodology for comparing two IVTFL variables and subsequently, put forward the concept of max-min principle for defining the lower and upper value of the interval linguistic game problem. However, in the absence of pure strategies, we designed a new approach for evaluating the optimal strategies and value of the game. We envision that the proposed method can easily be applied to large scale interval linguistic game problems, manufacturing companies, large scale decision-making problems where the existing players (or decision makers) have conflicting objectives.

Acknowledgment. This work was financially supported by Delhi Technological University Ref. No. DTU/IRD/619/2019/2107 and Ref. No. DTU/Maths/387/2019-20/2108.

References

1. Herrera, F., Martinez, L.: A 2-tuple fuzzy linguistic representation model for computing with words. IEEE Trans. Fuzzy Syst. **8**(6), 746–752 (2000)
2. Singh, A., Gupta, A., Mehra, A.: Matrix games with 2-tuple linguistic information. Ann. Oper. Res. (2018). https://doi.org/10.1007/s10479-018-2810-6
3. Singh, A., Gupta, A.: Matrix games with interval-valued 2-tuple linguistic information. Games **9**(3), 62 (2018)
4. Singh, A., Gupta, A., Mehra, A.: An AHP-PROMETHEE II method for 2-tuple linguistic multicriteria group decision making. In: 2015 4th International Conference on Reliability, Infocom Technologies and Optimization (ICRITO) (Trends and Future Directions), pp. 1–6. IEEE (2015). https://doi.org/10.1109/ICRITO.2015.7359374
5. Xu, Z.: Deviation measures of linguistic preference relations in group decision making. Int. Omega **33**(3), 249–254 (2005)
6. Von Neumann, J., Morgenstern, O.: Theory of Games and Economic Behavior, 2nd edn. Princeton University Press, Princeton (1947)
7. Zhang, H.: The multiattribute group decision making method based on aggregation operators with interval-valued 2-tuple linguistic information. Math. Comput. Model. **56**(1–2), 27–35 (2012)
8. Moore, R.E.: Interval Analysis, vol. 4. Prentice-Hall, Englewood Cliffs (1966)
9. Moore, R.E.: Methods and Applications of Interval Analysis. Society for Industrial and Applied Mathematics, Philadelphia (1979)
10. Sengupta, A., Pal, T.K.: A-index for ordering interval numbers. In: Indian Science Congress, pp. 3–8 (1997)
11. Ishibuchi, H., Tanaka, H.: Multiobjective programming in optimization of the interval objective function. Eur. J. Oper. Res. **48**(2), 219–225 (1990)
12. Wu, H.C.: On interval-valued nonlinear programming problems. J. Math. Anal. Appl. **338**(1), 299–316 (2008)
13. Dubois, D., Prade, H.: Ranking fuzzy numbers in the setting of possibility theory. Inf. Sci. **30**(3), 183–224 (1983)
14. Chanas, S., Kuchta, D.: Multiobjective programming in optimization of interval objective functions - a generalized approach. Eur. J. Optional Res. **94**(3), 594–598 (1996)
15. Bodenhofer, U.: Orderings of fuzzy sets based on fuzzy orderings part I: the basic approach. Mathw. Soft Comput. **15**(2), 201–218 (2008)
16. Bodenhofer, U.: Orderings of fuzzy sets based on fuzzy orderings part II: generalizations. Mathw. Soft Comput. **15**(3), 219–249 (2008)
17. Sakawa, M., Nishizaki, I.: Max-min solutions for fuzzy multiobjective matrix games. Fuzzy Sets Syst. **67**(1), 53–69 (1994)
18. Campos, L.: Fuzzy linear programming models to solve fuzzy matrix games. Fuzzy Sets Syst. **32**(3), 275–289 (1989)
19. Bector, C.R., Chandra, S.: Fuzzy Mathematical Programming and Fuzzy Matrix Games, vol. 169. Springer, Berlin (2005). https://doi.org/10.1007/3-540-32371-6
20. Allahdadi, M., Nehi, H.M.: The optimal solution set of the interval linear programming problems. Optim. Lett. **7**(8), 1893–1911 (2013)
21. Avşar, Z.M., Baykal-Gürsoy, M.: A note on two-person zero-sum communicating stochastic games. Oper. Res. Lett. **34**(4), 412–420 (2006)
22. Kurano, M., Yasuda, M., Nakagami, J.I., Yoshida, Y.: An interval matrix game and its extensions to fuzzy and stochastic games (2002). http://www.math.s.chiba-u.ac.jp/~yasuda/accept/SIG2.pdf. Accessed 5 June 2018
23. Arfi, B.: Linguistic fuzzy-logic game theory. J. Confl. Resolut. **50**(1), 28–57 (2006)

24. Arfi, B.: Linguistic fuzzy-logic social game of cooperation. Rat. Soc. **18**(4), 471–537 (2006)
25. Zhou, F., Huang, G.H., Chen, G.X., Guo, H.C.: Enhanced-interval linear programming. Eur. J. Oper. Res. **199**(2), 323–333 (2009)
26. Barron, E.N.: Game Theory: An Introduction, vol. 2. Wiley, Hoboken (2013)

Set-Membership Computation
of Integrals with Uncertain Endpoints

Olivier Mullier[(✉)][iD] and Julien Alexandre dit Sandretto[iD]

ENSTA Paris, 828 bd des maréchaux, 91120 Palaiseau, France
{mullier,alexandre}@ensta.fr

Abstract. An efficient guaranteed method for the computation of the integral of a nonlinear continuous function between two interval endpoints is proposed. This computation can be of interest for the computation of global optimization problems where such integrals occur like in robotics. The method results in the computation of the minimum and maximum of these integrals and provides the endpoints at stake. The complexity of the resulting algorithms is discussed, it depends on the number of roots of the function to be integrated. The computation is illustrated on several examples.

Keywords: Integral · Set-membership computation · Interval methods

1 Introduction

Numerical integration is one of the fundamental tools of scientific computation. It occurs in many domains and providing a reliable result to such problem is important. We can cite, for example, the computation of a validated simulation of ODEs [2,9] or for global optimization with continuous objective function [6]. It has applications in robotics as well [3,12].

The numerical computation of integrals has been intensively studied and the same happened for a validated computation. In [1], the validated computation of an integral where both endpoints are reals is considered using quadrature (see also [5]). In [11], the problem is tackled for piece-wise analytic functions.

For integrals where uncertainties happen in the endpoints defining them or if one wants to produce a set of possible integrals for which the endpoints take their value in an interval, there is fewer studies. We can nonetheless cite [4]. In there, the notion of integrals with intervals is defined and a formulation of the problem when the two interval endpoints are disjoint is given.

A simple algorithm in the general case remains to be defined, and it is the purpose of the present work. The subject of this work will not be to treat directly the guaranteed numerical computation of integrals where the endpoints are real even if it is mandatory when extending to set of endpoints. Indeed, a study of the general case with interval endpoints provides a decomposition of the problem that make the previous case appearing. Then any method for this matter can be

Y. D. Sergeyev and D. E. Kvasov (Eds.): NUMTA 2019, LNCS 11974, pp. 169–181, 2020.
https://doi.org/10.1007/978-3-030-40616-5_13

used in order to produce eventually the computation of an integral with unknown endpoints. We already mentioned (linear) quadrature and the interested reader to this particular point can read, for example, [10].

Our work makes use of interval analysis and the analysis of the roots of the function we integrate. It then requires the function to be analytic in the general case or make use of the computation when endpoints are known exactly to detect the parts where the integral is positive and the ones where it is negative. Our work is presented as follows: the next section provides the mathematical background for the understanding of this work, Sect. 3 is dedicated to the main result that is the computation of an integral with interval endpoints, some experiments make possible discussion on the presented method and the complexity of its associated algorithms in Sect. 4 before we conclude.

2   The Interval Integrals

In this Section, we recall the notions on interval analysis required in the following of the article and introduce the *interval integral*, an integral with interval endpoints.

2.1   Interval Analysis

Interval analysis [8] is well suited when dealing with computation involving sets of values or when handling uncertainties. Its goal is to produce an outer-approximation of a desired computation in a sound manner. We denotes hereafter an interval with brackets: $[x] = [\underline{x}, \overline{x}]$ with $\underline{x} \leqslant \overline{x}$ the lower and upper bounds of the interval. Any interval lies in the set of intervals $\mathbb{IR} = \{[x] = [\underline{x}, \overline{x}] \mid \underline{x}, \overline{x} \in \mathbb{R}, \ \underline{x} \leqslant \overline{x}\}$. For higher dimensions, we deal with Cartesian product of intervals $[\mathbf{x}] \in \mathbb{IR}^n$ which are named boxes.

As depicted in the fundamental theorem of interval analysis (see [7]), the evaluation of an arithmetic expression using intervals leads to an outer-approximation of the resulting set of values for this expression whatever the values considered in the intervals. This result can be extended to functions dealing with intervals we then call interval function or interval extension of a function thenceforth they verify the fundamental theorem.

We can cite classical ways to design such interval extension like the natural extension [8] which replaces the operations on reals by their interval counterparts using interval arithmetic or the mean value extension [8] which linearizes the function around its mean value. Such interval extension can easily be designed to produce a validated computation of integrals.

2.2   Validated Computation of Integrals

A simple way to produce an interval extension of the integral of a function is to extend the composite midpoint rule to the intervals. Using

$$\int_a^b f(x)dx \in \mathrm{wid}([a, b])\, [f]\, ([a, b]) \tag{1}$$

with $[f] : \mathbb{IR} \to \mathbb{IR}$ an interval extension of the function f and $\mathrm{wid}([a, b]) = b - a$ is the width of the interval $[a, b]$, one can cut the interval $[a, b]$ into n intervals and the computation of $\int_a^b f(x)dx$ can be then

$$\int_a^b f(x)dx \in \frac{\mathrm{wid}([a, b])}{n} \sum_{k=1}^n [f]\left(\left[a + (k - 1)\frac{b - a}{n}, a + k\frac{b - a}{n}\right]\right) \quad (2)$$

2.3   Integrals with Interval Endpoints

We denote an integral with interval endpoints as an *interval integral* and it is defined as follows:

Definition 1 (Interval integral). *Let* $f : \mathbb{R} \to \mathbb{R}$, *a continuous function and* $[x_1]$, $[x_2] \in \mathbb{IR}$ *two intervals. The interval integral of* f *with* $[x_1]$ *and* $[x_2]$ *as endpoints is denoted* $\int_{[x_1]}^{[x_2]} f(x)dx$ *and corresponds to the set*

$$\int_{[x_1]}^{[x_2]} f(x)dx = \left\{ \left. \int_{x_1}^{x_2} f(x)dx \; \right| \; x_1 \in [x_1], x_2 \in [x_2] \right\}. \quad (3)$$

This set considers all the integrals with the endpoints taken in the intervals $[x_1]$ and $[x_2]$. The following property is useful to decompose the computation of an interval integral.

Property 1. For an interval $[x_1] = \left[\underline{x_1}, \overline{x_1}\right]$ and $\tilde{x}_1 \in [x_1]$, an integral interval can be subdivided as follows

$$\int_{[x_1]}^{[x_2]} f(x)dx = \int_{\left[\underline{x_1}, \tilde{x}_1\right]}^{[x_2]} f(x)dx \bigcup \int_{\left[\tilde{x}_1, \overline{x_1}\right]}^{[x_2]} f(x)dx \quad (4)$$

and the same subdivision applies for $[x_2]$.

How to handle this set to produce an outer approximation using interval analysis is discussed in the next section.

3   Main Result

When dealing with the computation of the set in Eq. (3), three cases can occur whether the intervals $[x_1]$ and $[x_2]$ are disjoint, intersect or one is included. In the following, each case is discussed.

3.1   The Interval Endpoints Are Disjoint

The first case occurs when the two interval endpoints do not intersect. In the following, we will consider, without loss of generality, that for the computation of the integral $\int_{[x_1]}^{[x_2]} f(x)dx$, the endpoints $[x_1]$ and $[x_2]$ respect the constraint $\overline{x_1} < \underline{x_2}$. If not, we just have to consider the computation of $-\int_{[x_2]}^{[x_1]}$ and the constraint still apply for $\overline{x_2} < \underline{x_1}$. Using this assumption, the considered integrals are always with the first endpoint being smaller than the second endpoint and the integral cannot go backward.

As introduced in [4], An interval integral as defined in Definition 1 where the endpoints are disjoint can be decomposed as follows

$$\int_{[x_1]}^{[x_2]} f(x)dx = \int_{[x_1]}^{\overline{x_1}} f(x)dx + \int_{\overline{x_1}}^{\underline{x_2}} f(x)dx + \int_{\underline{x_2}}^{[x_2]} f(x)dx \qquad (5)$$

where 2 subcases of interval integrals appear and a more classical integral with real endpoints. Using Eq. (5), we can compute the minimum using

$$\min \int_{[x_1]}^{[x_2]} f(x)dx = \min \int_{[x_1]}^{\overline{x_1}} f(x)dx + \int_{\overline{x_1}}^{\underline{x_2}} f(x)dx + \min \int_{\underline{x_2}}^{[x_2]} f(x)dx \qquad (6)$$

and

$$\max \int_{[x_1]}^{[x_2]} f(x)dx = \max \int_{[x_1]}^{\overline{x_1}} f(x)dx + \int_{\overline{x_1}}^{\underline{x_2}} f(x)dx + \max \int_{\underline{x_2}}^{[x_2]} f(x)dx \qquad (7)$$

for the maximum since the minimum (or maximum) operator can be here distributed. Then computing the interval integral requires the computation of each integral in the right hand side of this equation, in particular the two interval integrals occurring.

Computing $\int_{[x]}^{\overline{x}} f(x)dx$

To produce the minimum and the maximum of this integral, we have to consider the parts where sub-integrals are positive and parts where they are negative. The change between positiveness and negativeness of the integral occurs at x being a root of x (such that $f(x) = 0$). Computing the minimum and maximum then requires to produce the set of roots

$$\mathcal{X}^* = \{x \in [x] : f(x) = 0\} \qquad (8)$$

The integral has to reach the endpoint $\overline{x}$ then all candidates to be minimum and maximum are the integrals

$$\left\{ \int_{x^*}^{\overline{x}} f(x)dx \, \middle| \, x^* \in \mathcal{X}^* \right\} \qquad (9)$$

If the set of roots is finite, we end up with a finite number of candidates to minimum and maximum when the general set contains an infinite number of integrals. Figure 1 illustrates an example of this computation where the set from Eq. (9) consists in 4 candidates. The sign of the integral starting at each $x^* \in \mathcal{X}^*$ dictates if the candidate is a minimum or a maximum (in Fig. 1, candidate 2 and 4 are candidates to be the minimum and candidates 1 and 3 can only be maximum).

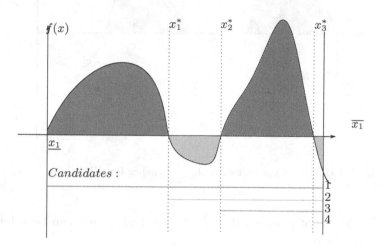

Fig. 1. Example of computation of $\int_{[\underline{x}]}^{\overline{x}} f(x)dx$ for $\mathcal{X}^* = \{x_1^*, x_2^*, x_3^*\}$ (blue: maximum; red: minimum). (Color figure online)

Computing $\int_{\underline{x}}^{[x]} f(x)dx$

In the case of the interval integral having only the endpoint as an interval, a dual method is applied from the previous example since the Second Fundamental Theorem of Calculus can be applied

$$\int_{\underline{x}}^{[x]} f(x)dx = -\int_{[x]}^{\underline{x}} f(x)dx \qquad (10)$$

and then it is the sign of integral prior to the endpoint x^* that defines the sub-integral to be candidate to minimum or maximum. Figure 2 provides an example of the computation of the minimum and the maximum.

The intervals 1 and 3 are candidates to be minimum and the intervals 2 and 4 to be maximum.

3.2   The Interval Endpoints Overlap

This case occurs when $\underline{x_1} \leqslant \underline{x_2} \leqslant \overline{x_1} \leqslant \overline{x_2}$. More interval integrals have to be considered since the interval integral can be backward with $x_1 > x_2$, $x_1 \in [x_1]$,

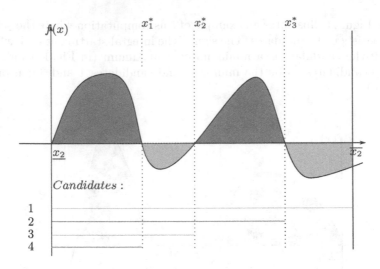

Fig. 2. Example of computation of $\int_{\underline{x_2}}^{[x_2]} f(x)dx$ for $\mathcal{X}_2^* = \{x_1^*, x_2^*, x_3^*\}$.

$x_2 \in [x_2]$. The set of integrals the interval integral defines can be subdivided (using Eq. (4)) since

$$\int_{[x_1]}^{[x_2]} f(x)dx = \int_{[\underline{x_1},\underline{x_2}]}^{[x_2]} f(x)dx \bigcup \int_{[\underline{x_2},\overline{x_1}]}^{[x_2]} f(x)dx \qquad (11)$$

$$= \int_{[\underline{x_1},\underline{x_2}]}^{[x_2]} f(x)dx \bigcup \int_{[\underline{x_2},\overline{x_1}]}^{[\underline{x_2},\overline{x_1}]} f(x)dx \bigcup \int_{[\underline{x_2},\overline{x_1}]}^{[\overline{x_1},x_2]} f(x)dx. \quad (12)$$

The first and the last interval integrals in Eq. (12) are of the same type as the one where endpoints are disjoint except that the integral can be equal to 0 when taking the same value for both endpoints.

Computing $\int_{[x]}^{[x]} f(x)dx$

We now discuss the middle interval integral occurring in the decomposition of Eq. (12). It is the case where both endpoints take their value in the same interval:

$$\int_{[x]}^{[x]} f(x)dx = \left\{ \int_{x_1}^{x_2} f(x)dx \mid x_1, x_2 \in [x] \right\}. \qquad (13)$$

The minimum and maximum will also be determined using the set of endpoints $\mathcal{X}^* = \{x \in [x] \mid f(x) = 0\}$. We have also to make the distinction between value in $\mathcal{X}^*$ for which $f'(x) > 0$ or $f'(x) < 0$. Then the computation of the minimum and maximum is stated by the following theorem.

Theorem 1. *Let* $f : \mathbb{R} \to \mathbb{R}$ *be a differentiable function. For* $[x] = [\underline{x}, \overline{x}]$ *an interval in* $\mathbb{IR}$, *we define*

$$I_{\max} = \int_{\mathcal{X}_+^* \cup \{\overline{x}\}}^{\mathcal{X}_-^* \cup \{\underline{x}\}} f(x)dx, \tag{14}$$

$$I_{\min} = \int_{\mathcal{X}_-^* \cup \{\overline{x}\}}^{\mathcal{X}_+^* \cup \{\underline{x}\}} f(x)dx \tag{15}$$

using the sets

$$\mathcal{X}_-^* = \{x \in [x] : f(x) = 0, f'(x) < 0\} \tag{16}$$
$$\mathcal{X}_+^* = \{x \in [x] : f(x) = 0, f'(x) > 0\}. \tag{17}$$

The minimum and the maximum of the set (13) *can be defined by*

$$\int_{[x]}^{[x]} f(x)dx = [-\max(-I_{\min}, I_{\max}), \max(-I_{\min}, I_{\max})] \tag{18}$$

3.3   The Starting Endpoint Is Included in the Ending Endpoint

When $[x_1] \subseteq [x_2]$, we have $\underline{x_2} \leqslant \underline{x_1} \leqslant \overline{x_1} \leqslant \overline{x_2}$ and the same decomposition as in Sect. 3.2 follows:

$$\int_{[x_1]}^{[x_2]} f(x)dx = \int_{[x_1]}^{[x_2,x_1]} f(x)dx \bigcup \int_{[x_1]}^{[x_1,\overline{x_2}]} f(x)dx \tag{19}$$

$$= \int_{[x_1]}^{[x_2,x_1]} f(x)dx \bigcup \int_{[x_1]}^{[x_1]} f(x)dx \bigcup \int_{[x_1]}^{[\overline{x_1},\overline{x_2}]} f(x)dx \tag{20}$$

so we go back to the already treated kind of interval integral that occurred in the previous cases.

Eventually, the computation of any interval integral only depends on the computation of the particular interval integrals

$$\int_{[x]}^{\overline{x}} f(x)dx \tag{21}$$

$$\int_{\underline{x}}^{[x]} f(x)dx \tag{22}$$

$$\int_{[x]}^{[x]} f(x)dx. \tag{23}$$

An implementation of the computation of those interval integrals is introduced in the next section.

4   Experiments

An implementation of the computation of the bounds of the set defined in
Eqs. (21)–(23) is introduced using interval analysis [7] and tested on several
examples.

4.1   Algorithms

We now introduce the algorithms that have been implemented in order to compute an interval outer approximation of an interval integral.

Algorithm for the Computation of $\int_{[x]}^{\overline{x}} f(x)dx$

The first algorithm treats the problem of computing $\int_{[x]}^{\overline{x}} f(x)dx$ (see
Algorithm 1).

Algorithm 1. COMPUTE_INTEGRAL1 compute an outer approximation of
the integral of $\int_{[x]}^{\overline{x}} f(x)dx$

Input: $[x]$ an interval
Input: f the function we want the integral
Output: the interval $[I_f] \supset \left\{ \int_x^{\overline{x}} f(x)dx : x \in [x] \right\}$

1 $l_{candidates} \leftarrow \emptyset$
2 $\mathcal{X}^* \leftarrow \{x \in [x] : f(x) = 0\}$
3 $[I_{current}] \leftarrow$ compute_integral$(f, \underline{x}, x_1^*)$
4 $l_{candidates} \leftarrow \{[I_{current}]\}$
5 **for** $i \leftarrow 1$ **to** $|\mathcal{X}^*| - 1$ **do**
6 $\quad$ $[I_{current}] \leftarrow$ compute_integral(f, x_i^*, x_{i+1}^*)
7 $\quad$ **foreach** $candidate \in l_{candidates}$ **do**
8 $\quad\quad$ $\lfloor$ candidate $\leftarrow$ candidate $+ [I_{current}]$
9 $\quad$ $l_{candidates} \leftarrow \{l_{candidates}, [I_{current}]\}$
10 $[I_{current}] \leftarrow$ compute_integral$(f, x_{|\mathcal{X}^*|}^*, \overline{x})$
11 **foreach** $candidate \in l_{candidates}$ **do**
12 $\quad$ $\lfloor$ candidate $\leftarrow$ candidate $+ [I_{current}]$
13 $l_{candidates} \leftarrow \{l_{candidates}, [I_{current}]\}$
14 $[I_{min}] \leftarrow \min(l_{candidates})$
15 $[I_{max}] \leftarrow \max(l_{candidates})$
16 **return** $[\underline{I_{min}}\overline{I_{max}}]$

The first step is to compute the set of roots $\mathcal{X}^* = \{x \in [x] : f(x) = 0\}$
(Line 2). Then the set of candidates is computed incrementally over the interval $[x]$ by computing each integral between two contiguous roots in the set $\mathcal{X}^*$.
Indeed, as we can see in Fig. 1, each integral between two elements of $\mathcal{X}^*$ is the
beginning of a candidate to be the minimum or the maximum so the algorithm

needs to compute this integral, add it to the list of candidates and add it to the existing candidates. This is done in Lines 5 to 9. After that the integral from the last element of $\mathcal{X}^*$ and the upper bound of $[x]$ is added (Lines 10 to 12). The minimum and the maximum is one of these candidates.

Algorithm for the Computation of $\int_{\underline{x}}^{[x]} f(x)dx$

The algorithm for the computation of $\int_{\underline{x}}^{[x]} f(x)dx$ is somehow the same as the previous one (see Algorithm 2). The difference is now that each couple of contiguous points in the set $\mathcal{X}^*$ is the end of a candidate. The integrals are then computed from the end to the beginning of the set $\mathcal{X}^*$, added to the current existing candidates, and added to the list of candidates (Lines 5 to 9). Eventually the integral from the lower bound of $[x]$ to the first element of $\mathcal{X}^*$ is computed, added to the existing candidates and added to the list (Line 10 to 12).

Algorithm 2. COMPUTE_INTEGRAL2 compute an outer approximation of the integral of $\int_{\underline{x}}^{[x]} f(x)dx$

Input: $[x]$ an interval
Input: f the function we want the integral
Output: the interval $[I_f] \supset \left\{ \int_{\underline{x}}^{x} f(x)dx : x \in [x] \right\}$

1   $l_{candidates} \leftarrow \emptyset$
2   $\mathcal{X}^* \leftarrow \{x \in [x] : f(x) = 0\}$
3   $[I_{current}] \leftarrow$ compute_integral$(f, x^*_{|\mathcal{X}^*|}, \overline{x})$
4   $l_{candidates} \leftarrow \{[I_{current}]\}$
5   **for** $i \leftarrow |\mathcal{X}^*|$ *downto* $x^*_1 + 1$ **do**
6      $[I_{current}] \leftarrow$ compute_integral(f, x^*_{i-1}, x^*_i)
7   **foreach** *candidate* $\in l_{candidates}$ **do**
8   candidate $\leftarrow$ candidate $+ [I_{current}]$
9      $l_{candidates} \leftarrow \{l_{candidates}, [I_{current}]\}$
10   $[I_{current}] \leftarrow$ compute_integral$(f, \underline{x}, x^*_1)$
11   **foreach** *candidate* $\in l_{candidates}$ **do**
12   candidate $\leftarrow$ candidate $+ [I_{current}]$
13   $l_{candidates} \leftarrow \{l_{candidates}, [I_{current}]\}$
14   $[I_{min}] \leftarrow$ min$(l_{candidates})$
15   $[I_{max}] \leftarrow$ max$(l_{candidates})$
16   **return** $[I_{min} \overline{I_{max}}]$

Algorithm for the Computation $\int_{[x]}^{[x]} f(x)dx$

We now introduce Algorithm 3 to compute an interval integral with the same interval endpoints. As described in Theorem 1, every pair of elements in $\mathcal{X}^*$ can be the candidate to be a minimum or a maximum. In Algorithm 3, we start by computing the integral from the lower bound of $[x]$ to the first element of $\mathcal{X}^*$ (Line 3). Then all the other integrals between two elements of $\mathcal{X}^*$ are computed

Algorithm 3. COMPUTE_INTEGRAL3 compute an outer approximation of the integral of $\int_{[x]}^{[x]} f(x)dx$

Input: $[x]$ an interval
Input: f the function we want the integral
Output: the interval $[I_f] \supset \left\{ \int_{x_1}^{x_2} f(x)dx \: : x_1, x_2 \in [x] \right\}$

1 $l_{\text{candidates}} \leftarrow \emptyset$
2 $\mathcal{X}^* \leftarrow \{x \in [x] : f(x) = 0\}$
3 $[I_{\text{current}}] \leftarrow \text{compute\_integral}(f, \underline{x}, x_1^*)$
4 $l_{\text{candidates}} \leftarrow \{[I_{\text{current}}]\}$
5 **for** $i \leftarrow 1$ **to** $|\mathcal{X}^*| - 1$ **do**
6 $\quad$ $[I_{\text{current}}] \leftarrow \text{compute\_integral}(f, x_i^*, x_{i+1}^*)$
7 $\quad$ **foreach** *candidate from last iteration* $\in l_{candidates}$ **do**
8 $\quad\quad$ $l_{\text{candidates}} \leftarrow \{l_{\text{candidates}}, \text{candidate} + [I_{\text{current}}]\}$
9 $\quad$ $l_{\text{candidates}} \leftarrow \{l_{\text{candidates}}, [I_{\text{current}}]\}$
10 $[I_{\text{current}}] \leftarrow \text{compute\_integral}(f, x_{|\mathcal{X}^*|}^*, \overline{x})$
11 **foreach** *candidate from last iteration (Line 5)* $\in l_{candidates}$ **do**
12 $\quad$ $l_{\text{candidates}} \leftarrow \{l_{\text{candidates}}, \text{candidate} + [I_{\text{current}}]\}$
13 $l_{\text{candidates}} \leftarrow \{l_{\text{candidates}}, [I_{\text{current}}]\}$
14 $[I_{\min}] \leftarrow -\max(-\min(l_{\text{candidates}}), \max(l_{\text{candidates}}))$
15 $[I_{\max}] \leftarrow \max(-\min(l_{\text{candidates}}), \max(l_{\text{candidates}}))$
16 **return** $[\underline{I_{min}} \overline{I_{max}}]$

---

and added to the list of candidates (Line 5). The condition in the **foreach** loop (Lines 7 and 11) means that the computation between two contiguous elements x_i^* and x_{i-1}^* can only be added to integrals where the endpoint is x_i^*. Only the candidates from the previous iteration fulfill this requirement.

The three algorithms that have been introduced can be used to compute the result of any interval integration discussed in Sect. 3. For example, we can use Algorithms 1 and 2 to compute the case where $[x_1]$ and $[x_2]$ do not intersect (see Algorithm 4). It simply corresponds to the decomposition given in Eq. (5).

Complexity

The complexity of Algorithms 1 and 2 is linear on the arity of the set $\mathcal{X}^*$. Since Algorithm 4 has no loop and simply uses Algorithms 1 and 2, it has the same complexity. Now for the case of the algorithm used for the computation of $\int_{[x]}^{[x]} f(x)dx$, we have more candidates to assume and the complexity is then factorial on the arity of the set $\mathcal{X}^*$ which corresponds to the number of candidates we need to compute.

Algorithm 4. COMPUTE_INTEGRAL_DISJOINT compute an outer approximation of the integral of $\left\{ \int_{x_1}^{x_2} f(x)dx : x_1 \in [x_1], x_2 \in [x_2] \right\}$

Input: $[x_1]$, $[x_2]$ two interval endpoints such that $[x_1] \cap [x_2] = \emptyset$
Input: f the function we want the integral
Output: the interval $[I_f] \supset \left\{ \int_{x_1}^{x_2} f(x)dx : x_1 \in [x_1], x_2 \in [x_2] \right\}$

1 $l_{candidates} \leftarrow \emptyset$

   /* Computation of $\int_{[x_1]}^{\overline{x_1}} f(x)dx$ */

2 $[I_{min}] \leftarrow \min(\text{compute\_integral1}(f, [x_1]))$
3 $[I_{max}] \leftarrow \max(\text{compute\_integral1}(f, [x_1]))$

   /* Computation of $\int_{\overline{x_1}}^{\underline{x_2}} f(x)dx$ */

4 $[I] \leftarrow \text{compute\_integral}(f, \overline{x_1}, \underline{x_2})$
5 $[I_{min}] \leftarrow [I_{min}] + [I]$
6 $[I_{max}] \leftarrow [I_{max}] + [I]$

   /* Computation of $\int_{\underline{x_2}}^{[x_2]} f(x)dx$ */

7 $[I] \leftarrow (\text{compute\_integral2}(f, [x_2]))$
8 $[I_{min}] \leftarrow [I_{min}] + [I]$
9 $[I_{max}] \leftarrow [I_{max}] + [I]$
10 **return** $[I_{min}, I_{max}]$

4.2 Examples

We now apply the previous algorithms in several cases using an implementation in C++ of them. When the computation of an integral with real endpoints is required, we simply use an interval version of the composite midpoint rule (see Sect. 2.2).

Example 1. The first example, from [4], is as follows: compute

$$\int_0^{[0,1]} \frac{dx}{1+x^2}. \tag{24}$$

The result is then

$$\left[\int_0^0 \frac{dx}{1+x^2}, \int_0^1 \frac{dx}{1+x^2} \right] \subset [0, 0.78543] \tag{25}$$

which is compatible with the exact result that is $\left[0, \frac{\pi}{4}\right]$.

Example 2. A second example illustrates the complexity over the number of elements in $\mathcal{X}^* = \{x \in [x] \mid f(x) = 0\}$. We want to compute the interval integral

$$\int_\alpha^{[\alpha,1]} \sin\left(\frac{1}{x}\right) dx, \alpha > 0. \tag{26}$$

The arity of $\mathcal{X}^*$ increases with $\alpha \to 0$.

Table 1. Results for the computation of the interval integral from Eq. (26).

α	$|\mathcal{X}^*|$	Minimum	Maximum	Computation time
10^{-1}	3	$\int_{0.1592}^{1} \sin\left(\frac{1}{x}\right) dx = 0.4815$	$\int_{0.3183}^{1} \sin\left(\frac{1}{x}\right) dx = 0.57774$	0.198216
10^{-2}	49	0.4815	0.57774	0.212816
10^{-3}	547	0.4815	0.57774	0.228498
10^{-4}	5640	0.4815	0.57774	0.492445

In Table 1 are represented the result of the computation of the interval integral in Eq. (26) for different values of the parameter α.

Example 3. The last example is the computation of the interval integral

$$\int_{[0,5]}^{[0,5]} x \sin x \, dx \tag{27}$$

In the interval $[0,5]$, two roots occur: 0 and π and the computation gives the results:

$$\left[\int_{3.14133}^{5} x \sin x \, dx, \int_{5}^{3.14133} x \sin x \, dx\right] \subset [-5.518, 5.52]. \tag{28}$$

Here since π is not representable, the method guarantees that the endpoints for the minimum are in the interval $[3.14133, 5]$.

5   Conclusion

In this work, we introduced algorithms for the computation of interval integrals, integrals with unknown endpoints for which their value is taken in an interval. These algorithm make use of interval analysis and the guaranteed computation of integrals with known endpoints to produce the minimum and the maximum of the corresponding set of integral the interval integral defines. These algorithms are simple to apply the moment we can compute the set of roots of the function: the value that make the function at stake in the integral equal to zero. The algorithms in the worst case scenario then have a factorial complexity on the number of the roots.

These algorithms can be embedded in any resolution of problems where interval integrals occur such as robotics problem like the optimal control problem. Future work will be to apply this method to such problems.

Acknowledgement. This research was supported by the "Complex Systems Engineering Chair - Ecole Polytechnique, THALES, DGA, FX, DASSAULT AVIATION, DCNS Research, ENSTA ParisTech, Télécom ParisTech, Fondation ParisTech and FDO ENSTA". This research is also partially funded by DGA MRIS.

References

1. Alefeld, G., Mayer, G.: Interval analysis: theory and applications. J. Comput. Appl. Math. **121**(1–2), 421–464 (2000)
2. Alexandre dit Sandretto, J., Chapoutot, A.: Validated explicit and implicit runge-kutta methods. Reliable Comput. **22**, 78–103 (2016). Electronic edition
3. Aubry, C., Desmare, R., Jaulin, L.: Loop detection of mobile robots using interval analysis. Automatica **49**(2), 463–470 (2013)
4. Corliss, G.F.: Computing narrow inclusions for definite integrals. In: Computer Arithmetic, pp. 150–179 (1987)
5. Corliss, G.F., Rall, L.B.: Adaptive, self-validating numerical quadrature. SIAM J. Sci. Stat. Comput. **8**(5), 831–847 (1987)
6. Hansen, E.: Global optimization using interval analysis–the multi-dimensional case. Numer. Math. **34**(3), 247–270 (1980)
7. Moore, R.E.: Interval Analysis, vol. 4. Prentice-Hall, Englewood Cliffs (1966)
8. Moore, R.E., Kearfott, R.B., Cloud, M.J.: Introduction to Interval Analysis, vol. 110. SIAM, Philadelphia (2009)
9. Nedialkov, N.S., Jackson, K.R., Corliss, G.F.: Validated solutions of initial value problems for ordinary differential equations. Appl. Math. Comput. **105**(1), 21–68 (1999)
10. Petras, K.: On the complexity of self-validating numerical integration and approximation of functions with singularities. J. Complex. **14**(3), 302–318 (1998)
11. Petras, K.: Self-validating integration and approximation of piecewise analytic functions. J. Comput. Appl. Math. **145**(2), 345–359 (2002)
12. Sandretto, J.A.D., Chapoutot, A., Mullier, O.: Optimal switching instants for the control of hybrid systems. In: 11th Summer Workshop on Interval Methods (SWIM) (2018)

Epidemic Spreading Curing Strategy Over Directed Networks

Clara Pizzuti$^{(\boxtimes)}$ (iD) and Annalisa Socievole (iD)

Institute for High Performance Computing and Networking (ICAR), National
Research Council of Italy (CNR), Via Pietro Bucci, 8-9C, 87036 Rende, CS, Italy
{clara.pizzuti,annalisa.socievole}@icar.cnr.it

Abstract. Epidemic processes on networks have been thoroughly investigated in different research fields including physics, biology, computer science and medicine. Within this research area, a challenge is the definition of curing strategies able to suppress the epidemic spreading while exploiting a minimal quantity of curing resources. In this paper, we model the network under analysis as a directed graph where a virus spreads from node to node with different spreading and curing rates. Specifically, we adopt an approximation of the Susceptible-Infected-Susceptible (SIS) epidemic model, the N-Intertwined Mean Field Approximation (NIMFA). In order to control the diffusion of the virus while limiting the total cost needed for curing the whole network, we formalize the problem of finding an Optimal Curing Policy (OCP) as a constrained optimization problem and propose a genetic algorithm (GA) to solve it. Differently from a previous work where we proposed a GA for solving the OCP problem on undirected networks, here we consider the formulation of the optimization problem for directed weighted networks and extend the GA method to deal with not symmetric adjacency matrices that are not diagonally symmetrizable.

Keywords: Epidemic spreading · NIMFA model · Directed networks · Genetic algorithms

1   Introduction

The spread and the permanence of viruses, both biological and digital, over a network represent a threat for society and organizations. Digital viruses mainly use the Internet as diffusion media and usually spread over telecommunication networks and social networks, while biological viruses propagate over contact networks of living beings through contacts. The first epidemic model emulating the diffusion of a virus between individuals of a population dates back to 1926 with the Kermack-McKendrick epidemic model [10] describing the interactions between individuals in susceptible, infected and immune states. In these last years, spreading processes continued to receive attention from researchers working in different research fields [12]. The most common applications of epidemic

© Springer Nature Switzerland AG 2020
Y. D. Sergeyev and D. E. Kvasov (Eds.): NUMTA 2019, LNCS 11974, pp. 182–194, 2020.
https://doi.org/10.1007/978-3-030-40616-5_14

models include information diffusion on social networks like Facebook or Twitter, diffusion of viruses in computer networks, propagation of infectious diseases in contact networks.

Independently of the kind of network where the epidemic spreads, the viral process can be formalized using the same network-based theoretical model. In fact, the entities experiencing the infection and their relationships can be formalized with a graph $G = (V, E)$ with $|V|$ nodes representing the entities involved and $|E|$ edges denoting the propagation of the virus from an infected node to a susceptible neighbor. In this context, to arrest the diffusion of a virus and make its pervasiveness as low as possible, the optimal distribution of resources (eg. medicines, medical staff, etc.) is fundamental. Thus, the development of policies aiming at controlling the spreading process with a fixed budget, referred as *Optimal Curing Policy* (OCP) is of paramount importance for public health and network security domains.

When the network resources are limited and the evolution of an epidemic has to be controlled, optimization techniques are usually adopted: distribute vaccines or antidotes by minimizing the costs for the medical cures [1,20,21], find the minimum number of nodes to protect with vaccines or the minimum number of links to remove [2,12], identify influential spreaders for maximizing the diffusion of information are just some examples. All these applications have in common the availability of a restricted budget and the need of methods able to provide fast and effective solutions.

In [17], we proposed a genetic algorithm, namely *OCPGA*, finding a minimal-cost curing strategy making the network virus-free in an undirected network where the virus spreads with a *Susceptible-Infected-Susceptible* (SIS) model. Due to the complexity of the model for large networks, we exploited one of its approximations, the heterogeneous version of the *N-Intertwined Mean-Field Approximation (NIMFA)* [24,25], where nodes have their own curing rates and the virus spreads to their neighbors with different infection rates.

In this paper, we propose to solve the optimization problem for directed weighted networks. Specifically, we extend the *OCPGA* method to deal with adjacency matrices that are not diagonally symmetrizable. The method, named *D-OCPGA (Directed OCPGA)* is validated through a comparison with the exact semidefinite programming solver *SDPT3* [23]. By testing the two algorithms over both real-world and synthetic networks, we find that *D-OCPGA* is able to outperform *SDPT3* in terms of total curing cost needed to make the network virus-free.

The paper is organized as follows. Section 2 describes the most relevant works on epidemic spreading in networks, and on genetic algorithms in combination with epidemic spreading. Section 3 introduces the OCP constrained minimization problem. Section 4 describes *D-OCPGA*, the genetic algorithm we propose for solving the OCP problem. In Sect. 5, we present the results of the performance comparison between *D-OCPGA* and *SDPT3* over real-world and synthetically generated networks. Finally, in Sect. 6, we draw the main conclusions.

2   Related Work

Epidemic spreading and infectious diseases have been studied in different research fields including physics, mathematics, computer science and epidemiology. A description of the state of the art on disease spreading models can be found in [12,15]. Pastor-Satorras and Vespignani [16] analyzed epidemic spreading on scale-free networks, Newman [11] on random graphs, Wang et al. [27] in contact networks.

In [1], Borgs et al. studied the problem of how to distribute antidotes to control the spread of an epidemic on a finite graph. Gourdin et al. [5] studied how to minimize the cost for curing a network when there is a given level of infection. Prakash et al. [19] analyzed the problem of properly distributing resources to nodes for minimizing the rate at which nodes infect each other.

In [20,21], Preciado et al. optimized the distribution of curing resources for controlling and protecting arbitrary networks from the diffusion of a virus by modifying the infection rates of the nodes.

Zhai et al. [28] analyzed several algorithms for epidemic evolution by proposing a framework for controlling the epidemic spread in broadcast networks.

More recently, Ottaviano et al. [13], focused on epidemic processes on networks organized in communities by studying an optimal policy for curing these types of network structures.

Regarding epidemic spreading and the use of genetic algorithms, so far, a few number of works have been proposed. Lahiri and Cebrian [7] proposed a *genetic algorithm diffusion model* (GADM) for static and dynamic social networks. Specifically, the authors defined a GA paired with specific forms of Holland's synthetic hyperplane-defined objective functions as a general diffusion model. GADM generates a spatially distributed population of chromosomes encoded with binary strings by exploiting the one-point crossover as genetic operator.

Liao et al. [9] focused on infectious diseases modeled through a stochastic ripple-spreading process emulating the effect of random mobility and contacts between individuals on the diffusion of a virus between them. The authors adopted a GA to tune the several parameters of the model.

Parousis-Orthodoxou and Vlachos [14], harnessed a GA for optimizing the distribution of vaccines on a SIR (Susceptible-Infected-Recovered) model. The objective function is the number of vaccines needed to have a minimal percentage of infected nodes, taking into account both the cost of the vaccine and the cost for the treatments.

Our work distinguishes from the work by Parousis-Orthodoxou and Vlachos for the model adopted, the SIS, for the spreading and curing rates that are nodal and link-dependent, respectively, and for the genetic operators employed. In addition, our scheme looks for an optimal curing strategy that is able to cure all the nodes (i.e., having all the nodes healthy).

In [2], Concatto et al. proposed a GA for minimizing the viral process of an infection by removing edges from the network graph. Here, the authors focused on the Min-SEIS-Cluster problem in which the SIS model is extended with the exposed (E) state, and nodes are organized in clusters where the epidemic spreads at a higher rate. Analogously to [14], this algorithm aims at minimizing the infected nodes while *D-OCPGA* completely cures the whole network with the applied curing policy.

3   The Optimal Curing Policy (OCP) Problem

The diffusion of a virus over a population of individuals is usually modeled through three different disease stages: *susceptible* (S), i.e. an individual can contract the infection, *infectious* (I), i.e. the infection has been contracted, and *recovered* (R), i.e. the individual recovered from the disease.

The Susceptible-Infected-Susceptible (SIS) is a type of epidemic model where an individual can pass from the state S to the state I and again to the state S. The states are modeled through a Bernoulli random variable $X_i \in \{0, 1\}$ which is $X_i = 0$ when the node is healthy and $X_i = 1$ when the node is infected [13]. The probability for a node of being in the *infected state* is $v_i(t) = \Pr[X_i(t) = 1]$, while the node is in the *healthy state* with probability $1 - v_i(t)$. To solve the SIS model, we need to compute $v_i(t)$ for each node.

In the homogeneous setting, each node is cured with the same *curing rate* δ and the *infection rate* β is the same for each link. In this situation, the *effective infection rate* is defined as $\tau = \beta/\delta$. In the heterogeneous case, on the contrary, the curing rate is node-specific thus each node i is recovered at rate δ_i, and also the infection rate is link-specific (i.e. β_{ij} can be different for each couple of connected nodes i and j).

In a network with N nodes, the SIS model can be described through a continuous Markov chain with 2^N states, corresponding to all the combinations of infected nodes [25]. After a certain time, the network converges to an *absorbing state* where the virus disappears. Moreover, the process is characterized by a phase transition τ_c, named *epidemic threshold*, a critical value for which if $\tau > \tau_c$, the infection becomes persistent, while if $\tau < \tau_c$, the virus extinguishes.

For networks with a high number of nodes, the exact solution of the Markovian chain can be obtained by solving a system of linear differential equations, whose number increases exponentially with the network size. Consequently, approximate models have been proposed [22,25], such as the *N-Intertwined Mean-Field Approximation* $(NIMFA)$, which substitutes the original 2^N linear differential equations with N non-linear differential equations.

In $NIMFA$, the probability of infection for a node i, $v_i(t)$, is modeled as:

$$\frac{dv_i(t)}{dt} = \sum_{j=1}^{N} \beta_{ij} v_j(t) - \sum_{j=1}^{N} \beta_{ij} v_i(t) v_j(t) - \delta_i v_i(t). \tag{1}$$

which can be rewritten as

$$\frac{dV(t)}{dt} = \bar{A} V(t) + F(V) \tag{2}$$

where $V(t)$ is the vector $V(t) = (v_1(t), v_2(t), ..., v_N(t))$, $\bar{A}$ is the matrix

$$\bar{A} = \begin{bmatrix} \delta_1 & \beta_{12} & \cdot\cdot & & \cdot & \beta_{1N} \\ \beta_{21} & \delta_2 & & & & \\ \cdot & & \cdot & & & \\ \cdot & & & \cdot & & \\ \cdot & & & & \cdot & \\ \beta_{N1} & \cdot & & \cdot\cdot & \beta_{NN-1} & \delta_N \end{bmatrix} \tag{3}$$

and $F(V)$ is the column vector having as i-th element

$$-\sum_{j=1}^{N} \beta_{ij} v_i(t) v_j(t) \tag{4}$$

In the heterogeneous setting, Ottaviano et al. [13] derived the epidemic threshold by relying on the work of Lajmanovich and Yorke [8]. By defining

$$r(\bar{A}) = max_{1 \le j \le N} Re(\lambda_j(\bar{A})) \tag{5}$$

where $Re(\lambda_j(\bar{A}))$ is the real part of the eigenvalues of $\bar{A}$, if $r(\bar{A}) \le 0$, then the virus dies out and this condition identifies the epidemic threshold.

Specifically, Ottaviano et al. [13] formalized the problem of suppressing the viral diffusion on a weighted network with a proper assignment of curing resources as follows. Let δ_i be the curing rate of node i and c_i the corresponding cost for recovering this node. The objective is to *minimize the total cost for curing the network* while making the infectious process die out. The total cost is defined as

$$U(\Delta) = \sum_{i=1}^{N} c_i \delta_i \tag{6}$$

where $\Delta = (\delta_1, \delta_2, ..., \delta_N)$ is the vector of the curing rates for each node to determine, knowing the cost c_i of the curing resources for a node i.

When the network is undirected and weighted, the adjacency matrix $A = (\beta_{ij})$ is symmetric ($\beta_{ij} = \beta_{ji}$) and, consequently, the values of the eigenvalues are real. In [13], Theorem 2.1 states that if $\lambda_{max}(A - diag(\Delta)) \le 0$ the viral infection is suppressed and all nodes are healthy. Thus, the epidemic threshold for the considered network is determined by the largest eigenvalue of $A - diag(\Delta)$.

However, for directed graphs, the weighted adjacency matrix A is not symmetric ($\beta_{ij} \ne \beta_{ji}$). Thus, for an arbitrary, strongly connected, directed weighted graph having a not symmetrizable[1] matrix A, instead of A, its Hermitian part $H = (A + A^T)/2$ needs to be considered, in this case obtaining only a suboptimal solution. In fact, it has been shown that $\lambda_{max}(A - diag(\Delta)) \le \lambda_{max}(H - diag(\Delta))$, thus the feasible region of the optimization problem using the Hermitian part of the matrix A, which is the region where $\lambda_{max}(H - diag(\Delta)) \le 0$,

---

[1] A matrix A is symmetrizable if there exists an invertible diagonal matrix D and symmetric matrix S such that $A = DS$.

is a subset of the feasible region of the original problem using A. This means that the optimal cost function value of the former problem is an upper bound of the original problem, thus, though the epidemic will go towards extinction, more effort will be necessary. The cost-optimal allocation of curing resources can thus be formulated as follows.

Problem *Optimal Curing Policy (OCP)*. Let $G = (V, E)$ be an undirected weighted graph with adjacency matrix A where the elements are not symmetric ($a_{ij} \neq \beta_{ji}$), thus meaning that node i can infect node j with rate β_{ij} while node j can infect i with a different rate β_{ji}. Let H be the Hermitian part of A, and $c_i > 0, i = 1, \ldots N$ the cost coefficients. The OCP problem can be formalized as the following nonlinear constrained optimization problem:

$$\text{minimize } U(\Delta)$$
$$\text{subject to } \lambda_{max}(H - diag(\Delta)) \leq 0$$
$$\Delta \geq 0$$

where $\Delta \geq 0$ is the curing vector to find.

Reformulated as a semidefinite programming problem (SDP) [26], solvable through an SDP solver like $SDPT3$ [23], the OCP problem is:

$$\text{minimize } U(\Delta)$$
$$\text{subject to } diag(\Delta) - H \geq 0$$
$$\Delta \geq 0$$

Since $diag(\Delta) \geq 0$ and the inequality sign in $diag(\Delta) - H \geq 0$, being $diag(\Delta) - H$ a matrix, means that it is semidefinite positive[2].

4   D-OCPGA: A Constrained Genetic Algorithm Solving the OCP Problem

We propose an extension of the constrained genetic algorithm $OCPGA$, which minimizes the total curing cost $U(\Delta)$ by evolving a population of individuals. Each individual is represented by a vector $\Delta = (\delta_1, \delta_2, \ldots, \delta_N)$, where δ_i is the curing rate for each node of the network assuming values in the interval $[x_i^l, x_i^u] = [0, 1]$. As the OCP formulation states, we need to check if a possible solution has the real part of the largest eigenvalue of $H - diag(\Delta)$ positive and the components δ_i fall within the bounds so that $x_i^l \leq \delta_i \leq x_i^u$.

D-$OCPGA$ receives in input the matrix $A = (\beta_{ij})$ of the spreading rates, the vector of curing costs $C = (c_1, c_2, \ldots, c_N)$ and the number T of iterations, then it performs the following steps:

1. compute $diag(\Delta)$, the diagonal matrix of the vector of curing costs;
2. compute H, the Hermitian part of the not symmetrizable matrix A;

---

[2] A semidefinite positive matrix $A \in R^{N \times N}$ is a symmetric matrix such that $x^T A x \geq 0$ for all the $x \in R^N$. Equivalently, all the eigenvalues of A are nonnegative.

3. compute the real part of the largest eigenvalue λ_{max} of $H - diag(\Delta)$;
4. for T iterations, run by using $U(\Delta)$ as fitness function to minimize subject to $\lambda_{max} \leq 0$ and $\Delta \geq 0$, applying crossover and mutation operators;

In output, the algorithm provides $\Delta^* = (\delta_1^*, \delta_2^*, ..., \delta_N^*)$, a curing vector having the lowest fitness function.

As crossover operator, we adopted the *simulated binary crossover (SBX)* proposed by Deb in [3]. As underlined by Deb, this crossover is able to manage the distance of the children from the parents generating feasible solutions. The spread of children is tuned using a distribution index η_c which is able to explore contiguous regions if the diversity among parents is sufficient. It exploits the diversity between the parents to drive the search towards certain regions. The *SBX* operator generates the children solutions $s^{(1)}$ and $s^{(2)}$ from the two feasible parents $x^{(1)}$ and $x^{(2)}$ as

$$s^{(1)} = 0.5 \left[(x^{(1)} + x^{(2)} - \bar{\beta}[x^{(2)} + x^{(1)}]) \right] \tag{7}$$

$$s^{(2)} = 0.5 \left[(x^{(1)} + x^{(2)} + \bar{\beta}[x^{(2)} + x^{(1)}]) \right] \tag{8}$$

where

$$\bar{\beta} = \begin{cases} (\alpha u)^{1/(\eta_c+1)} & \text{if } u \leq 1/\alpha \\ (\frac{1}{2-\alpha u})^{1/(\eta_c+1)} & \text{otherwise} \end{cases} \tag{9}$$

with $\alpha = 2 - \beta^{-(\eta_c+1)}$, u assumes a random number in the interval $[0, 1]$, η_c, if small, generates children solutions distant from the parents while for large values allows neighbor children and

$$\beta = 1 + \frac{2}{s^{(2)} - s^{(1)}} min[(x^{(1)} - x^l), (x^u - x^{(2)})]$$

This formulation guarantees that the children solutions fall within the fixed range $[x^l, x^u]$.

As mutation operator, since the problem is constrained, we selected the *mutation feasible* from the Global Optimization Toolbox of Matlab. This operator generates feasible mutants by choosing random directions that satisfy the bounds and the linear constraints.

5   Experimental Evaluation

In this section, we detail the experiments performed to test the effectiveness of *D-OCPGA*. We tested the algorithm both on real-world and synthetic networks whose topological characteristics are summarized in Table 1. Specifically, we compared the performance of *D-OCPGA* with those found by the *SDPT3* solver.

D-OCPGA has been implemented in Matlab v2015b. To run *SDPT3*, we used the *CVX* package which helps in specifying constraints and objectives and thus solving convex programs [6] by using the standard Matlab expression syntax. The genetic parameters have been set using a trial-and-error procedure on the benchmark datasets. We fixed crossover fraction to 0.9, mutation rate to 0.2, population size and number of generations to 1000 for networks with a number of nodes lower than 128, and to 2000 for the other networks. Finally, as suggested in [3], $\eta_c = 1$.

The infection spreading rates of each node, have been obtained by randomly generating rates in the range $[0, 1]$ and then by multiplying them for 10^{-3}. The spreading rates β_{ij} for each link have been fixed by generating random values in the range $[0, 10^{-3}]$. The following subsections detail the features of the dataset used and the results of the experimentation.

5.1 Datasets

Real-World Networks

- **Internet Backbones.** From the repository *Internet Topology Zoo*[3], we selected 5 Internet Backbone networks, namely *Bell South, OTEGlobe, ITC Deltacom, ION,* and *US Carrier* with different features in terms of number of nodes, average clustering coefficient and density. Each node which corresponds to a BGP (Border Gateway Protocol) router is usually connected to one or two routers. Such bidirectional networks are of particular interest for analyzing the OCP problem since are often subject to attacks like blackholing or traffic redirection that provoke instability.
- **Friendship Networks.** Friendship networks are examples of social networks composed by several ego networks (one central node directly connected to other alters/friends forming a star topology) connected between them through common friends. These networks are usually characterized by the spreading of fake news/comments to friends. We start analyzing two types of Facebook friendship networks, the *Ego 3980* and the *Ego 686*[4], representing the egos (i.e., social profiles) of two Facebook users. Since Facebook friendships are bidirectional, we also analyze three unidirectional friendship networks taken from the KONECT repository[5], namely the *HighSchool* and the *Residence Hall*. The *HighSchool* network contains self-declared friendships between boys in a small highschool in Illinois in two different time steps, for this reason we consider two snapshots of the network, the *HighSchool 1* and the *HighSchool 2*. Similarly, the *Residence Hall* contains self-declared friendship data between residents of a residence hall located on the Australian National University campus.

---

[3] http://www.topology-zoo.org/.
[4] https://snap.standford.edu/data/egonets-Facebook.html.
[5] http://konect.uni-koblenz.de.

Table 1. Features of the topologies considered in terms of number of nodes (N), average degree ($<k>$), average clustering coefficient ($<C>$) and density (D). For the synthetic networks, the measures refer to an average over 10 network samples of the same network class.

Network type	Network name	N	$<k>$	$<C>$	D
Backbone	Bell South	51	1.294	0.081	0.052
	OTE Globe	93	1.108	0.011	0.024
	ITC Deltacom	113	1.425	0.053	0.025
	ION	125	1.168	0.006	0.019
	US Carrier	158	1.196	0.002	0.015
Friendship	Facebook Ego 3980	52	5.625	0.462	0.11
	Facebook Ego 686	168	19.714	0.534	0.118
	HighSchool 1	73	3.328	0.353	0.046
	HighSchool 2	73	3.602	0.312	0.05
	Residence Hall	217	12.314	0.379	0.057
Synthetic	Erdős-Rényi	128	5.23	0.054	0.041
	Watts-Strogatz	128	6	0.109	0.047
	Bárabasi-Albert	128	3.954	0.132	0.031

Synthetic Networks

- **Erdős-Rényi random networks.** These networks are generated from an initial set of N isolated nodes that are then connected between them with a probability p_c. Since a threshold for the connectivity of Erdős-Rényi networks is $p_c \approx \ln(N)/N$ for large N, here, we set $p_c = 2\ln(N)/N$ to be sure to obtain a connected bidirectional graph. Erdős-Rényi networks are usually adopted for modeling with a good accuracy peer-to-peer and ad-hoc networks.
- **Watts-Strogatz small-world networks.** Watts-Strogatz networks are highly organized in clusters with nodes easily reachable in few hops by the other nodes. These networks can be created from an initial ring lattice of N nodes where each node is afterward linked to k nodes by rewiring each edge with probability p. In our simulations, we generate bidirectional Watts-Strogatz networks setting $k = 6$ and $p = 0.5$. Watts-Strogatz networks are commonly exploited to model Bluetooth or Wi-Fi contact networks.
- **Bárabasi-Albert power law networks.** These networks well model social networks, the Internet and the World Wide Web. The main characteristic of such kind of networks is the so-called *preferential attachment* feature, i.e. nodes prefer to connect to high-degree nodes. Starting from m_0 nodes, Bárabasi-Albert networks can be generated as follows. At every time step, a new node is connected to $m \leq m_0$ nodes with a probability proportional to the degree of the existing nodes. Here we set $m_0 = 5$ and $m = 2$ to obtain bidirectional Bárabasi-Albert networks.

Table 2. Performance comparison between $SDPT3$ and $D\text{-}OCPGA$ values of objective function and number of generations of $D\text{-}OCPGA$ needed to equal $SDPT3$ for equal costs (e) and random costs (r).

Network	$U(\Delta)^e_{SDPT3}$	$U(\Delta)^e_{D\text{-}OCPGA}$	#Gen^e	$U(\Delta)^r_{SDPT3}$	$U(\Delta)^r_{D\text{-}OCPGA}$	#Gen^r
Bell South	0.065	0.029	48.1	0.031	0.015	31.5
OTEGlobe	0.095	0.031	22.7	0.037	0.011	17
ITC Deltacom	0.16	0.074	25.2	0.066	0.036	33.8
ION	0.153	0.055	26	0.063	0.024	28.9
US Carrier	0.174	0.075	85.2	0.078	0.023	23.3
Ego 3980	0.144	0.113	160.4	0.065	0.049	80.5
Ego 686	1.678	1.329	150.3	0.788	0.624	120.9
High School 1	0.124	0.08	156.1	0.05	0.032	64.1
High School 2	0.134	0.086	48.3	0.05	0.033	124.2
Residence Hall	1.34	1.257	202.3	0.591	0.499	185.4
Erdős-Rényi	0.622	0.539	244.3	0.256	0.244	64.6
Watts-Strogatz	0.385	0.291	181.1	0.165	0.131	80.1
Bárabasi-Albert	0.247	0.165	54.3	0.107	0.077	276.9

5.2 Results

Table 2 shows the results obtained by the comparison between $D\text{-}OCPGA$ with the classical $SDPT3$ solver applied to the semidefinite programming based version of the OCP problem. Besides the value of the objective function obtained by the two methods, we also specify the number of iterations necessary to $D\text{-}OCPGA$ to obtain values of the objective function lower than those obtained by $SDPT3$. For $D\text{-}OCPGA$, the method has been executed 10 times over a particular network and the average values of the results have been indicated. For the synthetic networks, each result refers to the mean values over 10 different graph realizations of a network type.

We considered both the situation in which all the nodes have equal unitary costs and the setting in which each node has its own curing cost (i.e. we assign random costs to nodes). Observe that, for the bidirectional networks, in order to make them asymmetric, we fixed different spreading rates for each given link.

For the first group of real-world networks, the Internet Backbones, $D\text{-}OCPGA$ is always able to outperform $SDPT3$. This holds both with unitary costs and random costs. It is worth noting that when the curing costs are random, since $\delta_i \leq 1$, the objective function values $U(\Delta)$ are lower. $D\text{-}OCPGA$ shows to be much more effective than $SDPT3$ especially in the unitary costs case. On the *US Carrier* network, for example, the total curing cost for $SDPT3$ is 0.174 and for $D\text{-}OCPGA$ is 0.075, while for random costs, $SDPT3$ obtains 0.078 and $D\text{-}OCPGA$ 0.023. Observe that very few generations of GA are required to achieve $SDPT3$ values. For the *OTE Globe* network, for example, on average 22 generations are required for the unitary costs and 17 generations for the random costs over the 1000 generations we fixed.

For the real-world friendship networks, we found similar results. On the *Ego 3980*, for instance, $D\text{-}OCPGA$ outperforms $SDPT3$ obtaining a fitness value of 0.113 instead of 0.144, when unitary costs are considered, and 0.049 instead of 0.065 with random costs. On the *Ego 686* with equal costs, the difference between *D-OCPGA* and SPDT3 performance is even more marked. It is interesting the epidemic behavior of the *HighSchool* dataset over the two timestamps of friendship networks, *HighSchool 1* and *HighSchool 2*. Note that for a given optimization method, the values of the objective function are similar. For *D-OCPGA* with unitary costs, for example, *HighSchool 1* achieves a total curing cost of 0.08 and *HighSchool 2* of 0.086. A similar behavior can be observed for the random costs. We thus conclude that the epidemic processes over the two network timestamps are similar. Overall, comparing all the friendship networks to the Internet Backbones, we observe that the number of generations of *D-OCPGA* required to outperform *SDPT3* are higher. This is probably due to the higher average degree of the networks that facilitates the diffusion of the epidemic thus complicating the search of an optimal curing strategy. Finally, over the synthetic networks, *D-OCPGA* again outperforms *SDPT3* in all the network scenarios.

6   Conclusion

The *D-OCPGA* method has been proposed as a constrained genetic algorithm able to find an *Optimal Curing Policy* (OCP) in directed networks subject to a virus spreading modeled as Susceptible-Infected-Susceptible (SIS) epidemic process. As in [17], where we proposed a GA method for undirected networks solving the OCP problem, the method exploits the N-Intertwined Mean-Field Approximation (NIMFA) of the SIS spreading process to find curing rates for the nodes that minimize the cost needed for completely curing the network. Specifically, we extended the *GA* method to deal with not symmetric adjacency matrices. A thorough experimentation on both real-world and synthetic networks demonstrated that *D-OCPGA* finds solutions whose curing cost is lower than that obtained by the *SDPT3* solver over the semidefinite programming formulation of the OCP problem. In [18] a self-adaptive *SBX* crossover operator [4] has been investigated for the *OCPGA* method and showed to obtain better results than the *SBX* crossover. Future work will experiment this crossover operator also for directed networks, will concentrate on the effect of different strategies for setting the initial population, different mutation operators and the extension of the method to networks organized in communities.

References

1. Borgs, C., Chayes, J., Ganesh, A., Saberi, A.: How to distribute antidote to control epidemics. Random Struct. Algorithms **37**(2), 204–222 (2010)
2. Concatto, F., Zunino, W., Giancoli, L.A., Santiago, R., Lamb, L.C.: Genetic algorithm for epidemic mitigation by removing relationships. In: Proceedings of the Genetic and Evolutionary Computation Conference, pp. 761–768. ACM (2017)
3. Deb, K.: An efficient constraint handling method for genetic algorithms. Comput. Methods Appl. Mech. Eng. **186**, 311–338 (2000)

4. Deb, K., Jain, H.: Self-adaptive parent to mean-centric recombination for real-parameter optimization. Tech. rep., Indian Institute of Technology Kanpur (2011)
5. Gourdin, E., Omic, J., Van Mieghem, P.: Optimization of network protection against virus spread. In: Proceedings of the 8th International Workshop on Design of Reliable Communication Networks (DRCN), 2011, pp. 659–667 (2011)
6. Grant, M., Boyd, S., Ye, Y.: CVX: Matlab software for disciplined convex programming (2008). https://doi.org/10.1155/2013/506240. 11 pages, Article ID 506240
7. Lahiri, M., Cebrian, M.: The genetic algorithm as a general diffusion model for social networks. In: AAAI (2010)
8. Lajmanovich, A., Yorke, J.A.: A deterministic model for gonorrhea in a nonhomogeneous population. Math. Biosci. **28**(3), 221–236 (1976)
9. Liao, J.Q., Hu, X.B., Wang, M., Leeson, M.S.: Epidemic modelling by ripple-spreading network and genetic algorithm. Math. Probl. Eng. **2013** (2013)
10. McKendrick, A.: Applications of mathematics to medical problems. Proceedings Edinb. Math Soc. **14**, 98–130 (1926)
11. Newman, M.: Spread of epidemic disease on networks. Phys. Rev. E **66**(1), 016128 (2002)
12. Nowzari, C., Preciado, V.M., Pappas, G.J.: Analysis and control of epidemics: a survey of spreading processes on complex networks. IEEE Control Syst. **36**(1), 26–46 (2016)
13. Ottaviano, S., De Pellegrini, F., Bonaccorsi, S., Van Mieghem, P.: Optimal curing policy for epidemic spreading over a community network with heterogeneous population. J. Complex Netw. **6**(5), 800–829 (2018)
14. Parousis-Orthodoxou, K., Vlachos, D.: Evolutionary algorithm for optimal vaccination scheme. J. Phys. Conf. Ser. **490**, 012027 (2014). IOP Publishing
15. Pastor-Satorras, R., Castellano, C., Mieghem, P.V., Vespignani, A.: Epidemic processes in complex networks. Rev. Mod. Phys. **87**(3), 925–979 (2015)
16. Pastor-Satorras, R., Vespignani, A.: Epidemic spreading in scale-free networks. Phys. Rev. Lett. **86**(14), 99–108 (2014)
17. Pizzuti, C., Socievole, A.: A genetic algorithm for finding an optimal curing strategy for epidemic spreading in weighted networks. In: Proceedings of the Genetic and Evolutionary Computation Conference, GECCO 2018, pp. 498–504. ACM, New York (2018)
18. Pizzuti, C., Socievole, A.: Optimal curing strategy enhancement of epidemic processes with self-adaptive SBX crossover. In: Cagnoni, S., Mordonini, M., Pecori, R., Roli, A., Villani, M. (eds.) WIVACE 2018. CCIS, vol. 900, pp. 151–162. Springer, Cham (2019). https://doi.org/10.1007/978-3-030-21733-4_12
19. Prakash, B.A., Adamic, L., Iwashyna, T., Tong, H., Faloutsos, C.: Fractional immunization in networks. In: Proceedings of the SIAM Data Mining Conference, pp. 659–667 (2013)
20. Preciado, V.M., Zargham, M., Enyioha, C., Jadbabaie, A., Pappas, G.J.: Optimal vaccine allocation to control epidemic outbreaks in arbitrary networks. In: Proceedings of the 52nd IEEE Conference on Decision and Control, CDC 2013, December 10–13, 2013, Firenze, Italy, pp. 7486–7491 (2013)
21. Preciado, V.M., Zargham, M., Enyioha, C., Jadbabaie, A., Pappas, G.J.: Optimal resource allocation for network protection against spreading processes. IEEE Trans. Control Netw. Syst. **1**(1), 99–108 (2014)
22. Sahneh, F.D., Scoglio, C., Van Mieghem, P.: Generalized epidemic mean-field model for spreading processes over multilayer complex networks. IEEE/ACM Trans. Netw. **21**(5), 1609–1620 (2013)

23. Tütüncü, R.H., Toh, K.C., Todd, M.J.: Solving semidefinite-quadratic-linear programs using SDPT3. Math. Program. **95**(2), 189–217 (2003)
24. Van Mieghem, P., Omic, J.: In-homogeneous virus spread in networks. arxiv:1306.2588 (2013)
25. Van Mieghem, P., Omic, J., Kooij, R.: Virus spread in networks. IEEE/ACM Trans. Netw. **17**(1), 1–14 (2009)
26. Vandenberghe, L., Boyd, S.: Semidefinite programming. SIAM Rev. **38**(1), 49–95 (1996)
27. Wang, Y., Chakrabarti, D., Wang, C., Faloutsos, C.: Epidemic spreading in real networks: an eigenvalue viewpoint. In: Proceedings of International Symposium on Reliable Distributed Systems (SRDS), pp. 25–34 (2003)
28. Zhai, X., Zheng, L., Wang, J., Tan, C.W.: Optimization algorithms for epidemic evolution in broadcast networks. In: 2013 IEEE Wireless Communications and Networking Conference (WCNC), pp. 1540–1545. IEEE (2013)

Learning Aerial Image Similarity
Using Triplet Networks

Vytautas Valaitis[1(✉)], Virginijus Marcinkevicius[2], and Rokas Jurevicius[2]

[1] Vilnius University Institute of Computer Science,
Didlaukio str. 47, 08303 Vilnius, Lithuania
vytautas.valaitis@mif.vu.lt
[2] Vilnius University Institute of Data Science and Digital Technologies,
Akademijos str. 4, 08412 Vilnius, Lithuania
{virginijus.marcinkevicius,rokas.jurevicius}@mii.vu.lt

Abstract. Unmanned aerial vehicles (UAV) faces localization challenges in satellite navigation systems denied environments. Images taken
from on-board cameras can be used to compare against orthophotographical map to support visual localization algorithms. Image similarity estimation can be achieved calculating various similarity metrics. Pearson
correlation was found to be the best choice for evaluating areal images
similarity in our experiments. Still is not robust against image displacement caused by aircraft frame movement. We propose a new architecture
of triplet neural network to learn image similarity measure. The proposed
architecture incorporates VGG16 network base layers. Top layer structure, loss function and performance metrics being suggested by authors.
Images were matched to the maps from satellite photo. The matching
results from proposed neural network architecture were compared and
evaluated against Pearson correlation.

Keywords: Image similarity · Triplet loss · Neural networks · UAV
localization

1 Introduction

Unmanned aerial vehicles (UAV) faces localization challenges in satellite navigation systems denied environments. Conventional autopilot systems fail to navi-
gate safely if the GPS signal is lost, jammed or unavailable. UAV should estimate
its position without the need for external signals. Visual odometry, Simultaneous
Localization and Mapping (SLAM), or map-based localization techniques can be
used to process aerial imagery from a downward looking camera on-board UAV
may be used to solve the pose estimation problem. Visual odometry and SLAM
has shown astonishing results while performing flights in indoors or near-ground
altitudes (<100 m). While Visual odometry and SLAM methods do not require
an apriori known map of the environment (map-less methods), these algorithms
are prone to errors over long distance flights (>1 km). The accuracy of these

Y. D. Sergeyev and D. E. Kvasov (Eds.): NUMTA 2019, LNCS 11974, pp. 195–207, 2020.
https://doi.org/10.1007/978-3-030-40616-5_15

methods for low-altitude flights (>100 m) is not well studied since the GPS signal is usually available in this altitude and the problem of signal jamming and spoofing is receiving attention only in recent years. Map-based techniques can reduce the errors for long distance flights compared to map-less systems. Map-based techniques takes an image from on-board camera and compares against orthophotographical map to search for the most similar location on map and localize the UAV.

Similarity between two images represented as a numerical sequence can be calculated using a similarity function that quantifies the similarity between the images. Usually, image is represented as a vectors of numerical values, each value representing the intensity of a pixel (1). Therefore similarity function can be defined trough a distance metric (lower the distance grater the similarity).

$$x = x_1, x_2, ..., x_N, \tag{1}$$

where N is the dimension of a vector x and x_1 to x_N represents different pixels of the image.

There are various distance based similarity metrics, such as Euclidean distance [5], Pearson correlation [5], Root mean squares [5], Pattern intensity [3], RBF kernel [6] or Mutual information [7].

Image similarity estimation becomes more difficult as images are represented as multiple matrices (a tensor). Colored images are often represented via three or four different matrices (channels). Therefore, images usually are converted to grey scale and transformed to vectors, before calculating distance based similarity. Some information is inevitably lost during the process.

In our previous works it was shown [14], that Pearson correlation is not robust against image displacement caused by aircraft frame movement. The 5° error of aircraft heading angle change causes 35% decreased in correlation measure compared to the image with no rotational error. On the other hand, as image resolution is usually high, the distance-based metrics (e.g. Mean squares) struggle from the Curse of dimensionality. In our experiments, to get reliable correlation measure, high resolution images were downsized to 150×150 pixels. High resolution is defined by the capability of on board camera. It ranges from 640×480 pixels of some global shutter or thermal cameras, up to 4K to 8K for general use cameras.

Therefore, to be able to compare two high resolution images that could be rotated more than 5° or taken at different time, we build and train a model which is able to learn image similarity metric. The learning can be formulated in terms of ranking similarity learning. In ranking similarity learning, the goal is to learn a distance function d such that for any triplet of an images (im^a, im^p, im^n) it obeys (2):

$$d(im^a, im^p) > d(im^a, im^n), \tag{2}$$

where im^a denotes an anchor image, im^p an positive image (similar to the anchor), and im^n represents a negative (or different) image.

Recently, many deep neural networks were developed (e.g. AlexNet [17], Resnet [11], Inception [23]) for image classification task, one of classical networks is VGG16 [3]. Therefore, we used pre-trained lower layers of this network as feature extractor for the similarity learning. We make an assumption that if these classifiers have good discriminate properties, they should embed images in space where similar images are close to each other. This property should hold for images on which these networks were not yet trained. As experiments of this paper showed this assumption is valid and a simple network build on these features can be used as the similarity metric.

The paper structure is as follows: we present current state of similarity learning in Sect. 2, proposed deep neural network architecture is presented in Sect. 3, experiment results are shown in Sect. 4, conclusions were drawn in Sect. 5.

2   Related Works

Finding similarity between two images is a relevant task in image registration, image mosaicing, template matching, map-based robot localization, and other applications. This task extends beyond of image classification task and deals with a set of very similar images, otherwise often considered to be of the same class. Some examples are: search-by-example technique - used by many search engines to find very similar images; face recognition - most prominent example being Google's FaceNet [21]; landmark recognition - ability to identify current geographical position based of surroundings; in this paper examined - aerial images similarity.

Image similarity estimation can be achieved by calculating Pearson correlation, Mean squares, Pattern intensity or Mutual information, or other similarity metrics. To increase the accuracy of chosen metric some image prepossessing techniques, often including dimensionality reduction, can be used. The simplest techniques are images resizing and conversion to a lower dimensionality space (e.g. RGB to 8 bit gray scale color conversion). Advanced techniques uses feature extraction methods such as SIFT [18] or HOG [8].

One common approach to the problem is the category-level image similarity. It extends a classification algorithm and considers tho images with the probability of belonging to the same class to be similar [10,24].

With invention of Convolutional Neural Networks and hardware availability to evaluate deep models in a feasible amount of time lots of Deep Learning techniques emerged for image manipulation. A lot of these techniques emerged from ImageNet challenges [9].

Our proposed method requires images to be represented as visual embeddings. Visual embedding is a high-dimensional vector representation of an image which captures semantic similarity. This technique calculates the similarity in a semantic feature space instead of pixel (color) intensity space. This way, the problem of high dimensionality is solved by letting a neural network to find the feature space for image similarity calculation.

The embeddings are produced by deep neural networks in the intermediate layers. No hand tuning or hand-crafted metrics are required, model learns these representations by minimizing the loss function by examples from the training data.

Andreeva et al. [1] used VGG16 from FC6 layer (only top model removed) to get the embeddings. And technique called Locality-sensitive hashing (LSH) to get similarities between embeddings. Other widely used deep neural network architectures to get the embeddings are Inception [23], AlexNet [17], ResNet [12]. It is a similar approach to ones used with textual data, to transform it to k-dimensional space and capture semantic similarity, e.g. word2vec [19], fasttext [4], glove [20]. The choice of neural network model has no fundamental difference from implementation or model usage perspectives. These model are often trained using transfer learning. This way, initial weights are taken from a pretrained model, most often on ImageNet dataset. Additional layers are added and trained based on custom loss function. The pre-trained layers are frozen and their weights are not changed, only the added layers are trained.

Two main types of neural networks architectures for similarity learning are Siamese neural networks [16] and, in this paper discussed, triplet neural networks. Both of these architectures build two or three neural networks witch the same weights and optimize it's loss function to put similar images close to each other in the embeddings space, though triplet neural networks utilizes additional negative (non-similar) example to boost the embedding performance.

Triplet loss was used by Google scientists to create a cutting edge face recognition algorithm - FaceNet [21], at the time of publication it reduced the error rate of face recognition in comparison to the best published result by 30% on Labeled Faces in the Wild (LFW) [13] and YouTube Faces DB dataset [27]. While working on neural network that can easily detect failure in Telecom Operators networks Marc-Olivier Arsenault defined the triplet loss function (5) that prevents loss from going below zero [2]. This improved performance of his model on his dataset, compared to loss function defined by FaceNet paper [21]. Histogram loss was proposed by Ustinova and Lempitsky [25] and showed that such operations can be performed in a simple and piecewise-differentiable manner using 1D histograms with soft assignment operations.

Going deeper into triplet learning process observation can be made, that by reusing an already good classifier (e.g. VGG16 or AlexNet) to find the similarity between very similar images (that is the name fine grained similarity), only edge case examples from the training data can be used to increase networks performance and save some computational resources.

3 Network Architecture

A triplet-based general network architecture is shown in the Fig. 1. This network takes a triplet (Anchor image, Positive image, Negative image) or in a short form (im^a, im^p, im^n) as an input, which are fed independently into three identical deep neural networks $f(.)$ with shared architecture and parameters. The deep neural network consist of all layers taken from pre-trained VGG16 without very last layer and the proposed custom trainable layers (see Fig. 1). The network $f(.)$ computes the embedding of an image $im_i : f(im_i) \in (R)^d$, where d is the dimension of the feature embedding. After application of the deep neural network on a triplet of images, the embeddings $(f(im_i^a), f(im_i^p), f(im_i^n))$ are retrieved from output of the neural network. To effectively train the top layer of the network triplet, the loss function from Eq. (5) is used.

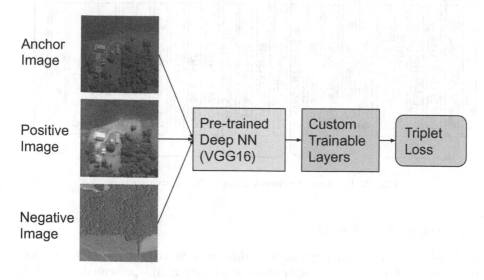

Fig. 1. General model.

Four configurations of the general model by including different number of VGG16 network layers were implemented. First configuration has all VGG16 layers except the last one (Output layer). This configuration is called ModelNN or ModelNN(0). The second configuration does not contain 4 top layers of VGG16 network - ModelNN(-4). And similarly ModelNN(-8) is VGG16 network without 8 top layers. Lastly, ModelNN(-12) has only 6 bottom layers of the VGG16 network. ModelNN(-12) is a network that has the lowest number of VGG16 layers and gives good classification results for image triplets.

Detailed model network ModelNN(-8) architecture is presented in the Fig. 2. All the VGG16 layers of the model network uses original weights, since modification of these weights reduced overall accuracy of the similarity estimates. This is most likely caused due to the small size of the dataset used for training and high number of trainable variables in VGG16. This leads to a very fast over-fitting of the model network. The custom layers of the model consists of one fully connected layer of sizes $28 \times 28 \times 1$ and a flatten layer to get vector for the triplet loss calculation. The fully connected layer size changes depending on the output of the last VGG16 layer. As lower layers of VGG16 network have larger output dimensions, therefore by omitting more VGG16 layers, we get more learnable weights in ModelNN custom layers.

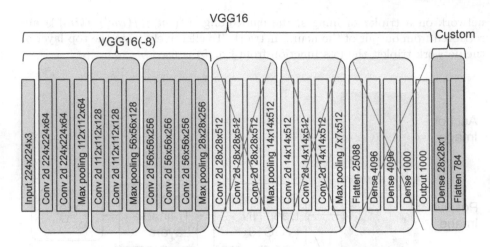

Fig. 2. The model network ModelNN(-8) architecture.

3.1 Triplet Loss Function

To be able to learn similarity metric in this research we looked for effective loss function. A one of the most popular triplet loss function (3) is presented bellow [26]:

$$loss = \sum_{i=1}^{N_{tr}} max(d(f_i^a, f_i^p) - d(f_i^a, f_i^n) + margin, 0), \qquad (3)$$

where d is some distance metric such that $d(f_i^a, f_i^p) < d(f_i^a, f_i^n)$, triplet (f_i^a, f_i^p, f_i^n) consist of an anchor image, a positive image and a negative image, respectively and N_{tr} is a number of triplets. $margin > 0$ is a gap parameter that regularizes the gap between the distance of the two image pairs: (f_i^a, f_i^p) and (f_i^a, f_i^n).

Closely related to the Eq. (3) is an Eq. (4) [21]. The Eq. (4) uses squared Euclidean distance as distance metric and $\alpha > 0$ as margin between two pairs of images and sum absolute values. This triplet loss function was successfully used for training deep neural networks for face verification and recognition.

$$loss = \sum_{i=1}^{N_{tr}} [||f_i^a - f_i^p||_2^2 - ||f_i^a - f_i^n||_2^2 + \alpha]_+, \qquad (4)$$

where triplet (f_i^a, f_i^p, f_i^n) consist of an anchor image, a positive image, and a negative image respectively, α is some small positive number (e.g. $e = 10^{-6}$) and N_{tr} is a number of triplets.

Histogram loss was proposed by Ustinova and Lempitsky [25] and showed that such operations can be performed in a simple and piecewise-differentiable manner using 1D histograms with soft assignment operations.

This paper uses triplet modified loss function [2]:

$$loss = -\sum_{i=1}^{N_{tr}} [ln(-\frac{\sum_{j=1}^{N}(f_{ij}^a - f_{ij}^p)^2}{\beta} + 1 + \epsilon) + ln(-\frac{N - \sum_{j=1}^{N}(f_{ij}^a - f_{ij}^n)^2}{\beta} + 1 + \epsilon)],$$

(5)

where f_{ij}^a is ModelNN output for an anchor image, f_{ij}^p and f_{ij}^n are ModelNN outputs for positive and negative image respectably. N is output vectors dimension, N_{tr} is a number of triplets in the training batch, ϵ is some small positive number (e.g. $e = 10^{-6}$). The loss function is equal to 0 when both positive and negative images are at the maximum distance. The loss (5) can be averaged over number of triplet to keep its value not depended on the N_{tr}.

To measure the similarity between some two images im_1 and im_2, we use modified loss function (5). Instead of Eq. (6) can be used standard Euclidean distance, but our similarity metrics gives lower values of the distances.

$$sim(im_1, im_2) = -ln(-\frac{\sum_{j=1}^{N}(f_{1j} - f_{2j})^2}{\beta} + 1 + \epsilon),$$

(6)

where $f_{1j} = f(im_1)_j$, $f_{2j} = f(im_2)_j$, N is number of dimensions and $j \in [1..N]$.

4 Experiments

The triplet network is implemented using Tensorflow 1.3.1 framework, Python 3.7 and a NVIDIA GeForce RTX 2080 Ti Graphics Card was used to train the network.

4.1 Image Dataset for Training and Testing the Neural Network

For this research special Aerial Imagery dataset [15] was used, which consists of 113474 images captured from the different UAV flights in a robotics simulator. All images are of 640×480 resolution and are rectified to do not contain camera distortion. For this experiment, a special subset of the dataset was created. It was created by performing flights and recording aerial images on two maps simultaneously, which were created a few years apart. The subset contains 1188 of image triplets (anchor image(im^a), positive image(im^p), negative image(im^n)) for training and 105 image triplets for validation (e.g. Fig. 3) and 64 for the testing. The anchor and positive images are aerial images of the same place but at a different time, the negative images are of randomly selected regions from the map that was used to create the anchor images. The input dimensions for the VGG16 network are 224×224, therefore the triplets images are resized to the input dimensions. In case of K-fold cross validation, the test and training datasets parts were joined before performing the cross validation splitting.

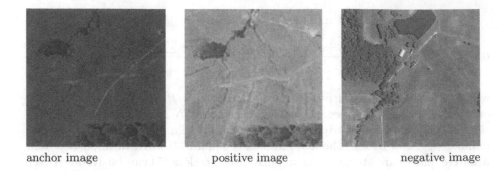

anchor image   positive image   negative image

Fig. 3. Example of a triplet of images.

4.2 Accuracy Measure

The goal of ranking similarity learning is to learn the distance function d such that for any triplet of images (im^a, im^p, im^n) it obeys $d(im^a, im^p) > d(im^a, im^n)$. Accuracy metric (7) from the evaluation of binary classifiers can be used to evaluate the accuracy of ModelNN.

$$accuracy = \frac{\sum True\ positive + \sum True\ negative}{\sum Total\ population}. \tag{7}$$

As it is very rare to $d(im^a, im^p)$ be equal to $d(im^a, im^n)$ we can assume that $\sum True\ positive$ is equal to $\sum True\ negative$. Therefore, we can count only $True\ positive$ triplets and divide by total number of triplets N_{tr} in the training dataset. The final Eq. (8) is used to evaluate the accuracy of the ModelNN.

$$accuracy = \frac{\sum_{d(im^a, im^p) > d(im^a, im^n)} 1}{N_{tr}}. \tag{8}$$

4.3 Pearson Correlation as Similarity Metric

In this research, Pearson correlation is used as a baseline, since it was the most suitable similarity metric found for calculating similarity of UAV images in our previous research [14]. Using Pearson correlation as the classifier on all training dataset (Fig. 4), an accuracy of 0.884 was achieved. Repeating the experiment on images transformed to gray scale, the accuracy slightly increased to 0.888. This result show that Pearson correlations struggle from the Curse of dimensionality, and lowering the dimensions of the images gives better results. A triplet (im^a, im^p, im^n) is considered a true positive in Eq. (8) if $pearson(im^a, im^p) > person(im^a, im^n)$.

4.4 Learning Metric on Full VGG16 Model

Grid search with K-fold cross validation of 4 folds was used to tune hyper-parameters to improve the model performance.

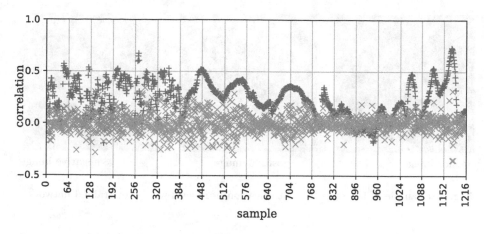

Fig. 4. Pearson correlation: blue points are representing correlation between im^a and im^p images, orange points are representing correlation between im^a and im^n images. (Color figure online)

Table 1. Dependency of ModelNN(0) average accuracy (in %) on number of epochs and batch size.

Batch size	1	4	16	50
3	99.7 (0.21)	99.92 (0.13)	99.78 (0.25)	100.0 (0.0)
6	99.25 (0.43)	99.92 (0.13)	99.92 (0.13)	99.92 (0.13)
12	98.22 (1.03)	99.85 (0.26)	99.92 (0.13)	99.92 (0.13)
24	96.48 (1.36)	98.98 (0.62)	99.92 (0.13)	99.92 (0.13)
48	96.2 (0.69)	99.12 (0.69)	100.0 (0.0)	99.92 (0.13)
96	90.82 (3.24)	98.4 (0.82)	99.78 (0.25)	100.0 (0.0)

Table 1 depicts iterations and batch size influence on accuracy of full VGG16 model with custom top layers or ModelNN(0). Standard deviation of the accuracy is presented in the parentheses. The table shows that ModelNN(0) accuracy on the test dataset increases with increasing number of the epochs, and decreases with increasing batch size. ModelNN(0) has only 49 trainable parameters, as number of images is greater than 1000, this avoids over-fitting the neural network and achieved high accuracy shows that ModelNN(0) learns to discriminate similar images from not one. Nevertheless, we can observe signs of over-fitting when we pass 16 epochs of training (e.g. batch size 48). Therefore, we suggest to train the ModelNN for 16 epochs.

Figure 5 presents a triplet of images with wrong classification due to over-trained ModelNN(0).

Looking at similarity metric distribution over image triplet (Fig. 6) we can observe that ModelNN(0) can learn to separate positive images from negative ones. The Fig. 6 has 4 different regions, this is because we used 4 folds cross

anchor image   positive image   negative image

Fig. 5. Example of false positive classification of images with over-trained network.

validation and trained 4 different networks. From right side of the figure we can conclude that, similar images in our dataset have very low distance (or high similarity) values (6). Assumption can be made that two images are similar if the similarity value is lower than 0.5, we get accuracy of 98%.

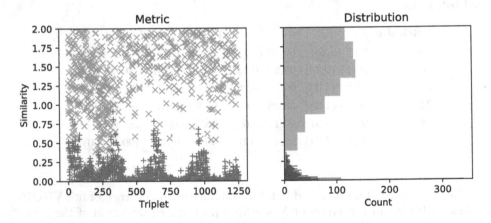

Fig. 6. K-fold cross-validation ModelNN(0) trained with 16 epochs.

4.5 VGG16 Depth Influence

Relation between different modifications of the ModelNN(k) and number of training epochs is presented in Table 2. Number of omitted layers k is shown in the first column of the Table 2. All experiment were run using cross validation of 4 folds. The result shows that ModelNN with lower number of VGG16 layers tend to learn slower, and to achieve same accuracy need more training epochs. Therefore, we suggest to use ModelNN(0) or ModelNN(−4) configuration for image similarity estimation.

As Fig. 6 shows, similarity values $sim(im^a, im^p)$ and $sim(im^a, im^n)$ are quite well separate from each other. Table 3 shows that the same tendency is applicable

Table 2. Average accuracy dependency on number of layers and epochs. Batch size is equal to 48

ModelNN(k)	1	4	16	50
0	96.2 (0.69)	99.12 (0.69)	100.0 (0.0)	99.92 (0.13)
−4	96.18 (3.01)	99.45 (0.79)	100.0 (0.0)	99.92 (0.13)
−8	88.88 (8.21)	99.12 (0.88)	99.52 (0.82)	99.92 (0.13)
−12	79.7 (9.12)	91.38 (10.45)	99.2 (0.67)	100.0 (0.0)

to other ModelNN configurations. Min-max of the similarity values for similar images are presented in each odd row of the Table 3 and the values for not similar images are in each even row.

Table 3. Dependency of min and max similarity (6) values on number of layers and epochs. The batch size is 48.

ModelNN(k)	1	4	16	50
0	0.0–0.99	0.0–1.25	0.0–0.77	0.0–0.8
	0.13–2.75	0.16–5.11	0.46–7.4	0.45–10.09
−4	0.01–0.92	−0.0–0.94	0.0–0.53	0.0–0.68
	0.34–2.24	0.27–4.18	0.62–13.82	0.51–13.82
−8	0.1–0.78	0.0–0.84	0.0–0.88	0.0–0.57
	0.25–1.21	0.2–2.57	0.15–4.27	0.23–5.05
−12	0.21–0.75	0.12–0.69	0.03–1.41	0.01–0.8
	0.37–0.97	0.2–1.7	0.28–3.0	0.41–2.94

5   Conclusions

This paper presents an investigation of image similarity metric learning to estimate image similarity of aerial images from UAV flights. We developed ModelNN neural network based on VGG16 with additional custom layer and a modified triple loss function. ModelNN was able to learn image similarity with the accuracy greater than 99%. All configurations of ModelNN with different VGG16 depth were able to learn images similarity, but ModelNN with smaller number of VGG16 layers requires higher number of training epochs. Therefore, we recommend to use ModelNN(0) for image similarity estimation, as it is the most accurate and can be trained quickly.

We proposed the similarity metric based on the image embeddings from ModelNN, and if the similarity value $sim(im_i, im_j) <= 0.5$, we can confidently state that images are similar. From the similarity function properties, it follows that the images are getting more similar when $sim(im_i, im_j)$ is getting closer to 0. The similarity of value 2 and higher can be used as a threshold to confidently reject the hypothesis that images are similar.

6 Future Works

If the future we are planing to investigate how affine transformations can influence our ModelNN performance. More complicated custom layer architecture may be required. Another approach is to use one of the deterministic optimization methods to optimize triplet loss function, as proposed, e.g., in [22].

References

1. Andreeva, E., Ignatov, D.I., Grachev, A., Savchenko, A.V.: Extraction of visual features for recommendation of products via deep learning. In: van der Aalst, W.M.P., et al. (eds.) AIST 2018. LNCS, vol. 11179, pp. 201–210. Springer, Cham (2018). https://doi.org/10.1007/978-3-030-11027-7_20
2. Arsenault, M.O.: Lossless triplet loss. A more efficient loss function for Siamese NN (2018)
3. Banerjee, P., Bhunia, A.K., Bhattacharyya, A., Roy, P.P., Murala, S.: Local neighborhood intensity pattern: a new texture feature descriptor for image retrieval. CoRR (2017)
4. Bojanowski, P., Grave, E., Joulin, A., Mikolov, T.: Enriching word vectors with subword information. arXiv preprint arXiv:1607.04606 (2016)
5. Boyd, S.: Introduction to Applied Linear Algebra Vectors, Matrices, and Least Squares. Cambridge University Press, Cambridge (2018)
6. Chang, Y., Hsieh, C., Chang, K., Ringgaard, M., Lin, C.: Training and testing low-degree polynomial data mappings via linear SVM. J. Mach. Learn. Res. **11**, 1471–1490 (2010)
7. Church, K.W., Hanks, P.: Word association norms, mutual information, and lexicography. Comput. Linguist. **16**(1), 22–29 (1990)
8. Dalal, N., Triggs, B.: Histograms of oriented gradients for human detection. In: International Conference on Computer Vision & Pattern Recognition (CVPR 2005), vol. 1, pp. 886–893. IEEE Computer Society (2005)
9. Deng, J., Dong, W., Socher, R., Li, L.J., Li, K., Fei-Fei, L.: ImageNet: a large-scale hierarchical image database. In: CVPR 2009 (2009)
10. Hadsell, R., Chopra, S., LeCun, Y.: Dimensionality reduction by learning an invariant mapping. In: 2006 IEEE Computer Society Conference on Computer Vision and Pattern Recognition (CVPR 2006), vol. 2, pp. 1735–1742. IEEE (2006)
11. He, K., Zhang, X., Ren, S., Sun, J.: Deep residual learning for image recognition. CoRR (2015)
12. He, K., Zhang, X., Ren, S., Sun, J.: Deep residual learning for image recognition. In: Proceedings of the IEEE Conference on Computer Vision and Pattern Recognition, pp. 770–778 (2016)
13. Huang, G.B., Mattar, M., Berg, T., Learned-Miller, E.: Labeled faces in the wild: a database for studying face recognition in unconstrained environments. In: Workshop on Faces in 'Real-Life' Images: Detection, Alignment, and Recognition (2008)
14. Jurevičius, R., Marcinkevičus, V.: Application of vision-based particle filter and visual odometry for UAV localization. In: WSCG 2017: Short Communications Proceedings: The 25th International Conference in Central Europe on Computer Graphics, Visualization and Computer Vision 2016 in co-operation with EUROGRAPHICS: University of West Bohemia, Plzen, Czech Republic, 29 May–2 June 2017, pp. 67–71. Václav Skala - UNION Agency (2017)

15. Jurevičius, R., Marcinkevičius, V., Šeibokas, J.: Robust GNSS-denied localization for UAV using particle filter and visual odometry. Mach. Vis. Appl. **30**(7–8), 1181–1190 (2019). https://doi.org/10.1007/s00138-019-01046-4. ISSN 0932-8092. eISSN 1432-1769

16. Koch, G., Zemel, R., Salakhutdinov, R.: Siamese neural networks for one-shot image recognition. In: ICML Deep Learning Workshop, vol. 2 (2015)

17. Krizhevsky, A., Sutskever, I., Hinton, G.E.: Imagenet classification with deep convolutional neural networks. In: Pereira, F., Burges, C.J.C., Bottou, L., Weinberger, K.Q. (eds.) Advances in Neural Information Processing Systems 25, pp. 1097–1105. Curran Associates, Inc., New York (2012)

18. Lowe, D.G., et al.: Object recognition from local scale-invariant features. In: ICCV, vol. 99, no. 2, pp. 1150–1157 (1999)

19. Mikolov, T., Sutskever, I., Chen, K., Corrado, G.S., Dean, J.: Distributed representations of words and phrases and their compositionality. In: Burges, C.J.C., Bottou, L., Welling, M., Ghahramani, Z., Weinberger, K.Q. (eds.) Advances in Neural Information Processing Systems 26, pp. 3111–3119. Curran Associates, Inc., New York (2013)

20. Pennington, J., Socher, R., Manning, C.D.: Glove: global vectors for word representation. In: EMNLP (2014)

21. Schroff, F., Kalenichenko, D., Philbin, J.: FaceNet: a unified embedding for face recognition and clustering. CoRR (2015)

22. Sergeyev, Y.D., Kvasov, D.E., Mukhametzhanov, M.S.: On the efficiency of nature-inspired metaheuristics in expensive global optimization with limited budget. Sci. Rep. **8**, article 453 (2018)

23. Szegedy, C., et al.: Going deeper with convolutions. CoRR (2014)

24. Taylor, G.W., Spiro, I., Bregler, C., Fergus, R.: Learning invariance through imitation. In: CVPR 2011, pp. 2729–2736. IEEE (2011)

25. Ustinova, E., Lempitsky, V.: Learning deep embeddings with histogram loss. In: Lee, D.D., Sugiyama, M., Luxburg, U.V., Guyon, I., Garnett, R. (eds.) Advances in Neural Information Processing Systems 29, pp. 4170–4178. Curran Associates, Inc., New York (2016)

26. Wang, J., et al.: Learning fine-grained image similarity with deep ranking. CoRR (2014)

27. Wolf, L., Hassner, T., Maoz, I.: Face recognition in unconstrained videos with matched background similarity. IEEE (2011)

In the Quest for Invariant Structures Through Graph Theory, Groups and Mechanics: Methodological Aspects in the History of Applied Mathematics

Sandra Visokolskis[1]([⊠]) [ID] and Carla Trillini[2] [ID]

[1] National University of Cordoba, Cordoba, Argentina
sandraviso@gmail.com
[2] National University of Villa Maria, Cordoba, Argentina

Abstract. The purpose of this paper is to analyze a geometrical case study as a sample of an intended methodology based on invariant theory's strategies, which have been developed particularly throughout the nineteenth century as one of the cornerstones of mathematics [15, p. 41], and whose resolution was reached by means of a combination of different disciplines: graph theory, mechanics and group theory, among others.

This case study presents the "perfect squared rectangle problem", that is an exhaustive classification of the dissection of a rectangle into a finite number of unequal squares. Despite its simplicity, in both description and mathematical resolution, it provides plausible elements of generalization from "the 'applied field' of mathematics" [8, p. 658], as a special case of applied mathematical toolkit [1, p. 715], related to the practice of invariant strategies that remain fixed through changes.

Keywords: Invariants · Graph theory · Geometry

1 Introduction

One of the cornerstones of mathematics is *invariance*, i.e. patterns of regularity that characterize permanent situations when a transformation occurs, stable properties of objects that remain constant despite changes in the system to which they belong. Objects and their properties may be invariant under specific changes that occur in the object or with the object. Therefore, an invariant relationship remains stable or unchanged regardless of the occurrence of other changes. When talking about identity, congruence, isomorphism, cycles, constancy, symmetry or periodicity, we refer to a situation in which a particular property remains permanent despite changes, either during the transition from one situation to the next, or during the process of transformation between two components of it. Invariance holds over a range of changes, and it is recognized as an intrinsic feature, a common property.

Thus, given a domain of mathematical objects and after applying some transformation on them, it turns out that some of the properties remain invariant. Turning our focus

© Springer Nature Switzerland AG 2020
Y. D. Sergeyev and D. E. Kvasov (Eds.): NUMTA 2019, LNCS 11974, pp. 208–222, 2020.
https://doi.org/10.1007/978-3-030-40616-5_16

back on such invariance producing transformations, these become entities of another level, susceptible of constituting a new field of study.

The advantage of knowing many invariant functions is that we can use them to completely characterize equivalent objects from different domains. Two objects are equivalent if they can be transformed into each other by means of making a suitable change in the mapping that connects them. Equivalence makes both objects indistinguishable. If we can determine completely the equivalent indistinguishable objects of some domain among the invariants functions, these invariants form the basic building blocks, which can be used to construct some kind of objects of mathematical interest. Conversely, and equally important, two equivalent objects must necessarily have the same invariants.

This is the sense in which it is characterized, within a specific domain, the "fundamental equivalence problem", i.e. to determine whether two mathematical objects can be transformed into each other by a suitable change of some kind. Therefore, a solution to this problem will allow the characterization of all the equivalent entities, as long as two objects can be identified under a specific transformation. The characterization of this class of objects can generate a new theory in mathematics.

Therefore, our aim is set at characterizing a methodology that not only involves developing a variety of techniques that will allow one to handle some or all of the problems of specific kinds in a wide variety of mathematical contexts, but also generating a strategy that allows the emergence of the totality of objects of a certain type. This is where the notion of invariance is relevant, since the construction of invariants and their description are used to characterize the equivalent objects and then completely solve the equivalent problem. Such equivalent objects brings to the discovery of indistinguishabilities, which, once they are fully determined and defined, can conform accepted theories.

The work of many mathematicians includes the determination of the totality of objects that meet specific conditions, a characterization of all the elements within a class that need to be defined. For instance, the search for symmetries of a geometric object ended by describing the notion of "group", which in turn allowed the classification of differential equations and variational problems. The search for invariants was influential in the process of developing the notion of group.

In the following sections, we will establish how different ways of understanding the notion of invariance have led to characterize the indistinguishability of objects in different areas of mathematics. In the case of geometry, where we will mainly focus, the search for invariants was the crucial step to find indistinguishabilities. That is to say, some type of equivalence that would allow us to characterize the type of geometry one was working with, and, consequently, the theory that came along with it. Therefore, it is worth emphasizing the heuristic and creative aspects that drive the search for invariants, and not just their justificatory processes, which must then be put into action to convert the findings into thoroughly consolidated results within some appropriate deductive system.

In this paper, the search for invariants will be analyzed from a methodological point of view, in order to characterize strategies present in mathematics since its inception. The history of Western mathematics has a long tradition, originated in Greece, on a *top-down* methodological characterization, described from theories conformed by true, evident, eternal and immutable basic elements: *Elements* of Euclid is a tangible sample of such top-down style. From such theories, applications can be built in various fields, internal as well as external to mathematics.

The Euclidean top-down style has been used in several domains: in logic for example (Aristotle's syllogistic), in probability (Suppes' axioms for propensities and foundations of objective probability [16]), in foundations of mathematics (Russell and Whitehead's *Principia Mathematica*), in biology (Woodger's axiomatic methodology [23]), in psychology (Hull's principles of behavior as well as Rottmayer's formal theory of perception), in physics (McKinsey, Sugar and Suppes' axiomatics foundations of classical particle mechanics); even in philosophy, as it was the case with Spinoza's ethics, written following traditional axiomatic schemes.

But also, and especially in Modernity, many other mathematical developments respect a *bottom-up* methodology, where it is the practice itself that leads to the generation of theories. Mathematicians such as Lagrange, Gauss, Laplace and Euler show mathematical developments arising from the interaction with empirical problems from other disciplines than the so-called "pure mathematics", which have usually received the label of "mixed" or "applied" mathematics.

This bottom-up style, even at the dawn of the nineteenth century, has been nourished by another cognitive strategy that we will refer to as "*transductive* methodology" [18–21], which will take a central role in this work. Our main objective will be to determine how this type of methodology is implemented in a good number of applied mathematics developments. The intended methodology is based on invariant theory's strategies, and have been developed particularly throughout the nineteenth century as one of the cornerstones of mathematics:

> "[Invariant theory] has had as deep and lasting influence on the development of mathematics, to the point that seldom in history has an international community of scholars [from England, Germany, France, Italy and America] felt so united lay a common scientific ideal" [15, p. 41].

Consequently, in section two of the article, we develop the transductive methodological proposal, for which we introduce a brief historical overview of the notion of invariance in mathematics, which will allow us to understand how this transductive style is typical in a great number of applied mathematics related works. Next, in section three, we analyze a geometric case study based on the proposed methodology, which highlights the importance of invariants in applied mathematics.

This case study has its origin in geometry, and presents the "perfect squared rectangle" problem, that is an exhaustive classification of the dissection of a rectangle into a finite number of unequal squares. Despite its simplicity, in both description and mathematical resolution, it provides plausible elements of generalization from "the 'applied field' of mathematics" [8, p. 658], as a special case of applied mathematical toolkit [1, p. 715], related to the practice of invariant strategies that remain fixed through changes.

2   Invariant Transductive Methodology: A Proposal

For the purpose of introducing our proposal, which is based on the theory of invariants, as we mentioned in the introduction, we carry out a historical outline of the emergence of such theoretical developments. We will concentrate on an analysis of different invariance principles that were adopted effectively by several geometers, mainly in the nineteenth century. This condensed journey will allow us to appreciate the relevance of problem solving strategies that rely on the search for patterns of regularity, and their applications in other domains different from those from which they arose.

Within the great variety of instances of the notion of invariance in mathematics, there is a case that should be highlighted and that owes its beginnings to quadratic binary forms[1] and to the problem of representing integers by them (as well as a to a type of indistinguishability based on the use of the discriminant to match equivalent forms): the classical invariant theory.

The roots of invariant theory can be traced back at least to Lagrange in 1788, on the one hand, and to Gauss in 1801, on the other. In the case of Carl Friedrich Gauss (1777–1855), his *Disquisitione arithmeticae* of 1801, at the turn of nineteenth century, had observed a special case of algebraic invariance, the discriminant of a binary quadratic form. Indeed, given a quadratic binary form f with integer coefficients $f(x, y) = ax^2 + 2bxy + cy^2$ with $a, b, c \in \mathbb{Z}$, an homogeneous polynomial of degree 2 (quadratic) in 2 (binary) unknowns, let T be a (non-singular[2]) linear transformation affecting f, that is $T(x', y') = (mx' + ny', m'x' + n'y')$ with $m, n, m', n' \in \mathbb{Z}$. Due to the application of T to f, Gauss get a new binary quadratic form:

$$f(T(x', y')) = Ax'^2 + 2Bx'y' + Cy'^2 \text{ where}$$

$A = am^2 + 2bmm' + cm'^2$, $B = amn + b(mn' + nm') + cm'n'$, $C = an^2 + 2bnn' + cn'^2$. Gauss observed that the discriminant $b^2 - 4ac$ of the original form f satisfied the following relation to the discriminant of the transformed form: $B^2 - 4AC = \Delta^2 . (b^2 - 4ac)$. Hence, the discriminant of the original form f was altered uniformly by a factor, which depends on the coefficients m, n, m', n' included in the transformation T. More precisely, a factor that is equal up to a power (2) of the determinant of the linear transformation T, that is $\left(mn' - nm' \right)^2$. In brief, the discriminant $b^2 - 4ac$ is an *invariant* of the binary form f, under a non-singular linear transformation T.

Now, regarding Joseph-Louis Lagrange (1736–1813), in his 1788 two-volume book *Mécanique analytique*, he worked on a specific physical problem concerning the motion of various kinds of bodies representing kinetic energy. This representation was realized by a reduction from a quadratic form to a sum of squares; in other words, by some linear substitution that then allowed a diagonalization of the quadratic form, via what we now call an orthogonal transformation.

Lagrange found that the coefficients of the transformation satisfied some vanishing conditions, which remain permanent through the algebraic operations. This aspect caught Boole's attention. Working now on Lagrange ideas, George Boole (1815–1864) isolated the phenomenon of invariance while working on the resolution of the general

---

[1] The word "quadratic" refers here to the degree of homogeneity of the variables of the form (so each term of the form has degree two); whereas the adjective "binary" indicates the number of variables involved in the form.

[2] This means that the determinant Δ is non zero: $\Delta = mn' - m'n \neq 0$.

problem of determining algebraic relationships among the coefficients of homogeneous polynomials of degree n in m unknowns, which remain *invariant* under a non-singular linear transformation. Boole eliminated the restrictions that Gauss imposed on the coefficients. More precisely, Boole called *invariant* any expression in the coefficients of a binary form, which varies only in a factor depending on the linear transformation. However, if the expression not only involves the coefficients but also the variables of the form, it is called a *covariant*. After that, two British algebraists, Arthur Cayley and James Joseph Sylvester, carried out the study of invariants and covariants.

Before discussing the work of these two English mathematicians, the following observation should be made regarding Boole's methodological work: his main goal was not the finding of invariants. This task ended up being a consequence of his greater concern for solving the equivalence problem, an indistinguishability problem. In order to solve the equivalence problem between two n-ary forms, in his two first papers [2, 3] Boole proceeded by setting to zero one particular invariant: the discriminant.

To understand this procedure, let us consider a very simple situation, the discriminant of the following quadratic polynomial (in one variable): $p(x) = x^2 + 2bx + c$ with b, $c \in \Re$. Nowadays, the simplest method to acquire the discriminant consists in applying Bhaskara's formula, with discriminant $D = b^2 - c$, which, in this case has three possibilities: $D > 0$ (two different real roots of the equation), $D = 0$ (double equal real root), and $D < 0$ (two different conjugate complex roots). However, what Boole noticed is that having a double root is *invariant* under translation, that is, if we consider the polynomial $p(x + d) = (x + d)^2 + 2b(x + d) + c$ where d is constant, then its corresponding discriminant $D' = b^2 - c$ would be the same as D.

Here, we are looking for an invariant whose vanishing expresses the condition that if the polynomial p has a double root, so does the polynomial p', result of the translation due to a change of variables $T(x) = x' = x + d$. However, the way Boole solved this problem in a general style implied the utilization of partial derivatives to *eliminate* the variables from the given polynomial. Again, using a very simple case, the one we saw previously from Gauss, but now without the restrictions over the variables and parameters, let $f(x, y) = ax^2 + 2bxy + cy^2$ with a, b, $c \in \Re$. We now calculate the partial derivatives of f with respect to x and y, and equating them to zero, we obtain: $\frac{\partial f}{\partial x}(x, y) = 2ax + 2bx = 0$ and $\frac{\partial f}{\partial y}(x, y) = 2bx + 2cy = 0$. The *elimination* of x and y from these equations yielded the expression $D(f) = b^2 - ac$, which was the relation he was looking for between the coefficients of f. If we now apply a linear transformation T as before (but now under $\Re$), we obtain the transformed binary form $g(x', y') = Ax'^2 + 2Bx'y' + Cy'^2$ with A, B, $C \in \Re$. Calculating the corresponding partial derivatives of g relative to x and y, we find that its discriminant is a multiple of the previously discriminant $D(f)$, altered by a factor which depends only on the coefficients included in the linear transformation, by elimination of the variables: $D'(g) = k.D(f)$.

What Boole ends up achieving is a *method of elimination*, that, eventually but not primarily, works by capturing one type of invariance as well as covariance. More precisely, this method, among equations in partial derivatives or differentials of the forms, instead of allowing to capture invariants, what is actually allowed is the application of his method to obtain substitutions that transform a pair of forms into an equivalent pair of forms. His goal, then, was to produce "important step[s] in the theory of the linear transformation of functions of two variables" [4, p. 98]. Nevertheless, who took over the task that Boole initiated is Cayley, taking it one step further: it aims at the isolation of many types of invariance (and covariance), that is, it is oriented towards the independence of a theory of invariants. As Paul R. Wolfson says:

> "Cayley [had discovered] that there were invariants (and covariants) other than those defines by Boole [that led Cayley] to refer to this property as 'the characteristic property'. By focusing on expressions that satisfy the 'characteristic property', Cayley was shifting the direction of research. Even though Boole had shown that his D enjoyed this property, he had not highlighted it as the defining relation of an invariant. Indeed, he would have seen no need to do so, for he was considering only D and related functions, anyway. Rather than study functions with a certain property, Boole had employed the relation to obtain the proportions" [22, p. 44].

Moreover, in later works [4, p. 95] to his classic articles of 1841 and 1842 [2, 3], Boole himself recognized that Cayley had detected other invariants in addition to the discriminants of n-ary forms, what Boole called 'constant functions': "There exist other functions than $D(f)$ [f n-ary form; D discriminant] possessing those [invariance] properties which I had regarded as peculiar to it" [4, p. 95]. Wolfson says: "After that, the hunt was on, not merely for new invariants, but for what we would now call generators and relations in the ring of invariants" [22, p. 44]. Then he declares:

> "By contrast to Boole's direct attack on the equivalence problem, (...) his principal aims in (1841b and 1842) had been, first to determine when two pairs of forms are equivalent, and second, if they are indeed equivalent, to determine those substitutions which take the first pair to the second (...) Cayley shifted attention to the production and study of the invariants (and covariants) themselves." [22, p. 45]

Note that, while Boole dealt with the fundamental problem of equivalence (i.e., to determine whether two pairs of forms can be transformed into each other, that is, one replaceable by the other, by a suitable change of variables), Cayley focused on the study of invariants and covariants. Thus, the equivalence problem and the invariance search are two sides of the same coin, that is, the search for an intelligible and complete characterization of mathematical theory. This is the kind of indistinguishability that invariant properties can preserve. In the situation we explained *ut supra*, where $D' = k.D$, two homogeneous functions become similar functions, as their discriminants are equal up to a power of a constant.

In 1846, Arthur Cayley (1821–1895) published a paper where his ideas about invariance were very clear. His purpose was, in his own words: "(...) to find all derivatives [i.e., *invariants*] of any number of functions, which have the property of preserving their form unaltered after any linear transformation of the variables" [6, p. 104]. The project began by a quest to find *all* invariants of any number of forms; but then he moved on to looking for what we now call the "generator and relations in the rings of invariants" [22, p. 44], as we said *ut supra*.

Arthur Cayley has employed a calculational oriented methodology: he intended and achieved to formulate several efficient algorithms for explicitly exhibiting invariants and covariants. However, this project finally stumbled upon many technical difficulties. The methods employed were based on what was then called the symbolic or *umbral* notation (*umbrae*, "shadows" of specifiable coefficients). Above all, it seemed impossible to convey any sense of the subject in nontechnical terms.

By 1850, Cayley met James Joseph Sylvester (1819–1897), when both of them, due to different motives, were initiating a career in law, besides their formation in mathematics. Sylvester brought new perspectives on the incipient theory of invariants. They both contributed to its inception, applying similar calculational techniques. Is worth noting that this theory too was the joint work of mathematicians from several countries. In England, besides those two just named, there were others: Young, Turnbull, and Salmon. In France: Hermite, Jordan, Laguerre. In Italy: Capelli, Brioschi, Trudi. In Germany: Aronhold, Clebsch, Gordan, Grassmann, Lie, Study. In America: Glenn, Dickson, Bell and later Weyl. Karen Hunger Parshall gave a very precise characterization of the cooperative work done by Cayley and Sylvester:

"Each of [the] texts [from Cayley and Sylvester] read, in a real sense, like a cookbook for the proper preparation of invariants and covariants. In the absence both of the necessary theoretical underpinnings and of a sufficiently general notation, the British school's techniques did not lend themselves to proving existence theorems (…) This is not to say that the non-symbolic approach of Cayley and Sylvester did not have spectacular successes. It enabled its adherents to calculate the invariants and covariants for binary forms up to the eighth degree and to determine the syzygies, or dependences, between them. They catalogued their results in massive tables, the very construction of which generated important discoveries in combinatorics and in the theory of symmetric forms" [13, p. 186].

The problem with this extreme technical style is the loss of intelligibility of mathematical expressions. If what is spoken here refers to the semantic content of the theories, Hermann Weyl (1885–1955), in his book *The Classical Groups* sought to highlight this aspect about invariants applied to geometry. There, Weyl indicated that the computation needs to express the "geometric facts", that is, facts about space that are independent of the choice of a coordinate system. There was a search for a translation of invariant algebraic equations expressed in terms of tensors into geometric facts, which could capture irreducible components under changes of coordinates. This task–one of the great advances in mathematics of all times, according to Gian-Carlo Rota [15] - involves the understanding of the indistinguishable ideas behind technical theory. It was discovered around the turn of twentieth century almost simultaneously by Issai Schur and Alfred Young. It is important to note that the emphasis on a theoretical description of invariance that shows the indistinguishabilities of each theory, was previously carried out successfully by the German school; especially in the field of geometry, by Felix Klein (1849–1925), and in the constitution of Modern Algebra, by David Hilbert (1862–1943).

An important observation deserves to be brought up: while Boole, in his investigations put the emphasis on the set of transformations or substitutions that can produce equivalences between two pairs of n-ary forms, the theory of algebraic invariants cultivated in England, especially by Cayley and Sylvester, explored those relations or properties that remain permanent under a set of transformations. Another point in this brief summary of ours should be directed toward developments involved in geometry research. In this sense, both the problem of equivalence treated by Boole and the theory of invariants from the English school of thought point towards geometry–with the works of Felix Klein and Hilbert-, where either the invariant properties or the set of transformations may be taken to characterize the geometry. The foregoing notes had consisted of a condensed summary of the classical theories of invariants.

Next, we will turn to specific cases where we will not refer to "theories" of invariance, but to ways of applying invariance without constituting theories in the style just described. These cases are Peacock's algebra and, finally, the projective geometry of Poncelet.

George Peacock formulated in 1830 a general principle that became influential in the later nineteenth and early twentieth centuries. He called it the *Principle of the Permanence of Equivalent Forms*, and stated it as follows:

"(...) If we discover an equivalent form in Arithmetical Algebra or any other subordinate science, when the symbols are general in form though specific in their nature, the same must be an equivalent form, when the symbols are general in their nature as well in their form" [14, p. 104].

An important feature of this principle was the origin of its ideas. In this regard, Peacock believed and defended the idea that algebra was a pure science, unlike what other prevailing approaches about the role of algebra in Cambridge society, like Peacock's Trinity College colleague and friend William Whewell, who throughout the 1820's did not consider the study of abstract algebra as an independent discipline.

Peacock's defense was supported by the introduction of the aforementioned principle, which was based on Peacock's philosophy of what he called "suggestion". Kevin Lambert argues that Peacock came about his principle through a historical investigation, an interpretation of his philosophy of suggestion taken from Natural History. Indeed, the Cambridge Philosophical Society, with researchers as John Stevens, Edward Daniel Clarke and Adam Sedgwick, professors of mineralogy and geology respectively oriented this society's goal to introduce "subjects of natural history to (...) Cambridge students" [10 p. 282]. In this context and under the influence of natural history, Peacock initiated research on the history of arithmetic, defending in several texts an empirical foundation for symbolic algebra.

He had the idea from a prior practice, an ancient and universal practical reasoning that transcended culture, a science of "suggestion" that ruled his search for the origins of algebraic thought. In other words, "practice suggests the abstract system". In this sense, by applying an ethnographic investigation, he found that, since the beginning of times, the practical problem of counting objects existed. These first practices of counting would *suggest* numeral language, that in turn would suggest new practices of calculation, and therefore, new symbols for those practices. Consequently, it would be a shift from the operations themselves. In this manner, the development of counting, the first stage of Peacock's history, would suggest arithmetic. Arithmetic is the science of suggestion for arithmetical algebra, which in turn acts as the science of suggestion for symbolic algebra, the last stage.

Here we find a process of abstraction that has to do with Peacock's idea of "suggestion". First, we have a suggestive discipline, e.g. arithmetical algebra, helping in the development of another branch of science, e.g. symbolical algebra. Thus, symbolic algebra is suggested or derived from some extensions we make over the properties of the operations from the former discipline. This process of abstraction operates as a heuristic strategy, where we take the rules of operation of arithmetic as conjectures or suggestions for developing an unrestricted symbolical algebra. Either Peacock's *Treatise* of 1830 or his *Report to the BAAS* in 1833 were texts that he wrote stripped of any reference to the historical developments of 1820s. These former papers were his philosophical justification not only for his principle of permanence of forms, but also for his project of a narrative structure for a story of the progress of algebraic reasoning, starting from counting objects and finishing with the emergence of symbolic algebra.

Jean Victor Poncelet was a mathematician who made a serious attempt to justify the introduction of imaginary, singular or improper points into his formulation of synthetic projective geometry. Due to this lack of meaning of this employed symbolism, there was a possibility to introduce a principle that could extend the scope of the theoretical statements about entities and contexts not previously addressed, since they are considered non real elements and prohibited methodologies, respectively, by the standard norms of the mathematics of that time.

Regarding this matter, Poncelet distinguished three types of correlation between two figures when one of them is obtained from the other by what he called a "general correlation": (1) a *direct* correlation, if the figures involved are composed of the same number of parts similarly placed, (2) an *indirect or inverse* correlation, when the parts of the correlative figures are in different order, differently placed, though the general relations remain the same, and (3) an *ideal* correlation, when certain distances and points cease to exist in a geometrical manner.

For each type of correlation, there must be an *invariance* of certain abstract relations stipulated in the initial conditions for the configuration, so that each figure in the series of figures obtained by gradual transformations must be an instance of these relations. This is what Poncelet called the "Principle of continuity or permanence of the mathematical relations".

We can see that for Poncelet, the geometric diagrams, which supposedly constitute the subject matter of geometry, were not necessarily real or actual configurations, with real existence, limited by visual perception or imagination. These figures are variable entities, abstract signs that could assume different values subject to certain rules of combination, and also could be left uninterpreted, as indeterminate magnitudes. In the case of ideal transformations, they could assume infinitely small or great values; they could be imaginary elements, improper points, things that could not be visualized (at least in a standard description).

Although Poncelet "would hardly have assented to the view that the task of the pure geometer is the exploration of the mutual interrelation of signs governed by specified rules of operation, irrespective of the 'interpretations' or 'meaning' which may be assigned to them" [12, p. 206], nonetheless, he understood that this use of the principle of continuity gave rise to an advance in the legitimate study of the relations between the transformed and the original figures, as legitimate as any deductive demonstrative procedure.

The work style, both of Peacock based on his notion of suggestion, and of Poncelet applying a questioned principle on ideal entities, are usually acceptable in the context of discovery. The main reason behind this lies in the type of argumentation implicit in these principles of permanence and continuity. Indeed, as it was stated in the introduction (Sect. 1), we can distinguish at least three types of argumentative inferences that configure work methodologies: top-down (deductive reasoning that goes from generic to generic, or applies generic to particular), bottom-up (reasoning which infers from particular to generic), and what we called "transductive" reasoning (which infers from particular to particular). This last term was coined by Gammerman, Vovk, and Vapnik [9] and applies to processes that try to match a current particular case with a familiar similar one, in order to transfer properties from one to the other, the known and familiar case to the unknown and problematic case that is sought to be solved.

By adopting a transductive inference, an analogy is constructed, described through a function from a domain A (the problem to be solved) to an image B (another problem already solved previously). This analogy allows us to explore how the problem A might be by comparing it with problem B, and how it might work if it were like the analogical problem B. Since we know a solution of B, all we have to do is transfer it to A. Therefore, the analogy must reflect the invariant structure that A has in terms of some invariance present in problem B and similar to that of A.

Thusly, we can portray the characteristics of the starting set A in terms of some already-formed and known set B with which we are familiar. In doing so, imagining some aspect or property of the domain A in terms of something else, we are able to think about the original problem from their transductive model. This allows solving the initial problem in terms of a solution that is already known in the analogous previously solved problem. The evocation of the familiar and known in advance problem B, leads to solving problem A, given the similarity between A and B. In this way, there is a connection–once unthinkable and surprising- between A and B, in making the terms from A to the analogical model B, and vice versa as the analogical terms fit back to problem A.

The advantage that problem A acquires when interacting with problem B consists in the creation of a new cognitive scheme to characterize A in terms of B, embedding A into B and redirecting A to the solution of B, which is based on its latent invariants, to finish capturing the invariants in A that allow its solution.

In the next section, we developed a case study that similarly connects a problem A of a geometric nature with another problem B belonging to the field of graph theory. In turn, B is analogically linked to a third problem C related to electrical circuits. The analogy between the three problems leads to solve A in terms of B first, and then in terms of C, allowing a characterization of the solution of A and A itself in physical terms.

3   The Perfect Squared Rectangle Problem: A Case Study

In this section, we put into practice the constructed and adopted transductive method-ology so as to allow analogical transferences of invariant structures between any two domains.

The problem to be solved is centered on the analysis of geometric tessellations through graph theory and planar electrical networks. A tessellation consists in a refinement of a specific surface by a set of pieces assembled together such that they are all together without spaces in between or causing overlaps. In our case, we consider flat tiles and, within those, flat geometric shapes. More precisely, we deal with "perfect squared rectangles". A rectangle is said to be squared into a finite number n of figures if it is tiled into n squares of sizes $s_1, s_2, \ldots, s_n$, all integer numbers, being n the order of the tiling. A squared rectangle is called perfect if the squares in the tiling are all of different sizes, and it is called simple if no square of the same size is repeated. Thus, the "perfect squared rectangle" problem (to which we will refer as problem A and will be the one we will deal with from here onward) is as follows: given a rectangle, is it possible to tile its surface with different squares? Are there perfect rectangles of all order, or what orders are there? And, are these unique, under such conditions?

This type of problem goes back at least to 1902, when Ernest H. Dudeney published in a magazine the solution to the puzzle concerning Lady Isabel's rectangular casquet with a square lid, which contained in this lid, a pattern of subdivisions into a rectangle and a number of squares, all of different sizes. The puzzle was designed to answer how many different squares there were on the lid, what those sizes were, and how the squares and the rectangle must fit together.

The following year, Max Dehn [7] proved that a rectangle can be squared if and only its sides are commensurable (using counting numbers), i.e. if the sides are integral multiples of each other. Therefore, solutions should be presented with integer lengths. In 1925, Moroń [11] found the 32 by 33 simple perfect rectangle of order 9, the first one published. In 1940, two important papers were published: on the one hand, R. Sprague proved that each rectangle with commensurable sides has a perfect squaring, and has infinitely many totally distinct perfect squarings. On the other hand, Leonard Brooks, Cedric Smith, Arthur Stone and William Thomas Tutte [5], in their attempt to demonstrate the uniqueness of the dissection of a square into smaller squares, all unequal, i.e. the Lusin's conjecture, came to work with perfect rectangles:

"After some practice we found that the construction of such rectangles was not difficult. Our method was to draw a rectangle dissected into smaller rectangles, and to pretend that these rectangles were badly drawn squares. On this assumption the relative sizes of the squares could be found by solving algebraic equations describing how the squares had to fit together (...) We amassed quite a respectable catalogue of perfect rectangles (...) Alas, no perfect square made its way into our catalogue" [17, pp. 2–3].

Thanks to this last work, the above-mentioned problem ends up being solved by graph theory and electrical networks. Below, we present a particular rectangle (Fig. 1) different from the one studied by Brooks et al., as a generalizable case, which allows us to appreciate the herein proposed transductive methodology.

The idea is to establish a correspondence between the original problem A and two equivalent problems: B represents the analogy in terms of graph theory, and C represents the translation from A to the field of electrical networks. This produces the parallelism inserted in Table 1:

Table 1. Three analogical problems.

Problem A	Problem B	Problem C
Geometry	Graph Theory	Electrical Networks
Squared rectangle	Connected planar graph	Electric circuit
[Figure 1]	[Figure 2]	[Figure 3]
Horizontal line segments	Vertices (nodes)	Terminals (dots)
Squares (that have two horizontal lines as boundaries)	Edges	Wires[a] (lines connecting two of the dots)
Side-length of the square (that the lines represent)	The length of the lines connecting two nodes	Current in a wire

[a] Assuming that the wires have electrical resistance of value 1, then Ohm's law implies that the value of the wire will be equal to the intensity of the current flowing through it.

Brooks et al. explain it so:

"The horizontal lines in the squared rectangle correspond to the terminals of the network, and the squares correspond to the wires joining them. The current in a wire is measured by the side-length of the corresponding square, and its direction is downward in the rectangle. The top edge of the squared rectangle corresponds to the positive pole, the terminal at which current enters the network. The bottom edge likewise corresponds to the negative pole, the terminal from which the current leaves" [17, pp. 3–4].

Figure 1 describes a tessellation of a squared rectangle: the variables x_1 to x_9 represent the side-lengths of the squares inside the rectangle. Linear relationships between the variables are established. The horizontal relationships represent the first law of Kirchhoff (the sum of the values of the wires that fit into a node is equal to the sum of the values of the wires that leave it): $x_2 = x_4 + x_5$, $x_6 = x_3 + x_5$, $x_1 + x_4 = x_7 + x_8$, $x_9 = x_6 + x_8$. On the other hand, Kirchhoff's second law states that the sum of the currents for the

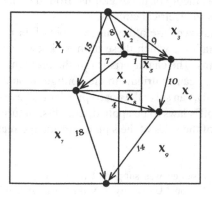

Fig. 1. Perfect squared rectangle.

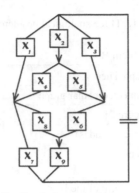

Fig. 2. Graph theoretical interpretation.

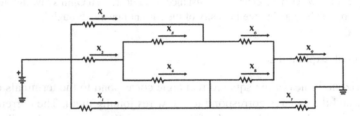

Fig. 3. Electrical interpretation.

entire circuit has to be zero. Referring to the Figs. 1, 2 and 3, the following equations are constructed: $x_1 = x_2 + x_4, x_7 = x_8 + x_9, x_4 + x_8 = x_5 + x_6, x_3 = x_2 + x_5$. The results obtained are: $x_1 = 15, x_2 = 8, x_3 = 9, x_4 = 7, x_5 = 1, x_6 = 10, x_7 = 18, x_8 = 4, x_9 = 14$, as shown in Fig. 1, finally producing a squared perfect rectangle of dimensions 32 by 33.

4 Conclusion

Brooks, Smith, Stone and Tutte's paper succeeded in separating the topological part of the problem, related to the theory of linear graphs, from the metrical part, associated to the theory of current flow in electrical circuits.

Consequently, the analyzed case offers an example of an invariant structure under-lying the three representations, which these authors knew how to recognize, beyond expressing it in other terms. The characteristics of the problem help to understand how the detection of an analogy can contribute to the configuration of a pattern of regularity that, when applied correctly, can solve the posed problem. The discovery of the electrical analogy allowed Kirchhoff's laws to be applied, and this, in turn, allowed these authors to solve the dimensions of the sides of the squares inside the rectangle, and many other problems.

Acknowledgements. This research was supported by the Research Group of Creativity and Innovation in Mathematics, National University of Villa Maria, Argentina.

References

1. Archibald, T.: Transmitting disciplinary practice in applied mathematics? Textbooks 1900-1910. In: Epple, M., Hoff Kjeldsen, T., Siegmund-Schultze, R. (eds.) Mathematisches Forschungsinstitut Oberwolfach, Annual Report, vol. 12, pp. 714–719, Germany (2013)
2. Boole, G.: Exposition of a general theory of linear transformations. Part I. Cambridge Math. J. **3**, 1–20 (1841)
3. Boole, G.: Exposition of a general theory of linear transformations. Part II. Cambridge Math. J. **3**, 106–119 (1842)
4. Boole, G.: On the general theory of linear transformations. Cambridge Dublin Math. J. **6**, 87–106 (1851)
5. Brooks, R., Smith, C., Stone, S., Tutte, W.T.: The dissection of rectangles into squares. Duke Math. J. **7**, 312–340 (1940)
6. Cayley, A.: On linear transformations. Cambridge Dublin Math. J. **1**, 104–122 (1846)
7. Dehn, M.: Über die Zerlegung von Rechtecken in Rechtecke. Math. Ann. **57**, 314–332 (1903)
8. Epple, M., Hoff Kjeldsen, T., Siegmund-Schultze, R.: From 'mixed' to 'applied' mathematics: tracing an important dimension of mathematics and its history. In: Mathematisches Forschungsinstitut Oberwolfach, Annual Report, vol. 12, pp. 657–660. Germany (2013)
9. Gammerman, A., Vovk, V., Vapnik, V.: Learning by transduction. In: Cooper, G.F., Moral, S. (eds.) Proceedings of the Fourteenth Conference on Uncertainty in Artificial Intelligence, pp. 148–155. Morgan Kaufmann, San Francisco (1998)
10. Lambert, K.: A natural history of mathematics. George peacock and the making of English algebra. Isis **104**(2), 278–302 (2013)
11. Moroń, Z.: O rozkladach prostokatów na kwadraty. Przleglad. Matem.-Fizyczny **3**, 152–153 (1925)
12. Nagel, E.: The formation of modern conceptions of formal logic in the development of geometry. In: Nagel, E. (ed.) Teleology Revisited and Other Essays in the Philosophy and the History of Science, pp. 195–259. Columbia University Press, New York (1979)
13. Parshall, K.H.: Toward a history of nineteenth-century invariant theory. In: Rowe, D.E., McCleary, J. (eds.) The History of Modern Mathematics. Ideas and their Reception, vol. I, pp. 157–206. Academic Press and Harcourt Brace Jovanovich Publishers, Boston (1989)
14. Peacock, G.: A Treatise on Algebra. J. & J. J. Deighton, Cambridge, C. J. G. & F. Rivington and Whittaker, Teacher & Arnot, London (1830)
15. Rota, G.-C.: What is invariant theory, really? In: Crapo, H., Senato, D. (eds.) Algebraic Combinatorics and Computer Science. A tribute to Gian-Carlo Rota, pp. 41–56. Springer, Italy (1998). https://doi.org/10.1007/978-88-470-2107-5_4
16. Suppes, P.: New foundations of objective probability. Axioms for propensities. Stud. Log. Found. Math. **74**, 515–529 (1973)
17. Tutte, W.T.: Graph Theory As I Have Known It. Clarendon Press/Oxford University Press, Oxford/New York (1998)
18. Visokolskis, A.S.: El fenómeno de la transducción en la matemática. Metáforas, analogías y cognición. In: Pochulu, M., Abrate, R., Visokolskis, A.S. (eds.) La metáfora en la educación. Descripción e implicaciones, pp. 37–53. Eduvim, Villa María (2009)
19. Visokolskis, A.S.: La noción de análisis como descubrimiento en la historia de la matemática. Propuesta de un modelo de descubrimiento creativo. Ph.D. Doctoral Dissertation. National University of Cordoba, Cordoba, Argentina (2016)
20. Visokolskis, A.S., Carrión, G.: Creative insights: dual cognitive processes in perspicuous diagrams. In: Sato, Y., Shams, Z. (eds.) Proceedings of the International Workshop on Set Visualization and Reasoning SetVR 2018, Set Visualization and Reasoning, pp. 28–43, Edinburgh (2018)

21. Visokolskis, A.S.: Filosofía de la creatividad en contextos matemáticos. Espacios transductivos como alternativa al dilema de Boden. Paper Presented at XX Jornadas Rolando Chuaqui Kettlun, Santiago, Chile, 27–30 August 2019
22. Wolfson, P.R.: George Boole and the origins of invariant theory. Hist. Math. **35**, 37–46 (2008)
23. Woodger, J.H.: The Axiomatic Method in Biology. Cambridge University Press, Cambridge (1937)

Generalizations of the Intermediate Value Theorem for Approximating Fixed Points and Zeros of Continuous Functions

Michael N. Vrahatis$^{(\boxtimes)}$ (iD)

Department of Mathematics, University of Patras, 26110 Patras, Greece
`vrahatis@math.upatras.gr`

Abstract. Generalizations of the traditional intermediate value theorem are presented. The obtained generalized theorems are particular useful for the existence of solutions of systems of nonlinear equations in several variables as well as for the existence of fixed points of continuous functions. Based on the corresponding criteria for the existence of a solution emanated by the intermediate value theorems, generalized bisection methods for approximating fixed points and zeros of continuous functions are given. These bisection methods require only algebraic signs of the function values and are of major importance for tackling problems with imprecise (not exactly known) information.

Keywords: Bolzano theorem · Bolzano-Poincaré-Miranda theorem · Intermediate value theorems · Existence theorems · Bisection methods · Fixed points · Nonlinear equations

1 Introduction

A system of n nonlinear equations in n real unknowns,

$$
\begin{aligned}
f_1(x_1, x_2, \ldots, x_n) &= 0, \\
f_2(x_1, x_2, \ldots, x_n) &= 0, \\
&\vdots \\
f_n(x_1, x_2, \ldots, x_n) &= 0,
\end{aligned}
\tag{1}
$$

may be represented in the real n-dimensional vector space $\mathbb{R}^n$ as follows:

$$
F_n(x) = \theta^n,
\tag{2}
$$

where $F_n = (f_1, f_2, \ldots, f_n) \colon \mathcal{D} \subset \mathbb{R}^n \to \mathbb{R}^n$ is a nonlinear mapping and $\theta^n = (0, 0, \ldots, 0)$ is the origin of $\mathbb{R}^n$. The problem of solving the Eq. (2) is to find a *zero* $x^* = (x_1^*, x_2^*, \ldots, x_n^*) \in \mathcal{D}$ for which $F_n(x^*) = \theta^n$. Similarly, the problem of finding a *fixed point* of F_n in $\mathcal{D} \subset \mathbb{R}^n$ is to find a point $x^* \in \mathcal{D}$ which satisfies the equation $F_n(x^*) = x^*$. Obviously, the problem of finding a fixed point is equivalent to the problem of solving the Eq. (2) by considering the mapping

© Springer Nature Switzerland AG 2020
Y. D. Sergeyev and D. E. Kvasov (Eds.): NUMTA 2019, LNCS 11974, pp. 223–238, 2020.
https://doi.org/10.1007/978-3-030-40616-5_17

$\Phi_n = I_n - F_n$ (where I_n indicates the identity mapping) instead of F_n and solving the equation $\Phi_n(x) = \theta^n$, instead of the Eq. (2).

Many problems require solution of systems of equations for which Newton's method and the related class of algorithms [15] fail due to nonexistence of derivatives or poorly behaved partial derivatives. Also, Newton's method as well as Newton's-like methods often converge to a solution x^* of $F_n(x) = \theta^n$ almost independently of the initial guess, while $F_n(x) = \theta^n$ may have several solutions, all of which are desired for the application [28]. Because of this reason, generalized bisection methods have been investigated. According to these methods one establishes the existence of at least one solution of the Eq. (2) in a given domain using a specific criterion for the existence of a solution. These kind of criteria can be obtained using the conditions of various "existence theorems" (intermediate value theorems). Once we have obtained a domain for which the criterion of the existence is fulfilled, we are able to obtain upper and lower bounds for solution values. To this end, by computing a sequence of bounded domains of decreasing diameters, we are able to obtain a region with arbitrarily small diameter that contains at least one solution of the Eq. (2).

These methods require only algebraic signs of the function values. The algebraic sign is the smallest amount of information (one bit of information) necessary for the purpose needed. Thus, the methods that require only algebraic signs are of major importance for tackling problems with imprecise (not exactly known) information. This kind of problems occurs in various scientific fields including mathematics, economics, engineering, computer science, biomedical informatics, medicine and bioengineering, among others. This is so, because, in a large variety of applications, precise function values are either impossible or time consuming and computationally expensive to obtain. One such application is provided in [28]. This application concerns the computation of all the periodic orbits (stable and unstable) of any period and accuracy which occur, among others, in the study of beam dynamics in circular particle accelerators like the Large Hadron Collider (LHC) machine at the European Organization for Nuclear Research (CERN). In this application, the method which is presented in [24] and is implemented in [25] is used. Furthermore, these methods are particularly useful for tackling various problems where the corresponding functions take very large and/or very small values.

2   Background Material

Notation 1. We denote by ϑA the boundary of a set A, by $\mathrm{cl}A$ its closure, by $\mathrm{int}A$ its interior, by $\mathrm{card}\{A\}$ its cardinality (i.e., the number of elements in the set A) and by $\mathrm{co}A$ its convex hull (i.e., the set of all finite convex combinations of elements of A).

Notation 2. We shall frequently use the index sets $N^n = \{0, 1, \ldots, n\}$, $N^n_{\neg 0} = \{1, 2, \ldots, n\}$ and $N^n_{\neg i} = \{0, 1, \ldots, i-1, i+1, \ldots, n\}$. Furthermore, for a given set $I = \{i, j, \ldots, \ell\} \subset N^n$ we denote by $N^n_{\neg I}$ or equivalently by $N^n_{\neg ij\ldots\ell}$ the set $\{k \in N^n \mid k \notin I\}$.

Definition 1. For any positive integer n, and for any set of points $V = \{v^0, v^1, \ldots, v^n\}$ in some linear space which are affinely independent (i.e., the vectors $\{v^1 - v^0, v^2 - v^0, \ldots, v^n - v^0\}$ are linearly independent) the convex hull $co\{v^0, v^1, \ldots, v^n\} = [v^0, v^1, \ldots, v^n]$ is called the n-*simplex with vertices* $v^0, v^1, \ldots, v^n$. For each subset of $(m + 1)$ elements $\{\omega^0, \omega^1, \ldots, \omega^m\} \subset \{v^0, v^1, \ldots, v^n\}$, the m-simplex $[\omega^0, \omega^1, \ldots, \omega^m]$ is called an m-*face* of $[v^0, v^1, \ldots, v^n]$. In particular, 0-faces are vertices and 1-faces are edges. The m-faces are also called *facets* of the n-simplex. An m-face of the n-simplex is called the *carrier* of a point p if p lies on this m-face and not on any sub-face of this m-face.

Notation 3. We denote the n-simplex with set of vertices $V = \{v^0, v^1, \ldots, v^n\}$ by $\sigma^n = [v^0, v^1, \ldots, v^n]$. Also, we denote the $(n-1)$-simplex that determines the i-th $(n-1)$-face of σ^n by $\sigma_{\neg i}^n = [v^0, v^1, \ldots, v^{i-1}, v^{i+1}, \ldots, v^n]$. Furthermore, for a given index set $I = \{i, j, \ldots, \ell\} \subset N^n$ with cardinality $card\{I\} = \kappa$, we denote by $\sigma_{\neg I}^n$ or equivalently by $\sigma_{\neg ij \ldots \ell}^n$ the $(n - \kappa)$-face of σ^n with vertices $v^m, m \in N_{\neg I}^n$.

Definition 2 [23,26]. The *diameter* of an m-simplex σ^m in $\mathbb{R}^n$, $m \leqslant n$, denoted by $diam(\sigma^m)$, is defined to be the length of the longest edge (1-face) of σ^m while the *microdiameter*, $\mu diam(\sigma^m)$, of σ^m is defined to be the length of the shortest edge of σ^m.

Definition 3. Let $\sigma^m = [v^0, v^1, \ldots, v^m]$ be an m-simplex in $\mathbb{R}^n$, $m \leqslant n$. Then the *barycenter* of σ^m denoted by K is the point $K = (m+1)^{-1} \sum_{i=0}^m v^i$ in $\mathbb{R}^n$.

Remark 1. By convexity it is obvious that the barycenter of any m-simplex σ^m in $\mathbb{R}^n$ is a point in the relative interior of σ^m.

Definition 4. An n-simplex is *oriented* if an order has been assigned to its vertices. If $\langle v^0, v^1, \ldots, v^n \rangle$ is an orientation of $\{v^0, v^1, \ldots, v^n\}$ this is regarded as being the same as any orientation obtained from it by an even permutation of the vertices and as the opposite of any orientation obtained by an odd permutation of the vertices. We shall denote oriented n-simplices by $\sigma^n = \langle v^0, v^1, \ldots, v^n \rangle$, and we shall write, for example, $\langle v^0, v^1, v^2, \ldots, v^n \rangle = -\langle v^1, v^0, v^2, \ldots, v^n \rangle = \langle v^2, v^0, v^1, \ldots, v^n \rangle$. The *boundary* $\vartheta \sigma^n$ of an oriented n-simplex $\sigma^n = \langle v^0, v^1, \ldots, v^n \rangle$ is given by $\vartheta \sigma^n = \sum_{i=0}^n (-1)^i \langle v^0, v^1, \ldots, v^{i-1}, v^{i+1}, \ldots, v^n \rangle$. The oriented $(n-1)$-simplex $\langle v^0, v^1, \ldots, v^{i-1}, v^{i+1}, \ldots, v^n \rangle$ will be called the ith *face* of σ^n.

Definition 5. An n-dimensional *polyhedron* Π^n is a union of a finite number of oriented n-simplices σ_i^n, $i = 1, 2, \ldots, k$ such that the σ_i^n have pairwise-disjoint interiors. We write $\Pi^n = \sum_{i=1}^k \sigma_i^n$ and $\vartheta \Pi^n = \sum_{i=1}^k \vartheta \sigma_i^n$.

Definition 6. Let $\psi \in \mathbb{R}$, then the *sign (or signum) function*, denoted by sgn, maps ψ to the set $\{-1, 0, 1\}$ as follows:

$$\operatorname{sgn} \psi = \begin{cases} -1, & \text{if } \psi < 0, \\ 0, & \text{if } \psi = 0, \\ 1, & \text{if } \psi > 0. \end{cases} \tag{3}$$

Furthermore, for any $a = (a_1, a_2, \ldots, a_n) \in \mathbb{R}^n$ the *sign* of a, denoted $\operatorname{sgn} a$, is defined as $\operatorname{sgn} a = (\operatorname{sgn} a_1, \operatorname{sgn} a_2, \ldots, \operatorname{sgn} a_n)$.

3   Bolzano Intermediate Value Theorem

The fundamental and pioneering Bolzano's theorem states the following [2,7]:

Theorem 1 (Bolzano's theorem). *If* $f : [a, b] \subset \mathbb{R} \to \mathbb{R}$ *is a continuous function and if it holds that* $f(a) f(b) < 0$, *then there is at least one* $x \in (a, b)$ *such that* $f(x) = 0$.

This theorem is also called *intermediate value theorem* since it can be easily formulated as follows:

Theorem 2 (Bolzano's intermediate value theorem). *If* $f : [a, b] \subset \mathbb{R} \to \mathbb{R}$ *is a continuous function and if* y_0 *is a real number such that:*

$$\min\{f(a), f(b)\} < y_0 < \max\{f(a), f(b)\},$$

then there is at least one $x_0 \in (a, b)$ *such that* $f(x_0) = y_0$.

Remark 2. Obviously, Theorem 2 can be deduced from Theorem 1 by considering the function $g(x) = f(x) - y_0$.

Remark 3. The first proofs of the above theorem, given independently by Bolzano in 1817 [2] and Cauchy in 1821 [4], were crucial in the procedure of *arithmetization of analysis*, which was a research program in the foundations of mathematics during the second half of the 19th century.

Based on the hypotheses of Theorem 1, a simple and very useful criterion for the existence of a zero of a continuous mapping $f : [a, b] \subset \mathbb{R} \to \mathbb{R}$ in some interval (a, b) is the following *Bolzano's existence criterion*:

$$f(a) \, f(b) < 0, \tag{4}$$

or equivalently:

$$\operatorname{sgn} f(a) \operatorname{sgn} f(b) = -1, \tag{5}$$

where sgn denotes the sign function (3).

Remark 4. The Bolzano existence criterion is well-known and widely used and it can be generalized to higher dimensions, see [27,30] (cf. Sects. 4 and 5). Note that when the condition (4) (or the condition (5)) is not fulfilled, then in the interval (a, b) either no zero exists or there are zeros for which the sum of their multiplicities is an even number (e.g., two simple zeros, one double and two simple zeros, one triple and one simple zeros etc.).

The well-know and widely applied *bisection method* is based on the Bolzano existence criterion in order to approximate a zero of a continuous function $f : [a, b] \subset \mathbb{R} \to \mathbb{R}$ in a given interval (a, b). A simplified version described in [24] is the following:

$$x^{p+1} = x^p + c \operatorname{sgn} f(x^p) \, / \, 2^{p+1}, \quad p = 0, 1, \ldots, \tag{6}$$

where $x^0 = a$ and $c = \operatorname{sgn} f(a) \, (b - a)$. Instead of the iterative formula (6) we can also use the following [24]:

$$x^{p+1} = x^p - \hat{c} \operatorname{sgn} f(x^p) \, / \, 2^{p+1}, \quad p = 0, 1, \ldots, \tag{7}$$

where $x^0 = b$ and $\hat{c} = \operatorname{sgn} f(b) \, (b - a)$.

The sequences (6) and (7) converge with certainty to a zero $r \in (a, b)$ if for some x^p it holds that:

$$\operatorname{sgn} f(x^0) \operatorname{sgn} f(x^p) = -1, \quad \text{for} \quad p = 1, 2, \ldots.$$

Furthermore, the number of iterations ν required to obtain an approximate zero r^* such that $|r - r^*| \leqslant \varepsilon$ for some $\varepsilon \in (0, 1)$ is given by:

$$\nu = \left\lceil \log_2(b - a) \, \varepsilon^{-1} \right\rceil, \tag{8}$$

where $\lceil x \rceil = \operatorname{ceil}(x)$ denotes the ceiling function that maps a real number x to the least integer greater than or equal to x.

Remark 5. The reasons for choosing the iterative schemes (6) and (7) are that:

1. They converge with certainty within the given interval (a, b).
2. They are globally convergent methods in the sense that they converge to a zero from remote initial guesses.
3. Using the relation (8) we may predetermine the number of iterations that are required for the attainment of an approximate zero to a given accuracy.
4. They have a great advantage since they are worst-case optimal. That is, they possess asymptotically the best possible rate of convergence in the worst case [20]. This means that they are guaranteed to converge within the predefined number of iterations, and, moreover, no other method has this important property.
5. They require only the algebraic signs of the function values to be computed, as is evident from (6) and (7); thus they can be applied to problems with imprecise function values.

For applications of the iterative schemes (6) and (7) we refer the interested reader, among others, to [5, 8, 9, 34, 35].

4 Bolzano-Poincaré-Miranda Intermediate Value Theorem

A straightforward generalization of Bolzano's intermediate value theorem to continuous mappings of an n-cube (parallelotope) into $\mathbb{R}^n$ was proposed (without proof) by Poincaré in 1883 and 1884 in his work on the *three body problem* [16,17]. This theorem, now known as Bolzano-Poincaré-Miranda theorem, states that [13,22,27]:

Theorem 3 (Bolzano - Poincaré - Miranda theorem). *Suppose that* $P = \{x \in \mathbb{R}^n \mid |x_i| < L,\ for\ 1 \leqslant i \leqslant n\}$ *and let the mapping* $F_n = (f_1, f_2, \ldots, f_n)\colon P \to \mathbb{R}^n$ *be continuous on the closure of* P *such that* $F_n(x) \neq \theta^n = (0, 0, \ldots, 0)$ *for* x *on the boundary of* P, *and*

(a) $f_i(x_1, x_2, \ldots, x_{i-1}, -L, x_{i+1}, \ldots, x_n) \geqslant 0, \qquad for \quad 1 \leqslant i \leqslant n,$
(b) $f_i(x_1, x_2, \ldots, x_{i-1}, +L, x_{i+1}, \ldots, x_n) \leqslant 0, \qquad for \quad 1 \leqslant i \leqslant n.$

Then, there is at least one $x \in P$ *such that* $F_n(x) = \theta^n$.

Theorem 3 it has come to be known as "Miranda's theorem" since in 1940 Miranda [13] proved that it is equivalent to the traditional Brouwer fixed point theorem [3]. It is worthy to mention that the Bolzano-Poincaré-Miranda theorem is closely related to important theorems in analysis and topology and constitutes an invaluable tool for verified solutions of numerical problems by means of interval arithmetic. For a short proof and a generalization of the Bolzano-Poincaré-Miranda theorem using topological degree theory we refer the interested reader to [27]. In addition, for generalizations with respect to an arbitrary basis of $\mathbb{R}^n$ that eliminate the dependence of the Bolzano-Poincaré-Miranda theorem on the standard basis of $\mathbb{R}^n$ see [6,27]. For various interesting relations between the theorems of Bolzano-Poincaré-Miranda, Borsuk, Kantorovich and Smale with respect to the existence of a solution of a system of nonlinear equations, we refer the interested reader to [1].

The conditions of the Bolzano-Poincaré-Miranda theorem give an invaluable existence criterion for a solution of the Eq. (2) where $F_n = (f_1, f_2, \ldots, f_n)\colon P \subset \mathbb{R}^n \to \mathbb{R}^n$ is continuous.

Remark 6. Similarly to Bolzano's criterion, the Bolzano - Poincaré - Miranda criterion requires only the algebraic sings of the function values to be computed on the boundary of the n-cube P. On the other hand, for general continuous functions, in contrary to Bolzano's criterion, the hypotheses (a) and (b) are not always fulfilled or it is impossible to be verified for a given n-cube P.

Next, the characteristic polyhedron criterion and the characteristic bisection method are briefly presented. These approaches, in contrary to Bolzano - Poincaré - Miranda criterion require only the algebraic sings of the function values to be computed on the vertices of the considered polyhedron.

There are various generalized bisection methods that require the computation of the topological degree in order to localize a solution of the Eq. (2) (see, e.g., [11,23]). We shall allow us to briefly discuss a few basic concepts regarding topological degree theory. To this end, suppose that a function $F_n = (f_1, f_2, \ldots, f_n) \colon \mathrm{cl}\mathcal{D}_n \subset \mathbb{R}^n \to \mathbb{R}^n$ is defined and twice continuously differentiable in an open and bounded domain $\mathcal{D}_n$ of $\mathbb{R}^n$ with boundary $\vartheta\mathcal{D}_n$. Suppose further that the solutions of the equation $F_n(x) = p$, where $p \in \mathbb{R}^n$ is a given vector, are not located on $\vartheta\mathcal{D}_n$, and that they are simple, i.e., the determinant, $\det J_{F_n}$, of the Jacobian matrix of F_n at these solutions is non-zero.

Definition 7. The *topological degree of F_n at p relative to $\mathcal{D}_n$* is denoted by $\deg[F_n, \mathcal{D}_n, p]$ and is defined by the following sum:

$$\deg[F_n, \mathcal{D}_n, p] = \sum_{x \in F_n^{-1}(p) \cap \mathcal{D}_n} \mathrm{sgn} \det J_{F_n}(x), \tag{9}$$

where sgn denotes the sign function (3).

Remark 7. The topological degree can be generalized when the function is only continuous [15]. Furthermore, if $\mathcal{D}_n = \mathcal{D}_n^1 \cup \mathcal{D}_n^2$ where $\mathcal{D}_n^1$ and $\mathcal{D}_n^2$ have disjoint interiors and $F_n(x) \neq \theta^n$ for all $x \in \vartheta\mathcal{D}_n^1 \cup \vartheta\mathcal{D}_n^2$, then the topological degree is additive, i.e.:

$$\deg[F_n, \mathcal{D}_n, \theta^n] = \deg[F_n, \mathcal{D}_n^1, \theta^n] + \deg[F_n, \mathcal{D}_n^2, \theta^n]. \tag{10}$$

The topological degree is invariant under changes of the vector p in the sense that, if $q \in \mathbb{R}^n$ is any vector, then it holds that [15]:

$$\deg[F_n, \mathcal{D}_n, p] \equiv \deg[F_n - q, \mathcal{D}_n, p - q],$$

where $F_n - q$ denotes the mapping $F_n(x) - q$, $x \in \mathcal{D}_n$. Thus, for simplicity reason, we consider the case where the topological degree is defined at the origin $\theta^n = (0, 0, \ldots, 0)$ in $\mathbb{R}^n$.

The topological degree $\deg[F_n, \mathcal{D}_n, \theta^n]$ can be represented by the Kronecker integral which is defined as follows:

$$\deg[F_n, \mathcal{D}_n, \theta^n] = \frac{\Gamma(n/2)}{2\pi^{n/2}} \int\int_{\vartheta\mathcal{D}_n} \cdots \int \frac{\sum_{i=1}^{n} A_i dx_1 \cdots dx_{i-1} dx_{i+1} \cdots dx_n}{\left(f_1{}^2 + f_2{}^2 + \cdots + f_n{}^2\right)^{n/2}}, \tag{11}$$

where Γ denotes the gamma function and A_i define the following determinants:

$$A_i = (-1)^{n(i-1)} \det \left[F_n \quad \frac{\partial F_n}{\partial x_1} \quad \cdots \quad \frac{\partial F_n}{\partial x_{i-1}} \quad \frac{\partial F_n}{\partial x_{i+1}} \quad \cdots \quad \frac{\partial F_n}{\partial x_n} \right],$$

where $\frac{\partial F_n}{\partial x_k} = \left(\frac{\partial f_1}{\partial x_k}, \frac{\partial f_2}{\partial x_k}, \ldots, \frac{\partial f_n}{\partial x_k} \right)$ is the kth column of the determinant $\det J_{F_n}$ of the Jacobian matrix J_{F_n}.

The important Kronecker's theorem [15] states that the equation $F_n(x) = \theta^n$ has at least one zero in $\mathcal{D}_n$ if $\deg[F_n, \mathcal{D}_n, \theta^n] \neq 0$. To this end, several methods for the computation of the topological degree have been proposed in the past few years (see, e.g., [11,22]). One such method is the fundamental and pioneering Stenger's method [22] that in some classes of functions is an almost optimal complexity algorithm (see, e.g., [14,20,22]). The accurate computation of topological degree using Stenger's or other related methods [11], is based on suitable assumptions, including appropriate representation of the boundary of $\mathcal{D}_n$. Specifically, if the boundary of $\mathcal{D}_n$ can be "sufficiently refined" then Stenger's method gives the value of the topological degree.

Definition 8 [11,22,33]. Let Π^n be an n-polyhedron. Let $F_n = (f_1, f_2, \ldots, f_n)\colon \Pi^n \subset \mathbb{R}^n \to \mathbb{R}^n$ be continuous with $\theta^n \notin F_n(\vartheta \Pi^n)$. If $n = 1$, $\vartheta \Pi^1$ is said to be *sufficiently refined relative to* $\operatorname{sgn} F_1$, if $0 \notin F_1(\vartheta \Pi^1)$. If $n > 1$, $\vartheta \Pi^n$ is said to be *sufficiently refined relative to* $\operatorname{sgn} F_n$, if $\vartheta \Pi^n$ has been subdivided so that it may be written as a union of a finite number of $(n-1)$-dimensional regions $Q_1^{n-1}, Q_2^{n-1}, \ldots, Q_m^{n-1}$, each consisting of a union of a finite number of $(n-1)$-simplices with pairwise disjoint $(n-1)$-dimensional interiors and having the following properties:

(a) the interiors of the Q_i^{n-1} are pairwise disjoint and each Q_i^{n-1} is connected;
(b) for each region Q_i^{n-1}, there exists at least one component of F_n, (for example f_{r_i}), that does not vanish on it;
(c) if $f_{r_i} \neq 0$ on Q_i^{n-1}, then ϑQ_i^{n-1} is sufficiently refined relative to $\operatorname{sgn} F_{n-1}^{r_i}$ where $F_{n-1}^{r_i} = (f_1, f_2, \ldots, f_{r_i-1}, f_{r_i+1}, \ldots, f_n)$.

As we have already mentioned previously, once we have obtained a domain for which the value of the topological degree relative to this domain is nonzero, we are able to obtain upper and lower bounds for solution values. To this end, by computing a sequence of bounded domains with nonzero values of topological degree and decreasing diameters, we are able to obtain a region with arbitrarily small diameter that contains at least one solution of the Eq. (2). However, although the nonzero value of topological degree plays an important role in the existence of a solution of the Eq. (2), the computation of this value is a time-consuming procedure. The bisection method, on the other hand, which is briefly described below, avoids all calculations concerning the topological degree by implementing the concept of the *characteristic n-polyhedron criterion* for the existence of a solution of the Eq. (2) within a given bounded domain. This criterion is based on the construction of a *characteristic n-polyhedron* [24,25,33]. To define a characteristic n-polyhedron (n-dimensional convex polyhedron) we construct the *n-complete $2^n \times n$ matrix* $\mathcal{M}_n$ whose rows are formed by all possible combinations of -1 and 1. To this end we compute the *n-binary $2^n \times n$ matrix* $\mathcal{M}_n^* = [e_{ij}^*]_{i,j=1}^{2^n,n}$ where e_{ij}^* is the jth digit of the n-digit binary representation of the number $(i-1)$ counting the left-most digit first. Then the elements of $\mathcal{M}_n = [e_{ij}]_{i,j=1}^{2^n,n}$ are given by $e_{ij} = 2e_{ij}^* - 1$.

Suppose now that $\Pi^n = \langle V_1, V_2, \ldots, V_{2^n} \rangle$ is an oriented (i.e., an orientation has been assigned to its vertices) n-dimensional convex polyhedron with 2^n vertices, $V_i \in \mathbb{R}^n$, and let $F_n = (f_1, f_2, \ldots, f_n)\colon \Pi^n \subset \mathbb{R}^n \to \mathbb{R}^n$ be a continuous mapping. Then,

Definition 9. The $2^n \times n$ matrix $\mathcal{S}(F_n; \Pi^n)$ whose entries in the k-th row are the corresponding coordinates of the vector:

$$\operatorname{sgn} F_n(V_k) = \left(\operatorname{sgn} f_1(V_k), \operatorname{sgn} f_2(V_k), \ldots, \operatorname{sgn} f_n(V_k)\right), \qquad (12)$$

will be called *matrix of signs associated with F_n and Π^n*, where $\operatorname{sgn} \psi$ defines the sign function (3).

Definition 10. An n-polyhedron Π^n is called *characteristic n-polyhedron relative to F_n*, iff the matrix $\mathcal{S}(F_n; \Pi^n)$ is identical with the matrix $\mathcal{M}_n$, after some permutation of its rows.

Definition 11. A polyhedron which is a convex hull of 2^{n-1} vertices of a characteristic n-polyhedron Π^n relative to F_n, will be called *r-side of Π^n* and will be noted by P_r, $r = 1, 2, \ldots, n$ iff for all its vertices V_k, $k = 1, 2, \ldots, 2^{n-1}$ the corresponding vectors $\operatorname{sgn} F_n(V_k)$ have their r-th coordinate equal to each other. Moreover, if this common r-th element is -1 or 1 then the P_r will be called *negative* or *positive r-side* correspondingly.

Lemma 1 [33]. *In each characteristic n-polyhedron relative to F_n there are n positive and n negative sides. Moreover, each side P_r of a characteristic n-polyhedron Π^n relative to $F_n = (f_1, f_2, \ldots, f_n): \Pi^n \subset \mathbb{R}^n \to \mathbb{R}^n$ is itself a characteristic $(n-1)$-polyhedron relative to $F_{n-1}^r = (f_1, f_2, \ldots, f_{r-1}, f_{r+1}, \ldots, f_n): P_r \to \mathbb{R}^{n-1}$.*

If the boundary $\vartheta \Pi^n$ of a characteristic polyhedron Π^n can be sufficiently refined then there is (at least) one zero within Π^n. More specifically, the following theorem holds:

Theorem 4 [33]. *Let $\mathcal{V} = \langle V_i \rangle_{i=1}^{2^n}$ and $\mathcal{P} = \{P_i\}_{i=1}^{2n}$ be the ordered set of vertices and the set of the sides, respectively, of a characteristic n-polyhedron Π^n relative to continuous $F_n: \Pi^n \subset \mathbb{R}^n \to \mathbb{R}^n$ for which $\theta^n \notin F_n(\vartheta \Pi^n)$. Suppose that $S = \{S_{i,j}\}_{i=1, j=1}^{2n, \, j_i}$ is a finite set of $(n-1)$-dimensional oriented simplices which lie on $\vartheta \Pi^n$ with the following properties:*

(a) $\vartheta \Pi^n = \sum_{i=1}^{2n} \sum_{j=1}^{j_i} S_{i,j}$,
(b) the interiors of the members of S are disjoint,
(c) these simplices make $\vartheta \Pi^n$ sufficiently refined relative to $\operatorname{sgn}(F_n)$, and
(d) the vertices of each simplex $S_{i,j}$ are a subset of vertices of P_i.

Then, it holds that $\deg[F_n, \Pi^n, \theta^n] = \pm 1$.

Remark 8. The above result implies the existence of at least one solution of the Eq. (2) within Π^n. For more details on how to construct a characteristic n-polyhedron and locate a desired solution see [24,25,28]. The characteristic polyhedron can be considered as a translation of the Poincaré-Miranda hypercube [22,27].

Next, we describe a generalized bisection method. This method combined with the above mentioned criterion, produces a sequence of characteristic polyhedra of decreasing size always containing the desired solution. We call it *Characteristic Bisection*. This version of bisection does not require the computation of the topological degree at each step, as others do [11,23]. It can be applied to problems with imprecise function values, since it depends only on their signs. The method simply amounts to constructing another refined characteristic polyhedron, by bisecting a known one, say Π^n. To do this, we compute the midpoint M of the longest edge $\langle V_i, V_j \rangle$, of Π^n (where the distances are measured in Euclidean norms). Then we obtain another characteristic polyhedron, Π_*^n, by comparing the sign, $\mathrm{sgn} F_n(M)$, of $F_n(M)$ with that of $F_n(V_i)$ and $F_n(V_j)$ and substituting M for that vertex for which the signs are identical [24,25,28]. Then we select the longest edge of Π_*^n and continue the above process. If the assumptions of Theorem 4 are satisfied, one of the $\mathrm{sgn} F_n(V_i)$, $\mathrm{sgn} F_n(V_j)$ coincides with $\mathrm{sgn} F_n(M)$, otherwise, we continue with another edge.

Theorem 5 [33]. *Let Π^n be a characteristic n-polyhedron whose longest edge length is $\Delta(\Pi^n)$. Then, the minimum number ζ of bisections of the edges of Π^n required to obtain a characteristic polyhedron Π_*^n whose longest edge length satisfies $\Delta(\Pi_*^n) \leqslant \varepsilon$, for some accuracy $\varepsilon \in (0, 1)$, is given by*

$$\zeta = \left\lceil \log_2 \left(\Delta(\Pi^n)\, \varepsilon^{-1} \right) \right\rceil . \tag{13}$$

Remark 9. Notice that ζ is independent of the dimension n and that the bisection algorithm has the same number of iterations as the bisection in one-dimension which is optimal and possesses asymptotically the best rate of convergence [19].

5   Intermediate Value Theorem for Simplices

In [30] the intermediate value theorem for simplices is proposed. The obtained proof is based on the Knaster-Kuratowski-Mazurkiewicz covering principle [12] (cf. Lemma 2 below). Also, in [31] two short proofs of this theorem are given which are based on Sperner covering principles (cf. Lemmas 3 and 4 below).

Lemma 2 (Knaster-Kuratowski-Mazurkiewicz (KKM Lemma)). *Let C_i, $i \in N^n = \{0, 1, \ldots, n\}$ be a family of $(n + 1)$ closed subsets of an n-simplex $\sigma^n = [v^0, v^1, \ldots, v^n]$ in $\mathbb{R}^n$ satisfying the following hypotheses:*

(a) $\sigma^n = \bigcup_{i \in N^n} C_i$ and
(b) For each $\emptyset \neq I \subset N^n$ it holds that $\bigcap_{i \in I} \sigma_{\neg i}^n \subset \bigcup_{j \in N_{\neg I}^n} C_j$.

Then, it holds that $\bigcap_{i \in N^n} C_i \neq \emptyset$.

Remark 10. It is worthy to mention that, the three fundamental and pioneering classical results, namely, the Brouwer fixed point theorem [3], the Sperner lemma [21], and the KKM lemma [12] are mutually equivalent in the sense that each one can be deduced from another. Furthermore, Scarf proposed a method for approximating a fixed point of a continuous function from a unit simplex into itself [18]. This approach is considered as the first constructive proof to Brouwer's fixed point theorem. Scarf's method is based on a simplicial subdivision (triangulation) of the given simplex and it uses a labeling of the vertices of the simplicial subdivision.

Definition 12. A system (family) of subsets of a set A whose union is A is called a *covering* of A. The *order* of a finite system of sets is the greatest integer k for which the system has k elements with nonempty intersection. A system of sets is said to be *simple* if every two elements of the system are distinct. A covering is called an *ε-covering* if the finite system of sets of this covering are of diameter less than $\varepsilon > 0$.

A similar to KKM covering principle was proposed by Sperner [21]:

Lemma 3 (Sperner covering principle). *Let C_i, $i \in N^n$ be a family of $(n+1)$ closed subsets of an n-simplex $\sigma^n = [v^0, v^1, \ldots, v^n]$ in $\mathbb{R}^n$ satisfying the following hypotheses:*

(a) $\sigma^n = \bigcup_{i \in N^n} C_i$ and
(b) $\sigma^n_{-i} \cap C_i = \emptyset$, $\forall i \in N^n$.

Then, it holds that $\bigcap_{i \in N^n} C_i \neq \emptyset$.

A similar result is the following:

Lemma 4 (Sperner covering principle). *Let C_i, $i \in N^n$ be a family of $(n+1)$ closed subsets of an n-simplex $\sigma^n = [v^0, v^1, \ldots, v^n]$ in $\mathbb{R}^n$ satisfying the following hypotheses:*

(a) $\sigma^n = \bigcup_{i \in N^n} C_i$ and
(b) $\sigma^n_{-i} \subset C_i$, $\forall i \in N^n$.

Then, it holds that $\bigcap_{i \in N^n} C_i \neq \emptyset$.

Next, we give the intermediate value theorem for simplices [30].

Theorem 6 (Intermediate value theorem for simplices [30]). *Assume that $\sigma^n = [v^0, v^1, \ldots, v^n]$ is an n-simplex in $\mathbb{R}^n$. Let $F_n = (f_1, f_2, \ldots, f_n) \colon \sigma^n \to \mathbb{R}^n$ be a continuous function such that $f_j(v^i) \neq 0$, $\forall j \in N^n_{-0} = \{1, 2, \ldots, n\}$, $i \in N^n = \{0, 1, \ldots, n\}$ and $\theta^n = (0, 0, \ldots, 0) \notin F_n(\vartheta\sigma^n)$ (i.e. F_n does not vanish on the boundary $\vartheta\sigma^n$ of σ^n). Assume that the vertices v^i, $i \in N^n$ are reordered such that the following hypotheses are fulfilled:*

$$(a) \quad \operatorname{sgn} f_j(v^j) \operatorname{sgn} f_j(x) = -1, \quad \forall x \in \sigma^n_{-j}, \ j \in N^n_{-0}, \tag{14}$$

$$(b) \quad \operatorname{sgn} F_n(v^0) \neq \operatorname{sgn} F_n(x), \quad \forall x \in \sigma^n_{-0}, \tag{15}$$

where $\operatorname{sgn} F_n(x) = \left(\operatorname{sgn} f_1(x), \operatorname{sgn} f_2(x), \ldots, \operatorname{sgn} f_n(x)\right)$ and σ^n_{-i} denotes the face opposite to vertex v^i. Then, there is at least one point $x \in \operatorname{int} \sigma^n$ such that $F_n(x) = \theta^n$.

Remark 11. The only computable information required by the hypotheses (14) and (15) of Theorem 6 is the algebraic sign of the function values on the boundary of the n-simplex σ^n. Thus, Theorem 6 is applicable whenever the signs of the function values are computed correctly. Theorem 6 has been applied for the localization and approximation of fixed points and zeros of continuous mappings using a simplicial subdivision of a simplex [31].

Next, we present a generalized method of bisection for simplices.

Definition 13 [10]. Let $\sigma_0^m = \langle v^0, v^1, \ldots, v^m \rangle$ be an oriented m-simplex in $\mathbb{R}^n$, $m \leqslant n$, suppose that $\langle v^i, v^j \rangle$ is the longest edge of σ_0^m and let $\Upsilon = (v^i \mid v^j)/2$ be the midpoint of $\langle v^i, v^j \rangle$. Then the *bisection* of σ_0^m is the order pair of m-simplices $\langle \sigma_{10}^m, \sigma_{11}^m \rangle$ where:

$$\sigma_{10}^m = \langle v^0, v^1, \ldots, v^{i-1}, \Upsilon, v^{i+1}, \ldots, v^j, \ldots, v^m \rangle,$$
$$\sigma_{11}^m = \langle v^0, v^1, \ldots, v^i, \ldots, v^{j-1}, \Upsilon, v^{j+1}, \ldots, v^m \rangle.$$

The m-simplices σ_{10}^m and σ_{11}^m will be called *lower simplex* and *upper simplex* respectively corresponding to σ_0^m while both σ_{10}^m and σ_{11}^m will be called *elements of the bisection* of σ_0^m. Suppose that $\sigma_0^n = \langle v^0, v^1, \ldots, v^n \rangle$ is an oriented n-simplex in $\mathbb{R}^n$ which includes at least one solution of the Eq. (2). Suppose further that $\langle \sigma_{10}^n, \sigma_{11}^n \rangle$ is the bisection of σ_0^n and that there is at least one root of the system (2) in some of its elements. Then this element will be called *selected n-simplex produced after one bisection* of σ_0^n and it will be denoted by σ_1^n. Moreover if there is at least one solution of the system (2) in both elements, then the selected n-simplex will be the lower simplex corresponding to σ_0^n. Suppose now that the bisection is applied with σ_1^n replacing σ_0^n giving thus the σ_2^n. Suppose further that this process continues for p iterations. Then we call σ_p^n the selected *n-simplex produced after p iterations of the bisection of σ_0^n* .

Theorem 7 [23]. *Suppose that $\sigma^m = [v^0, v^1, \ldots, v^m]$ is an m-simplex in $\mathbb{R}^n$, $m \leqslant n$. Let K be the barycenter of σ^m and let K_i be the barycenter of the i-th face $\sigma_{\neg i}^m = [v^0, v^1, \ldots, v^{i-1}, v^{i+1}, \ldots, v^m]$ of σ^m then the following relationships hold for all $0 \leqslant i \leqslant m$,*

(a) The points v^i, K and K_i are collinear points,

(b) $\|K - v^i\|_2 = \dfrac{m}{m+1} \left(\dfrac{1}{m} \displaystyle\sum_{\substack{j=0 \\ j \neq i}}^{m} \|v^i - v^j\|_2^2 - \dfrac{1}{m^2} \sum_{\substack{p=0 \\ p \neq i}}^{m-1} \sum_{\substack{q=p+1 \\ q \neq i}}^{m} \|v^p - v^q\|_2^2 \right)^{1/2},$

(c) $\|K - K_i\|_2 = m^{-1}\|K - v^i\|_2.$

Definition 14 [26]. The *barycentric radius* $\beta(\sigma^m)$ of an m-simplex σ^m in $\mathbb{R}^n$ is the radius of the smallest ball centered at the barycenter of σ^m and containing the simplex. The barycentric radius $\beta(A)$ of a subset A of $\mathbb{R}^n$ is the supremum of the barycentric radii of simplices with vertices in A.

Remark 12. The length of the barycentric radius $\beta(\sigma^m)$ of an m-simplex σ^m in $\mathbb{R}^n$, $m \leqslant n$, is $\max_i \|K - v^i\|_2$.

Theorem 8 [26]. *Any m-simplex $\sigma^m = [v^0, v^1, \ldots, v^m]$ in $\mathbb{R}^n$, $m \leqslant n$ is enclosable by the spherical surface S_β^{m-1} with radius $\beta(\sigma^m)$ given by:*

$$\beta(\sigma^m) = \frac{1}{m+1} \max_i \left(m \sum_{\substack{j=0 \\ j \neq i}}^{m} \|v^i - v^j\|_2^2 - \sum_{\substack{p=0 \\ p \neq i}}^{m-1} \sum_{\substack{q=p+1 \\ q \neq i}}^{m} \|v^p - v^q\|_2^2 \right)^{1/2}.$$

Remark 13. The barycentric radius $\beta(\sigma^n)$ of a n-simplex σ^n in $\mathbb{R}^n$ can be used to estimate error bounds for approximate fixed points or approximate roots of mappings in $\mathbb{R}^n$, by approximating a fixed point or a root by the barycenter of σ^n. Note that the computation of $\beta(\sigma^n)$ requires only the lengths of the edges of σ^n, which are also required in order to compute the diameter $\mathrm{diam}(\sigma^n)$ of σ^n. Furthermore, since the distance of the barycenter K of an n-simplex $\sigma^n = [v^0, v^1, \ldots, v^n]$ in $\mathbb{R}^n$ from the barycenter K_i of the i-th face $\sigma_{-i}^n = [v^0, v^1, \ldots, v^{i-1}, v^{i+1}, \ldots, v^n]$ of σ^n is equal to $\|K - v^i\|_2/n$ [23, 26], then using Theorem 8 we can easily compute the value of $\gamma(\sigma^n) = \min_i \|K - K_i\|_2/\mathrm{diam}(\sigma^n)$. The value $\gamma(\sigma^n)$ can be used to estimate the thickness $\theta(\sigma^n)$ of σ^n, that is:

$$\theta(\sigma^n) = \min_i \left\{ \min_{x \in \sigma_{-i}^n} \|K - x\|_2 \right\} / \mathrm{diam}(\sigma^n).$$

In general, the thickness $\theta(\sigma^n)$ is important to piecewise linear approximations of smooth mappings and, in general, to simplicial and continuation methods for approximating fixed points or roots of systems of nonlinear equations.

Theorem 9 [10]. *Suppose that σ_0^m is an m-simplex in $\mathbb{R}^n$ and let σ_p^m be any m-simplex produced after p bisections of σ_0^m. Then*

$$\mathrm{diam}(\sigma_p^m) \leqslant \left(\sqrt{3}/2 \right)^{\lfloor p/m \rfloor} \mathrm{diam}(\sigma_0^m), \tag{16}$$

where $\mathrm{diam}(\sigma_p^m)$ and $\mathrm{diam}(\sigma_0^m)$ are the diameters of σ_p^m and σ_0^m respectively and $\lfloor p/m \rfloor$ is the largest integer less than or equal to p/m.

Theorem 10 [23, 29]. *Suppose that σ_0^m, σ_p^m, $\mathrm{diam}(\sigma_0^m)$ and $\mathrm{diam}(\sigma_p^m)$ are as in Theorem 9 and let K_p^m be the barycenter of σ_p^m. Then for any point T in σ_p^m the following relationship is valid*

$$\|T - K_p^m\|_2 \leqslant \frac{m}{m+1} \left(\sqrt{3}/2 \right)^{\lfloor p/m \rfloor} \mathrm{diam}(\sigma_0^m). \tag{17}$$

Definition 15. Let σ^n be an n-simplex in $\mathbb{R}^n$ and let $\mathrm{diam}(\sigma^n)$ and $\mu\mathrm{diam}(\sigma^n)$ be the diameter and the microdiameter of σ^n respectively. Suppose that r is a solution of the Eq. (2) in σ^n. Then we define the barycenter K^n of σ^n to be an *approximation* of r and the quantity

$$\varepsilon(\sigma^n) = \frac{n}{n+1} \left((\operatorname{diam}(\sigma^n))^2 - \frac{n-1}{2n} (\mu \operatorname{diam}(\sigma^n))^2 \right)^{1/2}, \tag{18}$$

to be an *error estimate* for K^n.

Theorem 11 [23,29]. *Suppose that σ_p^n is the selected n-simplex produced after p bisections of an n-simplex σ_0^n in $\mathbb{R}^n$. Let r be a solution of the Eq. (2) which is included in σ_p^n and that K_p^n and $\varepsilon(\sigma_p^n)$ are the approximation of r and the error estimate for K_p^n respectively. Then the following hold:*

(a) $\quad \varepsilon(\sigma_p^n) \leqslant \dfrac{n}{n+1} \left(\sqrt{3}/2 \right)^{\lfloor p/n \rfloor} \operatorname{diam}(\sigma_0^n),$

(b) $\quad \varepsilon(\sigma_p^n) \leqslant \left(\sqrt{3}/2 \right)^{\lfloor p/n \rfloor} \varepsilon(\sigma_0^n),$

(c) $\quad \lim\limits_{p \to \infty} \varepsilon_p = 0,$

(d) $\quad \lim\limits_{p \to \infty} K_p^n = r.$

6 Synopsis

The paper presents, among some new results, an overview on generalizations of the intermediate value theorem for approximating fixed points and zeros of continuous functions. The presented generalized theorems are particular useful for the existence of solutions of systems of nonlinear equations in several variables as well as for the existence of fixed points of continuous functions. Based on the corresponding criteria for the existence of a solution emanated by the intermediate value theorems, generalized bisection methods for approximating fixed points and zeros of continuous functions are given. These bisection methods require only algebraic signs of the function values and are of major importance for tackling problems with imprecise (not exactly known) information.

Acknowledgment. The author would like to thank the anonymous reviewers for their helpful comments.

References

1. Alefeld, G., Frommer, A., Heindl, G., Mayer, J.: On the existence theorems of Kantorovich, Miranda and Borsuk. Electron. Trans. Numer. Anal. **17**, 102–111 (2004)
2. Bolzano, B.: Rein analytischer Beweis des Lehrsatzes, dass zwischen je zwei Werten, die ein entgegengesetztes Resultat gewähren, wenigstens eine reelle Wurzel der Gleichung liege. Prague (1817)
3. Brouwer, L.E.J.: Über Abbildungen von Mannigfaltigkeiten. Math. Ann. **71**, 97–115 (1912)

4. Cauchy, A.-L.: Cours d'Analyse de l'École Royale Polytechnique, Paris (1821). (Reprinted in Oeuvres Completes, series 2, vol. 3)
5. Grapsa, T.N., Vrahatis, M.N.: Dimension reducing methods for systems of nonlinear equations and unconstrained optimization: a review. In: Katsiaris, G.A., Markellos, V.V., Hadjidemetriou, J.D. (eds.) Recent Advances in Mechanics and Related Fields - Volume in Honour of Professor Constantine L. Goudas, pp. 343–353. University of Patras Press, Patras (2003)
6. Heindl, G.: Generalizations of theorems of Rohn and Vrahatis. Reliable Comp. **21**, 109–116 (2016)
7. Jarník, V.: Bernard Bolzano and the foundations of mathematical analysis. In: Bolzano and the Foundations of Mathematical Analysis, pp. 33–42. Society of Czechoslovak Mathematicians and Physicists, Prague (1981)
8. Kavvadias, D.J., Makri, F.S., Vrahatis, M.N.: Locating and computing arbitrarily distributed zeros. SIAM J. Sci. Comput. **21**(3), 954–969 (1999)
9. Kavvadias, D.J., Makri, F.S., Vrahatis, M.N.: Efficiently computing many roots of a function. SIAM J. Sci. Comput. **27**(1), 93–107 (2005)
10. Kearfott, R.B.: A proof of convergence and an error bound for the method of bisection in $\mathbb{R}^n$. Math. Comp. **32**(144), 1147–1153 (1978)
11. Kearfott, R.B.: An efficient degree-computation method for a generalized method of bisection. Numer. Math. **32**, 109–127 (1979)
12. Knaster, B., Kuratowski, K., Mazurkiewicz, S.: Ein Beweis des Fixpunkt-satzes für n-dimensionale Simplexe. Fund. Math. **14**, 132–137 (1929)
13. Miranda, C.: Un' osservatione su un theorema di Brouwer. Bollettino dell'U.M.I. **3**, 5–7 (1940)
14. Mourrain, B., Vrahatis, M.N., Yakoubsohn, J.C.: On the complexity of isolating real roots and computing with certainty the topological degree. J. Complex. **18**(2), 612–640 (2002)
15. Ortega J.M., Rheinboldt, W.C.: Iterative Solution of Nonlinear Equations in Several Variables. Classics in Applied Mathematics, vol. 30. SIAM, PA (2000)
16. Poincaré, H.: Sur certaines solutions particulières du problème des trois corps. Comptes rendus de l'Académie des Sciences Paris **91**, 251–252 (1883)
17. Poincaré, H.: Sur certaines solutions particulières du problème des trois corps. Bull. Astronomique **1**, 63–74 (1884)
18. Scarf, H.: The approximation of fixed points of a continuous mapping. SIAM J. Appl. Math **15**(5), 1328–1343 (1967)
19. Sikorski, K.: Bisection is optimal. Numer. Math. **40**, 111–117 (1982)
20. Sikorski, K.: Optimal Solution of Nonlinear Equations. Oxford University Press, New York (2001)
21. Sperner, E.: Neuer Beweis für die Invarianz der Dimensionszahl und des Gebietes. Abh. Math. Sem. Hamburg **6**, 265–272 (1928)
22. Stenger, F.: Computing the topological degree of a mapping in $\mathbb{R}^n$. Numer. Math. **25**, 23–38 (1975)
23. Vrahatis, M.N.: An error estimation for the method of bisection in $\mathbb{R}^n$. Bull. Greek Math. Soc. **27**, 161–174 (1986)
24. Vrahatis, M.N.: Solving systems of nonlinear equations using the nonzero value of the topological degree. ACM Trans. Math. Softw. **14**, 312–329 (1988)
25. Vrahatis, M.N.: CHABIS: a mathematical software package for locating and evaluating roots of systems of nonlinear equations. ACM Trans. Math. Softw. **14**, 330–336 (1988)
26. Vrahatis, M.N.: A variant of Jung's theorem. Bull. Greek Math. Soc. **29**, 1–6 (1988)

27. Vrahatis, M.N.: A short proof and a generalization of Miranda's existence theorem. Proc. Amer. Math. Soc. **107**, 701–703 (1989)
28. Vrahatis, M.N.: An efficient method for locating and computing periodic orbits of nonlinear mappings. J. Comput. Phys. **119**, 105–119 (1995)
29. Vrahatis, M.N.: Simplex bisection and Sperner simplices. Bull. Greek Math. Soc. **44**, 171–180 (2000)
30. Vrahatis, M.N.: Generalization of the Bolzano theorem for simplices. Topol. Appl. **202**, 40–46 (2016)
31. Vrahatis, M.N.: Intermediate value theorem for simplices for simplicial approximation of fixed points and zeros. Topol. Appl. (2019). Accepted for Publication
32. Vrahatis, M.N., Androulakis, G.S., Manoussakis, G.E.: A new unconstrained optimization method for imprecise function and gradient values. J. Math. Anal. Appl. **197**(2), 586–607 (1996)
33. Vrahatis, M.N., Iordanidis, K.I.: A rapid generalized method of bisection for solving systems of non-linear equations. Numer. Math. **49**, 123–138 (1986)
34. Vrahatis, M.N., Ragos, O., Skiniotis, T., Zafiropoulos, F., Grapsa, T.N.: RFSFNS: a portable package for the numerical determination of the number and the calculation of roots of Bessel functions. Comput. Phys. Commun. **92**, 252–266 (1995)
35. Zottou, D.-N.A., Kavvadias, D.J., Makri, F.S., Vrahatis, M.N.: MANBIS—A C++ mathematical software package for locating and computing efficiently many roots of a function: theoretical issues. ACM Trans. Math. Softw. **44**(3), 1–7 (2018). Article no. 35

Modelling Population Size Using Horvitz-Thompson Approach Based on the Zero-Truncated Poisson Lindley Distribution

Ratchaneewan Wongprachan[✉][iD]

Division of Statistics, Maejo University, Sansai, Chiang Mai 50290, Thailand
ratchaneewan@mju.ac.th

Abstract. Capture-recapture analysis is applied to estimate population size in ecology, biology, social science, medicine, linguistics and software engineering. The Poisson distribution is one of the simplest models for count data and appropriate for homogeneous populations. On the other hand, it is found to underestimate the counts for overdispersed data. In this study, population size estimation using the mixture of Poisson and Lindley distribution is proposed. It can exhibit overdispersed, equidispersed and underdispersed data. Additionally, it is able to present count data with long tail. As a result of the problem of unobserved individuals, the zero-truncated Poisson Lindley distribution is considered. The parameter of distribution can be estimated using the maximum likelihood estimation. The Horvitz-Thompson estimator based on the zero-truncated Poisson Lindley distribution for modelling the population size is investigated in this study. Point and interval estimation of the target population are presented. The technique of conditioning is used for variance estimation of the population size. Relative bias, relative variance and relative mean square error are used for measuring the accuracy of the estimator. The simulation results show that the Horvitz-Thompson estimator under the zero-truncated Poisson Lindley distribution provides a good fit when compared to the zero-truncated Poisson distribution.

Keywords: Poisson-Lindley distribution · Zero-truncated distribution · Horvitz-Thompson estimators · Variance estimation

1 Introduction

Estimating the unknown population size based on the idea of capture-recapture approach have been investigated in many fields. In ecology, this concept is applied to estimate the total number of species and the number of wildlife [6,8]. In medicine, it is used to estimate the number of hidden diseases. In linguistics, there are many words in novels and a linguist is interested in estimating the total vocabularies of authors [12]. In software engineering, estimating the number of faults in software inspection is useful for software quality improvements. In social

© Springer Nature Switzerland AG 2020
Y. D. Sergeyev and D. E. Kvasov (Eds.): NUMTA 2019, LNCS 11974, pp. 239–254, 2020.
https://doi.org/10.1007/978-3-030-40616-5_18

science, the problem of the number of illegal immigrants, the number of drug users, the number of forced labours and the number of domestic violence are discussed.

Let X_i denote the frequency with which individual i observed from the population of size N, for $X_i = 0, 1, 2, \ldots$ and $i = 0, 1, 2, \ldots, N$. Suppose that there are K individuals in sample from M observations, the number of individuals observed exactly x times is f_x for $x = 1, 2, \ldots, m$ and m is the maximum number of times any individual observed. Therefore, we have $K = \sum_{x=1}^{m} f_x$ and $M = \sum_{x=1}^{m} x f_x$.

1.1   Illegal Immigrants Data in the Netherlands

Illegal immigrants data in the Netherlands was recorded by the police in 1967. The number of illegal immigrants expelled ineffectively is 1880. Some of them are apprehended more than once. The frequencies of apprehension are shown in Table 1.

Table 1. Illegal immigrants data [10]

x	1	2	3	4	5	6	K
f_x	1645	183	37	13	1	1	1880

1.2   Heroin User Data in Bangkok

Capture-recapture experiment was studied to estimate the heroin user population in Bangkok in 2004 [2]. It found 9302 drug users treated at the centres. The number of contacts of heroin users from health treatment centres is observed and presented in Table 2.

Table 2. Heroin user data in Bangkok [2]

x	1	2	3	4	5	6	7	8	9	10	11	12	13
f_x	2176	1600	1278	976	748	570	455	368	281	254	188	138	99

x	14	15	16	17	18	19	20	21	K
f_x	67	44	34	17	3	3	2	1	9302

1.3   Forced Labour Data

The number of forced labour is a hidden data hard to estimate. Forced labour worldwide data in 2012 were observed. From report, it is found 5491 forced labours worldwide. The frequency of forced labour are shown in Table 3 [3].

Table 3. Forced labour data [3]

x	1	2	3	4	5	6	7	8	9	10	11	K
f_x	4069	1186	167	46	10	7	3	1	0	1	1	5491

1.4 Domestic Violence in the Netherlands

In 2009, it is found 17662 domestic violence identified in the Netherlands 2009. The number of domestic violence reported are shown in Table 4.

Table 4. Domestic violence in the Netherlands [11]

x	1	2	3	4	5	6	7	8	9	K
f_x	15169	1957	393	99	28	8	6	1	1	17662

The basic idea of the population size can be determined by both parts of observed and unobserved data. Nevertheless, the terms unobserved individual is unknown, f_0. Hence, the zero-truncated model is considered for the distribution of X_i or X in this study.

Modelling capture-recapture count data under the Poisson distribution is the simplest model for estimating N. It is a traditional model introduced for equidispersed data, the variance is equal to the mean. However, in practice, using the homogeneous Poisson model is not appropriate because of different probabilities of an individual being observed. This leads to a heterogeneous population which is over or under dispersed. When fitting a model under the Poisson distribution, it leads to an underestimate of the population size for overdispersed cases in which the variance is greater than the mean [4,5].

Then, a mixed Poisson distribution is considered for heterogeneous populations. It has been studied to fit count data for both overdispersion and underdispersion. Some example of Poisson mixture model are Poisson-Lindley [9, 18], Poisson-gamma [5,8,16], Generalized Poisson-Lindley [15], Poisson-Inverse Gaussian [13] Poisson-Weighted Exponential [19] and Poisson-Tweedie distributions [7].

Here, the problem about population size estimation is investigated using the Horvitz-Thompson (HT) estimator. Improving the HT estimator of N based on the zero truncated mixed Poisson model has been studied such as the zero truncated Poisson Gamma distribution known as the zero truncated negative binomial distribution. This estimator is proposed for overdispersed data in [5], [16]. However, it might be found numerical problem in optimization.

In this study, the focus is on the HT estimator based on the mixture of Poisson model with one parameter. The HT estimator under the zero truncated Poisson Lindley (ZTPL) distribution is developed and compared to the zero truncated Poisson model. The Horvitz-Thompson estimator based on the ZTPL distribution is presented with both point and interval estimation of N. The PL distribution and its properties are reviewed in Sect. 2. Point and interval for the population size estimator is discussed and variance estimation of N by conditioning approach is derived in Sect. 3. Simulation study is presented in Sect. 4. An application is shown in Sect. 5 and conclusion is in Sect. 6.

2 The Poisson Lindley Distribution and the Zero-Truncated Model

The mixture model based on the Poisson and Lindley distribution was developed in 1970 to model count data [17] and is named the Poisson Lindley distribution (PL). Let X denote a random variable which follows the Poisson distribution with parameter λ. The probability mass function is given by

$$g(x|\lambda) = \frac{e^{-\lambda}\lambda^x}{x!}, \tag{1}$$

where $x = 0, 1, 2, \ldots$ and $\lambda > 0$. When the parameter λ generated by the Lindley distribution with the probability density function

$$h(\lambda) = \frac{\theta^2}{\theta + 1}(1 + \lambda)e^{-\theta\lambda}, \tag{2}$$

where $\lambda > 0$ and $\theta > 0$, the random variable X has the PL distribution and the probability mass function of the PL distribution is given by

$$
\begin{aligned}
\Pr(X = x) &= \int_0^\infty g(x|\lambda)\, h(\lambda)d\lambda \\
&= \int_0^\infty \frac{e^{-\lambda}\lambda^x}{x!} \cdot \frac{\theta^2}{\theta + 1}(1 + \lambda)e^{-\theta\lambda}d\lambda \\
&= \frac{\theta^2}{(\theta + 1)^2}\left\{\frac{x + 1}{(\theta + 1)^{x+1}} + \frac{1}{(\theta + 1)^x}\right\} \\
&= \frac{\theta^2(x + \theta + 2)}{(\theta + 1)^{x+3}},
\end{aligned} \tag{3}
$$

where $x = 0, 1, 2, \ldots$ and θ is the Poisson-Lindley parameter for $\theta > 0$ [17].

Figure 1 shows the probability mass function of the PL distribution based on some values of the parameter θ. When some individuals are not identified, the zero count data is missing. The zero-truncated distribution is considered for the distribution of X, and the probability function is defined by a conditional probability on $X > 0$

$$\Pr(X = x|x > 0) = \frac{\Pr(X = x)}{\Pr(X > 0)} = \frac{\Pr(X = x)}{1 - \Pr(X = 0)}, \tag{4}$$

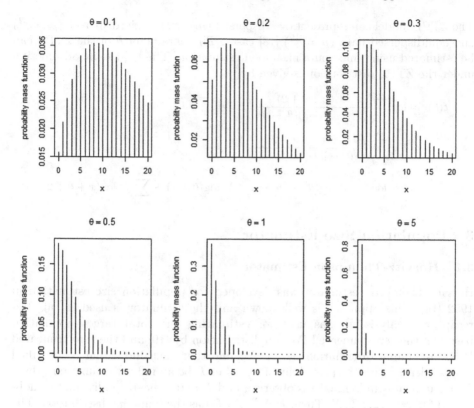

Fig. 1. Probability mass function for the Poisson-Lindley Distribution with the parameter $\theta = 0.1, 0.2, 0.3, 0.5, 1, 5$.

which can be written as

$$p_x^+ = \frac{p_x}{1 - p_0}, \qquad x = 1, 2, \ldots \tag{5}$$

Hence, the probability mass function of a random variable X under the zero-truncated Poisson Lindley distribution (ZTPL) can be defined by

$$p_x^+ = \frac{\theta^2(x + \theta + 2)}{(\theta^2 + 3\theta + 1)(\theta + 1)^x}, \tag{6}$$

where $x = 1, 2, \ldots$ and $\theta > 0$. The mean and variance of the X are given by

$$\mu = \frac{(\theta + 1)^2(\theta + 2)}{\theta(\theta^2 + 3\theta + 1)} \tag{7}$$

and

$$\sigma^2 = \frac{(\theta + 1)^2(\theta^3 + 6\theta^2 + 10\theta + 2)}{\theta^2(\theta^2 + 3\theta + 1)^2}. \tag{8}$$

The ZTPL model can represent overdispersed ($\mu < \sigma^2$), underdispersed ($\mu > \sigma^2$) and equidispersed data ($\mu = \sigma^2$) [18] cases. The parameter θ of the ZTPL can be estimated using maximum likelihood estimation. The log-likelihood function under the ZTPL distribution is given by

$$
\begin{aligned}
\ell(\theta) &= \log \prod_{x=1}^{K} \frac{\theta^2(x+\theta+2)}{(\theta^2+3\theta+1)(\theta+1)^x} \\
&= \log \prod_{x=1}^{m} \left(\frac{\theta^2(x+\theta+2)}{(\theta^2+3\theta+1)(\theta+1)^x} \right)^{f_x} \\
&= 2K \log \theta - K \log(\theta^2+3\theta+1) - M \log(\theta+1) + \sum_{x=1}^{m} f_x \log(x+\theta+2). \quad (9)
\end{aligned}
$$

3   Population Size Estimator

3.1   Horvitz-Thompson Estimator

Horvitz-Thompson estimator was developed for population size estimation in 1952 [14]. This approach is well known and the estimator is used in capture-recapture analysis. It was used for estimating the total target population based on the zero-truncated Poisson distribution by [10] and the zero-truncated Poisson-Gamma distribution by [5]. Assume that each individual is identified independently with the probability $1 - p_0$. Let I_i be an indicator function, where $I_i = 1$ indicates individual i is observed and $I_i = 0$ indicates individual i unobserved for $i = 1, 2, \ldots, N$. Then, $K = \sum_{i=1}^{N} I_i$ has the binomial distribution. The population size can be estimated by $\sum_{i=1}^{N} I_i [1 - \Pr(X = 0)]^{-1}$ and can written as

$$
\widehat{N} = \frac{K}{1 - p_0}, \quad (10)
$$

where $K = \sum_{i=1}^{N} I_i$ and $1 - p_0 = 1 - \Pr(X = 0)$, are the number of observed individuals and the probability of an individual is observed, respectively. Hence, considering the zero-truncated distribution, it leads to unknown probability p_0. Therefore, the estimated p_0 is required for modelling the size of target populations. The maximum likelihood estimation (MLE) is a well known statistical approach for estimating the unknown parameter. In this study, the log-likelihood function in [9] is maximized for estimating the parameter (θ) of the ZTPL distribution. After that, $\widehat{\theta}$ is replaced in [3] for $p_0(\widehat{\theta})$. Finally, we have the HT estimator as a point estimator of N under the ZTPL distribution.

For the interval estimator of N, the normal approximation is considered for the confidence interval for the population size under the large enough sample condition. The 95% confidence interval is given by [5, 10].

$$
\widehat{N} \pm 1.96 \sqrt{\mathrm{Var}(\widehat{N})}. \quad (11)
$$

3.2 Variance Estimation of N

The confidence interval for N requires the variance term. In this study, the technique of conditioning proposed by [1] is used to estimate the variance estimation of N which can be obtained as

$$\mathrm{Var}(\widehat{N}) = \mathrm{E}[\mathrm{Var}(\widehat{N}|K)] + \mathrm{Var}(\mathrm{E}[\widehat{N}|K]). \tag{12}$$

Considering the first term in (12), $\mathrm{E}[\mathrm{Var}(\widehat{N}|K)]$ can be estimated by $\mathrm{Var}(\widehat{N}|K)$ using the δ method. Under the HT estimator and the ZTPL distribution with parameter θ, this results in

$$\widehat{\mathrm{Var}}(\widehat{N}|K) = K^2 \left(\frac{\partial}{\partial \theta} \frac{1}{g(\theta)} \right)^T \mathrm{Var}(\theta) \left(\frac{\partial}{\partial \theta} \frac{1}{g(\theta)} \right), \tag{13}$$

where $g(\theta) = 1 - p_0$ and $p_0 = \dfrac{\theta^2(\theta+2)}{(\theta+1)^3}$ is the probability for unobserved individual based on the PL distribution and the variance of θ can be estimated by

$$\widehat{\mathrm{Var}}(\theta) = - \left(\frac{\partial^2 \ell}{\partial \theta^2} \right)^{-1},$$

where ℓ is the log likelihood defined in (9), the first derivative and the second derivative of the log-likelihood with respect to θ can be written as

$$\frac{\partial \ell}{\partial \theta} = \frac{2K}{\theta} + \sum_{x=1}^{m} \frac{f_x}{x+\theta+2} - \frac{K(2\theta+3)}{\theta^2+3\theta+1} - \frac{M}{\theta+1}$$

$$\frac{\partial^2 \ell}{\partial \theta^2} = \frac{2K}{\theta^2} + \sum_{x=1}^{m} \frac{f_x}{(x+\theta+2)^2} + \frac{2K}{\theta^2+3\theta+1} - K \left[\frac{2\theta+3}{\theta^2+3\theta+1} \right]^2 - \frac{M}{(\theta+1)^2}.$$

Then, the first term of variance in (12) can be evaluated as

$$\widehat{\mathrm{Var}}(\widehat{N}|K) = \left[\frac{K\theta(\theta+1)^2(\theta+4)}{(\theta^2+3\theta+1)^2} \right]^2 \widehat{\mathrm{Var}}(\theta). \tag{14}$$

For the second term, assume that $\mathrm{E}[\widehat{N}|K]$ can be estimated by $\widehat{N}$ and K is the number of observed individuals which follows the binomial distribution. Then,

$$\mathrm{Var}(\mathrm{E}[\widehat{N}|K]) \approx \mathrm{Var}(\widehat{N})$$
$$= \mathrm{Var}(\frac{K}{1-p_0})$$
$$= \frac{K\theta^2(\theta+2)(\theta+1)^3}{(\theta^2+3\theta+1)^2}. \tag{15}$$

Finally, the variance of $\widehat{N}$ in (12) can be estimated using (14) and (15) which is given as

$$\mathrm{Var}(\widehat{N}) = \left[\frac{K\theta(\theta+1)^2(\theta+4)}{(\theta^2+3\theta+1)^2} \right]^2 \widehat{\mathrm{Var}}(\theta) + \frac{K\theta^2(\theta+2)(\theta+1)^3}{(\theta^2+3\theta+1)^2}. \tag{16}$$

4   Simulation Study

The HT estimator based on the zero-truncated Poisson (ZTP) and the zero-truncated Poisson Lindley (ZTPL) distribution are compared to Chao estimator which is known as the lower bound of the estimated population size

$$\widehat{N}_{chao} = K + \frac{f_1^2}{2f_2}$$

and the estimated variance of $\widehat{N}$ is given by [1]

$$\widehat{\mathrm{Var}}(\widehat{N}_{chao}) = \frac{1}{4}\frac{f_1^4}{f_2^3} + \frac{f_1^3}{f_2^2} + \frac{1}{2}\frac{f_1^2}{f_2} - \frac{1}{4}\frac{f_1^4}{K f_2^2} - \frac{1}{2}\frac{f_1^4}{f_2(2K f_2 + f_1^2)}.$$

In a simulation study, $\widehat{N}$ and $\widehat{N} \pm 1.96\sqrt{\mathrm{Var}(\widehat{N})}$ for all estimators have been investigated as follows

1. Population of size $N = 250, 500, 1000, 2000$ are generated using the negative binomial distribution with parameter (p, k)

$$p_x = \frac{\Gamma(x+k)}{\Gamma(x+1)\Gamma(k)}p^k(1-p)^x, \qquad x = 0, 1, 2, \ldots, k,$$

   where $p = \dfrac{k}{k+\mu}$, $\mu = 0.5, 1, 1.5, 2, 4, 10$, $k = 0.5, 0.75, 1, 1.25, 1.5, 2$ and repeated 10000 times.
2. Likelihood function is constructed using the ZTP and ZTPL distribution before optimizing to estimate the parameter of the distribution.
3. Chao estimator and the HT estimator under the ZTP and the ZTPL model are evaluated.
4. Finally, the estimators are measured in terms of relative bias

$$\mathrm{RBIAS}(\widehat{N}) = \frac{1}{N}[\mathrm{E}(\widehat{N}) - N],$$

   relative variance

$$\mathrm{RVar}(\widehat{N}) = \frac{1}{N^2}\mathrm{Var}(\widehat{N}),$$

   and relative root mean square error

$$\mathrm{RRMSE}(\widehat{N}) = \frac{1}{N}\sqrt{\mathrm{Var}(\widehat{N}) + (\mathrm{bias}(\widehat{N}))^2}.$$

Considering the count data from the negative binomial distribution, the lower bound of the number of population in the simulation is approximated using the Chao estimator. The Horvitz-Thompson estimator based on the ZTP and ZTPL distributions is compared to the Chao estimator. Figure 2 represents the estimated population size for small and large sample size, $N = 250$ and $N = 1000$. It is clear that the estimator based on the ZTPL model outperforms ZTP

(a) N=250, k=1.25

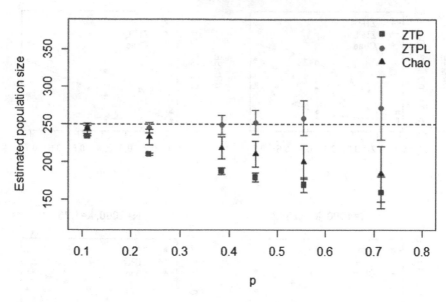

(b) N=1000,k=1.25

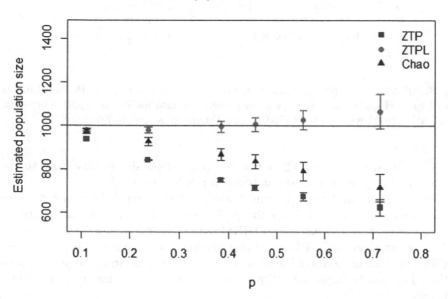

Fig. 2. Comparison of estimated population size using the HT estimator based on the Poisson (ZTP) and Poisson-Lindley (ZTPL) model and Chao estimator when generating data from the negative binomial distribution (p, k) with $N = 250, 1000$ and repeated 10000 times. The vertical lines at each dot represent the standard deviation of population size.

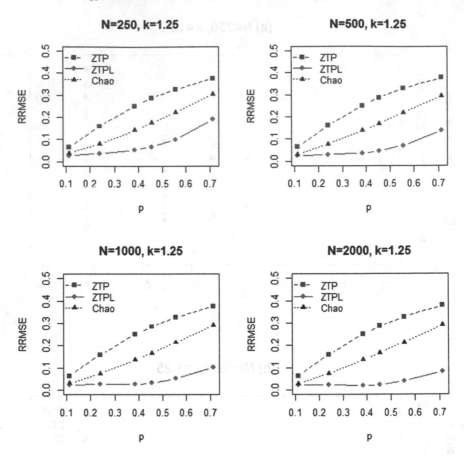

Fig. 3. RRMSE of the HT estimator based on the Poisson (ZTP) and Poisson-Lindley (ZTPL) model and Chao estimator when generating data under the negative binomial distribution (p, k) with $N = 250, 500, 1000, 2000$ and repeated 10000 times.

and Chao with small bias. The estimated population size is very close to the true number of population, especially when p is less than 0.5.

On the other hand, the ZTP and Chao estimators result in underestimates for all situations. Figure 3 shows the performance of the estimator in term of the relative root mean square error (RRMSE). The results indicate that the HT estimator based on the ZTPL model give the highest accuracy with the smallest RRMSE for various values of the parameter p. When N is large, the performance of the ZTPL can be improved, whereas the ZTP and Chao estimators give larger RRMSE.

Table 5 represents the accuracy of the ZTP, ZTPL and Chao estimator with the RBias. The results indicate that RBias for the ZTPL estimator is close to zero and gives the smallest RBias. The ZTPL estimator provides more accuracy, especially when $k < 1.5$. Table 6 shows that the RVar of $\widehat{N}$. It is found that the variance of the estimator decreases when μ is larger. Although the ZTP estimator gives the smaller RVar, it does not work well for the heterogeneous population and does not outperform the ZTPL and Chao estimators.

Considering the accuracy and precision in term of the RRMSE in Table 7, the ZTPL estimator provides a good fit with smallest RRMSE, particularly for $k < 1.5$. For example, when $N = 1000$, $\mu = 1$ and $k = 1.25$, RRMSE the ZTPL estimator is 0.0076 while for the ZTP estimator it is 0.0509. For RBias of the estimator, the ZTPL and ZTP estimators provide RRMSE = 2.0281 and 0.3533, respectively. When $k > 1.5$, it is found that Chao estimator can approximate the population size better than the ZTPL estimator.

Table 5. RBias for population size estimators, repeated 10000 times.

μ	k	$N = 250$			$N = 500$		
		Chao	ZTP	ZTPL	Chao	ZTP	ZTPL
1	1	−0.2389	−0.3689	−0.0513	−0.2446	−0.3705	−0.0542
	1.25	−0.2018	−0.3234	0.0322	−0.2087	−0.3262	0.0271
	1.5	−0.1735	−0.2883	0.0974	−0.1816	−0.2912	0.0915
	2	−0.1371	−0.2360	0.1947	−0.1432	−0.2388	0.1888
2	1	−0.1599	−0.2900	−0.0688	−0.1636	−0.2906	−0.0700
	1.25	−0.1278	−0.2502	−0.0037	−0.1317	−0.2502	−0.0041
	1.5	−0.1053	−0.2190	0.0478	−0.1084	−0.2196	0.0464
	2	−0.0773	−0.1763	0.1188	−0.0807	−0.1774	0.1169
4	1	−0.0950	−0.1943	−0.0692	−0.0974	−0.1943	−0.0697
	1.25	−0.0682	−0.1586	−0.0202	−0.0718	−0.1591	−0.0211
	1.5	−0.0522	−0.1334	0.0148	−0.0551	−0.1338	0.0142
	2	−0.0333	−0.1005	0.0611	−0.0349	−0.1005	0.0608
μ	k	$N = 1000$			$N = 2000$		
		Chao	ZTP	ZTPL	Chao	ZTP	ZTPL
1	1	−0.2471	−0.3714	−0.0561	−0.2484	−0.3717	−0.0569
	1.25	−0.2101	−0.3264	0.0263	−0.2120	−0.3273	0.0247
	1.5	−0.1830	−0.2913	0.0911	−0.1843	−0.2922	0.0894
	2	−0.1450	−0.2403	0.1860	−0.1469	−0.2409	0.1849
2	1	−0.1651	−0.2911	−0.0711	−0.1660	−0.2911	−0.0711
	1.25	−0.1329	−0.2507	−0.0050	−0.1337	−0.2507	−0.0054
	1.5	−0.1108	−0.2205	0.0449	−0.1112	−0.2205	0.0448
	2	−0.0818	−0.1774	0.1168	−0.0825	−0.1776	0.1163
4	1	−0.0988	−0.1943	−0.0700	−0.0991	−0.1942	−0.0699
	1.25	−0.0728	−0.1591	−0.0214	−0.0732	−0.1590	−0.0215
	1.5	−0.0556	−0.1338	0.0138	−0.0561	−0.1338	0.0139
	2	−0.0359	−0.1005	0.0606	−0.0364	−0.1005	0.0606

Table 6. RVar for population size estimators, repeated 10000 times.

μ	k	$N = 250$			$N = 500$		
		Chao	ZTP	ZTPL	Chao	ZTP	ZTPL
1	1	1.8505	0.2891	1.7055	1.6856	0.2801	1.6684
	1.25	1.9472	0.3718	2.1042	1.7743	0.3590	2.0509
	1.5	2.0334	0.4469	2.4554	1.8422	0.4296	2.3840
	2	2.1102	0.5737	3.0326	1.9433	0.5519	2.9455
2	1	0.9040	0.0071	0.5599	0.8351	0.0559	0.5535
	1.25	0.8741	0.0721	0.6624	0.8046	0.0709	0.6566
	1.5	0.8372	0.0857	0.7515	0.7831	0.0841	0.7431
	2	0.7785	0.1067	0.8838	0.7329	0.1049	0.8752
4	1	0.4806	0.0062	0.1891	0.4411	0.0061	0.1877
	1.25	0.4196	0.0080	0.2150	0.3805	0.0078	0.2136
	1.5	0.3630	0.0094	0.2347	0.3321	0.0093	0.2337
	2	0.2895	0.0116	0.2630	0.2684	0.0114	0.2614
μ	k	$N = 1000$			$N = 2000$		
		Chao	ZTP	ZTPL	Chao	ZTP	ZTPL
1	1	1.6125	0.2753	1.6479	1.5745	0.2724	1.6356
	1.25	1.7081	0.3533	2.0281	1.6719	0.3501	2.0143
	1.5	1.7783	0.4238	2.3618	1.7480	0.4201	2.3461
	2	1.8873	0.5431	2.9107	1.8460	0.5392	2.8953
2	1	0.8086	0.0551	0.5492	0.7928	0.0549	0.5483
	1.25	0.7818	0.0702	0.6525	0.7700	0.0699	0.6510
	1.5	0.7566	0.0832	0.7386	0.7476	0.0830	0.7374
	2	0.7128	0.1044	0.8726	0.7032	0.1038	0.8698
4	1	0.4207	0.0059	0.1866	0.4135	0.0059	0.1866
	1.25	0.3659	0.0077	0.2126	0.3595	0.0076	0.2123
	1.5	0.3229	0.0091	0.2328	0.3164	0.0091	0.2326
	2	0.2580	0.0113	0.2608	0.2525	0.0113	0.2607

5   Application with Real Data Examples

In this section, the Horvitz-Thompson estimator based on the ZTPL distribution is applied to some real data examples. The result of applying the ZTPL estimator to real data can be seen in Table 8. Point and interval estimation are evaluated for the data. Additional, Akaike information criterion (AIC) is used for model selection, $AIC = 2k - 2\ln(L)$, where L is the likelihood function and k is the number of parameters.

Table 7. RRMSE for population size estimators, repeated 10000 times.

μ	k	$N = 250$			$N = 500$		
		Chao	ZTP	ZTPL	Chao	ZTP	ZTPL
1	1	0.2539	0.3705	0.0972	0.2514	0.3713	0.0792
	1.25	0.2202	0.3257	0.0972	0.2170	0.3273	0.0695
	1.5	0.1955	0.2914	0.1389	0.1915	0.2927	0.1146
	2	0.1651	0.2408	0.2237	0.1561	0.2411	0.2038
2	1	0.1708	0.2904	0.0835	0.1686	0.2908	0.0775
	1.25	0.1408	0.2507	0.0516	0.1377	0.2505	0.0365
	1.5	0.1202	0.2198	0.0727	0.1154	0.2200	0.0603
	2	0.0954	0.1775	0.1328	0.0893	0.1780	0.1242
4	1	0.1046	0.1943	0.0745	0.1018	0.1944	0.0723
	1.25	0.0796	0.1587	0.0356	0.0769	0.1591	0.0295
	1.5	0.0646	0.1335	0.0340	0.0608	0.1338	0.0259
	2	0.0476	0.1008	0.0692	0.0419	0.1006	0.0649
μ	k	$N = 1000$			$N = 2000$		
		Chao	ZTP	ZTPL	Chao	ZTP	ZTPL
1	1	0.2503	0.3718	0.0692	0.2500	0.3719	0.0637
	1.25	0.2141	0.3269	0.0522	0.2139	0.3275	0.0402
	1.5	0.1878	0.2920	0.1032	0.1866	0.2926	0.0957
	2	0.1513	0.2415	0.1937	0.1500	0.2414	0.1888
2	1	0.1675	0.2912	0.0748	0.1672	0.2912	0.0730
	1.25	0.1358	0.2508	0.0260	0.1351	0.2510	0.0188
	1.5	0.1141	0.2207	0.0525	0.1128	0.2206	0.0488
	2	0.0860	0.1777	0.1205	0.0846	0.1778	0.1181
4	1	0.1009	0.1943	0.0713	0.1002	0.1942	0.0706
	1.25	0.0752	0.1591	0.0259	0.0744	0.1590	0.0238
	1.5	0.0584	0.1338	0.0206	0.0575	0.1338	0.0176
	2	0.0393	0.1006	0.0627	0.0381	0.1005	0.0617

According to illegal immigrants data, the results show that the lower bound of the estimated illegal immigrants is 9270 using the Chao estimator. The HT estimator under the ZTP and the ZTPL estimator can be estimated as $\widehat{N}_{ZTP} = 7080$, $\widehat{N}_{ZTPL} = 13334$, respectively.

For the heroin user data, the number of heroin users using Chao estimator is $\widehat{N}_{chao} = 10782$, while the ZTP and ZTPL model can approximate the number of heroin users as $\widehat{N}_{HT} = 9453$ and $\widehat{N}_{HT} = 11324$, respectively. When considering the AIC criterion, it is clear that the ZTPL outperforms the ZTP with smaller AIC as 42891.50 and the 95% confidence interval for the ZTPL is between 11213 and 11435.

For forced labour data, the size of force labour worldwide is estimated $\widehat{N}_{ZTPL} = 21879$. For domestic violence data, the size of domestic violence is approximated $\widehat{N}_{ZTPL} = 112502$. All application in real data sets, the ZTPL estimator gives a high estimated population size rather than Chao estimator. When comparing the ZTPL to the ZTP estimator, the ZTPL has a better fit compared to the ZTP ($AIC_{ZTPL} < AIC_{ZTP}$).

Table 8. Estimated total number of population

Data	$\widehat{N}_{HT}$	$\widehat{se}(\widehat{N}_{HT})$	95% CI	AIC
Illegal immigrants				
Chao	9270	635.65	8024–10516	-
ZTP($\widehat{\lambda} = 0.31$)	7080	365.75	6363–7797	1805.90
ZTPL($\widehat{\theta} = 6.88$)	13334	643.15	12073–14595	1769.10
Heroin user				
Chao	10782	80.21	10624–10940	-
ZTP($\widehat{\lambda} = 4.13$)	9453	12.84	9428–9479	50092.41
ZTPL($\widehat{\theta} = 0.49$)	11324	56.61	11213–11435	42891.50
Forced labour				
Chao	12470	288.40	11905–13035	-
ZTP($\widehat{\lambda} = 0.59$)	12326	240.17	11856–12797	8136.13
ZTPL($\widehat{\theta} = 3.66$)	21879	469.16	20960–22799	8086.10
Domestic violence				
Chao	76448	1579.65	73352–79544	-
ZTP($\widehat{\lambda} = 0.35$)	60214	941.76	58368–62059	18445.78
ZTPL($\widehat{\theta} = 6.14$)	112502	1686.10	109197–115807	18051.60

6   Conclusion

Capture-recapture approach is widely used for estimating the population size. The problem of unobserved data is an important consideration for population size estimation. Horvitz-Thompson estimator is a parametric approach that is applied to various distributions. The Poisson model is a fundamental distribution for count data. However, the mixture Poisson model is flexible for heterogeneity in the population. The Poisson Lindley distribution can represent underdispersed, equidispersed and overdispersed data. When applying the zero-truncated model, the Horvitz-Thompson estimator based on Poisson Lindley can approximate well and can be improved for large sample size. Whereas the zero-truncated Poisson distribution gives a poor fit. It underestimates the true population size in the simulation study. Hence, the Poisson Lindley distribution can be an alternative method for estimating the population size. In this study, maximizing the log likelihood function based on the distribution with one parameter converges to the correct parameter. In the future study, the aim is to improve the Horvitz-Thompson estimator using other distributions such as the Poisson mixture distribution with two parameters. However, the MLE would need to be evaluated numerically and may be more difficult.

Acknowledgment. This research was supported by Maejo University, Chiang Mai, Thailand.

References

1. Böhning, D.: A simple variance formula for population size estimators by conditioning. Stat. Methodol. **5**(5), 410–423 (2008). https://doi.org/10.1016/j.stamet.2007.10.001
2. Böhning, D., Suppawattanabodee, B., Kusolvisitkul, W., Viwatwongkasem, C.: Estimating the number of drug users in Bangkok 2001: a capture-recapture approach using repeated entries in one list. Eur. J. Epidemiol. **19**(12), 1075–1083 (2004). https://doi.org/10.2307/3583005
3. Böhning, D.: Ratio plot and ratio regression with applications to social and medical sciences. Stat. Sci. **31**(2), 205–218 (2016). https://doi.org/10.1214/16-STS548
4. Bunge, J., Barger, K.: Parametric models for estimating the number of classes. Biometrical J. **50**(6), 971–982 (2008). https://doi.org/10.1002/bimj.200810452
5. Cruyff, M.J., van der Heijden, P.G.: Point and interval estimation of the population size using a zero-truncated negative binomial regression model. Biometrical J. **50**(6), 1035–1050 (2008). https://doi.org/10.1002/bimj.200810455
6. Edwards, W.R., Eberhardt, L.: Estimating cottontail abundance from livetrapping data. J. Wildlife Manag. 87–96 (1967). https://doi.org/10.2307/3798362
7. El-Shaarawi, A.H., Zhu, R., Joe, H.: Modelling species abundance using the Poisson-Tweedie family. Environmetrics **22**(2), 152–164 (2011). https://doi.org/10.1002/env.1036
8. Fisher, R.A., Corbet, A.S., Williams, C.B.: The relation between the number of species and the number of individuals in a random sample of an animal population. J. Animal Ecol. 42–58 (1943). https://doi.org/10.2307/1411
9. Ghitany, M., Al-Mutairi, D.K., Nadarajah, S.: Zero-truncated Poisson-Lindley distribution and its application. Math. Comput. Simul. **79**(3), 279–287 (2008). https://doi.org/10.1016/j.matcom.2007.11.021
10. van der Heijden, P.G., Bustami, R., Cruyff, M.J., Engbersen, G., van Houwelingen, H.C.: Point and interval estimation of the population size using the truncated Poisson regression model. Stat. Model. **3**(4), 3305–322 (2003)
11. van der Heijden, P.G., Cruyff, M.J., Böhning, D.: Capture recapture to estimate criminal populations. In: Encyclopedia of Criminology and Criminal Justice pp. 267–276 (2014). https://doi.org/10.2307/41954244
12. Hidaka, S.: General type token distribution. Briometrika **101**(4), 999–1002 (2014). https://doi.org/10.1093/biomet/asu035
13. Holla, M.: On a Poisson-inverse Gaussian distribution. Metrika **11**(1), 115–121 (1967). https://doi.org/10.1007/bf02613581
14. Horvitz, D.G., Thompson, D.J.: A generalization of sampling without replacement from a finite universe. J. Am. Stat. Assoc. **47**(260), 663–685 (1952). https://doi.org/10.1080/01621459.1952.10483446

15. Mahmoudi, E., Zakerzadeh, H.: Generalized Poisson-Lindley distribution. Commun. Stat.-Theory Methods **39**(10), 1785–1798 (2010). https://doi.org/10.1080/03610920902898514
16. Rocchetti, I., Bunge, J., Böhning, D.: Population size estimation based upon ratios of recapture probabilities. Ann. Appl. Stat. **5**(2B), 1512–1533 (2011)
17. Sankaran, M.: The discrete Poisson-Lindley distribution. Biometrics 145–149 (1970). https://doi.org/10.2307/2529053
18. Shanker, R., Hagos, F., Sujatha, S., Abrehe, Y.: On zero-truncation of Poisson and Poisson-Lindley distributions and their applications. Biometrics Biostat. Int. J. **2**(6), 1–14 (2015)
19. Zamani, H., Ismail, N., Faroughi, P.: Poisson-weighted exponential univariate version and regression model with applications. J. Math. Stat. **10**(2), 148–154 (2014). https://doi.org/10.3844/jmssp.2014.148.154

Numerical Analysis of the Model of Optimal Consumption and Borrowing with Random Time Scale

Aleksandra Zhukova[1,2,3]([⊠]) [ID] and Igor Pospelov[1,2,3] [ID]

[1] Federal Research Center "Computer Science and Control" of RAS,
Vavilov st. 44, bld. 2, Moscow 119333, Russian Federation
sasha.mymail@gmail.com
[2] Moscow Institute of Physics and Technology, 9 Institutskiy per.,
Dolgoprudny, Moscow Region 141701, Russian Federation
[3] National Research University Higher School of Economics,
Myasnitskaya st. 20, Moscow 101000, Russian Federation

Abstract. This work is dedicated to modelling economic dynamics with random time scale. We propose a solution in the form a continuous time model where interactions of agents are random exchanges of finite portions of products and money at random points in time. In this framework, the economic agent determines the volume, but not the moments of the transactions and their order. The paper presents a correct formal description of optimal consumption and borrowing as a stochastic optimal control problem, which we study using the optimality conditions in the Lagrange's form. The solution appears to have a boundary layer near the end of planning horizon where the optimal control satisfies the specific functional equation. This equation was studied numerically using the functional Newton method adapted to a two-dimensional case.

Keywords: Time scale · Optimal control · Functional Newton method

1 Introduction

This work is dedicated to modelling economic dynamics with random time scale. The need for such model appears when a theoretical model formulated in continuous time is applied to numerical computations using the statistical data in discrete time. Observed data have certain characteristic period - hour, month, quarter, year, whereas the underlying economic processes, especially in what relates to financial transactions, inventory management and trade in durable goods, may not have periodicity at all, or have a characteristic time of change different from what is observed on the available data. For example, for financial accounts with equal balances at the end of period, the turnover during the period might differ tenfold.

© Springer Nature Switzerland AG 2020
Y. D. Sergeyev and D. E. Kvasov (Eds.): NUMTA 2019, LNCS 11974, pp. 255–267, 2020.
https://doi.org/10.1007/978-3-030-40616-5_19

Applied mathematical models of economy, nevertheless, have to combine the mathematical description of rational behavior of economic agents with the insights defined by the data. In this paper we try to adapt the standard models of optimal consumption of households in the presence of debt, to the aforementioned specifics of real-life transactions, by introducing uncertainty of the timing of deals. The studies of the consumption of households and their relation to debt attracted attention after the crisis of 2008 [9–12]. The empirical studies test the version of the model in continuos time or discrete time with fixed scale. To our knowledge, the random-timescale model of optimal consumption and borrowing by households was not studied before and its comparison to statistical data is a topic for further research.

The mathematical model in the presented paper represents the household's problem as a stochastic optimal control problem with Poisson uncertainty. It adds to the existing studies of similar economic models by the analysis of a case of high intensity of the random process and the finite horizon. For a short review see [14]. The typical approach to analytic solution for such problems is dynamic programming (see [13]), but we demonstrate that the Lagrange's method has its own benefits.

This paper concentrates on the specific feature of the stochastic optimal control model with finite time horizon - the presence of boundary layer, where the optimal solution is defined from the functional equation. This equation has a non-local form and is possible to analyze by numerical methods. We believe that this study might be used for the study of similar stochastic optimal control problems.

2   The Model

2.1   The Model of Optimal Consumption with Random Time Scale

The paper presents a correct formal description of optimal consumption and borrowing as a stochastic optimal control problem, which we study using the optimality conditions in the Lagrange's form. The baseline model is the finite-horizon problem of optimal consumption $C(t)$, borrowing $K(t)$ made as random moments of time appear to make the transaction.

$$\mathrm{E}\left[\int_0^T U\left(\frac{C(t)}{C_0}\right) e^{-\Delta t} d\eta(t) - W\left(A(T) - L(T)\right)\right] \to \max_{0 \leq L(t),\, 0 \leq A(t)} \quad (1)$$

The debt $L(t)$ changes whenever the consumer takes a loan $K(t)$. The bank accounts $A(t)$ change as the consumer buys the consumption goods in the amount $C(t)$ for the price $p_y(t)$, pay the interest at the rate $r_l(t)$ to the bank and the income from ownership of shares $Z_\pi(t)$ is paid to the consumer according to the equations

$$dL(t) = K(t)\, d\eta(t), \quad (2)$$

$$dA(t) = -p_y(t)\, C(t) d\eta(t) + K(t)\, d\eta(t) - r_l(t) L(t) dt + Z_\pi(t) dt. \quad (3)$$

The function $\eta(t)$ denotes the Poisson counting process $\eta(t, \omega)$ with ω omitted for simplicity and $\eta(t) = 1$ when an event of the Poisson flow of possible deals occurs in the small time interval $[t, t + dt]$ and zero otherwise. The intensity of the Poisson flow o events is assumed to be constant and we denote it by Λ.

The model assumes three constraints on the actions of the consumer. If the debt $L(t)$ is negative, the consumer might make it infinitely large negative number and from the Eq. (3) the consumption rate will also be infinitely large with positive probability. Therefore, the constraint $L(t) \geq 0$ is imposed. An additional constraint is made to guarantee the solvency of the consumer: $A(t) \geq 0$. The constraint $A(T) \geq L(T)$ is added to ensure that the consumer may pay out all the debt by the end of the planning period. See [2] for discussion of such conditions in applied economic models. Due to the specifics of the stochastic optimal control problem with the underlying Poisson process, one cannot guarantee the controlability of the processes of $A(t)$ and $L(t)$, because by the time T with positive probability no transaction might be possible. Therefore, in contrast with the continuous time model [2], for the inequality constraint we use the penalty function $W(\cdot)$ in the form

$$ W(x) = \frac{(min(x, 0))^2}{2\, a^2}, \ a \to 0. \tag{4} $$

The utility function $U(\cdot)$ might be any monotone concave function with $U'(0) = \infty$ and here it is $U(x) = ln(x)$. Due to the form of utility function, the consumption rate $C(t)$ may not be zero at any point, since a slight increase from zero will substantially increase the value of utility.

The processes $p_y(t), Z_\pi(t), r_l(t)$ are determined and external to the consumer. They are defined in the general equilibrium setting, where the consumer's model is a part of the household's block. The variables $C(t), L(t), A(t), K(t)$ are stochastic processes, left-continuous with defined right limit and adapted to the natural filtration generated by the random Poisson process $\eta(t)$. The stochastic optimal control problem is to find the optimal non-anticipating control $C(\cdot), K(\cdot)$, such that given the initial conditions $A(0), L(0)$, the processes $C(\cdot), K(\cdot), A(\cdot) \geq 0, L(\cdot) \geq 0$ satisfy the Eqs. (2 and 3) and the functional (1) achieves maximum.

2.2   The Solution to the Model

The analysis of the stochastic model, the sufficient optimality conditions and their solutions in the form of synthesis of optimal control for most of the phase space is provided in the paper [3]. We give the main results here to demonstrate the source of the equation we had to solve. In this paper we concentrate on the unusual problem that arises during analysis of the model for some values of phase variables. It requires solving a functional non-linear non-local equation that we made an attempt to analyze by asymptotic and numerical methods.

In [3] we formulate the sufficient optimality conditions that might be reduced to a system of differential equations with finite shifts and functional equations.

The approach uses the Lagrange's method, which is quite unusual for the stochastic optimal control problems in economic modelling [4–8]. The main idea is to add the dual variables to the constraints to construct the Lagrange functional. The dual variables are assumed to be right-continuous non-anticipating stochastic processes $\psi_1(t), \psi_3(t)$ for the equality constraints, and the dual variables to the inequalty constraints $A(\cdot) \geq 0, L(\cdot) \geq 0$ are assumed to be left-continuous non-anticipating stochastic processes $\phi_2(t), \phi_4(t)$.

The sufficient conditions for the optimal control in the form of the sythesis $C(t, A, L), K(t, A, L)$ is defined from the equations

$$\phi_2(t, A, L) L = 0, \quad \phi_4(t, A, L) A = 0, \tag{5}$$

$$\frac{\partial \psi_3(t, A, L)}{\partial t} = (r_l(t) L - Z_\pi(t)) \frac{\partial \psi_3(t, A, L)}{\partial A} - \phi_4(t, A, L)$$
$$- \frac{U'\left(\frac{C(t,A,L)}{C_0}\right) e^{-\Delta t} \Lambda}{C_0\, p_y(t)} + \psi_3(t, A, L)\,\Lambda, \tag{6}$$

$$\frac{\partial \psi_1(t, A, L)}{\partial t} = (r_l(t) L - Z_\pi(t)) \frac{\partial \psi_1(t, A, L)}{\partial A} - \phi_2(t, A, L)$$
$$+ r_l(t)\,\psi_3(t, A, L) + \psi_1(t, A, L)\,\Lambda + \frac{U'\left(\frac{C(t,A,L)}{C_0}\right) e^{-\Delta t} \Lambda}{C_0\, p_y(t)}, \tag{7}$$

$$\psi_3(t, K(t, A, L) + A - p_y(t) C(t, A, L), K(t, A, L) + L)$$
$$= \frac{U'\left(\frac{C(t,A,L)}{C_0}\right) e^{-\Delta t}}{C_0\, p_y(t)}, \tag{8}$$

$$\psi_1(t, K(t, A, L) + A - p_y(t) C(t, A, L), K(t, A, L) + L)$$
$$= -\frac{U'\left(\frac{C(t,A,L)}{C_0}\right) e^{-\Delta t}}{C_0\, p_y(t)}, \tag{9}$$

$$\psi_1(T, A, L) = W'(A - L),\ \psi_3(T, A, L) = -W'(A - L). \tag{10}$$

This is a system of partial differential equations with shifts and functional equations. For large values of Λ asymptotic methods allow one to obtain the approximate expression for the consumption $C(t, A, L)$ and $K(t, A, L)$ when $t \ll T$ [3]. But the terminal conditions (10) might be satisfied only if there exist a boundary layer in the vicinity of T. Changing the time variable t into

$$t = T - \frac{\theta}{\Lambda}, \tag{11}$$

$$C(t, A, L) = \frac{c((T - t)\Lambda, A, L)}{p_y(T)}, K(t, A, L) = k((T - t)\Lambda, A, L)$$

one obtains the approximate (of the order 1, with expressions of the order $O\left(\frac{1}{A}\right)$ omitted) version of the system. For this, firstly, the differential equations (6 and 7) are solved

$$\psi_1\left(T - \frac{\theta}{\Lambda}, A, L\right) = -\int_0^\theta \frac{e^{-\Delta T + x - \theta}}{c\left(x, A, L\right)} dx + W'\left(A - L\right) \tag{12}$$

$$\psi_3\left(T - \frac{\theta}{\Lambda}, A, L\right) = \int_0^\theta \frac{e^{-\Delta T + x - \theta}}{c\left(x, A, L\right)} dx - W'\left(A - L\right). \tag{13}$$

The solutions are substituted into the functional equations (8 and 9) taking into account the terminal conditions (10). The resulting two functional equations appear to be identical.

$$\int_0^\theta \frac{e^{-\Delta T + x}}{c\left(x, -c\left(\theta, A, L\right) + k\left(\theta, A, L\right) + A, k\left(\theta, A, L\right) + L\right)} dx$$

$$-W'\left(A - L - c\left(\theta, A, L\right)\right) = \frac{e^{\theta - \Delta T}}{c\left(\theta, A, L\right)}, \tag{14}$$

$$\int_0^\theta -\frac{e^{-\Delta T + x}}{c\left(x, -c\left(\theta, A, L\right) + k\left(\theta, A, L\right) + A, k\left(\theta, A, L\right) + L\right)} dx$$

$$+W'\left(-c\left(\theta, A, L\right) + A - L\right) e^{-\theta} = -\frac{e^{\theta - \Delta T}}{c\left(\theta, A, L\right)}. \tag{15}$$

Therefore, we consider only one equation in the following analysis. One may observe that the Eq. (14) might have a symmetric solution

$$c\left(\theta, A, L\right) = \sigma\left(\theta, A - L\right), \quad A - L = Y. \tag{16}$$

With such assumption, the Eq. (15) becomes more simple

$$\min\left(\frac{Y - \sigma\left(\theta, Y\right)}{a^2}, 0\right) = -\frac{e^{\theta - \Delta T}}{\sigma\left(\theta, Y\right)} + \int_0^\theta \frac{e^x}{\sigma\left(x, Y - \sigma\left(\theta, Y\right)\right)} dx e^{-\Delta T}. \tag{17}$$

By the change in variables

$$\sigma\left(\theta, Y\right) = f\left(e^\theta, \frac{Y e^{\frac{1}{2}\Delta T}}{a}\right) e^{-\frac{1}{2}\Delta T} a, \tag{18}$$

$$e^\theta = t, \quad Y = x e^{-\frac{1}{2}\Delta T} a, \tag{19}$$

the Eq. (17) becomes independent of parameters a, T and Δ.

$$\min\left(x - f\left(t, x\right), 0\right) = -\frac{t}{f\left(t, x\right)} + \int_{1}^{t} \frac{1}{f\left(\tau, x - f(t, x)\right)} d\tau. \tag{20}$$

Our attempts to find an analytical solution for this equation did not lead to any globally defined solution, but we suggest a method of numerical analysis of this equation, combined with asymptotic estimates where possible.

3 Numerical Solution

The Eq. (20) we are trying to solve has a specific form: it contains both $f(t, x)$ and $f\left(\tau, x - f(t, x)\right), \tau \in [1, t]$. Therefore, it is impossible to express $f(t, x)$ in terms of its values in some neighborhood of the point (t, x). The iterative procedure for computing the solution to such equation has to calculate $f(t, x)$ based on the whole sale of values of $f(t, x)$. We refer to this feature as nonlocality of the problem.

Since we are looking for the solution which is interpreted as the rate of consumption, the unknown function $f(t, x)$ is nonnegative (to be precise, positive, due to the properties of the utility function under the integral).

The equation defines the unknown function $f(t, x)$ in two regions $x \leq f(t, x)$ and $x \geq f(t, x)$ according to two equations

$$0 = -\frac{t}{f\left(t, x\right)} + \int_{1}^{t} \frac{1}{f\left(\tau, x - f(t, x)\right)} d\tau, \ x \leq f(t, x), \tag{21}$$

$$x - f(t, x) = -\frac{t}{f\left(t, x\right)} + \int_{1}^{t} \frac{1}{f\left(\tau, x - f(t, x)\right)} d\tau, \ x \geq f(t, x). \tag{22}$$

While the first equation (21) might have more than one solution (it is "almost" homogeneous), the second one seems to less likely to have, but it is a subject of additional research. We concentrated the numerical analysis on the second one for the additional reason of relatively simple boundary condition. As it will be demonstrated in the following section, the border $x = f(t, x)$ is a non-linear curve, so that if one would try to compute the solution to (21), they will only have this curve as a boundary condition and the values of $f(t, x)$ below this line as the values for integration of $f\left(\tau, x - f(t, x)\right)$. At the same time, the boundary condition for the solution to (22) is defined at $t = 1, \forall x$ and $x \rightarrow \infty, t \geq 1$.

3.1 The Problem

The problem is to numerically solve the functional equation

$$0 = -\frac{t}{f\left(t, x\right)} + \int_{1}^{t} \frac{t}{f\left(\tau, x - f(t, x)\right)} d\tau - x + f\left(t, x\right) \tag{23}$$

with the initial condition

$$f(1,x) = \frac{1}{2}x + \frac{\sqrt{x^2+4}}{2}. \tag{24}$$

The boundary condition was added by asymptotic approximation of the solution at $x \to -\infty$.

Asymptotic Approximation of the Solution. Nonnegativity of the unknown function $f(t,x)$ implies that at $x \to -\infty$, $f(t,x) > x$. Therefore, the asymptotic approximation for large x is obtained from the equation (22). If one assumes the solution to be asymptotically approximated by the asymptotic series

$$f(t,x) = \sum_{k=0}^{\infty} l_k(t)x^{-k}, \tag{25}$$

it becomes possible to derive the recursive expressions for the coefficients $l_k(t)$. We omit most of the derivations her, but provide the main results.

First of all, the coefficient $l_k(t) = 0$ because otherwise the Eq. (22) contains no terms matching the x in the left hand side of the equation

$$x - \sum_{k=0}^{\infty} l_k(t)x^{-k} = -\frac{t}{\sum\limits_{k=0}^{\infty} l_k(t)x^{-k}} + \int_1^t \frac{1}{\sum\limits_{k=0}^{\infty} l_k(\tau)\left(x - \sum\limits_{k=0}^{\infty} l_k(t)x^{-k}\right)^{-k}} d\tau. \tag{26}$$

Therefore, we look for

$$f(t,x) = \frac{1}{x}\sum_{k=0}^{\infty} a_k(t)x^{-k}, \tag{27}$$

Taking the exponents of the power series and combining the coefficients of the same terms, we rewrite the Eq. (22)

$$
\begin{aligned}
x - \frac{1}{x}\sum_{k=0}^{\infty} a_k(t)x^{-k} \\
= -tx\sum_{k=0}^{\infty} b_k(t)x^{-k} + \left(x - \frac{1}{x}\sum_{k=0}^{\infty} a_k(t)x^{-k}\right)\sum_{m=0}^{\infty}\int_1^t \beta_m(\tau,t)d\tau x^{-m}.
\end{aligned}
\tag{28}
$$

The coefficients $\beta_m(\tau,t)$ and $b_k(t)$ are defined iteratively

$$b_0(t) = \frac{1}{a_0(t)}, \quad b_m(t) = -\frac{1}{a_0(t)}\sum_{k=0}^{\infty} a_k(t)b_{m-k}(t), \tag{29}$$

$$\beta_0(\tau,t) = \frac{1}{\alpha_0(\tau,t)}, \quad \beta_m(\tau,t) = -\frac{1}{\alpha_0(\tau,t)}\sum_{k=1}^{m} \alpha_k(\tau,t)\beta_{m-k}(\tau,t) \tag{30}$$

$$\alpha_k(\tau,t) = -\frac{1}{a_0(t)}\sum_{n=0}^{\infty} a_n(\tau)c_{m-n,n}(t), \tag{31}$$

$$c_{0,n}(t) = 1, \; c_{m,n}(t)\, m = \sum_{k=0}^{m-2} (n(m+k)-k)\, a_{m-k-2}(t)\, c_k. \tag{32}$$

The Eq. (28) gives the integral equations for coefficients of the asymptotic expansion. The first 13 terms are obtained from relatively simple equations:

$$
\begin{aligned}
f(t,x) = &-\frac{1}{x} + \frac{\ln(t)+1}{x^3} + \frac{-5\ln(t)-2-2\,(\ln(t))^2}{x^5} \\
&+ \frac{23\ln(t)+22\,(\ln(t))^2+5\,(\ln(t))^3+5}{x^7} \\
&- \left(\frac{353}{2}\,(\ln(t))^2 + 105\ln(t) + 93\,(\ln(t))^3 + 14\,(\ln(t))^4 + 14\right)x^{-9} \\
&+ O\left(t,\frac{1}{x^{11}}\right).
\end{aligned}
\tag{33}
$$

The numerical computations in our work used the approximation till the 5th order for the value of $x = -50$.

3.2   The Interative Method

We represent the (21) as the functional equation

$$F(f) = 0 \tag{34}$$

and apply the functional Newton method to find its numerical solution.

The procedure was proposed by [1] for a similar problem of finding the function of one variable, but here we adapt it to the two-dimensional case. The iterative method consists in updating the approximation according to

$$f_n = f_{n-1} - \mathcal{D}^{-1}(f_{n-1})\,[F(f_{n-1})], \tag{35}$$

where by $\mathcal{D}(f_{n-1})$ we denote the inverse operator to the Frechet derivative of the operator F. In fact, the Frechet derivative has the form

$$
\mathcal{D}(f)[h] = -\frac{\left(t + \int_1^t \frac{D_2(f)(\tau, x - f(t,x))}{f(\tau, x - f(t,x))^2}\, d\tau\, f(t,x)^2 + f(t,x)^2\right)}{f(t,x)^2}\, h(t,x) \\
+ \int_1^t \frac{h(\tau, x - f(t,x))}{f(\tau, x - f(t,x))^2}\, d\tau.
\tag{36}
$$

We omitted the integral term $\int_1^t \frac{h(\tau, x - f(t,x))}{f(\tau, x - f(t,x))^2}\, d\tau$ to be able to compute the inverse of the operator in a simple form. Of course, this affects the precision, but it appear, the modified version of the method converges and the residual $F(f_n)$ becomes sufficiently small. We use the operator

$$
\mathcal{D}_0(f)[h] = -\frac{\left(t + \int_1^t \frac{D_2(f)(\tau, x - f(t,x))}{f(\tau, x - f(t,x))^2}\, d\tau\, f(t,x)^2 + f(t,x)^2\right)}{f(t,x)^2}\, h(t,x)
\tag{37}
$$

and build the iterative procedure on it

$$f_n = f_{n-1} - \mathcal{D}_0^{-1} (f_{n-1}) \left[F (f_{n-1}) \right].$$ (38)

The inverse operator is

$$\mathcal{D}_0^{-1} (f) [F] = - \frac{f (t, x)^2}{\left(t + \int_1^t \frac{D_2(f)(\tau, x - f(t,x))}{f(\tau, x - f(t,x))^2} d\tau f (t, x)^2 + f (t, x)^2 \right)} F (t, x).$$ (39)

3.3   The Results of Computation

The procedure starts with the first approximation

$$f_0 (t, x) = - (x - 1)^{-1}.$$ (40)

This function was selected among other because the iterations seem to converge faster with this starting point. The first two updated approximations might be found explicitly

$$F(f_0) = - (x - 1) t + x (t - 1) + \frac{t - 1}{x - 1} - t + 1 + x + (x - 1)^{-1},$$ (41)

$$f_1 (t, x) = - \frac{xt - t - 1}{t (x^2 - 2 x + 2)}.$$ (42)

The function f_2 is also found explicitly, but the expression is quite complex and we used the computer algebra Maple for symbolic operations.

The problem is complex and it appears to be convenient to split the grid into two areas: firstly, compute on the area $x \in [-50, 0], t \in [1, 11]$ and then use the computations as a boundary conditions for computing the numerical solution for the positive values of x: $x \in [0, 32], t \in [1, 11]$. We initialize the values of $f(t, x)$ by the values of $f_2(t, x)$.

The results of 18 iterations of the method suggest that the values of $f(t, x)$ for large negtive x do not vary as much as they vary for smaller negative x (see Fig. 1). Therefore, the grid was selected so that the points become more dense for smaller negative x.

The quality of approximation is measured as a pointwise difference between the left and the right hand side of the Eq. (21)

$$g_f (t, x) = \frac{t}{f (t, x)} - \int_1^t \frac{d\tau}{f (\tau, x - f (t, x))} + x - f (t, x).$$ (43)

The values of this indicator are presented on the Fig. 2 for $t = 1$ (red) and $t = 11$ (blue). Since the size of the set was of the order 10^{-2} we consider these values for the error to be satisfactory.

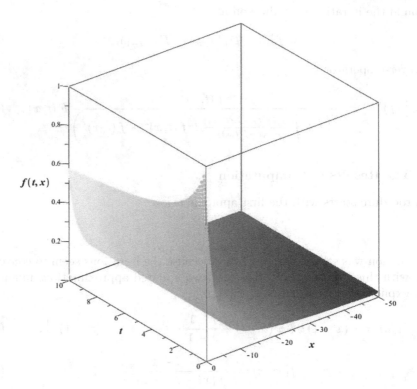

Fig. 1. The numerical approximation of $f(t,x)$ for $x \in [-50,0], t \in [1,11]$.

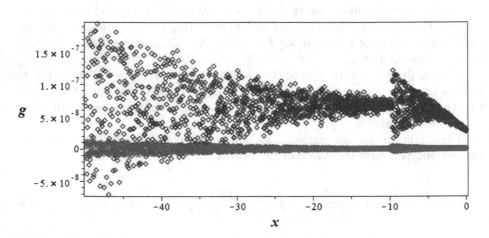

Fig. 2. The difference between the right and left hand side of numerical approximation of (21). (Color figure online)

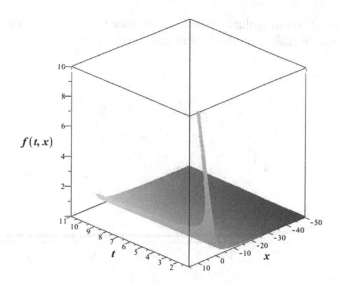

$f(t,x)$

Fig. 3. The numerical approximation of $f(t, x)$ for $x \in [-50, 12], t \in [1, 11]$

The Fig. 3 demonstrates the results of computation for both positive and negative values of x. There appears a border where the solution breaks. The reason is the presence of the border $\tilde{x}(t)$ where $0 = \tilde{x}(t) - f(t, \tilde{x}(t))$ and the equation changes the form from (22) to (21). The expression for this border might be given explicitly if one knows the values of $f(t, 0)$:

$$\tilde{x}(t) - f(t, \tilde{x}(t)) = -\frac{t}{\tilde{x}(t)} + \int_1^t \frac{1}{f(\tau, 0)} d\tau, \ \tilde{x}(t) = f(t, \tilde{x}(t)). \tag{44}$$

Therefore,

$$\tilde{x}(t) = \frac{t}{\int_1^t \frac{1}{f(\tau, 0)} d\tau}. \tag{45}$$

The problem of numerical computation above this border is the direction of our current research. The main challenge is that one can show that the border (45) comes close but never reaches the $t = 1$ asymptote (see Fig. 4). This means that the border is nonlinear for $x > 0$ and it is the only boundary condition to numerically compute the solution to the equation (21). The value at the point A on this figure uses the values at the interval CB below, which contains $f(t, x)$ from the areas above and below $\tilde{x}(t)$.

On the other hand, this effect has an interesting economic interpretation. Below the border, $f(t, x) > x$ which, in terms of A, L variables means that the consumption expenditures $p_y(t)C(t, A, L)$ are above the own money $A - L$. The closer is the end of the planning interval, the more the agent tries to spend instead of saving for the future.

At the same time the solution above the border uses all the values of $f(t, x)$ below this nonlinear border in the integral term. This creates a challenging problem.

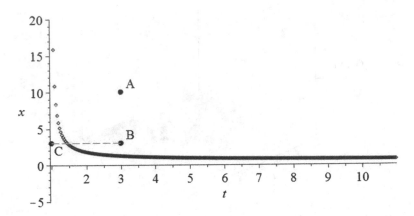

Fig. 4. The border $\tilde{x}(t)$ between the regions with $f(t, x) > x$ (below) and $f(t, x) > x$ (above).

However, it appears possible to find the asymptotic approximation to the solution for large positive x in the form of a series

$$f(t, x) = (\ln(t) + 1) x + (\ln(t) + 1) C_0 + e^{-\ln(t)^{-1}} C_1 x^{-1} + O(t, x^{-2}), \qquad (46)$$

where the constants $C_0, C_1, \ldots$ are defined from the correspondence between $f(t, x)$ for (t, x) above and below the border $\tilde{x}(t)$.

3.4 Conclusion

This paper presents the analysis of the model of optimal stochastic control of the processes of consumption and borrowing by a rational consumer. The solution to problem is found from a system of equations. Due to a finite planning horizon, there appears a boundary layer in the vicinity of the end of the planning time interval, $t \sim T$, where the solution is defined from the equation of a specific form (20). This equation might be solved by numerical methods, in particular, by the functional Newton method. For this, the boundary values are found by deriving the asymptotic behavior of the solution for large negative values of the argument. Another boundary condition comes from the terminal conditions of the underlying problem.

The results of computation are presented in this paper. The numerical solution appears to have a nonlinear form and the border that separates the solution defined by different segments in the left hand side of the Eq. (20) - where $f(t, x) > x$ and vice versa.

This paper concentrates on the specific feature of the stochastic optimal control model with finite time horizon - the presence of boundary layer, where the optimal solution is defined from the functional equation. This equation has a non-local form and is possible to analyze by numerical methods. We believe that this study might be used for the study of similar stochastic optimal control problems.

Acknowledgments.. The authors are very grateful to S. I. Bezrodnykh for valuable comments. This research was supported by the grant RFBR project 17-01-00588 A "Dynamic economic models with random time scale".

References

1. Bezrodnykh, S.I., Vlasov, V.I.: Analytic-numerial method for computation of interaction of physical fields in semiconductor diode. Math. Model. **27**(7), 15–24 (2015). (in Russian)
2. Pospelov, I.G., Pilnik, N.P.: On natural terminal conditions in models of intertemportal equilibrium. HSE Econ. J. **11**(1), 1–33 (2007). (in Russian)
3. Pospelov, I.G., Zhukova, A.A.: Model of optimal consumption with possibility of taking loans at random moments of time. HSE Econ. J. **22**(3), 330–361 (2018). (in Russian)
4. Pospelov, I.G., Zhukova, A.A.: Stochastic model of trade of non-liquid goods. Proc. MIPT **4**(2), 131–147 (2012). (in Russian)
5. Chow, G.C.: Dynamic Economics: Optimization by the Lagrange Method. Oxford University Press, Oxford (1997)
6. Chow, G.C.: Computation of optimum control functions by Lagrange multipliers. In: Belsley, D.A. (ed.) Computational Techniques for Econometrics and Economic Analysis, pp. 65–72. Springer, Dordrecht (1994). https://doi.org/10.1007/978-94-015-8372-5_4
7. Rong, S.: Optimization for a financial market with jumps by Lagrange's method. Pac. Econ. Rev. **4**(3), 261–276 (1999). https://doi.org/10.1111/1468-0106.00077
8. Rong, S.: Theory of Stochastic Differential Equations with Jumps and Applications: Mathematical and Analytical Techniques with Applications to Engineering. Springer, Boston (2006). https://doi.org/10.1007/b106901
9. Betti, G., Dourmashkin, N., Rossi, M., Ping Yin, Y.: Consumer over-indebtedness in the EU: measurement and characteristics. J. Econ. Stud. **34**(2), 136–156 (2007). https://doi.org/10.1108/01443580710745371
10. Keese, M.: Triggers and determinants of severe household indebtedness in Germany. Ruhr Economic Paper 150 (2009)
11. Kukk, M.: How did household indebtedness hamper consumption during the recession? Evidence from micro data. J. Comp. Econ. **44**(3), 764–786 (2016). https://doi.org/10.1016/j.jce.2015.07.004
12. Costa, S., Farinha, L.: Households' indebtedness: a microeconomic analysis based on the results of the households' financial and consumption survey. Financial stability report of Banco de Portugal (2012)
13. Oksendal, B.K., Sulem, A.: Applied Stochastic Control of Jump Diffusions. Springer, Berlin (2005). https://doi.org/10.1007/b137590
14. Posch, O., Trimborn, T.: Numerical solution of dynamic equilibrium models under Poisson uncertainty. J. Econ. Dyn. Control **37**(12), 2602–2622 (2013). https://doi.org/10.1016/j.jedc.2013.07.001

Short Papers

Short Papers

Conditions of the Stability Preservation Under Discretization of a Class of Nonlinear Time-Delay Systems

Alexander Yu. Aleksandrov[✉] [iD]

Saint Petersburg State University, 199034 Saint Petersburg, Russia
`a.u.aleksandrov@spbu.ru`

Abstract. Nonlinear differential systems with nonlinearities satisfying sector constraints and with constant delays are studied. Such systems belong to well-known class of Persidskii-type systems, and they are widely used for modeling automatic control systems and neural networks. With the aid of the Lyapunov direct method and original constructions of Lyapunov–Krasovskii functionals, we derive conditions of the stability preservation under discretization of the considered differential systems. The fulfilment of these conditions guarantees that the zero solutions of the corresponding difference systems are asymptotically stable for arbitrary values of delays. Moreover, estimates of the convergence rate of solutions are obtained. The proposed approaches are used for the stability analysis of a discrete-time model of population dynamics. An example is given to demonstrate the effectiveness of our results.

Keywords: Nonlinear system · Discretization · Stability · Convergence rate

1 Introduction

Preserving qualitative characteristics, when passing from differential equations to difference ones, is a fundamental and challenging research problem [1–3]. In many cases it is necessary to modify numerical schemes in order to preserve required characteristics. These modifications result in conservative computational schemes [2,4,5]. Such schemes, nowadays, are playing an important role in the design of reliable numerical methods in various areas of Science and Engineering (see, for instance, [2–4,6]).

However, it should be noted that the use of conservative schemes significantly complicates corresponding difference systems. Therefore, from a practical point of view, it is important to determine classes of systems for which discretization preserves qualitative properties without modifications of computational schemes.

In this contribution, a class of nonlinear difference systems with nonlinearities satisfying sector restrictions and with constant delays is studied. Such systems

Supported by the Russian Foundation for Basic Research, project no. 19-01-00146-a.

Y. D. Sergeyev and D. E. Kvasov (Eds.): NUMTA 2019, LNCS 11974, pp. 271–279, 2020.
https://doi.org/10.1007/978-3-030-40616-5_20

can be obtained as a result of application of the Euler numerical method to Persidskii-type differential systems [7].

It is assumed that zero solutions of the corresponding continuous-time systems are asymptotically stable for any values of delays. We are looking for conditions ensuring preservation of stability under discretization. A new construction of a Lyapunov–Krasovskii functional is proposed for the considered difference systems. It is shown that with the aid of such a functional, not only delay-independent asymptotic stability conditions, but also estimates of the convergence rate of solutions can be derived. In addition, the developed approaches are used for the stability analysis of an equilibrium position for a discrete-time model of population dynamics.

2   Preliminaries

Let R be the field of real numbers, R^n and $R^{n \times n}$ denote the vector spaces of n-tuples of real numbers and of $n \times n$ matrices respectively, $\| \cdot \|$ be the Euclidean norm of a vector. We use the notation $C^\top$ for the transpose of a matrix C and $C \succ 0$ ($C \prec 0$) to denote that the matrix C is positive (negative) definite.

Let $\mathrm{diag}\{\lambda_1, \ldots, \lambda_n\}$ be the diagonal matrix with the elements $\lambda_1, \ldots, \lambda_n$ along the main diagonal.

Consider a pair of matrices (A, B), where $A, B \in R^{n \times n}$.

Definition 1 (see [8]). *The pair (A, B) is diagonally Riccati stable if there exist diagonal matrices $P = \mathrm{diag}\{p_1, \ldots, p_n\}$ and $Q = \mathrm{diag}\{q_1, \ldots, q_n\}$ such that $P \succ 0$, $Q \succ 0$, $A^\top P + PA + Q + PBQ^{-1}B^\top P \prec 0$.*

Remark 1. The problem of diagonal Riccati stability was introduced in [8] and is motivated by the constructing diagonal Lyapunov–Krasovskii functionals for linear time-delay systems. Necessary and sufficient conditions for a given pair of matrices to be diagonally Riccati stable have been derived in [9]. Moreover, in [9, 10], some classes of matrices were determined for which simple and constructively verified conditions for diagonal Riccati stability can be obtained.

3   Statement of the Problem

Let the system of differential equations

$$\dot{x}(t) = A\Phi(x(t)) + B\Phi(x(t - \tau)) \tag{1}$$

be given. Here $x(t) = (x_1(t), \ldots, x_n(t))^\top$ is the state vector, A and B are constant matrices, τ is a constant nonnegative delay, $\Phi(x)$ is a separable nonlinearity: $\Phi(x) = (\varphi_1(x_1), \ldots, \varphi_n(x_n))^\top$, where scalar functions $\varphi_j(x_j)$ are continuous for $|x_j| < H$ ($0 < H \le +\infty$) and satisfy the sector conditions $x_j \varphi_j(x_j) > 0$ for $x_j \ne 0$, $j = 1, \ldots, n$.

Remark 2. The system (1) belongs to well-known class of Persidskii-type systems (see [7]). Such systems are widely used for modeling automatic control systems and neural networks [7,11].

We will assume that initial functions for (1) are chosen from the space $C([-\tau, 0], R^n)$ of continuous functions $\psi(\theta) : [-\tau, 0] \to R^n$ with the uniform norm $\|\psi\|_\tau = \sup_{\theta \in [-\tau, 0]} \|\psi(\theta)\|$. In addition, let x_t stand for the restriction of a solution $x(t)$ of (1) to the segment $[t - \tau, t]$, i.e., $x_t : \theta \to x(t + \theta)$, $\theta \in [-\tau, 0]$.
From the properties of $\Phi(x)$ it follows that (1) has the zero solution.

Assumption 1. Let the pair (A, B) be diagonally Riccati stable.

Remark 3. In [12] it is proved that, under Assumption 1, the zero solution of (1) is asymptotically stable for arbitrary constant nonnegative delay τ, and the system (1) admits a diagonal Lyapunov–Krasovskii functional.

In this paper, we consider the corresponding difference system

$$y(k + 1) = y(k) + h \left(A\Phi(y(k)) + B\Phi(y(k - l)) \right), \qquad (2)$$

where $y(k) \in R^n$, l is a nonnegative integer delay, $h > 0$ is a digitization step, $k = 0, 1, \ldots$. The system (2) can be obtained as a result of application of the Euler numerical method to system (1). Moreover, systems of the form (2) have own values, since they are used as discrete-time models of neural networks and digital filters (see [7]).
We will look for conditions ensuring preservation of stability, when passing from the continuous system (1) to its discrete-time counterpart (2). In addition, we will derive estimates of the convergence rate for solutions of (2).

Remark 4. It should be noted that stability of the corresponding delay-free difference system was studied in [13], whereas delay-independent stability conditions for systems of the form (2) were found in [14]. However, in [14] it was assumed that functions $\varphi_j(x_j)$ are essentially nonlinear, i.e., their expansions with respect to powers of state variables begin with powers of greater than one. In this paper, we do not impose such a restriction on $\varphi_j(x_j)$. Therefore, the approach proposed in [14] can not be applied to the stability analysis of (2).

4   Stability Conditions

To derive delay-independent asymptotic stability conditions for the difference system (2), we will use the discrete-time counterpart of the Lyapunov direct method and a special construction of a Lyapunov–Krasovskii functional.

Assumption 2. Functions $\varphi_j(x_j)$ satisfy a local Lipschitz condition, i.e., for any $H_1 \in (0, H)$, there exists a constant $L > 0$ such that $|\varphi_j(x_j') - \varphi_j(x_j'')| \le L|x_j' - x_j''|$ for $|x_j'| < H_1$, $|x_j''| < H_1$, $j = 1, \ldots, n$.

Theorem 1. *If Assumptions 1 and 2 are fulfilled, then there exists a number $h_0 > 0$ such that the zero solution of (2) is asymptotically stable for any $h \in (0, h_0)$ and for any nonnegative integer delay l.*

Proof. Let $y^{(k)} = \left(y^\top(k), y^\top(k-1), \ldots, y^\top(k-l)\right)^\top$ be the augmented state vector. Consider matrices $P = \mathrm{diag}\{p_1, \ldots, p_n\}$ and $Q = \mathrm{diag}\{q_1, \ldots, q_n\}$ satisfying Assumption 1. Choose a Lyapunov–Krasovskii functional candidate for (2) in the form

$$V\left(y^{(k)}\right) = 2 \sum_{i=1}^n p_i \int_0^{y_i(k)} \varphi_i(u)du + h \sum_{m=1}^l \Phi^\top(y(k-m))Q\Phi(y(k-m))$$

$$+ h \sum_{m=1}^l \lambda_m \|\Phi(y(k-m))\|^2, \tag{3}$$

where $\lambda_1, \ldots, \lambda_l$ are positive parameters.

Consider the difference $\Delta V = V\left(y^{(k+1)}\right) - V\left(y^{(k)}\right)$ of the functional (3) with respect to the system (2). We find that the difference satisfies

$$\Delta V = \sum_{i=1}^n p_i \int_{y_i(k)}^{y_i(k+1)} \varphi_i(u)du + h\Phi^\top(y(k))Q\Phi(y(k)) - h\Phi^\top(y(k-l))Q\Phi(y(k-l))$$

$$+ h\lambda_1 \|\Phi(y(k))\|^2 - h\lambda_l \|\Phi(y(k-l))\|^2 + h \sum_{m=2}^l (\lambda_m - \lambda_{m-1})\|\Phi(y(k+1-m))\|^2$$

$$= 2h\Phi^\top(y(k))P\left(A\Phi(y(k)) + B\Phi(y(k-l))\right) + h\Phi^\top(y(k))Q\Phi(y(k))$$

$$- h\Phi^\top(y(k-l))Q\Phi(y(k-l)) + 2\sum_{i=1}^n p_i \left(\varphi_i(y_i(k) + \xi_{ik}\Delta y_i(k)) - \varphi_i(y_i(k))\right) \Delta y_i(k)$$

$$+ h\lambda_1 \|\Phi(y(k))\|^2 - h\lambda_l \|\Phi(y(k-l))\|^2 + h \sum_{m=2}^l (\lambda_m - \lambda_{m-1})\|\Phi(y(k+1-m))\|^2$$

$$\leq -h(\alpha_1 - \lambda_1)\|\Phi(y(k))\|^2 + 2\sum_{i=1}^n p_i \left(\varphi_i(y_i(k) + \xi_{ik}\Delta y_i(k)) - \varphi_i(y_i(k))\right) \Delta y_i(k)$$

$$+ h \sum_{m=2}^l (\lambda_m - \lambda_{m-1})\|\Phi(y(k+1-m))\|^2 - h(\alpha_2 + \lambda_l)\|\Phi(y(k-l))\|^2.$$

Here α_1 and α_2 are positive constants, $\xi_{ik} \in (0,1)$, and $\Delta y_i(k) = y_i(k+1) - y_i(k)$.

Choose a number $H_1 \in (0, H)$ and find the corresponding Lipschitz constant L. We obtain that there exist numbers $\delta > 0$ and $\alpha_3 > 0$ such that

$$|(\varphi_i(y_i(k) + \xi_{ik}\Delta y_i(k)) - \varphi_i(y_i(k)))\Delta y_i(k)| \leq L\left(\Delta y_i(k)\right)^2$$

for $\left\|y^{(k)}\right\| < \delta$. Thus, if $0 < \lambda_l < \lambda_{l-1} < \ldots < \lambda_1$, and values of λ_1 and h are sufficiently small, then the estimate

$$\Delta V \leq -\alpha_4 h \sum_{m=0}^{l} \|\Phi(y(k-m))\|^2 \qquad (\alpha_4 = \text{const} > 0) \tag{4}$$

holds for $\left\|y^{(k)}\right\| < \delta$. Hence (see [7]), the zero solution of (2) is asymptotically stable for sufficiently small h and for any nonnegative integer delay l. $\qquad \square$

5 Convergence Rate of Solutions

Next, let us show that in the case where function $\Phi(x)$ satisfies some additional conditions, estimates of the convergence rate for solutions of (2) can be derived with the aid of the functional (3).

Assumption 3. Functions $\varphi_j(x_j)$ are represented in the form $\varphi_j(x_i) = \rho_j x_j^{\sigma_j} + \omega_j(x_j)$, where ρ_j are positive coefficients, $\sigma_j \geq 1$ are rationals with odd numerators and denominators, and $\omega_j(x_j)/x_j^{\sigma_j} \to 0$ as $x_j \to 0$, $j = 1, \ldots, n$.

Remark 5. Without loss of generality, we will assume that $\rho_j = 1$, $j = 1, \ldots, n$, and $\sigma_1 \leq \ldots \leq \sigma_n$.

Remark 6. If $\sigma_1 = \ldots = \sigma_n = 1$, then it is easy to prove that, under Assumptions 1–3 and for sufficiently small h, the zero solution of (2) is exponentially stable. Therefore, in what follows, we consider the case where $\sigma_n > 1$.

Let $y\left(k, k_0, y^{(k_0)}\right)$ denote a solution of (2) with initial conditions $k_0 \geq 0$, $y^{(k_0)} \in R^{n(l+1)}$.

Theorem 2. *Let Assumptions 1–3 be fulfilled and $\sigma_n > 1$. Then, there exists $h > 0$ such that, for any $h \in (0, h_0)$ and any nonnegative integer delay l, one can find positive numbers c_1, c_2 and $\tilde{\delta}$ such that*

$$\left|y_i\left(k, k_0, y^{(k_0)}\right)\right|^{\sigma_i+1} \leq c_1 \left\|y^{(k_0)}\right\|^{\sigma_1+1} \left(1 + c_2 \left\|y^{(k_0)}\right\|^{\nu(\sigma_1+1)} (k-k_0)\right)^{-\frac{1}{\nu}} \tag{5}$$

for $i = 1, \ldots, n$, $\left\|y^{(k_0)}\right\| < \tilde{\delta}$, $k_0 \geq 0$, $k \geq k_0$. Here $\nu = (\sigma_n - 1)/(\sigma_n + 1)$.

Proof. Consider the functional (3). Let parameters $h_0, \lambda_1, \ldots, \lambda_l, \delta$ be chosen in a such way that the inequality (4) hold for $\left\|y^{(k)}\right\| < \delta$.

Using Assumption 3, we obtain that if δ is sufficiently small, then

$$a_1 \sum_{i=1}^{n} \left(y_i^{\sigma_i+1}(k) + \sum_{m=1}^{l} y_i^{2\sigma_i}(k-m)\right) \leq V\left(y^{(k)}\right)$$
$$\leq a_2 \sum_{i=1}^{n} \left(y_i^{\sigma_i+1}(k) + \sum_{m=1}^{l} y_i^{2\sigma_i}(k-m)\right), \tag{6}$$

$$\Delta V \leq -a_3 \sum_{i=1}^{n} \sum_{m=0}^{l} y_i^{2\sigma_i}(k-m)$$

for $\|y^{(k)}\| < \delta$, where a_1, a_2, a_3 are positive constants. Hence, there exist numbers $a_4 > 0$ and $\tilde{\delta} > 0$ such that if $\|y^{(k_0)}\| < \tilde{\delta}$, $k_0 \geq 0$, then

$$V\left(y^{(k+1)}\right) \leq V\left(y^{(k)}\right) - a_4 \tilde{V}^{\nu+1}\left(y^{(k)}\right)$$

for $k \geq k_0$. Applying Lemma 1 from [15] and taking into account the estimates (6), we arrive at the inequalities (5).   □

6   A Discrete-Time Model of Population Dynamics

Let the difference system

$$z_i(k+1) = z_i(k) \exp\left(h\left(c_i + \sum_{j=1}^{n} a_{ij} f_j(z_j(k)) + \sum_{j=1}^{n} b_{ij} f_j(z_j(k-l))\right)\right), \quad (7)$$
$$i = 1, \ldots, n,$$

be given. The system describes interaction of n species in a biological community. Here $z_i(k)$ is the density of population i at the kth generation, functions $f_i(x_i)$ are defined for $x_i \geq 0$, a_{ij}, b_{ij}, c_i are constant coefficients, $i, j = 1, \ldots, n$, h is a positive parameter characterizing the transient time between two consecutive generations. System (7) is a discrete counterpart of a continuous generalized Lotka–Volterra ecosystem model [16, 17]. It is known [17, 18] that in the case where populations have non-overlapping generations, discrete-time models are more appropriate than continuous ones.

We consider functions $f_i(x_i)$, $i = 1, \ldots, n$, satisfying the following conditions, which are consistent with the standard assumptions made in [16, 18–20] and elsewhere:

(a) $f_i(0) = 0$;
(b) $f_i(x_i)$ is continuous and locally Lipschitz for $x_i \in [0, +\infty)$;
(c) $f_i(x_i)$ is a strictly increasing function for $x_i \geq 0$, and $f_i(x_i) \to +\infty$ as $x_i \to +\infty$.

By R_+^n we denote the nonnegative orthant of R^n, and int R_+^n being the interior of R_+^n. For biological reasons, we consider (7) in int R_+^n which is an invariant set for this system. In addition, let $A = \{a_{ij}\}_{i,j=1}^n$, $B = \{b_{ij}\}_{i,j=1}^n$.

Assumption 4. The system (7) admits an equilibrium position $\bar{z} = (\bar{z}_1, \ldots, \bar{z}_n)^\top \in$ int R_+^n.

Theorem 3. *If Assumptions 1 and 4 are fulfilled, then there exists a number $h_0 > 0$ such that the equilibrium position $\bar{z}$ of (7) is asymptotically stable for any $h \in (0, h_0)$ and for any nonnegative integer delay l.*

Proof. With the aid of the substitution $y_i(k) = \log(z_i(k)/\bar{z}_i)$, $i = 1,\ldots,n$, we transform (7) to a system of the form (2) with $\varphi_i(x_i) = f_i(\bar{z}_i e^{x_i}) - f_i(\bar{z}_i)$, $i = 1,\ldots,n$. To complete the proof, one should verify that all the conditions of Theorem 1 are fulfilled for the obtained system. □

Corollary 1. *If Assumptions 1 and 4 are fulfilled and functions $f_i(\bar{z}_i e^{x_i})$ are globally Lipschitz for $x_i \in (-\infty, +\infty)$, $i = 1,\ldots,n$, then there exists a number $h_0 > 0$ such that the equilibrium position $\bar{z}$ of (7) is globally asymptotically stable in int R_+^n for any $h \in (0, h_0)$ and for any nonnegative integer delay l.*

7   Example

Consider the case where $n = 4$ and the system (2) has the form

$$y_i(k+1) = y_i(k) - ha_i y_i(k) + hc_i \varphi(y_4(k)), \quad i = 1,2,3,$$

$$y_4(k+1) = y_4(k) - ha_4 \varphi(y_4(k)) + h\sum_{j=1}^{3} b_j y_j(k-l). \tag{8}$$

Here a_i, b_j, c_j are constant coefficients with $a_i > 0$, $i = 1,2,3,4$, $j = 1,2,3$, and $\varphi(x_4)$ is a sector nonlinearity that is continuous for $x_4 \in (-\infty, +\infty)$ and satisfies a local Lipschitz condition.

The system (8) is a discrete-time counterpart of a canonical form of the Yakubovich indirect control system with delay in the feedback law (see [7,11]).

With the aid of results of [10], it can be shown that the pair of matrices (A, B) corresponding to the system (8) is diagonally Riccati stable if and only if

$$a_4 > \frac{|b_1 c_1|}{a_1} + \frac{|b_2 c_2|}{a_2} + \frac{|b_3 c_3|}{a_3}. \tag{9}$$

Using Theorem 1, we obtain that, under the condition (9), there exists $h_0 > 0$ such that the zero solution of (8) is asymptotically stable for an arbitrary $h \in (0, h_0)$ and an arbitrary nonnegative integer delay l.

Assume that $a_1 = 0.7563$, $a_2 = 0.038$, $a_3 = 0.0163$, $b_1 = 0.2773$, $b_2 = 0.0468$, $b_3 = 0.0642$, $c_1 = 0.0264$, $c_2 = 0.1259$, $c_3 = 0.1348$. It should be noted that in [21,22] the corresponding continuous system with these values of coefficients was used as a mathematical model of a transport ship motion, and the problem of the course-keeping autopilot synthesis for the ship was studied.

In this case the condition (9) takes the form $a_4 > 0.6957$ (we give the result of approximate computations rounded to 10^{-4}).

8   Conclusion

In the present paper, using the discrete-time counterpart of the Lyapunov direct method, sufficient conditions of the delay-independent asymptotic stability are found for a class of nonlinear difference systems. In addition, with the aid of the constructed Lyapunov–Krasovskii functional, estimates for the convergence rate of solutions are obtained. An interesting direction of the future research is an application of the developed approaches to stability analysis of switched nonlinear difference systems with delay.

References

1. Dekker, K., Verwer, J.G.: Stability of Runge-Kutta methods for stiff nonlinear differential equations. North-Holland, Amsterdam, New York, Oxford (1984)
2. Mickens, R.E.: Applications of Nonstandard Finite Difference Schemes. World Scientific, Singapore (2000)
3. Stuart, A.M., Humphries, A.R.: Dynamical Systems and Numerical Analysis. Cambridge University Press, Cambridge (1996)
4. Patidar, K.G.: On the use of nonstandard finite difference methods. J. Differ. Equ. Appl. **11**(8), 735–758 (2005)
5. Sanz-Serna, J.M.: Symplectic integrators for Hamiltonian problems: an overview. Acta Numer. **1**, 243–286 (1992)
6. Zubov, V.I.: Conservative numerical methods for integrating differential equations in nonlinear mechanics. Doklady Math. **55**(3), 388–390 (1997)
7. Kazkurewicz, E., Bhaya, A.: Matrix Diagonal Stability in Systems and Computation. Birkhauser, Boston (1999)
8. Mason, O.: Diagonal Riccati stability and positive time-delay systems. Syst. Control Lett. **61**, 6–10 (2012)
9. Aleksandrov, A., Mason, O.: Diagonal Riccati stability and applications. Linear Algebra Appl. **492**, 38–51 (2016)
10. Aleksandrov, A., Mason, O., Vorob'eva, A.: Diagonal Riccati stability and the Hadamard product. Linear Algebra Appl. **534**, 158–173 (2017)
11. Letov, A.M.: Stability in Nonlinear Control Systems. Princeton University Press, Princeton (1961)
12. Aleksandrov, A.Y., Mason, O.: On diagonal stability of positive systems with switches and delays. Autom. Remote Control **79**(12), 2114–2127 (2018)
13. Aleksandrov, A.Yu., Chen, Y., Platonov, A.V., Zhang, L.: Stability analysis and uniform ultimate boundedness control synthesis for a class of nonlinear switched difference systems. J. Differ. Equ. Appl. **18**(9), 1545–1561 (2012)
14. Aleksandrov, A.Yu., Aleksandrova, E.B.: Delay-independent stability conditions for a class of nonlinear difference systems. J. Franklin Instit. **355**, 3367–3380 (2018)
15. Aleksandrov, A.Y., Zhabko, A.P.: On stability of solutions to one class of nonlinear difference systems. Siberian Math. J. **44**(6), 951–958 (2003)
16. Chen, F.D.: Permanence and global attractivity of a discrete multispecies Lotka-Volterra competition predator-prey systems. Appl. Math. Comput. **182**(1), 3–12 (2006)
17. Hofbauer, J., Hutson, V., Jansen, W.: Coexistence for systems governed by difference equations of Lotka-Volterra type. J. Math. Biol. **25**, 553–570 (1987)
18. Li, Y.: Dynamics of a discrete food-limited population model with time delay. Appl. Math. Comput. **218**(12), 6954–6962 (2012)
19. Aleksandrov, A.Yu., Aleksandrova, E.B., Platonov, A.V.: Ultimate boundedness conditions for a hybrid model of population dynamics. In: Proceedings of the 21st Mediterranean Conference on Control and Automation, 25–28 June 2013, Platanias-Chania, Crite, Greece, pp. 622–627. IEEE (2013)

20. Saker, S.H.: Oscillation and attractivity of discrete nonlinear delay population model. J. Appl. Math. Comput. **25**(1–2), 363–374 (2007)
21. Veremei, E.I.: Dynamical correction of control laws for marine ships' accurate steering. J. Mar. Sci. Appl. **13**, 127–133 (2014)
22. Aleksandrov, A.Yu., Aleksandrova, E.B., Zhabko, A.P.: Asymptotic stability conditions and estimates of solutions for nonlinear multiconnected time-delay systems. Circuits Syst. Signal Process **16**, 3531–3554 (2017)

Numerical Investigation of Natural Rough-Bed Flow

Giancarlo Alfonsi(iD), Domenico Ferraro(iD), Agostino Lauria$^{(\boxtimes)}$(iD),
and Roberto Gaudio(iD)

Department of Civil Engineering, University of Calabria, Via Bucci 42b,
87036 Rende, Cosenza, Italy
{giancarlo.alfonsi,domenico.ferraro,agostino.lauria,
roberto.gaudio}@unical.it

Abstract. The turbulent flow in natural rough-bed watercourses is a rather complex phenomenon, still poorly investigated. The majority of the existing works on this subject is of experimental nature, while the numerical ones are mostly related to artificially and regularly-roughened beds. In the present work a numerical investigation is carried out, in which the fully turbulent flow in an open channel is simulated, where the channel bottom is constituted by natural-pebble layers. In the numerical simulations, the Large Eddy Simulation (LES) approach is used, in conjunction with the Wall-Adapting Local Eddy viscosity (WALE) Sub-Grid Scale (SGS) closure model at Reynolds number 46,500 and Froude number 0.186. The Finite-Volume discretized governing equations are solved numerically by means of the InterFOAM solver, embedded in the OpenFOAM C++ digital library. In order to take into account the free-surface dynamics, the Volume of Fluid (VoF) method has been used. The results of the simulations are compared with those obtained in a companion experiment, mainly in terms of turbulence statistics of different order, obtaining a rather good agreement.

Keywords: Pebble bed flow · Large Eddy Simulation · Volume of Fluid

1 Introduction

The fluid-dynamic phenomena occurring in turbulent open-channel flows play a key role in rivers and watercourses, where the characteristic element is the presence of a macro-roughness bed of natural origin. In the recent years this subject has been faced by several researchers, both numerically and experimentally. In the experimental field, a noticeable research effort has been made as related to the turbulent characteristics of open-channel flows [1, 2]. In the last two decades, the spatial average method became popular, and was used in several works to capture the heterogeneous nature of the bed flow in terms of stress characteristics [3–7]. More recently, the low relative submergence condition raised the interest of the researchers, and some works have been published about the turbulence characteristics in this latter condition [8–12]. In the numerical field, the Large Eddy Simulation (LES) technique has been often used. Nevertheless, most of the numerical works are related to regular-roughness geometries. Stoesser and Nikora [13] performed

© Springer Nature Switzerland AG 2020
Y. D. Sergeyev and D. E. Kvasov (Eds.): NUMTA 2019, LNCS 11974, pp. 280–288, 2020.
https://doi.org/10.1007/978-3-030-40616-5_21

a LES calculation of a turbulent flow over square ribs mounted on a wall. Bomminayuni and Stoesser [14] executed a LES simulation of the flow over a channel bed artificially roughened with hemispheres. Fang et al. [15] investigated the case of the open-channel flow over three kinds of macro-roughened beds, as obtained with spheres with different sizes and arrangements. Omidyeganeh and Piomelli [16, 17] executed a simulation taking into account the macro-roughness through three-dimensional dunes at laboratory scale. As for nonregular roughness, Hardy et al. [18, 19] performed a simulation of the flow over a gravel surface by investigating the role of near-bed turbulence on the flow-structure development process. Stoesser [20] proposed a physically-realistic method for the LES of turbulent channel flows over a granular bed, in which the description of the bed roughness was obtained by means of a roughness-geometry function. Overall, these studies provided a relatively good insight of the near-wall turbulence, but the case of the naturally-rough bed flow needs to be further investigated. In the aforementioned works, the free surface is often assumed as a rigid frictionless boundary, so that its vertical movement is neglected. In the present work, the rigid lid condition is abandoned and a LES is performed of turbulent open-channel flow with a highly-rough bed of natural origin. A computational solver embedded in the OpenFOAM C++ digital library has been used, in conjunction with the WALE SGS closure model [21]. Moreover, the VoF technique has been adopted in order to follow the free-surface behavior, so that the effects of the moving surface are taken into account. The results of the simulations are compared with those of a companion experiment, and are presented in terms of turbulence statistics.

The experimental set-up of the companion experiment is here concisely summarized. The experiment was carried out in a 16 m long, 1 m wide, 0.8 m high tilting flume with rectangular cross-section. At the flume outlet, the water depth was regulated by an adjustable tailgate. In the downstream channel, a honeycomb was placed upstream from a Bazin weir, used to measure the discharge with an accuracy less than 2%. Natural nonuniform pebbles were spread onto the channel bed in four layers, with median size $d_{50} = 70$ mm. The 2.5 m long test section was located 10 m downstream of the inlet for an appropriate development of the boundary layer along the flume centerline [8, 9]. The measuring grid included 25 vertical-velocity profiles, 20 mm spaced along the streamwise direction, 5 mm and 3 mm spaced in the vertical direction near the free surface and the bottom, respectively. A down-looking four-beam acoustic Doppler velocimeter was used for the local pointwise measurement of the three instantaneous components of the fluid velocity. The instantaneous velocity components (streamwise u, spanwise v, and vertical w, with fluctuations u', v' and w', along the x, y and z axes, respectively) were measured in the x-z plane along the centerline. The data sampling rate and the sampling duration were 100 Hz and 300 s, respectively, which were found to be adequate to achieve the statistically time independent turbulent quantities [7]. The velocity measurements were taken from the lowest accessible position into the bed grain gaps up to an elevation of about 50 mm below the water surface, owing to the limitation of the ADV down-looking probe. The Signal-to-Noise Ratio (SNR) was kept equal to about 15 and the ellipsoid method proposed by Goring and Nikora [22] was applied for despiking. In Table 1 the characteristic parameters of the experiment considered in the present work for comparison with the numerical results are summarized, S being the longitudinal flume bottom slope, B the flume width, Q the discharge, h_w the water depth measured above the maximum grain-crest level, k_s the statistical roughness scale defined as the biggest gap into the granular bed, $\Delta = h_w/k_s$ the relative submergence, $Re = Uh_w/\nu$ the Reynolds number, U the bulk flow velocity, $Fr = U/(gh_w)^{1/2}$ the Froude number, and g the acceleration due to gravity. For the numerical simulations, the physical pebble-bed surface has been captured with a 3D Laser scanner (Minolta Vivid 300) with resolution of 0.1×0.1 mm^2, and used for the characterization of the bottom of the numerical channel.

Table 1. Characteristic parameters of the experimental and simulated flow case

S (%)	B (m)	Q (l/s)	h_w (m)	U (m/s)	ν (m^2/s)	k_s (m)	Δ	Re	Fr
0.4	1.00	46.5	0.185	0.251	10^{-6}	0.0573	3.13	46500	0.186

2 Computing Procedures

The filtered unsteady Navier-Stokes equations for incompressible fluids are considered (Einstein summation convention applies to repeated indices, $i, j = 1, 2, 3$):

$$\frac{\partial \bar{u}_i}{\partial t} + \frac{\partial}{\partial x_j}(\bar{u}_i \bar{u}_j) = -\frac{1}{\rho}\frac{\partial \bar{p}}{\partial x_i} - \frac{\partial \tau_{ij}}{\partial x_j} + \upsilon \frac{\partial^2 \bar{u}_i}{\partial x_j \partial x_j} \tag{1}$$

$$\frac{\partial \bar{u}_i}{\partial x_i} = 0 \tag{2}$$

where ρ is the fluid density, ν is the water kinematic viscosity, p is the fluid pressure, u_i (u, v, w) is i^{th} the fluid velocity component, x_1 (x), x_2 (y) and x_3 (z) are the streamwise, spanwise and vertical directions, respectively (overbars denote filtered quantities). The effect of the small turbulent scales is mirrored by the term $(\tau_{ij} = \bar{u}_i\bar{u}_j - \overline{u_i u_j})$, representing the subgrid-scale stress, that has to be modeled. The scales smaller than the grid size are not resolved, but accounted for through the sub-grid scale tensor τ_{ij}:

$$\tau_{ij} - \frac{1}{3}\tau_{kk}\delta_{ij} = -2\upsilon_t \bar{S}_{ij} \tag{3}$$

where $\bar{S}_{ij}$ is the deformation-rate tensor of the resolved field. In Eq. (3), ν_t is the SGS eddy viscosity, that is evaluated using the WALE SGS model [21]. The LES governing equations have been discretized by means of finite volumes. As for the computing domain (Fig. 1), a structured mesh has been built, where the dependent variables are stored at the cell center in a co-located arrangement. Both convective and diffusive terms are approximated with second-order central differences, while the Crank-Nicolson scheme is used in time. The Pressure Implicit with Split Operator (PISO) technique introduced by Issa [23] has been employed to couple pressure and velocity. The stability of the procedure has been ensured utilizing an adaptive time step with an initial value of 10^{-6} s, in conjunction with a mean Courant-Friedrichs-Lewy (CFL) number limit of 0.5. The governing equations have been solved numerically by means of the InterFOAM solver, embedded in the OpenFOAM C++ digital libraries. The InterFOAM code uses the VoF interface capturing approach [24, 25], that has been satisfactorily utilized in several flow cases [26–28]. Boundary conditions have been imposed of no-slip and zero wall-normal velocity at the pebble-bed surface. Periodic boundary conditions have been used in the streamwise and spanwise directions. A number of tests has been performed before reaching the final configuration of the grid (Table 2), employing an increasing number of points mainly along the normal direction (N_x, N_y and N_z are the grid node number in the ith direction x_i, with $i = 1, 2, 3$ for the x, y, z directions, respectively. Recall that $x_i^+ = x_i u_* / \upsilon$, $u_i^+ = \bar{u}_i / u_*$, being $u_* = \sqrt{\tau_w / \rho}$ the shear velocity). The grid is uniformly spaced in the x and y directions, and the grid spacing in terms of wall units is $\Delta x^+ = \Delta y^+ = 22$. Along the vertical direction ($\hat{z} = (z - z_c)/h_w$, where z_c is the maximum crest level), the grid is uniformly spaced among the pebbles ($-0.54 \leq \hat{z} \leq 0$, $\Delta z^+ = 1$), while in the main portion of the computing domain, in order to have a better spatial resolution near the wall and near the free surface, a grid stretching law of hyperbolic tangent type has been introduced [29]. After the insertion of appropriate initial conditions (an initial velocity profile evolving with time), the initial transient of the flow has been firstly simulated by means of the Reynolds-Averaged Navier-Stokes (RANS) equations coupled with the k-ω SST closure model [25]. Then, the RANS steady state has been mapped onto the LES domain and run for $15t$ (being $t = 5.4 h_w/U$ the flow-through time) for an appropriate turbulence development. Finally, the LES turbulent steady state has been run for $30t$ to build up the turbulent-flow database. A CPU-based computational system has been used for the simulations. The system included 1 worker node equipped with 2 eight-core Intel E5-2640 CPUs (for a total of 16 cores/16 threads at 2.0 GHz), 128 GB RAM at 1899 MHz, and 4 TB of disk space. The code has been parallelized through the OpenMPI implementation of the standard MPI. The computational domain has been decomposed into 16 sub-domains (4 along the spanwise direction and 4 along the vertical direction) with the Simple Geometric Domain Decomposition (SGDD) technique, for an appropriate balancing of the computational weight among the different processors.

The simulation has been preliminarily run on different machine configurations, and the results in terms of runtime in executing one full time step (Δt) of the calculation procedure on the computational grid are reported in Table 3 (I/O operations are not included). It can be noted that the execution time decreases, as expected, with the number of processors involved in the calculations. The numerical simulation of the case at hand has been run on the 16-core machine configuration ($N_{tot} = 8.4 \times 10^7$, $R_e = 46{,}500$, where N_{tot} is the total number of grid nodes). The elapsed computational time resulted in a total of about 1,500 h, which is a rather reasonable value for a relatively complex LES simulation using the worker node depicted before.

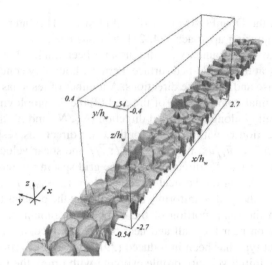

Fig. 1. Computing domain

Table 2. Computational grid

N_x	N_y	N_z	N_{tot}	Δx^+	Δy^+	$\Delta z^+_{(min)}$	$\Delta z^+_{(max)}$
1024	160	512	8.4×10^7	22	22	1.19	22

Table 3. Full-Δt runtime with different machine configurations (seconds per Δt)

CPU cores	2	4	8	16
Runtime	262.50	86.42	65.38	55.00

3   Results

Computed and measured velocity profiles in the computing domain are compared in Fig. 2, showing a good agreement along the entire water depth. Some discrepancies are visible only below the crest level, related to the ADV data, because of the difficulties in measuring inside the layers influenced by the bed roughness. In Fig. 3 the streamwise time-averaged velocity-vector distributions along the channel centerline are shown. Figure 3 shows the high resolution of the LES results, where a clear recirculation in the bed gaps ($1.4 < x/h_w < 1.6$) is observed.

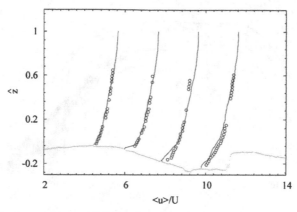

Fig. 2. Mean velocity profiles at different locations in the computing domain: open bullets represent experimental results, whereas solid lines represent numerical results.

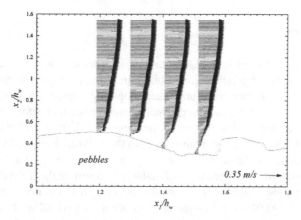

Fig. 3. Mean-velocity-vector distributions along the channel centreline: numerical values.

In real space a stringent way to check the existence of the Kolmogorov scaling is to consider the third-order structure functions [30] given by the compensated form [8–31]. In Fig. 4 a comparison between the turbulent kinetic energy (TKE) dissipation rate $\langle \varepsilon \rangle$, $\langle \cdot \rangle$ denoting the spatial averaging operator, from LES and experiment is shown at several vertical dimensionless elevations. Although the third-order statistics are sensible, the comparison between simulation and measurements is rather satisfactory. All the compensated laws in Fig. 4 show a plateau in the so-called inertial range, which can be fitted to obtain a precise value of the TKE dissipation rate, showing the correct simulation of the Kolmogorov inertial range.

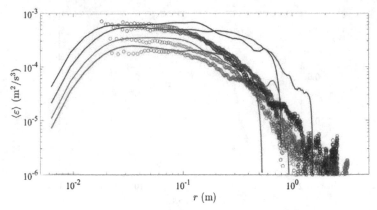

Fig. 4. TKE dissipation rate vs. space increment. Open bullets represent the experimental data, whereas solid lines are related to the LES results. Black colour represents $\langle \varepsilon \rangle$ ($\hat{z} = 0.2$), blue colour $\langle \varepsilon \rangle$ ($\hat{z} = 0.3$), green colour $\langle \varepsilon \rangle$ ($\hat{z} = 0.4$), red colour $\langle \varepsilon \rangle$ ($\hat{z} = 0.5$). (Color figure online)

4 Conclusions

Overall, the LES code used for the simulations in conjunction with the WALE SGS closure model has shown a rather good ability in correctly reproducing the turbulent phenomena at hand. The numerical approach adopted guarantees a resolved-scale resolution that is enough to glimpse into the inertial subrange and perform the analysis of turbulence proposed here. A first portion of the present work deals with the testing of the reliability of the LES calculations with respect to the measured dataset. Once confident about the convergence of the second-order moment analysis, as well as the more sensible third-order structures, we used the LES dataset to compute the TKE dissipation rate (including the smallest scales of the inertial subrange) that was not captured with the ADV. Moreover, the ADV down-looking configuration does not allow the measurements in the upper 50 mm of the water depth. Third-order structures clearly show the presence of an inertial subrange as related to both experimental and numerical results, giving a way to precisely compute the value of $\langle \varepsilon \rangle$.

References

1. Giménez-Curto, L.A., Lera, M.A.C.: Oscillating turbulent flow over very rough surfaces. J. Geophys. Res. Oceans **101**, 20745–20758 (1996)
2. Dittrich, A., Koll, K.: Velocity field and resistance of flow over rough surface with large and small relative submergence. Int. J. Sedim. Res. **12**, 21–33 (1997)
3. Nikora, V., Goring, D., McEwan, I., Griffiths, G.: Spatially averaged open-channel flow over rough bed. ASCE J. Hydraul. Eng. **127**, 123–133 (2001)
4. Nikora, V., McEwan, I., McLean, S., Coleman, S., Pokrajac, D., Walters, R.: Double averaging concept for rough-bed open-channel and overland flows: theoretical background. ASCE J. Hydraul. Eng. **133**, 873–883 (2007)
5. Manes, C., Pokrajac, D., McEwan, I.: Double-averaged open-channel flows with small relative submergence. ASCE J. Hydraul. Eng. **133**, 896–904 (2007)

6. Aberle, J., Koll, K., Dittrich, A.: Form induced stresses over rough gravel-beds. Acta Geophys. **56**, 584–600 (2008)

7. Dey, S., Das, R.: Gravel-bed hydrodynamics: double-averaging approach. ASCE J. Hydraul. Eng. **138**, 707–725 (2012)

8. Ferraro, D., Servidio, S., Carbone, V., Dey, S., Gaudio, R.: Turbulence laws in natural bed flows. J. Fluid Mech. **798**, 540–571 (2016)

9. Coscarella, F., Servidio, S., Ferraro, D., Carbone, V., Gaudio, R.: Turbulent energy dissipation rate in a tilting flume with a highly rough bed. Phys. Fluids **29**, 085101 (2017)

10. Cameron, S., Nikora, V., Stewart, M.: Very-large-scale motions in rough-bed open-channel flow. J. Fluid Mech. **814**, 416–429 (2018)

11. Padhi, E., Penna, N., Dey, S., Gaudio, R.: Hydrodynamics of water-worked and screeded gravel beds: a comparative study. Phys. Fluids **30**, 085105 (2018)

12. Padhi, E., Penna, N., Dey, S., Gaudio, R.: Spatially averaged dissipation rate in flows over water-worked and screeded gravel beds. Phys. Fluids **30**, 125106 (2018)

13. Stoesser, T., Nikora, V.I.: Flow structure over square bars at intermediate submergence: large eddy simulation study of bar spacing effect. Acta Geophys. **56**, 876–893 (2008)

14. Bomminayuni, S., Stoesser, T.: Turbulence statistics in an open-channel flow over rough bed. ASCE J. Hydraul. Eng. **137**, 1347–1358 (2008)

15. Fang, H., Han, X., He, G., Dey, S.: Influence of permeable beds on hydraulically macro-rough flow. J. Fluid Mech. **847**, 552–590 (2018)

16. Omidyeganeh, M., Piomelli, U.: Large-eddy simulation of three-dimensional dunes in a steady, unidirectional flow. Part 1. Turbulence statistics. J. Fluid Mech. **721**, 454–483 (2013)

17. Omidyeganeh, M., Piomelli, U.: Large-eddy simulation of three-dimensional dunes in a steady, unidirectional flow. Part 2. Flow structures. J. Fluid Mech. **734**, 509–534 (2013)

18. Hardy, R.J., Lane, S.N., Ferguson, R.I., Parsons, D.R.: Emergence of coherent flow structures over a gravel surface: a numerical experiment. Water Resour. Res. **43**, Article no. W03422 (2007)

19. Hardy, R.J., Best, J.L., Lane S.N., Carbonneau, P.E.: Coherent flow structures in a depth-limited flow over a gravel surface: the role of near-bed turbulence and influence of Reynolds number. J. Geophys. Res. Earth Surf. **114**, Article no. F011003 (2009)

20. Stoesser, T.: Physically-realistic roughness closure scheme to simulate turbulent channel flow over rough beds within the framework of LES. ASCE J. Hydraul. Eng. **136**, 812–819 (2010)

21. Nicoud, F., Ducros, F.: Subgrid-scale stress modelling based on the square of the velocity-gradient tensor. Flow Turbul. Combust. **62**, 183–200 (1999)

22. Goring, D.G., Nikora, V.I.: Despiking acoustic Doppler velocimeter data. ASCE J. Hydraul. Eng. **128**, 117–126 (2002)

23. Issa, R.I.: Solution of the implicitly discretized fluid flow equations by operator splitting. J. Comput. Phys. **62**, 40–65 (1986)

24. Hirt, C.W., Nichols, B.D.: Volume of fluid (VoF) method for the dynamics of free boundaries. J. Comput. Phys. **39**, 201–225 (1981)

25. Calomino, F., et al.: Experimental and numerical study of free-surface flows in a corrugated pipe. Water **10**, 638 (2018)

26. Alfonsi, G., Lauria, A., Primavera, L.: Proper orthogonal flow modes in the viscous-fluid wave diffraction case. J. Flow Vis. Image Process. **24**, 227–241 (2013)

27. Alfonsi, G., Lauria, A., Primavera, L.: The field of flow structures generated by a wave of viscous fluid around vertical circular cylinder piercing the free surface. Procedia Eng. **116**, 103–110 (2015)

28. Alfonsi, G., Lauria, A., Primavera, L.: Recent results from analysis of flow structures and energy modes induced by viscous wave around a surface-piercing cylinder. Math. Probl. Eng. **2017**, Article no. 5875948 (2017)

29. Alfonsi, G., Lauria, A., Primavera, L.: On evaluation of wave forces and runups on cylindrical obstacles. J. Flow Vis. Image Process. **20**, 269–291 (2013)
30. Frisch, U.: Turbulence. Cambridge University Press, Cambridge (1995)
31. Alfonsi, G., Ferraro, D., Lauria, A., Gaudio, R.: Large-eddy simulation of turbulent natural-bed flow. Phys. Fluids **31**(8), 085105 (2019)

A Dynamic Precision Floating-Point Arithmetic Based on the Infinity Computer Framework

Pierluigi Amodio[1], Luigi Brugnano[2], Felice Iavernaro[1]([✉]),
and Francesca Mazzia[3]

[1] Dipartimento di Matematica, Università di Bari, Bari, Italy
{pierluigi.amodio,felice.iavernaro}@uniba.it
[2] Dipartimento di Matematica e Informatica "U. Dini", Università di Firenze,
Firenze, Italy
luigi.brugnano@unifi.it
[3] Dipartimento di Informatica, Università di Bari, Bari, Italy
francesca.mazzia@uniba.it

Abstract. We introduce a dynamic precision floating-point arithmetic based on the Infinity Computer. This latter is a computational platform which can handle both infinite and infinitesimal quantities epitomized by the positive and negative finite powers of the symbol ①, which acts as a radix in a corresponding positional numeral system. The idea is to use the positional numeral system from the Infinity Computer to devise a variable precision representation of finite floating-point numbers and to execute arithmetical operations between them using the Infinity Computer Arithmetics. Here, numerals with negative finite powers of ① will act as infinitesimal-like quantities whose aim is to dynamically improve the accuracy of representation only when needed during the execution of a computation. An application to the iterative refinement technique to solve linear systems is also presented.

Keywords: Infinity Computer · Floating-point arithmetic · Conditioning · Iterative refinement

1 Introduction

The Infinity Computer paradigm, patented in EU, USA, and Russia (see for example [16]), is based on a positional numeral system with the infinite radix ① (called grossone) representing, by definition, the number of elements of the set of natural numbers $\mathbb{N}$ [11,14]. A number in this system is a linear combination of powers of ① with coefficients in the standard numeral system, such as

$$d_{p_m}①^{p_m} \ldots d_{p_1}①^{p_1} d_{p_0}①^{p_0} d_{p_{-1}}①^{p_{-1}} \ldots d_{p_{-k}}①^{p_{-k}},$$

This work was funded by the INdAM-GNCS 2018 Research Project "Numerical methods in optimization and ODEs".

© Springer Nature Switzerland AG 2020
Y. D. Sergeyev and D. E. Kvasov (Eds.): NUMTA 2019, LNCS 11974, pp. 289–297, 2020.
https://doi.org/10.1007/978-3-030-40616-5_22

with the usual meaning

$$d_{p_m}①^{p_m} + \ldots + d_{p_1}①^{p_1} + d_{p_0}①^{p_0} + d_{p_{-1}}①^{p_{-1}} + \ldots + d_{p_{-k}}①^{p_{-k}}.$$

The numerals $d_i \neq 0$ belong to a traditional numeral system and are called *grossdigits*, while numerals p_i are sorted in the decreasing order with $p_0 = 0$

$$p_m > p_{m-1} > \ldots > p_1 > p_0 > p_{-1} > \ldots p_{-(k-1)} > p_{-k},$$

and are called *grosspowers* (only finite grosspowers are considered in this contribution). Among the many fields of research this new methodology has been successfully applied, we mention *numerical differentiation and optimization* [4,12,17] and *numerical solution of differential equations* [1,6,9,10].[1] First results on handling ill-conditioning using the Infinity Computer may be found in [5,13].

Of particular interest in our study are grossnumbers consisting of a finite expansion of integer grosspowers such as, for example,

$$X = ①^P \sum_{j=0}^{T} x_j ①^{-j}, \qquad \text{with grossdigits} \quad x_j = \pm\beta^{p_j} \sum_{i=0}^{t} d_{ij}\beta^{-i}, \qquad (1)$$

where $P, p_j \in \mathbb{Z}$ and T, t are given positive integers, while β stands for the base of the traditional floating-point arithmetic system (usually $\beta = 2$).

This representation suggests interesting applications of the Infinity Computer if now ① identifies a suitable prescribed finite value. The idea is to exploit the grossdigits x_i in order to store a large number of significant digits in a dynamic manner during the execution of an algorithm. This means that the accuracy may be increased/decreased on demand during the flow of computations by automatically activating/deactivating a number of negative grosspowers. Taking aside the technical aspects related to the hardware implementation of the Infinity Computer, our study explores this path of investigation and is addressed to the accurate solution of ill-conditioned/unstable problems [3,7].

It should be noticed that, in principle, neither the user nor the programmer needs to know what the value of ① actually is. This assumption should be instead understood as an inherent feature of the machine architecture which, consistently with the Infinity Arithmetic methodology, will perceive the negative powers of ① as infinitesimal-like quantities if related to the classical floating-point system. Adopting this point of view, it turns out that changing the meaning of ① as we are going to do in the present work, does not affect that much the philosophical principles the grossone methodology is rooted in.

2   The Framework

In this section, we discuss an application of the Infinity Arithmetic system which consists in devising a floating-point variable precision arithmetic that will be

---

[1] For further references and applications see the survey [15].

used later to overcome the intrinsic loss of accuracy experienced when solving an ill-conditioned problem in the standard floating-point arithmetic.

The implementation of multiple and, in particular, variable precision arithmetic has been largely explored and successfully implemented, as is testified by the rich literature on this topic (see, for example, the survey paper [2] and reference therein). Here, however, we explore a further generalization of this paradigm in that the number of significant digits needs not to be a priori specified and maintained fixed but may be dynamically changed during the flow of computations. In particular, depending on the specific problem at hand and the algorithm used to solve it, the involved variables may allocate a different and variable amount of memory during the execution of the algorithm. The final goal is to control the error and make sure that the desired accuracy in the output data is achieved. This dynamic precision arithmetic is introduced as a natural byproduct of the Infinity Computer architecture and thus is expected to be easy to handle and to perform efficiently, once a hardware implementation of this latter will be available.

Representation of a Real Number. Consider first the problem of storing a real non-zero number $x = \pm\beta^p \sum_{i=0}^{\infty} d_i \beta^{-i} \in \mathbb{R}$, $d_0 \neq 0$, by preserving $N + 1 > 0$ significant digits. This task can be accomplished by setting in (1): $\text{①} \doteq \beta^{t+1}$ and assuming $p_j = p, j = 0, \ldots, T$, and $P = 0$.[2] The first $N + 1$ digits of x may be gathered in adjacent groups of $t + 1$ elements as follows:

$$x = \pm\beta^p \underbrace{d_0.d_1 \ldots d_t} \underbrace{d_{t+1} \ldots d_{2t+1}} \cdots \underbrace{d_{j(t+1)} \ldots d_{(j+1)(t+1)-1}} \cdots d_N. \qquad (2)$$

Then, assuming that $(N + 1) \leq (T + 1)(t + 1)$ and setting $d_i = 0$ for $i = N + 1, \ldots, (T + 1)(t + 1)$, the floating-point grossnumber representing x takes the form

$$\mathrm{fl}(x) = \pm\beta^p \sum_{j=0}^{T} \text{①}^{-j} \sum_{i=0}^{t} d_{j(t+1)+i} \beta^{-i}. \qquad (3)$$

Notice that all grossdigits share the same exponent p which, therefore, could be stored only once. We call (3) the *normalized* machine representation of x and its uniqueness comes from the uniqueness of the standard normalized notation (2).

Renormalization After a Computation. According to the Infinity Arithmetic methodology, the four basic operations over two grossnumbers follow the same rules of operations with polynomials. In fact, by definition, in this numeral system the radix ① is infinite while all digits d_{ij} are finite. For example, given the two grossnumbers (see (1))

$$X = x_1 \text{①}^1 + x_2 \text{①}^0 + x_3 \text{①}^{-1}, \qquad Y = y_1 \text{①}^0 + y_2 \text{①}^{-1} + y_3 \text{①}^{-2}$$

we get

$$X + Y = x_1 \text{①}^1 + (x_2 + y_1)\text{①}^0 + (x_3 + y_2)\text{①}^{-1} + y_3 \text{①}^{-2} \qquad (4)$$

---

[2] $\doteq$ denotes the identification operation, so the meaning of ① remains unaltered.

and

$$X \cdot Y = x_1 y_1 \mathbb{1}^1 + (x_1 y_2 + x_2 y_1)\mathbb{1}^0 + (x_1 y_3 + x_2 y_2 + x_3 y_1)\mathbb{1}^{-1}$$
$$+ (x_2 y_3 + x_3 y_2)\mathbb{1}^{-2} + x_3 y_3 \mathbb{1}^{-3}. \tag{5}$$

Now, due to the identification of $\mathbb{1}$ with a finite number and the assumption that all grossdigits must share the same exponent, it follows that, in general, the result of an operation will be not normalized and thus it is necessary to carry forward or backward some digits along the powers of $\mathbb{1}$ in order to obtain the result in the form (3). Without entering into details of this normalization procedure, which would go beyond the aims of this short paper, we consider an illustrative example.

Example 1. Set $\beta = 2$ (binary base) and $t = 3$ (four significant digits). Consider the sum of the two floating-point normalized grossnumbers

$$X = 2^0 \cdot (\mathbb{1}^0 1.101 + \mathbb{1}^{-1} 1.010 + \mathbb{1}^{-2} 1.111),$$
$$Y = 2^{-2} \cdot (\mathbb{1}^0 1.011 + \mathbb{1}^{-1} 1.110 + \mathbb{1}^{-2} 1.001).$$

The procedure, which moves along similar lines as for standard floating-point arithmetic, is summarized by the following steps of obvious meaning:

		$\mathbb{1}^0$	$\mathbb{1}^{-1}$	$\mathbb{1}^{-2}$	
(a) alignment	2^0	1.101	1.010	1.111	
	2^0	0.010	1.111	1.010	01
(b) sum with carrying	2^0	10.000	1.010	1.001	01
(c) normalization	2^1	1.000	0.101	0.101	

Notice that, as explained above, the Infinity Computer would perform the addition without carrying. This means that step (b) needs to be suitably arranged, also considering how to manage the rounding effects in each floating-point grossdigit. This aspects needs a specific study and is not addressed here since we just intend to show the general lines of our apprach. As a further remark, one clear advantage arising from the use of the Infinity Computer is that the computation of the grossdigits outcoming from basic operations such as (4) and (5) may be carried out in parallel.

Dynamic Precision Usage. Among the features offered by the computational platform based on the Infinity Computer, we assume that the user may decide how many infinitesimals stored in a variable should be involved in a given computation. If X is chosen as in (1), we denote by $X^{(q)}$ the grossnumber obtained by neglecting, in the sum, all the infinitesimals of order greater than q, that is, $X^{(q)} = \mathbb{1}^P \sum_{j=0}^{q} x_j \mathbb{1}^{-j}$. For example, choosing X and Y as in Example 1, we see that $X^{(0)} + Y^{(0)} = 2^1 \cdot 1.000$ would become the standard floating-point addition, and one can improve the accuracy by letting the subsequent infinitesimals come into play. This possibility may be exploited in a dynamical manner to overcome ill-conditioning issues associated with a given problem. The following example has a heuristic purpose in this direction.

Example 2. Set $\beta = 10$ (decimal base) and $t = 3$ (four significant digits), without rounding. Consider the three grossnumbers

$$X = 10^0 \cdot (①^0 1.234 + ①^{-1} 5.678 + ①^{-2} 9.012 + ①^{-3} 3.456),$$
$$Y = 10^0 \cdot (①^0 1.234 + ①^{-1} 4.444 + ①^{-2} 4.444 + ①^{-3} 4.444),$$
$$Z = 10^{-4} \cdot (①^0 1.230 + ①^{-1} 1.234 + ①^{-2} 1.234 + ①^{-3} 1.234).$$

storing three corresponding decimal numbers, say $x, y, z \in \mathbb{R}$ with 16 significant digits. Then $X^{(0)}, Y^{(0)}$ and $Z^{(0)}$ may be interpreted as the single precision representation of x, y, z while, on the other side, X, Y and Z are their quadruple precision approximations. Consider the problem of computing $w = (x - y) - z$ with four significant digits, which would suggest the use of single precision. Unfortunately, the first subtraction $x - y$ is ill-conditioned in single precision:

$$\left| \frac{(x - y) - (X^{(0)} - Y^{(0)})}{x - y} \right| = \left| \frac{(X - Y) - (X^{(0)} - Y^{(0)})}{X - Y} \right| = 1. \qquad (6)$$

The following scheme illustrates the procedure to obtain the correct result while minimizing the computational effort. It goes without saying that a cheap estimation of the relative error be available.[3] However, for simplicity of exposition, we evaluate the error by exploiting formulae similar to (6).

steps	error	action
(a) $X^{(0)} - Y^{(0)} = 0$	$1.9 \cdot 10^{-1}$	improve the accuracy
(b) $X^{(1)} - Y^{(1)} = 1.234 \cdot 10^{-4}$	$3.7 \cdot 10^{-4}$	accept the result
(c) $(X^{(1)} - Y^{(1)}) - Z^{(0)} = 4.000 \cdot 10^{-7}$	$7.7 \cdot 10^{-2}$	improve the accuracy
(d) $X^{(2)} - Y^{(2)} = 1.2344568 \cdot 10^{-4}$	$8.0 \cdot 10^{-9}$	
(e) $(X^{(2)} - Y^{(2)}) - Z^{(1)} = 4.333 \cdot 10^{-7}$	$5.1 \cdot 10^{-6}$	accept the result

Steps (a)–(b) produce four significant digits in the difference $X - Y$. However, a new cancellation phenomenon occurs at step (c). To overcome the loss of significant digits at this stage, a further improvement in accuracy of $X - Y$ is required (step (d)). The final step (e) reveals the coexistence of variables combined with different precisions.

In general, improving the accuracy of variables results in an increase of the overall computational complexity. However, it turns out that, in certain situations, algorithms may devised where only a marginal amount of computation needs to be performed with high accuracy. One such example is the iterative refinement and will be considered in the next section to illustrate the idea.

3   A Case Study

Citing Cleve Moler [8], *iterative refinement reduces the roundoff errors in the computed solution to a system of linear equations.* Starting from an initial

---

[3] For example, one could take the values $X^{(k+1)}$, $Y^{(k+1)}$ and $Z^{(k+1)}$ as a reference solution with respect to $X^{(k)}$, $Y^{(k)}$ and $Z^{(k)}$, and change the procedure accordingly.

approximated solution x_0 to a linear system $Ax = b$, this procedure consists of three steps iteratively executed. For $k = 0, 1, \ldots,$

step 1: compute the residual $r_k = b - Ax_k$;
step 2: solve the system $Ad_k = r_k$;
step 3: add the correction $x_{k+1} = x_k + d_k$.

In absence of roundoff errors the iteration would converge after one step to the true solution $x^* = A^{-1}b$. As is well-known, the use of finite precision arithmetic causes an amplification of the representation errors of the input data A and b proportional to the condition number $\kappa(A)$ of the coefficient matrix A. It turns out that, if step 1 is performed using a higher precision arithmetic with respect to the standard precision used at steps 2–3, the accuracy of the approximation may be significantly improved. In particular, denoting by ε_1 and ε_2 the round-off units defining the accuracy of the evaluations of steps 2–3 and step 1 respectively, in [8] it is shown that

$$\frac{||x_k - x^*||_\infty}{||x^*||_\infty} \leq (\sigma \kappa_\infty(A)\varepsilon_1)^k + \mu_1 \varepsilon_1 + n \mu_2 \kappa_\infty(A)\varepsilon_2 \tag{7}$$

where n is the dimension of A and σ, μ_1, μ_2 are suitable positive quantities with $\mu_1, \mu_2 = O(1/(1 - \sigma \kappa_\infty(A)\varepsilon_1))$. Consequently, under the assumption $0 < \sigma \kappa_\infty(A)\varepsilon_1 \ll 1$, we see that μ_1 and μ_2 become of the order of unity and the relative error approaches the size of the greatest round-off unit (that is ε_1 if we assume $\varepsilon_2 \ll \varepsilon_1$). Usually, the LU factorization with partial pivoting of matrix A is used at step 2 to reduce the computational effort associated with the linear systems to be solved at each iteration. During the execution of the algorithm on the Infinity Computer, one can control the convergence of the scheme by looking at the norms of the residuals computed at step 1:

- In the unfortunate event that $||r_k||$ diverges, the algorithm should improve the overall accuracy of step 2 by involving suitable negative powers of ①, thus reducing ε_1 (see (7)). One alternative we adopt in the example below is to compute the LU factorization of A with a higher accuracy and then truncate L and U to the roundoff level ε_1.
- In the case where $||r_k||$ stagnates before the error reaches the desired size, an improvement of the accuracy in performing step 1 is needed to reduce the value of ε_2. This is again accomplished by introducing new negative powers of ① in the representation of the data A and b and in the computation of the residual r_k.

In the following example, the Infinity Computer arithmetic has been emulated in the Matlab environment by using $\beta = 2$ and $t = 52$, which means that the grossdigits associated with ①^0 carry the 64-bit base-2 format of the IEEE 754 standard (double precision). This precision is doubled or tripled by involving the grossdigits associated with ①$^{-1}$ and ①$^{-2}$ respectively.

Example 3. Consider the system $Ax = b$, where A is the Vandermonde matrix of size 25 and coefficients $a_{ij} = (i - 1)^{j-1}$, $i, j = 1, \ldots, 25$, while $b = Ae$ with

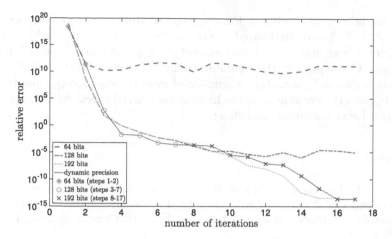

Fig. 1. Fixed versus dynamic precision implementation of the iterative refinement on a linear system with Vandermonde coefficient matrix of size 25.

$e = (1, \ldots 1)^{\top}$, so that $x^* := A^{-1}b = e$. The condition number of A is $\kappa_{\infty}(A) \approx 8.5 \cdot 10^{39}$. The iterative refinement procedure described above is executed starting from the initial guess $x_0 = (0, \ldots, 0)^{\top}$. We make the choice $\varepsilon_1 = \beta^{-t}/2$ (double precision) and we want to gain as many correct significant digits as possible while dynamically changing ε_2 in order to minimize the overall computational cost. A double precision accurate LU factorization with pivoting has been initially computed to solve the systems at step 2. As for the previous examples, we set $A^{(j)}$ and $b^{(j)}$ the truncations of A and b to the term j in their expansion along the negative grosspowers, and initially perform step 1 with $A^{(0)}$ and $b^{(0)}$ until an increase in $||r_k||$ is detected. In fact, the condition $||r_k|| > ||r_{k-1}||$ is symptomatic that a stagnation of the error is going to occur. This happens at the third iteration: here step 1 is run again with $A^{(1)}$ and $b^{(1)}$ (quadruple precision) in place of $A^{(0)}$ and $b^{(0)}$ and, consequently, the accuracy increases in the subsequent iterations. Notice that r_k has to be truncated at t digits, that is the standard double precision accuracy, before implementing step 2.

A further improvement in the accuracy of r_k is needed at the eighth iteration to avoid the saturation of the error at about 10^{-5}. Therefore $A^{(2)}$ and $b^{(2)}$ come into play at step 1 and assure a representation of A and b in sextuple precision (192 bits). This is enough to allow the error decrease at roundoff level ε_1 (see (7)). The results are illustrated in Fig. 1. The dashed, dash-dotted and dotted lines refer to the execution of the iterative refinement using fixed accuracy $\varepsilon_2 = \beta^{-t}/2, \beta^{-2t}/2, \beta^{-3t}/2$, respectively, and reveal the error levels it is possible to reach with these choices. In particular, we see that sextuple precision is needed at step 1 in order to obtain a double precision accurate solution. The solid line refers to the dynamic precision implementation of the procedure illustrated above. The errors at the first two iterations executed with $\varepsilon_2 = \varepsilon_1$ are labeled with asterisks. The errors in the subsequent five iterations, executed with $\varepsilon_2 =$

$\beta^{-2t}/2$ are labeled with circles. Finally, the remaining iterations are executed with $\varepsilon_2 = \beta^{-3t}/2$ and the related errors are labeled with crosses. We see that the error decreases until the saturation level of the corresponding precision mode is attained. Consequently, the dynamic change of the accuracy is finely tuned for this example and guarantees a number of total iterations very close to those needed by directly working with the highest considered precision but, evidently, requiring a lower computational effort.

References

1. Amodio, P., Iavernaro, F., Mazzia, F., Mukhametzhanov, M.S., Sergeyev, Y.D.: A generalized Taylor method of order three for the solution of initial value problems in standard and infinity floating-point arithmetic. Math. Comput. Simul. **141**, 24–39 (2016)
2. Bailey, D.H., Borwein, J.M.: High-precision arithmetic in mathematical physics. Mathematics **3**(2), 337–367 (2015)
3. Brugnano, L., Mazzia, F., Trigiante, D.: Fifty years of stiffness. In: Simos, T. (ed.) Recent Advances in Computational and Applied Mathematics, pp. 1–21. Springer, Dordrecht (2011). https://doi.org/10.1007/978-90-481-9981-5_1
4. De Cosmis, S., Leone, R.D.: The use of grossone in mathematical programming and operations research. Appl. Math. Comput. **218**(16), 8029–8038 (2012)
5. Gaudioso, M., Giallombardo, G., Mukhametzhanov, M.S.: Numerical infinitesimals in a variable metric method for convex nonsmooth optimization. Appl. Math. Comput. **318**, 312–320 (2018)
6. Iavernaro, F., Mazzia, F., Mukhametzhanov, M.S., Sergeyev, Y.D.: Conjugate-symplecticity properties of Euler-Maclaurin methods and their implementation on the infinity computer. Appl. Numer. Math. (2019). https://doi.org/10.1016/j.apnum.2019.06.011
7. Iavernaro, F., Mazzia, F., Trigiante, D.: Stability and conditioning in numerical analysis. J. Numer. Anal. Ind. Appl. Math. **1**(1), 91–112 (2006)
8. Moler, C.B.: Iterative refinement in floating point. J. ACM (JACM) **14**(2), 316–321 (1967)
9. Sergeyev, Y.D.: Solving ordinary differential equations by working with infinitesimals numerically on the infinity computer. Appl. Math. Comput. **219**(22), 10668–10681 (2013)
10. Sergeyev, Y.D., Mukhametzhanov, M.S., Mazzia, F., Iavernaro, F., Amodio, P.: Numerical methods for solving initial value problems on the infinity computer. Int. J. Unconvent. Comput. **12**(1), 3–23 (2016)
11. Sergeyev, Y.D.: A new applied approach for executing computations with infinite and infinitesimal quantities. Informatica **19**(4), 567–596 (2008)
12. Sergeyev, Y.D.: Higher order numerical differentiation on the infinity computer. Optim. Lett. **5**(4), 575–585 (2011)
13. Sergeyev, Y.D., Kvasov, D., Mukhametzhanov, M.S.: On strong homogeneity of a class of global optimization algorithms working with infinite and infinitesimal scales. Commun. Nonlinear Sci. Numer. Simul. **59**, 319–330 (2018)
14. Sergeyev, Y.D.: Numerical computations and mathematical modelling with infinite and infinitesimal numbers. J. Appl. Math. Comput. **29**(1–2), 177–195 (2009)

15. Sergeyev, Y.D.: Numerical infinities and infinitesimals: methodology, applications, and repercussions on two Hilbert problems. EMS Surv. Math. Sci. 4(2), 219–320 (2017)
16. Sergeyev, Y.D.: Computer system for storing infinite, infinitesimal, and finite quantities and executing arithmetical operations with them. USA patent 7,860,914 (2010)
17. Žilinskas, A.: On strong homogeneity of two global optimization algorithms based on statistical models of multimodal objective functions. Appl. Math. Comput. 218(16), 8131–8136 (2012)

Numerical Strategies for Solving Multiparameter Spectral Problems

Pierluigi Amodio$^{(\boxtimes)}$ ⓘ and Giuseppina Settanni ⓘ

Dipartimento di Matematica, Università degli Studi di Bari "Aldo Moro", Bari, Italy
{pierluigi.amodio,giuseppina.settanni}@uniba.it

Abstract. We focus on the solution of multiparameter spectral problems, and in particular on some strategies to compute coarse approximations of selected eigenparameters depending on the number of oscillations of the associated eigenfunctions. Since the computation of the eigenparameters is crucial in codes for multiparameter problems based on finite differences, we herein present two strategies. The first one is an iterative algorithm computing solutions as limit of a set of decoupled problems (much easier to solve). The second one solves problems depending on a parameter $\sigma \in [0,1]$, that give back the original problem only when $\sigma = 1$. We compare the strategies by using well known test problems with two and three parameters.

Keywords: Multiparameter spectral problems · High order methods · Finite difference schemes

1 Introduction

We consider a self-adjoint multiparameter regular spectral problem defined as

$$\left(p_i\left(t_i\right) y_i' \right)' + \left[q_i\left(t_i\right) + \sum_{j=1}^{N} r_{ij}\left(t_i\right) \lambda_j \right] y_i = 0 \quad \alpha_i \leq t_i \leq \beta_i \quad i = 1, 2, \ldots, N,$$

(1)

and separated boundary conditions

$$\begin{aligned} a_{i1}y_i(\alpha_i) + a_{i2}y_i'(\alpha_i) &= 0, & a_{i1}, a_{i2} \text{ real with } a_{i1}^2 + a_{i2}^2 > 0, \\ a_{i3}y_i(\beta_i) + a_{i4}y_i'(\beta_i) &= 0, & a_{i3}, a_{i4} \text{ real with } a_{i3}^2 + a_{i4}^2 > 0, \end{aligned}$$

where the number of eigenfunctions $N \geq 2$ and the functions p_i, r_{ij}, q_i are real and continuous on the interval $[\alpha_i, \beta_i]$ for all i, j.[1] Moreover, it is supposed

---

[1] In case of singular problem, the solution is not continuous at some endpoints but the strategies discussed below continue to work well.

This research was supported by the project "Equazioni di Evoluzione: analisi qualitativa e metodi numerici" of the Università degli Studi di Bari.

Y. D. Sergeyev and D. E. Kvasov (Eds.): NUMTA 2019, LNCS 11974, pp. 298–305, 2020.
https://doi.org/10.1007/978-3-030-40616-5_23

$p_i > 0$ on $[\alpha_i, \beta_i]$ for all i and $\det\{r_{ij}(t_i)\}_{i,j=1,\dots N} > 0$ for all $t_i \in [\alpha_i, \beta_i]$. Let the eigenparameters $\lambda_i^{(l_i)} := \lambda_i$ be such that the system (1) has non-trivial solution with l_i oscillations, $i = 1, \dots, N$. For ease of notation in the following we do not use the upper index even if we refer to specific eigenvalues.

This kind of spectral problems finds great application in the area of mathematical physics, for example in boundary value problems associated with the Lamè equation, the Laplace equation or the Helmholtz wave equation using separation of variables. Problem (1), also defined as a multiparameter Sturm-Liouville problem, see [19], has been discussed by different authors. In particular, different approaches for its solution have been analysed in [17], although a large study concerns the solution of the two-parameter eigenvalue problems, see [9,10,12,13,18]. An automatic algorithm based on the solution of a suitable initial value problems has been presented in [11] in order to obtain an estimation of the eigenvalues for both singular and non singular two-parameter Sturm-Liouville problems. The method uses the code SLEIGN based on the Prüfer transformation that handles with a differential initial value problem. Another approach discussed in [16] and used in the MATLAB toolbox MultiParEig is based on the spectral collocation method combined with the Sylvester-Arnoldi method.

The solution of classical Sturm-Liouville problems ($N = 1$) has been much more investigated and different numerical approaches exist; the authors of this note have proposed in [2,5,6,8] a matrix method able to compute an accurate estimation of the eigenvalues and the eigenfunctions. The HOFiD method follows the idea of BVMs (Boundary Value Methods) using finite difference schemes of high order, see [7]. An extension of this method HOFiD_SLP [8] has been also applied to solve two-parameter Sturm-Louville problems, most of them arising from the separation of variables in the Helmhotz equation [3,4]. In this case the coupled system needs an initial approximation of the eigenparameters, that are computed using the Prüfer transformation.

The aim of this work is to compute eigenparameters and eigenfunctions corresponding to a specific number of oscillations $(l_1, \dots, l_N)$ of multiparameter spectral problems pursuing the idea of the previous works [3,4] (see Sect. 2). Since the main difficulty is to get an initial estimation of these eigenparameters, we focus on the analysis of two different strategies allowing to overcome this drawback (Sects. 4 and 5). Both strategies require, before starting, initial approximations that are computed by a "frozen problem" described in Sect. 3. Finally, the considered strategies are compared by using known regular multiparameter problems in Sect. 6.

2   High Order Finite Differences

We use finite differences to obtain a discrete counterpart of the continuous problem. For this reason, we rewrite problem (2) in order to explicit the first and second derivatives

$$p_i(t_i) y_i'' + p_i'(t_i) y_i' + \left[q_i(t_i) + \sum_{j=1}^{N} r_{ij}(t_i) \lambda_j \right] y_i = 0, \quad i = 1, 2, \dots, N. \quad (2)$$

Then, we define a suitable discretization for each interval $[\alpha_i, \beta_i]$ and substitute continuous functions with the discrete ones

$$\text{diag}\,(P_i)\,Y_i^{(2)} + \text{diag}\left(P_i^{(1)}\right)Y_i^{(1)} + \text{diag}\left(Q_i + \sum_{j=1}^{N} R_{ij}\lambda_j\right)Y_i = 0, \quad (3)$$

where now $i = 1, 2, \ldots, N$ and $P_i, P_i^{(1)}, Q_i, R_{ij}, Y_i^{(2)}, Y_i^{(1)}, Y_i$ are vectors of the same length containing the discretizations in the mesh points of the respectively functions in (2).

Successively, we approximate both the derivative vectors by means of finite differences (linear combinations in the same points of Y_i). We use the fundamental idea of BVMs, that is a main formula with the best stability properties centred in the mid-point when possible, and different methods (with the same order of accuracy of the main method) in the initial and final points in order to obtain approximations in closed form. Therefore, $Y_i^{(2)} = A_i Y_i$ and $Y_i^{(1)} = B_i Y_i$, where A_i and B_i are essentially banded matrices (the bandwidth depend on the order of used formulas) containing the coefficients of the methods. Assembling the formulae we obtain an algebraic spectral problem with unknowns the N parameters λ_i and the associated eigenvectors Y_i that we rewrite in this form (as previously, $i = 1, 2, \ldots, N$)

$$\left(\text{diag}(P_i)A_i + \text{diag}\left(P_i^{(1)}\right)B_i + \text{diag}(Q_i)\right)Y_i = -\text{diag}\left(\sum_{j=1}^{N} R_{ij}\lambda_j\right)Y_i. \quad (4)$$

For $N = 1$ the problem reduces to the computation of eigenvalues and eigenvectors of a band matrix. For $N > 1$ it may be solved by means of a Newton method but, since the problem has infinite solutions, it requires a good initial guess to converge to the desired one. In the following sections we define two different strategies to obtain this starting value; both begin by computing the solution of the same frozen problem.

3   The Frozen Problem

In [1] it is suggested to compute a first approximation of the eigenparameters by simplifying the original problem in order to each equation in (2) becomes independent of the others. We consider functions $r_{i,j}(t_i)$ to be constant by choosing points $t_i^0 \in [\alpha_i, \beta_i]$ such that the frozen values $r_{i,j}^0 \equiv r_{i,j}(t_i^0)$ approximate $r_{i,j}(t_i)$. Substituting $\mu_i = \sum_{j=1}^{N} r_{i,j}^0 \lambda_j$ in (2) we obtain a set of N decoupled problems

$$p_i(t_i)y_i'' + p_i'(t_i)y_i' + [q_i(t_i) + \mu_i]\,y_i = 0, \quad i = 1, 2, \ldots, N, \quad (5)$$

that give (μ_i, y_i) by using the idea in [2,5]. Values λ_i^0, which represent the first approximation of the eigenparameters, are obtained by means of the solution of the linear system

$$
\begin{pmatrix}
r_{11}^0 & r_{12}^0 & \cdots & r_{1N}^0 \\
r_{21}^0 & r_{22}^0 & \cdots & r_{2N}^0 \\
\vdots & & & \\
r_{N1}^0 & r_{N2}^0 & \cdots & r_{NN}^0
\end{pmatrix}
\begin{pmatrix}
\lambda_1^0 \\
\lambda_2^0 \\
\vdots \\
\lambda_N^0
\end{pmatrix}
=
\begin{pmatrix}
\mu_1 \\
\mu_2 \\
\vdots \\
\mu_N
\end{pmatrix}.
$$

It is important to observe that these initial estimations of the eigenparameters could be far away from the exact ones, since they arise from a different problem.

4 Itevative Correction Strategy

Following the idea presented in [11], the first strategy continues to solve decoupled equations (much easier to solve) by using an iterative Gauss-Seidel–like procedure until a sufficiently accurate estimate of the eigenparameters is reached.

Given the solution of the frozen problem in Sect. 3, we define a sequence of Sturm-Liouville problems $(k = 1, 2, \dots)$

$$
p_i(t_i)y_i'' + p_i'(t_i)y_i' + [s_i(t_i) + r_{ii}(t_i)\lambda_i]\, y_i = 0, \qquad i = 1, 2, \dots, N, \tag{6}
$$

where $s_i(t_i) = q_i(t_i) + \sum_{j=1}^{i-1} r_{ij}(t_i)\lambda_j^k + \sum_{j=i+1}^{N} r_{ij}(t_i)\lambda_j^{k-1}$, which is used to compute λ_i^k, for $i = 1, 2, \dots, N$. Problem (6) are decoupled and may be solved, as the frozen problem, with techniques for the computations of spectral algebraic problems.

It is worth to note that this strategy converges for regular problems while it should be investigated for singular problems too.

5 Variation Strategy

Following the Abramov's idea in [1], we define a set of multiparameter problems depending on a parameter $\sigma \in [0, 1]$

$$
p_i(t_i)y_i'' + p_i'(t_i)y_i' + \left[q_i(t_i) + \sum_{j=1}^{N} r_{ij}(t_i,\sigma)\lambda_j^k\right] y_i = 0, \qquad i = 1, 2, \dots, N, \tag{7}
$$

where $r_{ij}(t_i, \sigma) = \sigma r_{ij}(t_i) + (1 - \sigma)r_{ij}(t_i^0)$, that, starting from the solution of the frozen problem $(\sigma = 0)$, leads to the solution of the original problem (2) $(\sigma = 1)$. The number of steps needed for σ to go from 0 to 1 depends on the problem to be solved. The solution obtained for $\sigma = \sigma_k$ after k steps is used as starting value for $\sigma_{k+1} > \sigma_k$, meanwhile it is necessary to check that the new eigenvectors have the right number of oscillations. If the Newton method computes a different solution, it is necessary to decrease the distance between

σ_k and σ_{k+1}, and try again. Therefore, the algorithm uses a stepsize variation strategy on the parameter σ.

We observe that problem (7) is solved by using the same strategy explained in Sect. 2. From a computational point of view, each step of this algorithm seems to be more expensive than the previous algorithm, since solving N decoupled problems is cheaper than solving one problem N times larger, however the variation strategy offers stronger guarantee to achieve the final solution.

6   Numerical Results

In this section we analyse two multiparameter spectral problems, both considering the separation of variables for solving boundary value problems associated with the Helmholtz equation; the only difference is that the used coordinates are elliptic for the Mathieu problem [14,16] and ellipsoidal for the Lamé problem [15]. In order to execute the previous strategies, we use $tol = 10^{-6}$ to estimate the eigenparameters in the iterative strategy. For both the examples we have plotted the frozen and the final solutions for each eigenvector (Figs. 1 and 3) and the speed of convergence in the computation of the eigenparameters for the two strategies (Figs. 2 and 4).

Mathieu Equations (see [14]). The two-parameter problem has been solved with $\beta_2 = \text{arccosh} \frac{c}{\sqrt{c^2-d^2}}$, where $c = 2$, $d = 1$, and the following boundary conditions $y_1'(0) = y_1(\pi/2) = y_2'(0) = y_2(\beta_2) = 0$. We have computed the solution with $l_1 = 2$ and $l_2 = 1$ oscillations by using $n_1 = n_2 = 50$ constant stepsizes. Since the frozen solution is similar to the problem solution, both strategies require few steps. We observe that, as usual, the variation approach computes completely different solutions for the intermediate problems.

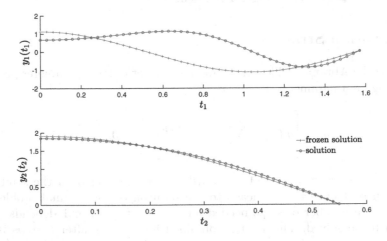

Fig. 1. Mathieu equations. Frozen and final solutions.

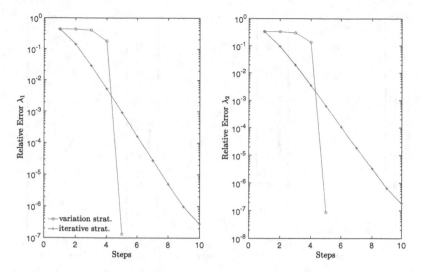

Fig. 2. Mathieu equations. Convergence to the eigenparameters for the two strategies.

Lamé Wave Equations (see [15]). The three-parameter problem has been solved with $d^2 = \xi - \rho^2$, $\xi = 3$ and $\rho^2 = 2$ and boundary conditions $y_1'(0) = y_1'(1) = y_2'(0) = y_2'(1) = y_3'(0) = y_3'(1) = 0$. We have solved the problem with $l_1 = 11$, $l_2 = 31$, $l_3 = 6$ oscillations by using $n_1 = 500$, $n_2 = 1000$, $n_3 = 400$ constant stepsize. There are two main difficulties in this problem: the required eigenfunctions have several oscillations, hence much more equispaced points needed with respect to the previous example; moreover, the solutions computed by the frozen

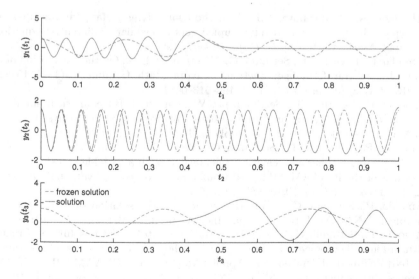

Fig. 3. Lamé wave equations. Frozen and final solutions.

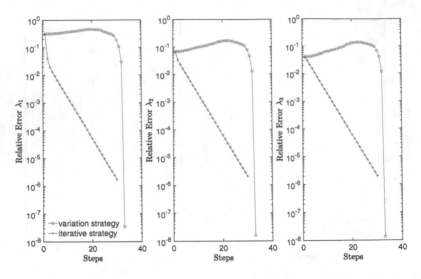

Fig. 4. Lamé wave equations. Convergence to the eigenparameters for the two strategies.

problem are quite different from the final ones, hence both the strategies require much more steps to reach the final solution.

We point out that the iterative approach is in general faster than the variational one and should be preferable. However we need to investigate more the convergence properties of such method.

References

1. Abramov, A.A., Ul'yanova, V.I.: A method for solving selfadjoint multiparameter spectral problems for weakly coupled sets of ordinary differential equations. Comput. Math. Math. Phys. **37**(5), 552–557 (1997)
2. Amodio, P., Levitina, T., Settanni, G., Weinmüller, E.B.: On the calculation of the finite Hankel transform eigenfunctions. J. Appl. Math. Comput. **43**(1–2), 151–173 (2013). https://doi.org/10.1007/s12190-013-0657-1
3. Amodio, P., Levitina, T., Settanni, G., Weinmüller, E.B.: Numerical simulation of the whispering gallery modes in prolate spheroids. Comput. Phys. Commun. **185**(4), 1200–1206 (2014). https://doi.org/10.1016/j.cpc.2013.12.012
4. Amodio, P., Levitina, T., Settanni, G., Weinmüller, E.B.: Whispering gallery modes in oblate spheroidal cavities: calculations with a variable stepsize. In: Proceedings of the ICNAAM-2014, AIP Conference Proceedings, vol. 1648, p. 150019 (2015). https://doi.org/10.1063/1.4912449
5. Amodio, P., Settanni, G.: A matrix method for the solution of Sturm-Liouville problems. JNAIAM. J. Numer. Anal. Ind. Appl. Math. **6**(1–2), 1–13 (2011)
6. Amodio, P., Settanni, G.: A stepsize variation strategy for the solution of regular Sturm-Liouville problems. In: Simos, T.E., Psihoyios, G., Tsitouras, C., Anastassi, Z. (eds.) Numerical Analysis and Applied Mathematics ICNAAM 2011, AIP Conference Proceedings, vol. 1389, pp. 1335–1338 (2011). https://doi.org/10.1063/1.3637866

7. Amodio, P., Settanni, G.: A finite differences MATLAB code for the numerical solution of second order singular perturbation problems. J. Comput. Appl. Math. **236**(16), 3869–3879 (2012). https://doi.org/10.1016/j.cam.2012.04.011

8. Amodio, P., Settanni, G.: Variable-step finite difference schemes for the solution of Sturm-Liouville problems. Commun. Nonlinear Sci. Numer. Simulat. **20**(3), 641–649 (2015). https://doi.org/10.1016/j.cnsns.2014.05.032

9. Arscott, F.M.: Two-parameter eigenvalue problems in differential equations. Proc. London Math. Soc. **14**(3), 459–470 (1964). https://doi.org/10.1112/plms/s3-14.3.459

10. Atkinson, F.V.: Multiparameter Eigenvalue Problems. Matrices and Compact Operators, vol. 1. Academic Press, New York (1972). Mathematics in Science and Engineering, vol. 82

11. Bailey, P.B.: The automatic solution of two-parameter Sturm-Liouville eigenvalue problems in ordinary differential equations. Appl. Math. Comput. **8**(4), 251–259 (1981). https://doi.org/10.1016/0096-3003(81)90021-7

12. Faierman, M.: On the distribution of the eigenvalues of a two-parameter system of ordinary differential equations of the second order. SIAM J. Math. Anal. **8**(5), 854–870 (1977). https://doi.org/10.1137/0508065

13. Faierman, M.: Distribution of eigenvalues of a two-parameter system of differential equations. Trans. Amer. Math. Soc. **247**, 86–145 (1979). https://doi.org/10.1090/S0002-9947-1979-0517686-7

14. Gheorghiu, C.I., Hochstenbach, M.E., Plestenjak, B., Rommes, J.: Spectral collocation solutions to multiparameter Mathieu's system. Appl. Math. Comput. **218**(24), 11990–12000 (2012). https://doi.org/10.1016/j.amc.2012.05.068

15. Levitina, T.V.: A numerical solution to some three-parameter spectral problems. Comput. Math. Math. Phys. **39**(11), 1715–1729 (1999)

16. Plestenjak, B., Gheorghiu, C.I., Hochstenbach, M.E.: Spectral collocation for multiparameter eigenvalue problems arising from separable boundary value problems. J. Comput. Phys. **298**, 585–601 (2015). https://doi.org/10.1016/j.jcp.2015.06.015

17. Sleeman, B.D.: Multi-parameter eigenvalue problems in ordinary differential equations. Bul. Inst. Politehn. Iaşi (N.S.) **17(21)**(3–4, sect. I), 51–60 (1971)

18. Sleeman, B.D.: The two parameter Sturm-Liouville problem for ordinary differential equations. Proc. Roy. Soc. Edinburgh, Sect. A **69**(2), 139–148 (1971)

19. Turyn, L.: Sturm-Liouville problems with several parameters. J. Differ. Equ. **38**(2), 239–259 (1980). https://doi.org/10.1016/0022-0396(80)90007-8

Algorithms of 3D Wind Field Reconstructing by Lidar Remote Sensing Data

Nikolay Baranov(✉) 🆔

Dorodnicyn Computing Centre, FRC CSC RAS, 40 Vavilova Street, 119333 Moscow, Russia
baranov@ians.aero

Abstract. In this paper, we analyzed the performance of wind vector field recovery from the wind lidar measurements. Wind lidar (LIDAR – Light Identification Detection And Ranging) remotely measures the wind radial speed by using the Doppler principle. Algorithms of the wind vector reconstruction using different versions of the least squares method are considered. In particular, the versions of weighted least squares (WLS) are considered, as well as the use of data spikes filtering procedures in the source data. The weights were calculated inversely with the local approximation error. As the initial data, the data of real measurements obtained in various wind conditions were used. The situations of a stationary wind field, a wind field with speed gusts, a wind field with fluctuations in direction, a wind field of variable speed and direction are considered. Lidar data were obtained for a region with a low-hilly terrain; therefore, even in the case of a stationary in time, the wind field was characterized by spatial heterogeneity. The questions of the use of regularization methods are considered. The analysis of the influence of the size of the averaging region on the quality of the recovery process was carried out.

Keywords: Remote sensing · Wind lidar · Wind field recovery

1 Introduction

Data on the spatial distribution of the wind speed and direction in the surface layer is an important factor for aviation. To obtain this data, Doppler wind lidars are scanning in PPI (Plan Position Indicator) mode with low elevation angles. However, only the wind component along the measurement direction is directly measured, and post-processing methods are required to determine wind speed and direction. For the tasks of wind profile measurements, methods for wind vector recovery are researched in sufficient detail. The impetus for the development of these methods is primarily the problem of wind energy. Most of these methods are variations of the VAD - Velocity Azimut Display method. The idea of the method is that with a conical scanning of a stationary wind field, the radial component changes according to a sinusoidal law. VAD is a retrieval technique that performs a harmonic analysis on radial wind data [1]. A huge amount of work has been devoted to various variations of this method, in particular [2, 3], as well as works of the author [4, 5], in which we are considered only the wind profile recovering.

© Springer Nature Switzerland AG 2020
Y. D. Sergeyev and D. E. Kvasov (Eds.): NUMTA 2019, LNCS 11974, pp. 306–313, 2020.
https://doi.org/10.1007/978-3-030-40616-5_24

VAD is not applicable for the tasks of spatial scanning of the wind field. In this case, VVP – Velocity Volume Processing is used. This method was developed in relation to Doppler Weather Radars. The analysis volume in the VVP technique is a 3D space containing several PPI observations. The hypothesis adopted for the analysis volume is that the wind field varies linearly and remains constant during radar scanning [6, 7].

Wind vector recovery allows to define wind field characteristics such as wind shear [8].

When the wind field recovering is according to lidar measurements, there are certain features that are primarily related to the fact that all the data belong to the same PPI observation. This scanning was performed with small elevation angles, which allows us to consider a problem of reconstructing a two-dimensional vector.

Lidar scanning is usually performed at low speed to provide a high angular resolution of the measurements. Therefore, the rate of data updating in a large scanning sector is low, and the wind field may change during this period. In this case, the problem of wind field reconstructing is different in that to determine the wind vector we can use scan data only in a small sector. This can be explained by the fact that t the surface wind is characterized by considerable variability in time and space, and for practical usage of interest is the wind field structure on a small scale. For a good conditionality of the equations system for determining wind parameters, it is necessary to increase the size of the averaging region. On the other hand, smoothing of the wind field occurs when we use measurement data from a large area to calculate the wind vector at a point. In addition, as shown in this study, spatial variations in the wind field can lead to incorrect solutions to the problem of recovering wind speed and direction.

2 Recovery Algorithms

Algorithms for recovering wind parameters at a given point minimize the root mean square residual of remote sensing data and theoretical values of radial projections of wind speed in a certain area. Suppose that the wind in a region is stable in time and space. We consider a two-dimensional approximation of the wind field, assuming that the contribution of the vertical velocity component is small. This hypothesis is valid when scanning is performed with a small elevation angle or when the vertical wind speed is small.

Let us to restore the wind speed vector $W_0 = (u_0, v_0)$ at the point r_0. Here u_0, v_0 are the Cartesian components of the wind speed in the scanning plane.

Suppose the radial projection of the wind speed V_j along the direction of measurement φ_j is measured at the points r_j. The points r_j are located in a neighborhood of the point r_0 at a distance of not more than d:

$$|r_j - r_0| \leq d.$$

The radial component of wind speed for the direction φ_j has the form

$$v_j = u_0\cos\varphi_j + v_0\sin\varphi_j, \tag{1}$$

and the mimimized function is

$$I_{OLS} = \sum_j (V_j - v_j)^2. \tag{2}$$

Minimizing the function (2) to determine the components u_0, v_0 is a classic ordinary problem of least squares (OLS). The weighted least squares (WLS) method, which uses instead of (2) weighted rms sums

$$I_{WLS} = \sum_j w_j (V_j - v_j)^2,$$

(3)

is a variation of the method OLS. Here w_j is the weighting coefficient.

The function (2) is identical to (3) when $w_j = 1$.

The solution to the minimization problem of (3) is calculated as a solution to the equations system of the form

$$\begin{pmatrix} \sum_j w_j \cos^2 \varphi_j & \sum_j w_j \cos \varphi_j \sin \varphi_j \\ \sum_j w_j \cos \varphi_j \sin \varphi_j & \sum_j w_j \sin^2 \varphi_j \end{pmatrix} \begin{pmatrix} u_0 \\ v_0 \end{pmatrix} = \begin{pmatrix} \sum_j w_j V_j \cos \varphi_j \\ \sum_j w_j V_j \sin \varphi_j \end{pmatrix},$$

or in the matrix form

$$S \begin{pmatrix} u_0 \\ v_0 \end{pmatrix} = Q.$$

(4)

The weighting coefficients are chosen inversely proportional to the residual of the measurements approximation by the relation (1), obtained as an OLS solution (2). In other words, let u_{OLS}, v_{OLS} is the solution, that minimizes (2). Define

$$\sigma_j^2 = (V_j - u_{OLS} \cos \varphi_j - v_{OLS} \sin \varphi_j)^2.$$

The weighting coefficients of (3) are equal to

$$w_j = \frac{\frac{\sigma_0^2}{\sigma_0^2 + \sigma_j^2}}{\sum_j \frac{\sigma_0^2}{\sigma_0^2 + \sigma_j^2}}.$$

(5)

The value $\sigma_0^2 > 0$ is a parameter.

Weights can also be determined as follows. Let σ_{lim}^2 is the maximum allowable value of the residual. Then we define alternative weights

$$w_j = \begin{cases} 1, if \sigma_j^2 \le \sigma_{lim}^2, \\ 0, if \sigma_j^2 > \sigma_{lim}^2. \end{cases}$$

(6)

Using of weight coefficients (6) actually means solving the problem with removing emissions. That is, having found the solution u_{OLS}, v_{OLS} of the problem (2), we perform the repeated solution on the reduced set of measurement points, excluding points for which the residual value exceeds the permissible level σ_{lim}^2.

The value σ_{lim}^2 is selected in proportion to the rms value of the residual:

$$\sigma_{lim}^2 = k \frac{1}{N-1} \sum_j \sigma_j^2.$$

The choice of the radius d for selecting the measurement points near the point r_0 is non-trivial. If distance $|r_0|$ is small, a large radius value leads to unnecessary averaging of the wind field. On the other hand, if the selection radius is small, then with a large distance to the recovery point, the recovery accuracy decreases.

The accuracy of the numerical solution of the equations system (4) is inversely proportional to the minimum eigenvalue of the matrix S [9]. The minimum eigenvalue of the matrix S is a function of the angles between the measurement directions. As an example, we consider a particular case of two measurement points, assuming that one of the directions coincides with the x-axis direction of the Cartesian coordinate system. Then system (4) has the form

$$\begin{pmatrix} 1 + \cos^2\varphi & \sin\varphi \\ \sin\varphi & \sin^2\varphi \end{pmatrix} \begin{pmatrix} u_0 \\ v_0 \end{pmatrix} = \begin{pmatrix} V_1 + V_2\cos\varphi \\ V_2\sin\varphi \end{pmatrix},$$

The minimum eigenvalue is approximately equal to

$$\lambda_{min} \approx \frac{1}{4}\cos^2\varphi\sin^2\varphi,$$

and the value $\sin\varphi$ is related to the parameters d, $|r_0|$ by the ratio

$$|\sin\varphi| \leq \frac{d}{|r_0|},$$

then we get the estimate

$$\lambda_{min} \leq \frac{1}{4}\left(\frac{d}{|r_0|}\right)^2.$$

Therefore, the problem of the rational choice of size of the recovering area as function of distance $d(|r_0|)$ is studied in addition to the main task. In this paper, the selection radius is considered as a linear function of distance

$$d(|r_0|) = d_0 + \alpha|r_0|. \tag{7}$$

Considering that, in the general, the matrix of system (4) is ill-conditioned, we use the Tikhonov regularization method [10] for its solving. Thus, we solve the system in the form

$$(S^* \cdot S + \mu I)\begin{pmatrix} u_0 \\ v_0 \end{pmatrix} = S^* \cdot Q, \tag{8}$$

where S^* is the transposed matrix, and $\mu > 0$ is the regularization parameter.

3 Result Analysis

For the analysis we use the remote sensing data from the Windex-5000 pulse wind lidar. Scanning was performed in PPI mode with an angular velocity 1°/s and elevation 6°.

The time of one measurement is equal to 1 s (accumulation time is equal to 1 s), then at a given scanning speed we have an angular resolution 1°. The distance resolution is equal to 40 m.

Let consider the scanning data presented in Fig. 1. Qualitative analysis of the data shows the presence of zones of moderate wind gain from 3 up to 6м/с. In addition, we can observe a typical S-shaped data structure, corresponding to the situation of changing of the wind direction. Note that there are areas of local velocity variations, such as, a region around a point (−500, 500), where a local decrease in velocity is observed (Fig. 2).

We show that in the areas of local gusts of winds, the OLS algorithm leads to inadequate recovery of the wind field. Local intensive changes in speed are interpreted as a change in direction. As a result, situations when in neighboring nodes (e.g., (−500, 800) and (−400, 900)) there are wind vectors with almost opposite directions arise. The use of the WLS (3), (5) algorithm without regularization (8) gives similar results (Fig. 3).

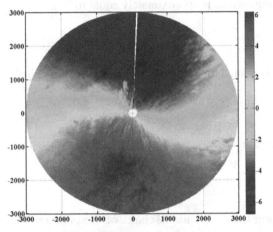

Fig. 1. Radial speed lidar data.

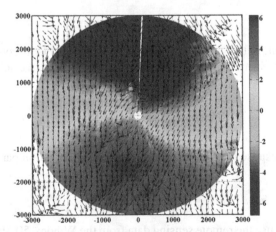

Fig. 2. Retrieved wind field. OLS method without regularization.

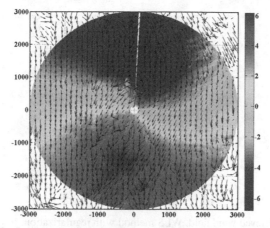

Fig. 3. Retrieved wind field. WLS method without regularization.

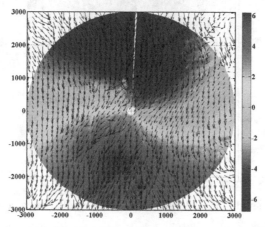

Fig. 4. Retrieved wind field. WLS method with regularization, $\mu = 0.00001$.

The use of regularization (8) for solving the system of Eq. (4) with small values of the regularization parameter gives a solution without obvious anomalies in the distribution of the wind direction (Fig. 4). However, as μ increases to 0.0001, the tendency for the solution to degeneration appears (Fig. 5). A further increase in μ to values of 0.0005 gives the actually degenerate solution (Fig. 6): the wind directions coincide with the direction of measurements, that is non-physical.

Figure 7 presents the recovery results for another wind situation. The distribution of the radial velocities of the wind field has a jet character: in the azimuthal direction the sections with low and high radial velocities alternate, that can be seen better at medium distances.

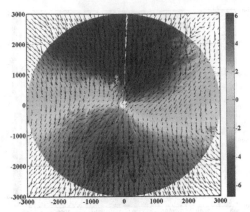

Fig. 5. Retrieved wind field. WLS method with regularization, $\mu = 0.0001$.

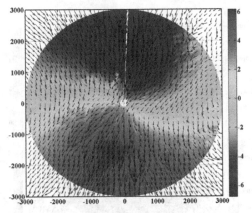

Fig. 6. Retrieved wind field. WLS method with regularization, $\mu = 0.0005$.

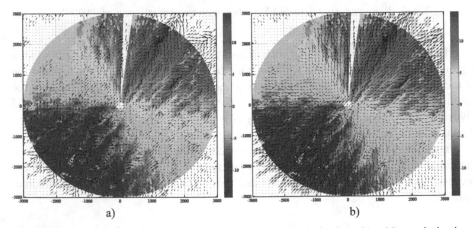

a)   b)

Fig. 7. Retrieved wind field. WLS method: (a) without regularization, (b) with regularization, $\mu = 0.00001$.

In this case, the WLS method without regularization gives in practic a virtually chaotic distribution of the wind direction in the zone of negative radial velocities (Fig. 7a). At the same time, the use of regularization (8) with a small $\mu = 0.00001$ ensures a qualitatively more appropriate recovery of the wind field (Fig. 7b).

4 Conclusions

The paper presents various options for solving the problem of wind field recovery using lidar remote sensing data. With a low rate of updating lidar data, the temporal variability of the wind field transforms to the spatial distribution of the measurement data. The feature of the solved problem is the requirement of high spatial resolution of the wind vector distribution. The complexity of its implementation is associated with a small amount of measurement data in a small recovery area and a small angular resolution.

The variations of the least squares method for the reconstruction of the wind vector are considered. The method is analogous to the VVP method. The parameters of the task are determined and the effect of some of them on the quality of wind field recovery is shown. Direct application of OLS in the case of spatially inhomogeneous lidar remote sensing data leads to qualitative errors in the recovery of the wind field. In particular, the spatial variability of the wind speed can transform during recovery into the variability of the wind direction.

The WLS method with the regularization algorithms provides a higher quality of recovering. However, the use of regularization requires the careful choice of the regularization parameter.

References

1. Lhermitte, R.M., Atlas, D.: Precipitation motion by pulse Doppler. In: Ninth Weather Radar Conference, Kansas City, American Meteorological Society, pp. 218–223 (1961, preprints)
2. Gao, J., Droegemeier, K.K., Gong, J., Xu, Q.: A method for retrieving mean horizontal wind profiles from single-doppler radar observations contaminated by aliasing. Mon. Weather Rev. **12**, 1399–1409 (2004)
3. Teschke, G., Lehmann, V.: Mean wind vector estimation using the velocity–azimuth display (VAD) method: an explicit algebraic solution. Atmos. Meas. Tech. **10**, 3265–3271 (2017)
4. Baranov, N., Petrov, G., Shiriaev, I.: Wind speed vector restoration algorithm. In: EPJ Web of Conferences, vol. 176, p. 06012 (2018)
5. Petrov, G., Baranov, N.: Data processing technique for the all-fiber wind profiler. In: Proceedings of SPIE 104290G (2017)
6. Shenghui, Z., Ming, W., Lijun, W., Chang, Z., Minghu, Z.: Sensitivity analysis of the VVP wind retrieval method for single-doppler weather radars. J. Atmos. Oceanic Technol. **31**, 1289–1300 (2014)
7. Li, N., Wei, M., Mu, X., Zhao, Ch.: A support vector machine-based VVP wind retrieval method. Atmos. Sci. Lett. **16**, 331–337 (2015)
8. Baranov, N., Lemishchenko, E.: Windshear identification algorithms by doppler pulse lidar. In: ITM Web of Conferences, vol. 24, p. 01011 (2018)
9. Bakhavalov, N.S.: Numerical Methods: Analysis, Algebra, Ordinary Differential Equations. MIR Publishers, Moscow (1977)
10. Tikhonov, A., Arsenin, V.Y.: Solution of Ill-Posed Problems. Winston & Sons, Washington (1977)

Stationarity Condition for Nonsmooth MPVCs with Constraint Set

David Barilla[1] , Giuseppe Caristi[1](✉) , and Nader Kanzi[2]

[1] Department of Economics, University of Messina, Via dei Verdi, 75, Messina, Italy
{dbarilla,gcaristi}@unime.it
[2] Department of Mathematics, Payam Noor University,
P.O. Box 19395-3697, Tehran, Iran
nad.kanzi@gmail.com

Abstract. We consider a nonsmooth optimization problem with vanishing constraints and constraint set. A new constraint qualification and a necessary condition for M-stationary of the problem are presented. Our results are formulated in terms of Mordukhoivich subdifferential.

Keywords: Stationary conditions and vanishing constraints ·
Nonsmooth optimization · Constraint qualification

1 Introduction

In this paper we study necessary and sufficient conditions for nonsmooth multiobjective mathematical programming problem with vanishing constraints and constrained set, which is defined as

$$
\begin{aligned}
\text{(MMPVCC):} \qquad \min \quad & \big(f_1(x), \ldots, f_p(x)\big) \\
\text{s.t.} \quad & g_i(x) \le 0, & i \in I_g := \{1, \ldots, s\}, \\
& h_i(x) = 0, & i \in I_h := \{1, \ldots, r\}, \\
& H_i(x) \ge 0, & i \in I := \{1, \ldots, m\}, \\
& H_i(x)G_i(x) \le 0, & i \in I, \\
& x \in \Omega,
\end{aligned}
$$

where, $f_j : \mathbb{R}^n \to \mathbb{R}$ ($j \in J := \{1, \ldots, p\}$), $g_i : \mathbb{R}^n \to \mathbb{R}$ ($i \in I_g$), $h_i : \mathbb{R}^n \to \mathbb{R}$ ($i \in I_h$), $H_i : \mathbb{R}^n \to \mathbb{R}$ ($i \in I$), and $G_i : \mathbb{R}^n \to \mathbb{R}$ ($i \in I$) are locally Lipschitz functions, and $\Omega \subseteq \mathbb{R}^n$ is a closed set.

If $p = 1$ and $\Omega = \mathbb{R}^n$, then (MMMPVCC) reduces to "mathematical programming with vanishing constraints" (MPVC in brief) which were introduced by Kanzow and his coauthors in 2007 [1,5] as an important extension of "mathematical programming with equilibrium constraints" (MPEC) [9,11], defined as

$$
\begin{aligned}
\min \quad & f_1(x) \\
\text{s.t.} \quad & g_i(x) \le 0 & i \in I_g, \\
& h_i(x) = 0 & i \in I_h, \\
& 0 \le H_i(x) \perp G_i(x) \ge 0, & i \in I,
\end{aligned}
$$

© Springer Nature Switzerland AG 2020
Y. D. Sergeyev and D. E. Kvasov (Eds.): NUMTA 2019, LNCS 11974, pp. 314–321, 2020.
https://doi.org/10.1007/978-3-030-40616-5_25

where the later condition means that $H_i(x) \geq 0$, $G_i(x) \geq 0$, and $H_i(x)G_i(x) = 0$ for all $i \in I$.

After defining the MPVC, finding the optimality conditions, named stationary conditions, for it become an interesting subject for some researchers; see [3–5] in smooth case and [7,8] in nonsmooth case.

It is worth mentioning that the feasible set of above problems are not convex, so their optimality conditions can formulate by different normal cones, including Clarke, Micheal-Penot, and Mordukhovich normal cones. Motivated to [3–5,9,11], we focus on Mordukhovich one. Also, the present paper considered the properly efficient solutions of (MMPVCC), and another kinds of efficiency do not consider here. Since there are several definitions for properly efficient solutions of multiobjective problems (see [2]), we consider the newest definition of them from [6].

The structure of subsequent sections of this paper is as follows: in Sect. 2, we define required definitions, theorems and relations of non-smooth analysis. In Sect. 3, we will introduce a constraint qualification for (MMPVCC) and present a Karusk-Kuhn-Tucker (KKT) type necessary condition for properly efficient solutions of the problem.

2   Notations and Preliminaries

In this section we present some preliminary results on nonsmooth analysis from [10].

The Mordukhovich subdifferential of locally Lipschitz function $\varphi : \mathbb{R}^p \to \mathbb{R}$ at $x_0 \in \mathbb{R}^p$ is defined as

$$\partial\varphi(x_0) := \limsup_{x \to x_0} \left\{ \xi \in \mathbb{R}^n \mid \liminf_{y \to x} \frac{\varphi(y) - \varphi(x) - \langle \xi, y - x \rangle}{\|y - x\|} \geq 0 \right\}.$$

We observe that for two locally Lipschitz functions φ_1 and φ_2 from $\mathbb{R}^p$ to $\mathbb{R}$, and for two arbitrary real numbers α and β, the following subadditive formula holds:

$$\partial(\alpha\varphi_1 + \beta\varphi_2)(x_0) \subseteq \alpha\partial\varphi_1(x_0) + \beta\partial\varphi_2(x_0). \tag{1}$$

Notice that the subdifferential $\partial\varphi(x_0)$ is always a compact (not necessarily convex) subset of $\mathbb{R}^p$.

Recall also that the normal cone of a closed subset $\mathcal{A} \subseteq \mathbb{R}^p$ at $x_0 \in \mathcal{A}$ is defined by $N(\mathcal{A}, x_0) := \partial\Theta_{\mathcal{A}}(x_0)$, where $\Theta_{\mathcal{A}}(.)$ denotes the indicator function of $\mathcal{A}$, i.e., $\Theta_{\mathcal{A}}(x) := 0$ for $x \in \mathcal{A}$, and $\Theta_{\mathcal{A}}(x) := +\infty$ otherwise. The following theorem will be useful in what follows.

Theorem 1. *If $x_0 \in V$ is a local minimizer of $\varphi : \mathbb{R}^p \to \mathbb{R}$ on $V \subseteq \mathbb{R}^p$, then one has*

$$0_p \in \partial\varphi(x_0) + N(V, x_0),$$

where, 0_p denotes the zero vector in $\mathbb{R}^p$.

Let $M : \mathbb{R}^r \rightrightarrows \mathbb{R}^s$ be a set-valued function, and

$$(\overline{y}, \overline{x}) \in \mathrm{Gph} M := \{(y, x) \in \mathbb{R}^r \times \mathbb{R}^s \mid x \in M(y)\}.$$

We say that M is calm at $(\overline{y}, \overline{x})$ if there exist some $L > 0$ and neighborhoods $\mathcal{X}$ and $\mathcal{Y}$ around $\overline{x}$ and $\overline{y}$, respectively, such that

$$d_{M(\overline{y})}(x) \leq L\|y - \overline{y}\|, \qquad \forall y \in \mathcal{Y}, \ \forall x \in \mathcal{X} \cap M(y),$$

where, $d_B(a) := \inf_{b \in B} \|a - b\|$ denotes the point-to-set distance between $a \in \mathbb{R}^s$ to $B \subseteq \mathbb{R}^s$ induced by standard norm $\|.\|$ on $\mathbb{R}^s$.

Also, we associate Mordukhovich's coderivative to M as $D^* M(\overline{y}, \overline{x}) : \mathbb{R}^s \rightrightarrows \mathbb{R}^r$ defined by

$$D^* M(\overline{y}, \overline{x})(x^*) := \{y^* \in \mathbb{R}^r \mid (y^*, -x^*) \in N(\mathrm{Gph} M, (\overline{y}, \overline{x}))\}.$$

If M is a single-valued, we simply write $D^* M(\overline{y})$ instead of $D^* M(\overline{y}; M\overline{x})$. For single-valued locally Lipschitz function h, it holds as

$$D^* h(\overline{y})(x^*) = \partial_M \langle x^*, h \rangle(\overline{y}), \tag{2}$$

where $\langle x^*, h \rangle(y) := \sum_{k=1}^{s} x_k^* h_k(y)$ for $x^* = (x_1^*, \ldots, x_s^*)$ and $h(y) = (h_1(y), \ldots, h_s(y))$.

Suppose that the set-valued mapping $\widetilde{M} : \mathbb{R}^l \rightrightarrows \mathbb{R}^k$ is defined as

$$\widetilde{M}(y) := \{x \in \widetilde{C} \mid \widetilde{g}(x) + y \in \widetilde{E}\}, \tag{3}$$

where the function $\widetilde{g} : \mathbb{R}^k \to \mathbb{R}^l$ is locally Lipschitz and $(\widetilde{C}, \widetilde{E}) \subseteq \mathbb{R}^k \times \mathbb{R}^l$ is closed. The following important theorems will be used in sequel.

Theorem 2 *[4, Theorem 4.1]. Consider the multifunction $\widetilde{M}$ given by (3) and a pair $(0, \overline{x}) \in \mathrm{Gph}\widetilde{M}$. If $\widetilde{M}$ is calm at $(0_l, \overline{x})$, then*

$$N(\widetilde{M}(0_l), \overline{x}) \subseteq \bigcup_{y^* \in N(\widetilde{E}, \widetilde{g}(\overline{x}))} D^* \widetilde{g}(\overline{x})(y^*) + N(\widetilde{C}, \overline{x}).$$

Theorem 3 *[4, Corollary 3.4]. Consider the set-valued function $\widehat{M} : \mathbb{R}^p \rightrightarrows \mathbb{R}^k$,*

$$\widehat{M}(y) := \{x \in \widehat{C} \mid \widehat{g}(x, y) \in \widehat{E}\},$$

where $\widehat{g} : \mathbb{R}^k \times \mathbb{R}^p \to \mathbb{R}^q$ is locally Lipschitz and $\widehat{E} \subseteq \mathbb{R}^q$, $\widehat{C} \subseteq \mathbb{R}^k$ are closed. Let $(\overline{y}, \overline{x}) \in \mathrm{Gph}\widehat{M}$ and $\widehat{C}$ be regular and semismooth at $\overline{x}$ (in the sense of [4, Definition 2.2]). Further, assume the following qualification condition holds,

$$\bigcup_{z^* \in N(\widehat{E}, \widehat{g}(\overline{x}, \overline{y})) \setminus \{0_q\}} [\partial \langle z^*, \widehat{g} \rangle(\overline{x}, \overline{y})]_x \cap -\mathrm{bd}\, N(\widehat{C}, \overline{x}) = \emptyset,$$

where $[\ ]_x$ denotes projection onto the x-component. Then $\widehat{M}$ is calm at $(\overline{y}, \overline{x})$. It is noteworthy that $\mathrm{bd}\, A$ denotes the topological bound of A.

3   Main Results

At starting point of this section, we denote the feasible set of (MMPVCC) by S, i.e.,

$$S := \Omega \cap \{x \in \mathbb{R}^n \mid g_i(x) \leq 0, \ i \in I_g; \ h_i(x) = 0, \ i \in I_h;$$

$$H_i(x) \geq 0, \ G_i(x)H_i(x) \leq 0, \ i \in I\}.$$

A feasible point $x_0 \in S$ is said to be properly efficient for (MMPVCC) if there exist positive numbers $\lambda_1, \ldots, \lambda_p$ such that

$$\sum_{j=1}^{p} \lambda_j f_j(x_0) \leq \sum_{j=1}^{p} \lambda_j f_j(x), \qquad \forall x \in S.$$

We fix a feasible point $\hat{x} \in S$, and consider following index sets:

$$I_{+0} := \{i \in I \mid H_i(\hat{x}) > 0, \ G_i(\hat{x}) = 0\},$$
$$I_{+-} := \{i \in I \mid H_i(\hat{x}) > 0, \ G_i(\hat{x}) < 0\},$$
$$I_{0+} := \{i \in I \mid H_i(\hat{x}) = 0, \ G_i(\hat{x}) > 0\},$$
$$I_{00} := \{i \in I \mid H_i(\hat{x}) = 0, \ G_i(\hat{x}) = 0\},$$
$$I_{0-} := \{i \in I \mid H_i(\hat{x}) = 0, \ G_i(\hat{x}) < 0\}.$$

Also, we set

$$I_g^0 := \{i \in I_g \mid g_i(\hat{x}) = 0\}.$$

Obviously, $I = I_0 \cup I_+$ in which $I_+ := I_{+0} \cup I_{+-}$ and $I_0 := I_{0+} \cup I_{00} \cup I_{0-}$.

The mention of M-stationary point was introduced for smooth MPVC in [3,4] and for nonsmooth MPEC in [9]. Now, we present a suitable extension of this concept for (MMPVCC).

Definition 1. *The feasible point $\hat{x}$ is said to be generalized M-stationary (G-M-stationary) point for (MMPVCC) if there exist $\lambda_j > 0$ for $j \in J$, $\lambda_i^g \geq 0$ for $i \in I_g^0$, $\lambda_i^h \in \mathbb{R}$ for $i \in I_h$, $\lambda_i^G \in \mathbb{R}$ for $i \in I$, and $\lambda_i^H \in \mathbb{R}$ for $i \in I$ such that*

$$0_n \in \sum_{j=1}^{p} \lambda_j \partial f_j(\hat{x}) + \sum_{i \in I_g^0} \lambda_i^g \partial g_i(\hat{x}) + \sum_{i=1}^{r} \lambda_i^h \partial h_i(\hat{x})$$

$$+ \sum_{i=1}^{m} \left[\lambda_i^G \partial G_i(\hat{x}) - \lambda_i^H \partial H_i(\hat{x}) \right] + N(\Omega, \hat{x}), \qquad (4)$$

$$\lambda_i^G \geq 0, \ i \in I_{00} \cup I_{+0}, \qquad \lambda_i^G = 0, \ i \in I_{0+} \cup I_{0-} \cup I_{+0}, \qquad (5)$$

$$\lambda_i^H \ \text{free}, \ i \in I_{00} \cup I_{0+}, \qquad \lambda_i^H \geq 0, \ i \in I_{0-}, \qquad (6)$$

$$\lambda_i^H = 0, \ i \in I_+, \qquad \lambda_i^H \lambda_i^G = 0, \ i \in I_{00}. \qquad (7)$$

To simplicity in writing, we rearrange the constraints of G_is and H_is such that the constraints with index $i \in I_{00}$ are first written, then the constraints with index $i \in I_{0+}$, then $i \in I_{0-}$, then $i \in I_{+0}$, and finally $i \in I_{+-}$. We keep this order throughout this paper. Also, we assume that

$$I_g^0 = \{1, \ldots, k\}, \quad k \le s.$$

Let the function $\Gamma : \mathbb{R}^n \to \mathbb{R}^{k+r+2m}$ be define by

$$\Gamma(x) := \big(g_1(x), \ldots, g_k(x), h_1(x), \ldots, h_r(x), G_1(x), H_1(x), \ldots, G_m(x), H_m(x)\big).$$

Consider following problem which has been parameterized respect to $y \in \mathbb{R}^{k+r+2m}$:

$$\hat{P}(y): \qquad \min f(x)$$
$$\text{s.t. } \Gamma(x) + y \in \mathfrak{D}$$
$$x \in \Omega,$$

in which

$$\mathfrak{D} := \mathbb{R}_-^k \times \{0_r\} \times \{(p, q) \in \mathbb{R}^m \times \mathbb{R}^m \mid q_i \ge 0 \text{ and } p_i q_i \le 0, \quad \forall i \in I\}.$$

$\hat{P}(0_{k+r+2m})$ is clearly (MMPVCC). We denote the feasible set of $\hat{P}(y)$ by $\hat{S}(y)$, i.e.,

$$\hat{S}(y) := \{x \in \Omega \mid \Gamma(x) + y \in \mathfrak{D}\}.$$

Note that $\hat{S} : \mathbb{R}^{k+r+2m} \rightrightarrows \mathbb{R}^n$ is a set-valued mapping. The following important theorem gives an upper estimate for $N(S, \hat{x})$.

Theorem 4. *Suppose that $\hat{x}$ is an feasible point for (MMPVCC) and the set-valued mapping $\hat{S}$ is calm at $(0_{k+r+2m}, \hat{x})$. Then*

$$N(S, \hat{x}) \subseteq$$
$$\bigcup_{\mu \in N(\mathfrak{D}, \Gamma(\hat{x}))} \left[\sum_{i=1}^k \mu_i^g \partial g_i(\hat{x}) + \sum_{i=1}^r \mu_i^h \partial h_i(\hat{x}) + \sum_{i=1}^m \left[\mu_i^H \partial H_i(\hat{x}) + \mu_i^G \partial G_i(\hat{x}) \right] \right]$$
$$+ N(\Omega, \hat{x}),$$

where, $\mu = \big(\mu_1^g, \ldots, \mu_k^g, \mu_1^h, \ldots, \mu_r^h, \mu_1^G, \mu_1^H, \ldots, \mu_m^G, \mu_m^H\big)$.

Proof. Taking in Theorem 2

$$\tilde{g}(x) = \Gamma(x), \qquad \widetilde{M} = \hat{S}, \qquad \widetilde{E} = \mathfrak{D}, \qquad \tilde{C} = \Omega,$$

we deduce that

$$N(S, \hat{x}) \subseteq \bigcup_{\mu \in N(\mathfrak{D}, \Gamma(\hat{x}))} D^* \Gamma(\hat{x})(\mu) + N(\Omega, \hat{x}). \tag{8}$$

By well-known equality (2), for each

$$\mu := \big(\mu_1^g, \ldots, \mu_k^g, \mu_1^h, \ldots, \mu_r^h, \mu_1^G, \mu_1^H, \ldots, \mu_m^G, \mu_m^H\big) \in \mathbb{R}^{k+r+2m}$$

we have

$$D^* \Gamma(\hat{x})(\mu) = \partial \langle \mu, \Gamma(.) \rangle(\hat{x}) =$$

$$= \partial \left[\sum_{i=1}^{k} \mu_i^g g_i + \sum_{i=1}^{r} \mu_i^h h_i + \sum_{i=1}^{m} \left(\mu_i^H H_i + \mu_i^G G_i \right) \right](\hat{x})$$

$$\subseteq \sum_{i=1}^{k} \mu_i^g \partial g_i(\hat{x}) + \sum_{i=1}^{r} \mu_i^h \partial h_i(\hat{x}) + \sum_{i=1}^{m} \left[\mu_i^H \partial H_i(\hat{x}) + \mu_i^G \partial G_i(\hat{x}) \right].$$

Thus, (8) implies that

$$N(S, \hat{x}) \subseteq$$

$$\bigcup_{\mu \in N(\mathfrak{D}, \Gamma(\hat{x}))} \left[\sum_{i=1}^{k} \mu_i^g \partial g_i(\hat{x}) + \sum_{i=1}^{r} \mu_i^h \partial h_i(\hat{x}) + \sum_{i=1}^{m} \left[\mu_i^H \partial H_i(\hat{x}) + \mu_i^G \partial G_i(\hat{x}) \right] \right]$$

$$+ N(\Omega, \hat{x}),$$

as required.

Same as everywhere in optimization theory, we require a constraint qualification for presenting a KKT type necessary condition for (MMPVCC). The concept of "No Nonzero Abnormal Multiplier Constraint Qualification" (NNAMCQ in brief) have been introduced by Ye [11] for smooth MPECs, and were extended for nonsmooth MPECs by Movahedian and Nobakhtian [9]. We generalize this constraint qualification for (MMPVCC) in following definition.

Definition 2. *We say that (MMPVCC) at $\hat{x}$ satisfies in "Generalized NNAMCQ" (GNNAMCQ in short) if there is no nonzero scalars $\alpha_i^g \in \mathbb{R}$ for $i \in I_g^0$, $\alpha_i^h \in \mathbb{R}$ for $i \in I_h$, $\alpha_i^G \in \mathbb{R}$ for $i \in I$, and $\alpha_i^H \in \mathbb{R}$ for $i \in I$ such that*

$$0_n \in \sum_{i \in I_g^0} \alpha_i^g \partial g_i(\hat{x}) + \sum_{i=1}^{r} \alpha_i^h \partial h_i(\hat{x}) + \sum_{i=1}^{m} \left[\alpha_i^G \partial G_i(\hat{x}) + \alpha_i^H \partial H_i(\hat{x}) \right] + N(\Omega, \hat{x}),$$

$$\alpha_i^g \geq 0, \ i \in I_g^0, \qquad \alpha_i^H \leq 0, \ i \in I_{0-},$$

$$\alpha_i^H = 0, \ i \in I_+, \qquad \alpha_i^H \alpha_i^G = 0, \ i \in I_{00},$$

$$\alpha_i^G \geq 0, \ i \in I_{00} \cup I_{+0}, \qquad \alpha_i^G = 0, \ i \in I_{0+} \cup I_{0-} \cup I_{+0}.$$

Now, we can state a necessary condition for G-M-stationarity of (MMPVCC).

Theorem 5. *Suppose that $\hat{x}$ is a properly efficient solution for (MMPVCC). If GNNAMCQ is satisfied at $\hat{x}$), then $\hat{x}$ is a G-M-stationary point for (MMPVCC).*

Proof. The properly efficiency of $\hat{x}$ concludes that there exist positive scalars $\lambda_j > 0$, for $j \in J$, such that $\hat{x}$ is a minimizer to the following weighted problem:

$$\min \sum_{j=1}^{p} \lambda_j f_j(x) \qquad \text{subject to} \qquad x \in S.$$

Thus, Theorem 1 implies that

$$0_n \in \partial\Big(\sum_{j=1}^{p}\lambda_j f_j\Big)(\hat{x}) + N(S,\hat{x}) \subseteq \sum_{j=1}^{p}\lambda_j\partial f_j(\hat{x}) + N(S,\hat{x}). \qquad (9)$$

We claim that the set-valued mapping $\widehat{S}$ is calm at $(0_{k+r+2m},\hat{x})$. For this end, we observe that

$$N(\mathfrak{D},\Gamma(\hat{x})) = N\Big(\mathbb{R}^k_- \times \{0_r\} \times \{(p,q) \in \mathbb{R}^m \times \mathbb{R}^m \mid q_i \geq 0 \text{ and } p_i q_i \leq 0, \quad \forall i \in I\}, \Gamma(\hat{x})\Big)$$

$$= N\Big(\mathbb{R}^k_-, 0_k\Big) \times N\Big(\{0_r\}, 0_r\Big)$$

$$\times N\Big(\{(p,q) \in \mathbb{R}^m \times \mathbb{R}^m \mid q_i \geq 0 \text{ and } p_i q_i \leq 0, \quad \forall i \in I\}, (G_1(\hat{x}), H_1(\hat{x}), \ldots, G_m(\hat{x}), H_m(\hat{x}))\Big). \qquad (10)$$

On the other hand, by [3, Lemma 3.2] we have

$$N\Big(\{(p,q) \in \mathbb{R}^m \times \mathbb{R}^m \mid q_i \geq 0 \text{ and } p_i q_i \leq 0, \quad \forall i \in I\}, (G_1(\hat{x}), H_1(\hat{x}), \ldots, G_m(\hat{x}), H_m(\hat{x}))\Big)$$

$$= \prod_{i \in I_{00}}\mathfrak{B} \times \prod_{i \in I_{0+}}(\{0\} \times \mathbb{R}) \times \prod_{i \in I_{0-}}(\{0\} \times \mathbb{R}_-) \times \prod_{i \in I_{+0}}(\mathbb{R}_+ \times \{0\}) \times \prod_{i \in I_{+-}}(\{0\} \times \{0\}),$$

where, $\mathfrak{B} := \{(r,s) \in \mathbb{R}^2 \mid r \geq 0, \ rs = 0\}$. Combining the latter equality and (10), we get

$$N(\mathfrak{D},\Gamma(\hat{x})) = \mathbb{R}^k_+ \times \mathbb{R}^r \times \prod_{i \in I_{00}}\mathfrak{B} \times \prod_{i \in I_{0+}}(\{0\} \times \mathbb{R}) \times \prod_{i \in I_{0-}}(\{0\} \times \mathbb{R}_-) \times \prod_{i \in I_{+0}}(\mathbb{R}_+ \times \{0\}) \times \prod_{i \in I_{+-}}(\{0\} \times \{0\}). \qquad (11)$$

This means that $\alpha := \big(\alpha_1^g, \ldots, \alpha_k^g \alpha_1^h, \ldots, \alpha_r^h, \alpha_1^G, \alpha_1^H, \ldots, \alpha_m^G, \alpha_m^H\big) \in N(\mathfrak{D}, \Gamma(\hat{x}))$ if and only if

$$\begin{cases} \alpha_i^g \geq 0, \ i \in I_g^0, & \alpha_i^H \leq 0, \ i \in I_{0-}, \\ \alpha_i^H = 0, \ i \in I_+, & \alpha_i^H \alpha_i^G = 0, \ i \in I_{00}, \\ \alpha_i^G \geq 0, \ i \in I_{00} \cup I_{+0}, & \alpha_i^G = 0, \ i \in I_{0+} \cup I_{0-} \cup I_{+0}. \end{cases}$$

Thus, the satisfying GNNAMCQ at $\hat{x}$ guaranties if $\alpha \in N(\mathfrak{D}, \Gamma(\hat{x}))$ and

$$\sum_{i \in I_g^0}\alpha_i^g \partial g_i(\hat{x}) + \sum_{i=1}^{r}\alpha_i^h \partial h_i(\hat{x}) + \sum_{i=1}^{m}\big[\alpha_i^G \partial G_i(\hat{x}) + \alpha_i^H \partial H_i(\hat{x})\big] \cap -N(\Omega, \hat{x}) \neq \emptyset,$$

then, $\alpha = 0_{k+r+2m}$. This implies that

$$\bigcup_{0_{k+r+2m} \neq \alpha \in N(\mathfrak{D}, \Gamma(\hat{x}))} [\partial\left(\langle\alpha, \Gamma(x) + y\rangle\right)(\hat{x}, 0_{k+r+2m})]_x \cap -N((\Omega, \hat{x}) = \emptyset,$$

where y is a variable in $\mathbb{R}^{k+r+2m}$. Thus, the hypotheses of Theorem 3 hold by taking $\widehat{g}(x,y) = \Gamma(x) + y$ and $\bar{y} := 0_{k+r+2m}$. Therefore, Theorem 3 implies that $\widehat{M}(=\widehat{S})$ is calm at $(0_{k+r+2m}, \hat{x})$, and our claim is proved. Finally, Theorem 4 and inclusion (9) yield the result.

References

1. Achtziger, W., Kanzow, C.: Mathematical programs with vanishing constraints: optimality conditions and constraint qualifications. Math. Program. **114**, 69–99 (2007)
2. Ehrgott, M.: Multicriteria Optimization. Springer, Berlin (2005). https://doi.org/10.1007/3-540-27659-9
3. Hoheisel, T., Kanzow, C.: Stationarity conditions for mathematical programs with vanishing constraints using weak constraint qualifications. J. Math. Anal. Appl. **337**, 292–310 (2008)
4. Hoheisel, T., Kanzow, C., Outrata, J.: Exact penalty results for mathematical programs with vanishing constraints. Nonlinear Anal. **72**, 2514–2526 (2010)
5. Hoheisel, T., Kanzow, C.: First- and second-order optimality conditions for mathematical programs with vanishing constraints. Appl. Math. **52**, 495–514 (2007)
6. Gopfert, A., Riahi, H., Tammer, C., Zalinescu, C.: Variational Methods in Partioal Ordered Spaces. Springer, New York (2003). https://doi.org/10.1007/b97568
7. Kazemi, S., Kanzi, N.: Constraint qualifications and stationary conditions for mathematical programming with non-differentiable vanishing constraints. J. Optim. Theory Appl. (2018). https://doi.org/10.1007/s10957-018-1373-7
8. Kazemi, S., Kanzi, N., Ebadian, A.: Estimating the Fréchet normal cone in optimization problems with nonsmooth vanishing constraints. Iran. J. Sci. Technol. Trans. A Sci. (2019). https://doi.org/10.1007/s40995-019-00683-8
9. Movahedian, N., Nobakhtian, S.: Necessary and sufficient conditions for nonsmooth mathematical programs with equilibrium constraints. Nonlinear Anal. **72**, 2694–2705 (2010)
10. Rockafellar, R.T., Wets, J.B.: Variational Analysis. Springer, Heidelberg (1998). https://doi.org/10.1007/978-3-642-02431-3
11. Ye, J.: Necessary and sufficient optimality conditions for mathematical programs with equilibrium constraints. J. Math. Anal. Appl. **307**, 350–369 (2005)

Molecular Dynamics Performance Evaluation with Modern Computer Architecture

Emanuele Breuza[1](ID), Giorgio Colombo[2](ID), Daniele Gregori[1](ID), and Filippo Marchetti[2,3](✉)(ID)

[1] E4 Computer Engineering, Viale Martiri della Libertà 66, 42019 Scandiano, Italy
{emanuele.breuza,daniele.gregori}@e4company.com
[2] Department of Drug Sciences, University of Pavia,
Viale Taramelli 12–14, 27100 Pavia, Italy
{g.colombo,filippo.marchetti}@unipv.it
[3] Department of Chemistry, University of Milan,
Via Venezian 21, 20133 Milano, Italy

Abstract. An important task of chemical biology is to discover the mechanism of recognition and binding between proteins. Despite the simplicity of the ligand-based model, fundamental mechanisms that regulate these interactions are poorly understood. An adequate equipment is mandatory to unravel this scientific challenge, not only through cost savings but also with high-quality results. With this in mind, we performed Molecular Dynamics simulations using the Gromacs package on two promising platforms: Cavium ThunderX2 ARM based cluster setup and shared-memory Intel based single-node machine. Aforementioned tests were also performed on common Intel based servers as a reference. Acquired results shown that shared-memory machine features the higest performance, although ARM and Intel clutsers are only slightly slower when more than four sockets are employed. During measurements, idle and job-execution consumptions were sampled in order to evaluate the energy required by a single simulation step. Results show that ARM and Intel servers are much less power-hungry with respect to shared-memory machine. The latter, on the other hand, features a decrement in power consumption when more resources are employed. Said unexpected behaviour is later discussed.

Keywords: Gromacs · High performance computing · Molecular Dynamics

1 Introduction

Ligand-based control of protein functional motions can provide fresh opportunities in the study of fundamental biological mechanisms and in the development of novel therapeutics. A prototypical example is represented by integrin $\alpha\nu\beta6$,

Y. D. Sergeyev and D. E. Kvasov (Eds.): NUMTA 2019, LNCS 11974, pp. 322–329, 2020.
https://doi.org/10.1007/978-3-030-40616-5_26

a transmembrane protein with promising anticancer properties. Integrins are heterodimeric cell adhesion receptors formed by a bilobular head and two legs that can assume two different configuration: closed (also called inactive) with bended legs and open (or active, Fig. 1) in which it is proned to bind partners [1,2].

A large number of biological pathways (e.g. cell migration or intracellular signal transduction) require the transition between close and open state. The malfunction of this machinery could lead to the pathogenesis of many diseases, as reported in ref. [3–6].

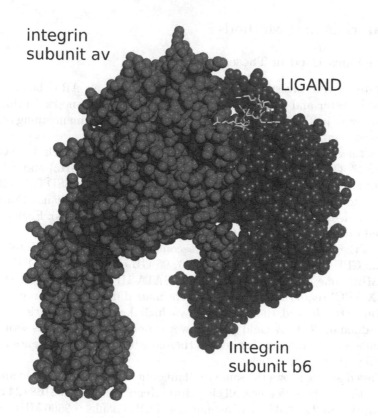

Fig. 1. Integrin model in open configuration. Color code: αv subunit in red, $\beta 6$ subunit in blue, ligand in yellow. (Color figure online)

Here we focus on the integrin $\alpha v \beta 6$, an epitelial specific class of integring that interact with TGF-β, i.e. the transforming growth factor. The head part of $\alpha v \beta 6$ was recently crystallized alone and in complex with TGF-β peptide by Dong et al. [7] providing a starting model for the protein-ligand interaction. The investigation of different activation/deactivation mechanisms of integrin is a rather challenging problem, usually addressed with molecular dynamics computational technique. Indeed, internal protein dynamics are strongly influenced by binded ligands, requiring extensive simulations.

The large system size (above 200000 atoms) drammatically increases CPU-time requirements to compute an integration step, leading to unfeasible computational costs. For this reasons, the research of efficient architecture represents an important step in the evolution of computational simulations towards real biological structures.

In this work, we measured the performances of Gromacs, [8] a popular Molecular Dynamics software, on different architectures in order to compare computational efficiency and power consumption.

2   Materials and Methods

2.1   Machines Used in These Simulations

Three systems were investigated in the present study: an ARM based and an Intel based cluster and a shared-memory Intel based single-node. During tests, all systems were installed in the E4 R&D facility capable of manteaning constant environment temperature and power delivery.

The former one consists of four 2U servers equipped with two Cavium ThunderX2® CN9980 processors (32 cores each @ 2.2–3.0 GHz) and a total of 256 GB of memory (16 DDR4 banks @ 2666 MHz). An Intel SSD (240 GB with SATA III interface) is dedicated to the local storage, while a Mellanox ConnetX-5 PCIe card is dedicated to network connectivity (link @ 100 Gb/s). Each machine is powered with a redundant 1200 W Platinum grade power supply.

Each of the four worker-nodes of the second cluster features two Intel Xeon® Gold 6130 CPUs (16 cores @ 2.1–3.7 GHz), 96 GB of memory (12 DDR4 banks @ 2400 MHz), one Intel SSD (960 GB with SATA III interface) and a Mellanox ConnectX-5 PCIe card (link @ 100 Gb/s). The main difference with previous setup is the form factor. Indeed, these nodes show a high density configuration and they share a redundant 2200 W Gold grade power supply. Nowadays, this setup is the most common in the modern datacenter, therefore it is taken as reference for comparisons reported in this article.

The last machine shows the same total amount of processors of previous clusters but with a major difference: all eight Intel Xeon® Platinum 8168 (24 cores @ 2.7-3.7 GHz) as well as 1536 GB of memory (48 DDR4 banks @ 2666 MHz) are presented as part of the same board to the OS. Comunications between sockets is handled by the QuickPath Interconnect bus (QPI) that features much lower latency and higher bandwidth with respect to a traditional network interface. This feature could result in a major advantage if handled properly (see Sect. 2.2). Two Intel SSD in mirror mode (480 GB with SATA III interface) and a 4800 W Titanium grade power supply are completing the configuration.

All investigated platforms were running CentOS Linux distribution (release 7.5). GCC [9] v. 6.4.0, OpenMPI [10] v 4.0.0 and FFTW [11] v. 3.3.8 were adopted to build Gromacs v. 2018.3 from source code. All firmware and drivers were updated to the latest version before testing.

2.2   Software Parameters and Test Execution

The investigated structure is the headpiece of integrin $\alpha\nu\beta6$ complex with TGF-β peptide (the pdb code is 4UM9). Said model was obtained using X-Ray single crystal crystallography and consequently refined to delete purely resolved parts. The total residue count is 987. Three Mg atoms were added in the ion binding sites and the complex was solvated with tip3p water molecules in an octahedron box (1.4 nm of margin), Na+ counterions were added to ensure electroneutrality. The final system is composed of 231756 atoms.

The molecular simulation was conducted with Gromacs package using the amberff99SB-ildn forcefield [12]. Parameters used for Mg ions were developped by Allner et al. [13] After an initial minimization, the system was equilibrated at a constant temperature of 300 K for 100 ps and, successively, at costant temperatue and pressure (300 K, 1 bar) for 100 ps. For temperature and pressure coupling, the modified Berendsen thermostat (v-rescale) and Parrinello-Rahman barostat were used [14,15] with τT of 0.1 ps and τP 2 ps. Electrostatic forces were evaluated by Particle Mesh Ewald method [16] and Lennard-Jones forces by a cutoff of 0.9 nm.

To execute Gromacs parallel code, we used `mpirun` command provided by OpenMPI package. More specifically, we employed available resources as follow: one MPI process per physical core and one OpenMP thread per MPI process. Consequently, the final command was

```
mpirun -np #_of_cores --machinefile hosts_file mdrun_mpi.
```

We took particular precautions to correctly distribute computational resources on shared-memory machine. More specifically, we adopted a slightly different command, i.e.

```
mpirun -np #_of_cores --map-by ppr:24:socket mdrun_mpi.
```

Said command allowed to fully occupy only requested sockets, mimicking a cluster setup. Indeed, `mpirun` by default would spread cores across all eight sockets of the single-node server, leading to a partial CPU allocation and therefore a much higer clock frequency compared to a cluster setup with the same amount of resources.

3   Results and Discussion

A molecular dynamics simulation consists in the integration of the Newton's law of dynamics for every atom in the system. Interactions between atoms are modelled with a force field potential. Therefore the system is represented by a large set of coupled differential equations that could not be integrated analytically but had to be solved numerically. Gromacs package implement various algorithms for the numerical integration with different order of precision, in our test we use the leap-frog algorithm [17].

The choice of the integration interval has to be balance between performance and integration errors, we set a time step of 0.002 ps. This should not be confused with the real CPU-time needed to compute system evolution. Indeed, the actual duration of a step is in the order of ten ms. Consequently, our results are expressed in Gromacs performances, i.e. the number of evolutionary steps (espressed in ns) during the 24 h.

Acquired results are reported in Fig. 2. Shared-memory machine turned out to be the fastest in our pool. This can be easily traced to the higher clock frequency of CPU. This being said, performance differences are significantly reduced with the increment of employed resources (i.e. number of sockets). As a matter of fact, ARM and Intel based servers turned out to be 40% and 20% slower than single-node platform when four and eight sockets are employed.

One could guess that this behavior could be related to (i) the excellent parallelizzation of Gromacs code and (ii) suboptimal memory allocation inside shared-memory machine. Indeed, network interfaces increase communication latency between processes executed on different servers, and consequently augmenting the simulation length. In single-node systems said interfaces are substituted with the much faster QPI busses, resulting in a substantial increment in performance. On the other hand, computational cores are handling both calculation processes and QPI communication messages (mostly related to allocated data in memory). Therefore, calculation processes are frequently interrupted by communication requests, slowing down the overall simulation.

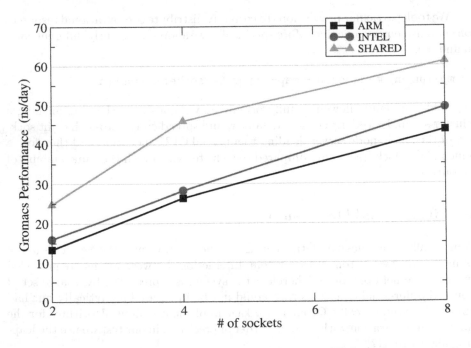

Fig. 2. Gromacs perfomance vs employed resources. Higher is better.

In recent years, attention to energy efficiency has become fundamental since acceptable total cost of ownership must include not only hardware purchase but also system power consumptions. Therefore, we measured energy requirements of investigated architectures during job executions.

Figure 3 shows acquired results scaled by idle consumptions. This correction is mandatory since compared machines feature different formfactors (i.e. high-density twin-square for Intel servers, 2U for ARM once and 7U for shared-memory) and consequently they adopt different power management strategies.

Shared-memory machine turned out to be the most power-hungry system in our pool. On the other hand, ARM and Intel based servers show a rather modest incremental trend. It must be noted that single-node server features a decrement in power consumption when more than two sockets were employed. This unexpected behavior could be caused by the fact that full load and idle consumptions does not change very much in shared-memory architecture, while Gromacs performance are almost doubled. Machine efficiency substantially drops when 8 sockets are employed (see Fig. 2). Consequently, energy required for a single simulation step turned out to be comparable with tests requiring 4 sockets.

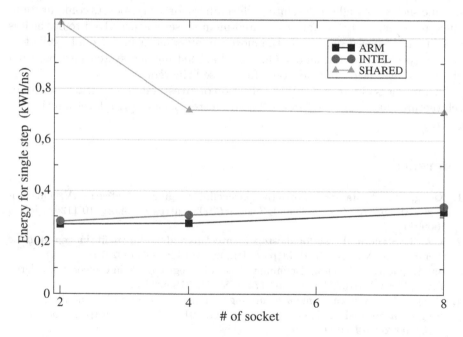

Fig. 3. Required energy to calcualte a single simulation step vs employed resources (i.e. sockets). Lower is better.

Recently the interest of HPC community in ARM processors is increasing and a comparison between ARM and x86 architectures were conducted [18], in our test we introduce also a comparison with a shared-memory machine and an evaluation of the energy consumption.

Moreover the improvements in hardware configuration led us to extend the length of molecular simulations arriving to simulation times in the order of μs, necessary for analyze this system as is previously done with an other type of integrin [19].

4   Conclusions

In the present work, we compared Gromacs performances on two rather new architectures with rising popularity in datacenter environment: ARM based clusters and shared-memory single-node machines. Also traditional Intel based cluster were considered for comparative purposes. Our tests focused on activation/deactivation mechanisms of integrin $\alpha\nu\beta6$, a rather large systems comparable to real size proteins.

We observe that integrin complexes could be efficiently simulated on all investigated platfoms. Indeed, all architectures were capable to simulate more than 40 ns/day. Intel and ARM servers feature a linear trend with employed resources (i.e. number of sockets) in Gromacs performance. This being said, shared-memory machine shows the highest absolute performances for every sockets configuration. Concerning energy comsumptions, shared-memory server turned out to be the less efficient. ARM based servers are the most energy efficient in our pool, while featuring respectable performances. One could argue that said systems are indeed a compelling alternative to traditional Intel based clusters.

Future works aim to expand the collection of investigated platforms with other architectures like AMD and Power based cluster or enterprise level multi-GPU server.

References

1. Xiong, J.P.: Crystal structure of the extracellular segment of integrin $\alpha V\beta3$ in complex with an Arg-Gly-Asp ligand. Science (2002). https://doi.org/10.1126/science.1069040
2. Xiao, T.: Structural basis for allostery in integrins and binding to fibrinogen-mimetic therapeutics. Nature (2004). https://doi.org/10.1038/nature02976
3. Bandyopadhyay, A., et al.: Defining the role of Integrin $\alpha\nu\beta6$ in Cancer. Curr. Drug Targ. (2009). https://doi.org/10.2174/138945009788680374
4. Munger, J.S.: A mechanism for regulating pulmonary inflammation and fibrosis: the integrin $\alpha\nu\beta6$ binds and activates latent TGF $\beta1$. Cell (1999). https://doi.org/10.1016/S0092-8674(00)80545-0
5. Xu, M., et al.: Epigenetic regulation of integrin $\beta6$ transcription induced by TGF-$\beta1$ in human oral squamous cell carcinoma cells. J. Cell. Biochem. (2018). https://doi.org/10.1002/jcb.26642
6. Morgan, M.R., et al.: Psoriasin (S100A7) associates with integrin $\beta6$ subunit and is required for $\alpha\nu\beta6$-dependent carcinoma cell invasion. Oncogene (2011). https://doi.org/10.1038/onc.2010.535
7. Dong, X., et al.: Structural determinants of integrin β-subunit specificity for latent TGF-β. Nat. Struct. Mol. (2014). https://doi.org/10.1038/nsmb.2905

8. Abraham, M.J.: GROMACS: high performance molecular simulations through multi-level parallelism from laptops to supercomputers. SoftwareX **1**, 19–25 (2015)
9. The GNU Compiler Collection Homepage. https://gcc.gnu.org/. Accessed 27 May 2019
10. OpenMPI Homepage. https://www.open-mpi.org/. Accessed 27 May 2019
11. FFTW Hoempage. http://www.fftw.org/. Accessed 27 May 2019
12. Abil, E.A., et al.: Motional timescale predictions by molecular dynamics simulations: case study using proline and hydroxyproline sidechain dynamics. Proteins (2014). https://doi.org/10.1002/prot.24350
13. Allner, et al.: Magnesium ion-water coordination and exchange in biomolecular simulations. J. Chem. Theory Comput. (2012). https://doi.org/10.1021/ct3000734
14. Bussi, G., et al.: Canonical sampling through velocity-rescaling. J. Chem. Phys. (2007). https://doi.org/10.1063/1.2408420
15. Parrinello, M., Rahman, A.: Polymorphic transitions in single crystals: a new molecular dynamics method. J. Appl. Phys. (1981). https://doi.org/10.1063/1.328693
16. Essman, U., et al.: A smooth particle mesh Ewald method. J. Chem. Phys. (1995). https://doi.org/10.1063/1.470117i
17. http://manual.gromacs.org/documentation/2019-rc1/reference-manual/algorithms/molecular-dynamics.html#the-leap-frog-integrator.   Accessed   10 Jan 2019
18. Smith S., et al.: A performance analysis of the first generation of HPC-optimized arm processors. Concurr. Pract. Exp. Comput. (2019). https://doi.org/10.1002/cpe.5110
19. Paladino, A., et al.: High affinity vs. native fibronectin in the modulation of $\alpha\nu\beta3$ integrin conformational dynamics: insights from computational analyses and implications for molecular design. Plos CB (2017). https://doi.org/10.1371/journal.pcbi.1005334

Artificial Neural Networks Training Acceleration Through Network Science Strategies

Lucia Cavallaro[1]([✉]) [iD], Ovidiu Bagdasar[1] [iD], Pasquale De Meo[2] [iD],
Giacomo Fiumara[3] [iD], and Antonio Liotta[4] [iD]

[1] University of Derby, Kedleston Road, Derby DE22 1GB, UK
l.cavallaro@derby.ac.uk
[2] Polo Universitario Annunziata, University of Messina, 98122 Messina, Italy
[3] MIFT Department, University of Messina, 98166 Messina, Italy
[4] Edinburgh Napier University, 10 Colinton Road, Edinburgh EH10 5DT, UK

Abstract. Deep Learning opened artificial intelligence to an unprecedented number of new applications. A critical success factor is the ability to train deeper neural networks, striving for stable and accurate models. This translates into Artificial Neural Networks (ANN) that become unmanageable as the number of features increases. The novelty of our approach is to employ Network Science strategies to tackle the complexity of the actual ANNs at each epoch of the training process. The work presented herein originates in our earlier publications, where we explored the acceleration effects obtained by enforcing, in turn, scale freeness, small worldness, and sparsity during the ANN training process. The efficiency of our approach has also been recently confirmed by independent researchers, who managed to train a million-node ANN on non-specialized laptops. Encouraged by these results, we have now moved into having a closer look at some tunable parameters of our previous approach to pursue a further acceleration effect. We now investigate on the revise fraction parameter, to verify the necessity of the role of its double-check. Our method is independent of specific machine learning algorithms or datasets, since we operate merely on the topology of the ANNs. We demonstrate that the revise phase can be avoided in order to half the overall execution time with an almost negligible loss of quality.

Keywords: Network Science · Artificial Neural Networks

1 Introduction

The idea to simulate the human brain behaviour is one of the top scientific trends today. In particular, Deep Learning strategies pave the way to an unprecedented number of new applications thanks to their ability to manage more complex architectures such as speech recognition [8], image [9] and signal [15] processing, and cyber-security [2]. Other applications which are gaining popularity are in the field of bio-medicine [3] and in drug discovery [4,14].

© Springer Nature Switzerland AG 2020
Y. D. Sergeyev and D. E. Kvasov (Eds.): NUMTA 2019, LNCS 11974, pp. 330–336, 2020.
https://doi.org/10.1007/978-3-030-40616-5_27

Despite their success, Deep Learning architectures suffer from important scalability issues: generally speaking, they translate into Artificial Neural Networks (ANN) that become unmanageable as the number of features increases.

While most current strategies focus on using more powerful hardware, our approach is to employ Network Science strategies to tackle the complexity of the actual ANNs iteratively, that is at each epoch of the training process.

The work presented herein originates in our earlier publication [13], a promising research avenue to speed up Neural Network training. There, we defined a new approach, called Sparse Evolutionary Training (SET), in which the acceleration effects obtained by enforcing, in turn, scale-freeness, small-worldness, and sparsity during the ANN training process are explored.

In the SET framework, an ANN is first initialized as a sparse weighted Erdős-Rényi graph in which the graph density is fixed (20% by default), and weights on edges are drawn from a normal distribution with mean equals to zero. In a second stage (called revision), null-edges (i.e. links with weight equal to zero) are iteratively replaced with non-zero weights with the twofold goal of reducing the loss on the training set and to keep the number of connections constant.

The efficiency of this approach has also been recently confirmed by independent researchers, who managed to train a million-node ANN on non-specialized laptops [12].

Encouraged by these results, we have now moved into looking at algorithm tuning parameters to pursue a further acceleration effect, with a negligible accuracy loss of the final model. The method is independent of specific machine learning algorithms or datasets since we operate merely on the topology of the ANNs. The focus is on the revision stage and on its impact on the training time over epochs. Noteworthy results have been achieved, such as an improvement of 50% in terms of time gain, as shown in Sect. 4.

The rest of the paper is organized as follows. Section 2, presents the related background. The adopted methodology is addressed in Sect. 3. Next, on Sect. 4, the results are discussed. In Sect. 5, our conclusions are drawn.

2   Background

This section briefly introduces the main concepts required to understand our work.

Note that, for the sake of simplicity, we use the words 'weight' and 'link' interchangeably. We demonstrate our approach in the context of the multilayer perceptron (MLP), a popular supervised model. MLP is a feed-forward Artificial Neural Network (ANN) composed by several hidden layers, forming a Deep Network, as shown in Fig. 1. Because of the intra-layer links flow, an MLP can be seen as a fully connected directed graph between the input and output layers.

Supervised learning involves observing several samples of a given dataset, which will be divided into 'training' and 'test' samples. While the former is used to train the neural network, the latter has the role of the litmus test, as it is

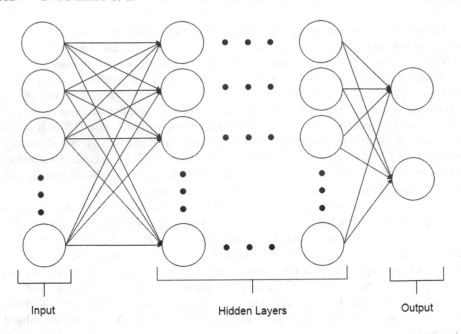

Fig. 1. Example of a generic multilayer perceptron graph with more than two hidden layers.

compared with the ANN predictions. For further details on Deep Learning, refer to [11, 7] .

The construction of a fully connected graph inevitably leads to higher computational costs, as the network grows. To overcome this issue, in our earlier SET framework [13], drawing inspiration from human brain models, we suggested to model an ANN topology as a weighted sparse Erdős-Rényi graph in which edges were randomly placed with nodes, according to a fixed probability [5,1,10].

Just like in [13], we set the edge probability as follows:

$$p\left(W_{ij}^{k}\right) = \frac{\epsilon(n^{k} + n^{k-1})}{n^{k}n^{k-1}}, \tag{1}$$

where $W^{k} \in R^{n^{k-1} \times n^{k}}$ is a sparse weight matrix between the k-th layer and the previous one, $\epsilon \in R^{+}$ is the sparsity parameter, and i, j are a pair of neurons; moreover, n^{k} is the number of neurons in the k-th layer [13].

As outlined in the previous section, this process led to forcing network sparsity. This trick was balanced by introducing the tunable revise fraction parameter ζ that has the role to confirm (or to change) the weights during the training process as explained below.

After that, the first input vector from training data to the network is presented and propagated through the ANN to obtain an output (forward propagation phase). Then, the error signal is computed by comparing the actual output to the target output (from the test set). Next, this error signal is propagated back through the network. This event is called back-propagation. During this phase, to minimise the overall error the weights are revised and adjusted. The combined stage of both one forward pass and one backward pass of all the training examples is called *epoch*. The feed-forward and back-propagation stages are repeated until the overall error is satisfactorily small [6].

At the end of each epoch, in [13] there is a weight adjustment phase. It consists of removing the closest-to-zero links in between layers plus a wider revising range (i.e. ζ). This parameter has the role to verify the correctness of the forced-to-be-zero weights.

Subsequently, new weighs are added randomly to exactly compensate the removed weights. Thanks to this procedure, the number of links between layers remains constant across different epochs, while no isolated neurons are allowed. The revise fraction, represented by the ζ parameter, is the one that we analyze herein.

3   Method

In this section, we illustrate the research questions and the strategy that we followed to achieve our goal.

To speed-up the training process, we examine the effects drawn by the variation of ζ during the evolutionary weight phase, at each epoch. We gradually reduce ζ with the aim to better understand the trade-off between speed-up and accuracy loss.

In [13] the default revise fraction was set to $\zeta = 0.3$. Yet, no further investigations on the sensitivity to ζ (in terms of accuracy and running time) was carried out. We set out to understand these effects, particularly how the revise step affects the training when ζ is substantially reduced.

Some obvious considerations of this problem are that a shorter execution time and a certain percentage of accuracy loss are expected, if we opt for smaller values of ζ. Nonetheless, this relationship is bound to be non-linear; thus, it is crucial to get to quantitative results, which is our aim herein.

Following a pragmatic approach, we have conducted our experiments using a well-known dataset, the Lung Cancer Data Set[1], which is widely used for its importance in medicine. This comprises 203 instances, and 3,312 input features, with five output classes.

Our ANN is composed of three hidden layers with 3,000 neurons per layer. The activation functions used by default are ReLu for the hidden layers, and Sigmoid for the output.

4   Results

In this section, we compare the results obtained varying the parameter ζ. We evaluate the training goodness in terms of the best balance between short execution time and high accuracy reached.

---

[1] http://featureselection.asu.edu/.

For brevity, only the most important outcomes are reported hereafter. We focused on the range of $0 \leq \zeta \leq 1$; furthermore, we enlarged the number of epochs range from the default value of 100 up to 150 with the aim to find the ending point of the transient phase. By combining these two tuning parameters we discovered that, with this dataset, the meaningful range is $0 \leq \zeta \leq 0.02$.

Figure 2 shows substantial accuracy fluctuations, but no well-defined transient phase for $\zeta > 0.02$. The default ζ value, used here as benchmark, shows an accuracy variation of more than 10% (e.g. accuracy from 82% to 97% at the 60th epoch, and the accuracy from 85% to 95% at the 140th epoch). Note that because the first 10 epochs are within the settling phase, the observations given below concern the following steps. Due to this uncertainty, and the absence of a transient phase, it is impossible to identify an absolute best stopping condition for the algorithm. For instance, at the 60th epoch an accuracy collapse from 97% to 82% was spotted, followed by an accuracy of 94% at the next step.

Considering a lower revise fraction, such as $\zeta \leq 0.02$, an improvement in stability and a loss in accuracy emerged, as was to be expected. In this scenario, defining an exit condition according to the accuracy trend over time is easier. In this case, there is no unexpected sharp decrease.

To quantify the amount of accuracy loss compared with the execution time gain, we refer to Table 1. This reports in percentage terms both the revise fraction and the accuracy, at the 150th epoch, as well as the highest accuracy reached during the whole simulation. Moreover, mean and confidence interval bounds are given.

What becomes evident is that the revise phase substantially affects execution time. The difference between the simulation with $\zeta = 1\%$ and $\zeta = 30\%$ is of 7 min approximately (29.36 vs 36.74). This is a small gap if compared with the 11 min between $\zeta = 0\%$ and $\zeta = 1\%$ (17.88 vs 31.46). Therefore, on average, the improvement achieved by using a higher revise fraction (as the default one is) has a gain of 3%.

Table 1. Evaluating parameters varying the revise fraction. From left: the revise fraction in percentage; the overall time expressed in minutes; the accuracy at the last epoch considered (i.e. 150th) in percentage; the highest accuracy reached during the simulation expressed in percentage; the accuracy mean during the simulation, and the confidence interval bounds. Notes that these last three parameters are computed after the first 10 epochs.

ζ (%)	Exec time (min)	Last Acc (%)	Max Acc (%)	Mean	Lower bound	Upper bound
30%	36.74	95.59%	97.06%	0.931164	0.926608	0.935720
2%	31.47	95.59%	95.59%	0.903630	0.900610	0.906649
1%	29.36	92.65%	94.12%	0.901648	0.898086	0.905210
0%	17.88	92.65%	94.12%	0.901856	0.897713	0.906000

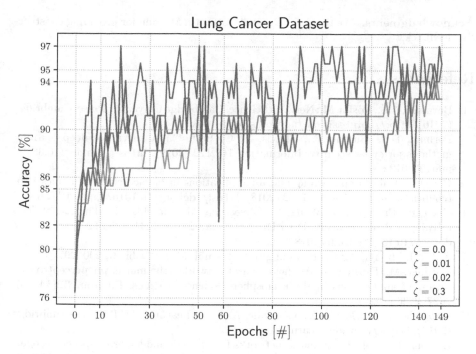

Fig. 2. Accuracy percentage over 150 epochs with $\zeta = 0\%$ in blue, $\zeta = 1\%$ in yellow, $\zeta = 2\%$ in green, and the default value $\zeta = 30\%$ in red as benchmark. (Color figure online)

5   Conclusions

In this paper, we moved a step forward from earlier work [13]. Not only our experiments confirmed the efficiency arising from training sparse neural networks. We managed to further exploit sparsity through a better tuned algorithm, obtaining a further speed up effect at a negligible accuracy loss.

Of course, the actual (quantitative) results will depend on the particular application domain, dataset and training algorithm. Yet, it is evident that network science algorithms that keep sparsity in ANNs are a promising direction and have considerable further potential. Our next priority will be to test out a broader range of training algorithms and datasets, venturing into more network science strategies.

In the future, multiple datasets will be considered in order to analyse the impact of ζ parameter on the training phase in different application domains. The results obtained will allow us to verify that our approach works well regardless of the reference scenario. The experiments will include also different distributions for the initial weight assignments, and the sparsity parameter ϵ will be examined. This last one improvement, in particular, leads to other important questions, like: *is there a critical threshold in pushing network sparsity any further?*

Acknowledgments. We thank Dr Decebal Costantin Mocanu for providing constructive feedback.

References

1. Barabási, A.L., Pósfai, M.: Network Science. Cambridge University Press, Cambridge (2016). http://barabasi.com/networksciencebook/
2. Berman, D.S., Buczak, A., Chavis, J., Corbett, C.: A survey of deep learning methods for cyber security. Information **10**, 122 (2019). https://doi.org/10.3390/info10040122
3. Cao, C., et al.: Deep learning and its applications in biomedicine. Genomics Proteomics Bioinform. **16**(1), 17–32 (2018). https://doi.org/10.1016/j.gpb.2017.07.003
4. Chen, H., Engkvist, O., Wang, Y., Olivecrona, M., Blaschke, T.: The rise of deep learning in drug discovery. Drug Discov. Today **23**(6), 1241–1250 (2018). https://doi.org/10.1016/j.drudis.2018.01.039
5. Erdös, P., Rényi, A.: On random graphs I. Publ. Math. Debr. **6**, 290–297 (1959)
6. Gardner, M., Dorling, S.: Artificial neural networks (the multilayer perceptron)–a review of applications in the atmospheric sciences. Atmos. Environ. **32**(14–15), 2627–2636 (1998)
7. Goodfellow, I.J., Bengio, Y., Courville, A.: Deep Learning. MIT Press, Cambridge (2016). http://www.deeplearningbook.org
8. Hinton, G., et al.: Deep neural networks for acoustic modeling in speech recognition: the shared views of four research groups. IEEE Signal Process. Mag. **29**(6), 82–97 (2012). https://doi.org/10.1109/MSP.2012.2205597
9. Krizhevsky, A., Sutskever, I., Hinton, G.E.: Imagenet classification with deep convolutional neural networks. Commun. ACM **60**(6), 84–90 (2017). https://doi.org/10.1145/3065386
10. Latora, V., Nicosia, V., Russo, G.: Complex Networks: Principles, Methods and Applications. Cambridge University Press, Cambridge (2017)
11. LeCun, Y., Bengio, Y., Hinton, G.: Deep learning. Nat. Cell Biol. **521**(7553), 436–444 (2015). https://doi.org/10.1038/nature14539
12. Liu, S., Mocanu, D.C., Matavalam, A., Pei, Y., Pechenizkiy, M.: Sparse evolutionary deep learning with over one million artificial neurons on commodity hardware. arXiv:1901.09181 (2019)
13. Mocanu, D.C., Mocanu, E., Stone, P., Nguyen, P., Gibescu, M., Liotta, A.: Scalable training of artificial neural networks with adaptive sparse connectivity inspired by network science. Nat. Commun. **9**, 2383 (2018). https://doi.org/10.1038/s41467-018-04316-3
14. Ruano-Ordás, D., Yevseyeva, I., Fernandes, V.B., Méndez, J.R., Emmerich, M.T.M.: Improving the drug discovery process by using multiple classifier systems. Expert Syst. Appl. **121**, 292–303 (2019). https://doi.org/10.1016/j.eswa.2018.12.032
15. Yu, D., Deng, L.: Deep learning and its applications to signal and information processing [exploratory DSP]. IEEE Signal Process. Mag. **28**(1), 145–154 (2011). https://doi.org/10.1109/MSP.2010.939038

Grossone Methodology for Lexicographic Mixed-Integer Linear Programming Problems

Marco Cococcioni[1]([⊠])[ID], Alessandro Cudazzo[1][ID], Massimo Pappalardo[1][ID], and Yaroslav D. Sergeyev[2,3][ID]

[1] University of Pisa, Pisa, Italy
{marco.cococcioni,massimo.pappalardo}@unipi.it, alessandro@cudazzo.com
[2] University of Calabria, Rende, Italy
yaro@dimes.unical.it
[3] Lobachevsky State University, Nizhni Novgorod, Russia

Abstract. In this work we have addressed lexicographic multi-objective linear programming problems where some of the variables are constrained to be integer. We have called this class of problems LMILP, which stands for Lexicographic Mixed Integer Linear Programming. Following one of the approach used to solve mixed integer linear programming problems, the branch and bound technique, we have extended it to work with infinitesimal/infinite numbers, exploiting the Grossone Methodology. The new algorithm, called GrossBB, is able to solve this new class of problems, by using internally the GrossSimplex algorithm (a recently introduced Grossone extension of the well-known simplex algorithm, to solve lexicographic LP problems without integer constraints). Finally we have illustrated the working principles of the GrossBB on a test problem.

Keywords: Multi-objective optimization · Lexicographic optimization · Mixed Integer Linear Programming · Numerical infinitesimals

1 Introduction

Engineering applications often lead to optimization problems where several objectives should be optimized. An important class of problem of this kind is the Lexicographic Mixed-Integer Linear Programming (LMILP), where the first objective is incomparably more important than the second, which, on its turn, is incomparably more important than the third one, and so on. An LMILP problem is also characterized by the fact that the feasibility domain is defined by a set of linear inequalities, with the addition of the integrality constraint on some or all the decision variables. Here the LMILP problem will be approached using a recently introduced computational methodology allowing one to work *numerically* with infinities and infinitesimals in a handy way (see for a detailed introduction surveys [11,13,14] and the book [10] written in a popular way). This computational methodology has already been successfully applied in optimization [2,3,7,8,16], game theory [4–6], and in many other fields, as

© Springer Nature Switzerland AG 2020
Y. D. Sergeyev and D. E. Kvasov (Eds.): NUMTA 2019, LNCS 11974, pp. 337–345, 2020.
https://doi.org/10.1007/978-3-030-40616-5_28

338 M. Cococcioni et al.

reported in [15]. This methodology uses a numeral system working with an infinite number called *Grossone*, expressed by the numeral ①, and introduced as the number of elements of the set of natural numbers (the non-contradictory of the methodology has been studied in [9]). This numeral system allows one to express a variety of numbers involving different infinite and infinitesimal parts and to execute operations with all of them in a unique framework.

In this work we propose a novel Branch-and-Bound (BB) algorithm using the Grossone framework to analyze the LMILP problem. The key idea is to scalarize the lexicographic objective functions using Grossone, in order to obtain a scalar function (actually, a scalar function taking Grossone-based values), following the same approach described in [2]. By removing the integrality constraints at each node, we are able to map an LMOMILP (Lexicographic Multi-Objective Mixed Integer Linear Programming) problem into an LP (Linear Programming)-like problem, which can be solved using the GrossSimplex algorithm [2]. The next step is to theoretically extend the BB algorithm to the case of function assuming Gross-scalar values. We have called this algorithm GrossBB. After providing the associated pruning and branching rules, we use this algorithm (coupled with the GrossSimplex one) to solve the LMILP problem.

2 Lexicographic Mixed-Integer Linear Programming

The LMILP problem is formalized as follow:

$$
\begin{aligned}
\text{lexmin} \quad & \mathbf{c}^{1\,T}\mathbf{x},\, \mathbf{c}^{2\,T}\mathbf{x},\, ...,\, \mathbf{c}^{r\,T}\mathbf{x} \\
\text{s.t.} \quad & \mathbf{A}\mathbf{x} \leqslant \mathbf{b}, \\
& \mathbf{x} = \begin{bmatrix} \mathbf{p} \\ \mathbf{q} \end{bmatrix} \qquad \mathbf{p} \in \mathbb{Z}^k, \qquad \mathbf{q} \in \mathbb{R}^{n-k}
\end{aligned} \qquad \mathcal{P}
$$

where $\mathbf{c}^i$, $i = 1, ..., r$, are column vectors $\in \mathbb{R}^n$, $\mathbf{x}$ is a column vector $\in \mathbb{R}^n$, $\mathbf{A}$ is a full-rank matrix $\in \mathbb{R}^{m \times n}$, $\mathbf{b}$ is a column vector $\in \mathbb{R}^m$. The notation *lexmin* in $\mathcal{P}$ denotes the *Lexicographic Minimum* and means that the first objective is much more important than the second, and so on: $\mathbf{c}^{1\,T}\mathbf{x} \gg \mathbf{c}^{2\,T}\mathbf{x} \gg ... \gg \mathbf{c}^{r\,T}\mathbf{x}$.

From problem $\mathcal{P}$, we can define the polyhedron with only linear constraints:

$$
S \equiv \{\mathbf{x} \in \mathbb{R}^n : \mathbf{A}\mathbf{x} \leqslant \mathbf{b}\}. \tag{1}
$$

and with that we can define the **relaxation** of a Lexicographic (mixed) integer linear problems arises by removing the integrality constraint of each variable:

$$
\begin{aligned}
\text{lexmin} \quad & \mathbf{c}^{1\,T}\mathbf{x},\, \mathbf{c}^{2\,T}\mathbf{x},\, ...,\, \mathbf{c}^{r\,T}\mathbf{x} \\
\text{s.t.} \quad & \mathbf{A}\mathbf{x} \leqslant \mathbf{b},
\end{aligned} \qquad \mathcal{R}
$$

$\mathcal{R}$ is called Lexicographic Multi-Objective Linear Programming (LMOLP) problem and can be solved as in [2].

Hereinafter we assume that S is bounded and non-empty. In next section we briefly introduce the Grossone approach, that will allow us to provide in Sec. 4 another formulation of problem $\mathcal{P}$, much easier to deal with.

3   The Grossone Methodology

As said before, in [10–14] a computational methodology working with an infinite unit of measure called Grossone and indicated by the numeral ① has been introduced as the number of elements of the set of natural numbers $\mathbb{N}$. On the one hand, this allows one to treat easily many problems related to the traditional set theory operating with Cantor's cardinals by computing the number of elements of infinite sets using ①-based numerals.

On the other hand, in the numeral system built upon Grossone, there is the opportunity to treat infinite and infinitesimal numbers in a unique framework and to work with all of them numerically, i.e., by executing arithmetic operations with floating-point numbers and the possibility to assign concrete infinite and infinitesimal values to variables.

A general way to express infinities and infinitesimals is also provided in [10–14] by using records similar to traditional positional number systems, but with the radix ①. A number $\tilde{c}$ in this new numeral system ($\tilde{c}$ will be called $Gross$-scalar from here on) can be constructed by subdividing it into groups of corresponding powers of ① and thus can be represented as

$$\tilde{c} = c_{p_m}①^{p_m} + \dots + c_{p_1}①^{p_1} + c_{p_0}①^{p_0} + c_{p_{-1}}①^{p_{-1}} + \dots + c_{p_{-k}}①^{p_{-k}},$$

where $m, k \in \mathbb{N}$, exponents p_i are called $Gross$-powers (they can be numbers of the type of $\tilde{c}$) with $p_0 = 0$, and $i = m, \dots, 1, 0, -1, \dots, -k$. Then, $c_{p_i} \neq 0$ called $Gross$-digits are finite (positive or negative) numbers, $i = m, \dots, 1, 0, -1, \dots, -k$. In this numeral system, finite numbers are represented by numerals with the highest $Gross$-power equal to zero, e.g., $-6.2 = -6.2①^0$.

4   LMILP Solved Using the GrossSimplex-Based GrossBB

First of all, let us show how the LMILP problem $\mathcal{P}$ can be rewritten, using Gross-numbers, in the following way:

$$
\begin{aligned}
\min \quad & \tilde{\mathbf{c}}^T\mathbf{x} \\
\text{s.t.} \quad & \mathbf{Ax} \leqslant \mathbf{b}, \\
& \mathbf{x} = \begin{bmatrix} \mathbf{p} \\ \mathbf{q} \end{bmatrix} \qquad \mathbf{p} \in \mathbb{Z}^k, \quad \mathbf{q} \in \mathbb{R}^{n-k}
\end{aligned}
\qquad \tilde{\mathcal{P}}
$$

where $\tilde{\mathbf{c}}$ is a column $Gross$-vector having n $Gross$-scalar components:

$$\tilde{\mathbf{c}} = \sum_{i=1}^{r} \mathbf{c}^i ①^{-i+1} \tag{2}$$

and $\tilde{\mathbf{c}}^T\mathbf{x}$ is the $Gross$-scalar obtained by multiplying the $Gross$-vector $\tilde{\mathbf{c}}$ by the purely finite vector $\mathbf{x}$:

$$\tilde{\mathbf{c}}^T\mathbf{x} = (\mathbf{c}^{1T}\mathbf{x})①^0 + (\mathbf{c}^{2T}\mathbf{x})①^{-1} + \dots + (\mathbf{c}^{rT}\mathbf{x})①^{-r+1}, \tag{3}$$

where (3) can be equivalently written in the extended form as:

$$\tilde{\mathbf{c}}^{\mathbf{T}}\mathbf{x} = (c_1^1 x_1 + ... + c_n^1 x_n)\textcircled{1}^0 + (c_1^2 x_1 + ... + c_n^2 x_n)\textcircled{1}^{-1} + ... + (c_1^r x_1 + ... + c_n^r x_n)\textcircled{1}^{-r+1}.$$

What makes this formulation attractive is the fact that its relaxed version (from the integrality constraint) is a Gross-LP problem [2], which can be effectively solved using a single run of the GrossSimplex algorithm. This means that the set of multiple objective functions is mapped into a single (Gross-) scalar function to be optimized. This opens the possibility to solve the integer-constrained variant of the problem using an adaptation of the BB algorithm (see Sect. 4.3), coupled with the GrossSimplex. Of course the GrossSimplex will solve problem $\tilde{R}$, the relaxed version of $\tilde{P}$:

$$\begin{aligned} \min \quad & \tilde{\mathbf{c}}^T \mathbf{x} \\ \text{s.t.} \quad & \mathbf{Ax} \leqslant \mathbf{b} \end{aligned} \qquad\qquad \tilde{R}$$

Next subsequent subsections provide the pruning and branching rules, then the GrossBB algorithm, i.e. a BB algorithm able to work with Gross-numbers.

4.1   Pruning Rules for the GrossBB

Theorem 1 (Pruning Rules for the GrossBB). *Let $\mathbf{x}_{opt}$ be the best upper bound found so far for $\tilde{P}$, and let be $\tilde{v}_S(\tilde{P}) = \tilde{\mathbf{c}}^T \mathbf{x}_{opt}$ be the current upper bound. Considering the current node $(\tilde{P}_c)$ and the associated problem $\tilde{P}_c$:*

1. *If the feasible region of problem $\tilde{P}_c$ is empty the sub-tree with root $(\tilde{P}_c)$ has no feasible solutions having values lower than $\mathbf{x}_{opt}$. So, we can prune this node.*
2. *If $\tilde{v}_I(\tilde{P}_c) \geqslant \tilde{v}_S(\tilde{P})$, then we can prune at node $(\tilde{P}_c)$, since the sub-tree with root $(\tilde{P}_c)$ cannot have feasible solutions having a value lower than $\tilde{v}_S(\tilde{P})$.*
3. *If $\tilde{v}_I(\tilde{P}_c) < \tilde{v}_S(\tilde{P})$ and the optimal solution $\bar{\mathbf{x}}$ of the relaxed problem $\tilde{R}_c$ is feasible for $\tilde{P}$, then $\bar{\mathbf{x}}$ is a better candidate solution for $\tilde{P}$, and thus we can update the value of $\mathbf{x}_{opt}$ ($\mathbf{x}_{opt} = \bar{\mathbf{x}}$) and that of the upper bound ($\tilde{v}_S(\tilde{P}) = \tilde{v}_I(\tilde{P}_c)$). Finally, prune this node, for the same reasons of the second rule.*

The correctness of the pruning rules above can be found in [1].

4.2   Branching Rule for the GrossBB

When the sub-tree below $\tilde{P}_c$ cannot be pruned (because it could contain better solutions), its sub-tree must be explored. Thus we have to branch the current node into P_l and P_r and to add these two new nodes to the tail of the queue of the sub-problems to be analyzed and solved (by the GrossSimplex).

4.3   Pseudo-Code for the GrossSimplex-Based GrossBB Algorithm

The steps of the branch and bound method for determining an optimal integer solution for a minimization model (with $\leqslant$ constraints) is summarized in Algorithm 1.

Algorithm 1. The GrossSimplex-based GrossBB Algorithm

Inputs: maxIter and a specific LMOMILP problem $|\tilde{P}|$, to be put in the root node $(\tilde{P})$
Outputs: $\mathbf{x}_{opt}$ (the optimal solution, a purely finite vector), $\tilde{f}_{opt}$ (the optimal value, a Gross-scalar)

Step 0.   Insert $|\tilde{P}|$ into a queue of the sub problems that must be solved. Put $\tilde{v}_S(\tilde{P}) = ①$, $\mathbf{x}_{opt} = [\ ]$, and $\tilde{f}_{opt} = ①$ or use a greedy algorithm to get an initial feasible solution.

Step 1a. If all the remaining leaves have been visited (empty queue), or the maximum number of iterations has been reached, or the $\tilde{\epsilon}$-optimality condition holds, then goto Step 4. Otherwise extract from the head of the queue the next problem to solve and call it $\tilde{P}_c$ (*current problem*). Remark: this policy of insertion of new problems at the tail of the queue and the extraction from its head leads to a *breadth-first* visit for the binary tree of the generated problems.

Step 1b. Solve $\tilde{R}_c$, the relaxed version of the problem $\tilde{P}_c$ at hand, using the Gross-Simplex and get $\bar{\mathbf{x}}$ and $\tilde{f}_c$ ($= \tilde{\mathbf{c}}^T \bar{\mathbf{x}}$):

$$[\bar{\mathbf{x}}, \tilde{f}_c, \texttt{emptyPolyhedron}] \leftarrow \texttt{GrossSimplex}(\tilde{R}_c)$$

Step 2a. If the LP solver has found that the polyhedron is empty, then prune the sub-tree of $(\tilde{P}_c)$ (according to Pruning Rule 1) by going to Step 1a (without branching $(\tilde{P}_c)$). Otherwise, we have found a new lower value for $\tilde{P}_c$:

$$\tilde{v}_I(\tilde{P}_c) = \tilde{f}_c$$

Step 2b. If $\tilde{v}_I(\tilde{P}_c) \geqslant \tilde{v}_S(\tilde{P})$, then prune the sub-tree under $\tilde{P}_c$ (according to Pruning Rule 2), by going to Step 1a (without branching $\tilde{P}_c$).

Step 2c. If $\tilde{v}_I(\tilde{P}_c) < \tilde{v}_S(\tilde{P})$ and all components of $\bar{\mathbf{x}}$ that must be integer are actually ϵ-integer (i.e., $\bar{\mathbf{x}}$ is feasible), then we have found a better upper bound estimate. Thus we can update the value of $\tilde{v}_S(\tilde{P})$ as:

$$\tilde{v}_S(\tilde{P}) = \tilde{v}_I(\tilde{P}_c).$$

In addition we set $\mathbf{x}_{opt} = \bar{\mathbf{x}}$ and $\tilde{f}_{opt} = \tilde{v}_I(\tilde{P}_c)$. Then we also prune the sub-tree under $(\tilde{P}_c)$ (according to Pruning Rule 3) by going to Step 1a (without branching $(\tilde{P}_c)$).

Step 3.   If $\tilde{v}_I(\tilde{P}_c) < \tilde{v}_S(\tilde{P})$ but *not* all components of $\bar{\mathbf{x}}$ that must be integer are actually ϵ-integer, we have to branch. Select the component $\bar{x}_t$ of $\bar{\mathbf{x}}$ having the greatest fractional part, among all the components that must be integer. Create two new nodes (i.e., problems) with a new constraint for this variable, one with a new $\leqslant$ constraint for the rounded down value of $\bar{x}_t$ and another with a new $\geqslant$ constraint for the rounded up value of $\bar{x}_t$. Let us call the two new problems $\tilde{P}_l$ and $\tilde{P}_r$ and put them at the tail of the queue of the problems to be solved, then goto Step 1a.

Step 4.   End of the algorithm.

---

5   A Numerical Illustration

In this work we have presented the GrossBB for lexmin problems. Clearly, using it we can also solve lexmax problems, by considering the opposite of the cost Gross-vector, i.e., by calling the lexmin-based GrossBB with $-\tilde{\mathbf{c}}$ instead of $\tilde{\mathbf{c}}$. The following lexmax-formulated problem is solved in this way.

Example: This problem is a variant to the two dimensional problem with three objectives and all integer variables described in [2] with a known solution.

$$\text{lexmax}\quad 8x_1 + 12x_2,\ 14x_1 + 10x_2,\ x_1 + x_2$$
$$\text{s.t.}\quad 2x_1 + 1x_2 \leqslant 12$$
$$2x_1 + 3x_2 \leqslant 210 + 2.5$$
$$4x_1 + 3x_2 \leqslant 270 \qquad\qquad |T|$$
$$x_1 + 2x_2 \geqslant 60$$
$$-200 \leqslant x_1, x_2 \leqslant +200,\ \mathbf{x} \in \mathbb{Z}^n$$

The polygon $\mathcal{S}$ associated to this problem is shown in Fig. 1. In black the integer points and in light grey the LP problem without integer constraints.
It can be seen that the first objective vector $\mathbf{c}^1 = [8, 12]^T$ is orthogonal to segment $[\alpha, \beta]$ ($\alpha = (0, 70.83), \beta = (28.75, 51.67)$) shown in the same figure. All the nearest integer points parallel to this segment are optimal for the first objective. Since the solution is not unique, the is the chance to try to improve the second objective vector ($\mathbf{c}^2 = [14, 10]^T$).
It is known that the optimal solution is $\bar{\mathbf{x}} = [28, 52]^T$ and $\tilde{\mathbf{c}}^T\mathbf{x} = 848①^0 + 912①^{-1} + 80①^{-2}$ is the optimal Gross-scalar.

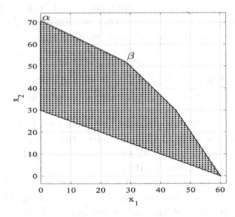

Fig. 1. This example is in two dimensions with three objectives. All the nearest integer points parallel to the segment $[\alpha, \beta]$ (there are many), are optimal for the first objective, while point $(28, 52)$ is the unique lexicographic optimum for the given problem (i.e., considering the second objective too). The third objective plays no role in this case.

Table 1. Iterations performed by GrossSimplex-based GrossBB Algorithm on test $|T|$

Iter.	result at node(iter)
Init:	- $\tilde{v_S}(\tilde{P}) = ①$
	- Queue len. 1 (add the root problem to the queue)
1	$\tilde{v_I}(\tilde{P_c})$: $-850①^0 - 919.167①^{-1} - 80.4167①^{-2}$. Queue length : 0
	- no pruning rules applied, branch $\tilde{P_c}$ in two sub-problems. Queue length: 2
	- $\tilde{\Delta} = 100①^0 + 100①^{-1} + 100①^{-2}$
2	$\tilde{v_I}(\tilde{P_c})$: $-850①^0 - 915.5①^{-1} - 80.25①^{-2}$. Queue length: 1
	- no pruning rules applied, branch $\tilde{P_c}$ in two sub-problems. Queue length: 3
	- $\tilde{\Delta} = 100①^0 + 100①^{-1} + 100①^{-2}$
3	$\tilde{v_I}(\tilde{P_c})$: $-846①^0 - 919.5①^{-1} - 80.25①^{-2}$. Queue length: 2
	- no pruning rules applied, branch $\tilde{P_c}$ in two sub-problems. Queue length: 4
	- $\tilde{\Delta} = 100①^0 + 100①^{-1} + 100①^{-2}$
4	prune node: rule 1, empty feasible region. Queue length: 3
5	$\tilde{v_I}(\tilde{P_c})$: $-850①^0 - 913.667①^{-1} - 80.1667①^{-2}$. Queue length: 2
	- no pruning rules applied, branch $\tilde{P_c}$ in two sub-problems. Queue length: 4
	- $\tilde{\Delta} = 100①^0 + 100①^{-1} + 100①^{-2}$
6	$\tilde{v_I}(\tilde{P_c})$: $-840①^0 - 920①^{-1} - 80①^{-2}$. Queue length: 3
	- A feasible solution has been found: $\mathbf{x}_{opt} = [30 \ 50]^T$
	- update $\tilde{v_S}(\tilde{P}) = \tilde{v_I}(\tilde{P_c})$, prune node: rule 3
	- $\tilde{\Delta} = 0.0119048①^0 - 0.00688406①^{-1} + 0.00208333①^{-2}$
7	$\tilde{v_I}(\tilde{P_c})$: $-844①^0 - 916①^{-1} - 80①^{-2}$. Queue length: 2
	- A feasible solution has been found: $\mathbf{x}_{opt} = [29 \ 51]^T$
	- update $\tilde{v_S}(\tilde{P}) = \tilde{v_I}(\tilde{P_c})$, prune node: rule 3
	- $\tilde{\Delta} = 0.354191①^0 - 0.127528①^{-1} + 0.104058①^{-2}$
8	$\tilde{v_I}(\tilde{P_c})$: $-850①^0 - 904.5①^{-1} - 79.75①^{-2}$. Queue length: 1
	- no pruning rules applied, branch $\tilde{P_c}$ in two sub-problems. Queue length: 3
	- $\tilde{\Delta} = 0.007109①^0 - 0.00254731①^{-1} + 0.00208333①^{-2}$
9	$\tilde{v_I}(\tilde{P_c})$: $-848①^0 - 912①^{-1} - 80①^{-2}$. Queue length: 2
	- A feasible solution has been found: $\mathbf{x}_{opt} = [28 \ 52]^T$
	- update $\tilde{v_S}(\tilde{P}) = \tilde{v_I}(\tilde{P_c})$, prune node: rule 3
	- $\tilde{\Delta} = 0.00235849①^0 - 0.00822368①^{-1} - 0.003125①^{-2}$
10	prune node: rule 1, empty feasible region. Queue length: 1
11	. $\tilde{v_I}(\tilde{P_c})$: $-850①^0 - 899①^{-1} - 79.5①^{-2}$. Queue length: 0
	- no pruning rules applied, branch $\tilde{P_c}$ in two sub-problems. Queue length: 2
	- $\tilde{\Delta} = 0.00235849①^0 - 0.0142544①^{-1} - 0.00625①^{-2}$
...	... ...
80	prune node: rule 1, empty feasible region. Queue length: 1
81	$\tilde{v_I}(\tilde{P_c})$: $-848①^0 - 714①^{-1} - 71①^{-2}$. Queue length: 0
	- $\tilde{v_I}(\tilde{P_c}) \geqslant \tilde{v_S}(\tilde{P})$ prune node: rule 2
Result	at iteration 81: optimization ended. Optimal solution found:
	$\mathbf{x}_{opt} = [28 \ 52]^T \qquad \tilde{f}_{opt} = -848①^0 - 912①^{-1} - 80①^{-2}$
	$\tilde{\Delta} = 0①^0 + 0①^{-1} + 0①^{-2}$

Table 1 provides a synthesis with the most interesting iterations performed by the GrossBB algorithm.

6   Conclusions

In this work, the LMILP problem has been formulated and a branch and bound algorithm using the powerful Grossone methodology from [10–14] has been proposed. The proposed algorithm has been called GrossBB and was coupled with the GrossSimplex solver developed in [2]. Its properties have been analyzed briefly and a numerical example showing a quite satisfactory behavior of the new method has been provided.

These preliminary results show a high potential of the GrossBB method combined with the GrossSimplex algorithm, and confirm that this research direction deserves a deeper investigation.

References

1. Cococcioni, M., Cudazzo, A., Pappalardo, M., Sergeyev, Y.D.: Solving the lexicographic mixed-integer linear programming problem using branch-and-bound and Grossone methodology. Comm. Nonlinear Sci. Numer. Simul., Submitted
2. Cococcioni, M., Pappalardo, M., Sergeyev, Y.D.: Lexicographic multi-objective linear programming using grossone methodology: theory and algorithm. Appl. Math. Comput. **318**, 298–311 (2018)
3. De Leone, R., Fasano, G., Sergeyev, Y.D.: Planar methods and grossone for the conjugate gradient breakdown in nonlinear programming. Comput. Optim. Appl. **71**, 73–93 (2018)
4. Fiaschi, L., Cococcioni, M.: Non-archimedean game theory: a numerical approach. Submitted
5. Fiaschi, L., Cococcioni, M.: Numerical asymptotic results in game theory using Sergeyev's infinity computing. Int. J. Unconventional Comput. **14**, 1–25 (2018)
6. Fiaschi, L., Cococcioni, M.: Generalizing pure and impure iterated prisoner's dilemmas to the case of infinite and infinitesimal quantities. In: Proceedings of the 3rd International Conference on "Numerical Computations: Theory and Algorithms". Springer Lecture Notes in Computer Science (2019)
7. Gaudioso, M., Giallombardo, G., Mukhametzhanov, M.S.: Numerical infinitesimals in a variable metric method for convex nonsmooth optimization. Appl. Math. Comput. **318**, 312–320 (2018)
8. Lai, L., Fiaschi, L., Cococcioni, M.: Solving mixed pareto-lexicographic multi-objective optimization problems: the case of priority chains. Submitted
9. Lolli, G.: Metamathematical investigations on the theory of grossone. Appl. Math. Comput. **255**, 3–14 (2015)
10. Sergeyev, Y.D.: Arithmetic of Infinity, 2nd edn. (2013). Orizz, Merid., CS (ed.)
11. Sergeyev, Y.D.: A new applied approach for executing computations with infinite and infinitesimal quantities. Informatica **19**(4), 567–596 (2008)
12. Sergeyev, Y.D.: Counting systems and the First Hilbert problem. Nonlinear Anal. Ser. A Theory Methods Appl. **72**(3–4), 1701–1708 (2010)

13. Sergeyev, Y.D.: Lagrange lecture: methodology of numerical computations with infinities and infinitesimals. Rend. del Seminario Matematico dell'Università e del Politecnico di Torino **68**(2), 95–113 (2010)
14. Calude, C.S., Dinneen, M.J. (eds.): UCNC 2015. LNCS, vol. 9252. Springer, Cham (2015). https://doi.org/10.1007/978-3-319-21819-9
15. Sergeyev, Y.D.: Numerical infinities and infinitesimals: methodology, applications, and repercussions on two Hilbert problems. EMS Surv. Math. Sci. **4**, 219–320 (2017)
16. Sergeyev, Y.D., Kvasov, D.E., Mukhametzhanov, M.S.: On strong homogeneity of a class of global optimization algorithms working with infinite and infinitesimal scales. Comm. Nonlinear Sci. Numer. Simul. **59**, 319–330 (2018)

Infinite Games on Finite Graphs Using Grossone

Louis D'Alotto[1,2][✉] [iD]

[1] York College, The City University of New York, Jamaica, Queens, NY 11451, USA
ldalotto@york.cuny.edu
[2] The Graduate Center, The City University of New York,
356 Fifth Avenue, New York City, NY 10016, USA

Abstract. In his seminal work, Robert McNaughton (see [1] and [7])
developed a model of infinite games played on finite graphs. This paper
presents a new model of infinite games played on finite graphs using the
Grossone paradigm. The new Grossone model provides certain advantages such as allowing for draws, which are common in board games,
and a more accurate and decisive method for determining the winner
when a game is played to infinite duration.

Keywords: Infinite games · Grossone · Finite automata

1 Introduction

This paper applies the theory of grossone (see [10, 13, 15–20]) to investigate games
of infinite duration with finitely many configurations. The games investigated
occur on finite graphs and are those with *perfect information*. That is, and
typically, a perfect information game is played on a board where a player moves
pieces subject to a given set of rules and each player knows everything important
to the game that has previously occurred.

We are all very familiar with finite board games such as *tic-tac-toe*, *chess*,
checkers, and *go*, to provide four examples. These are games of strategy, once the
specific positions are known. Of course we must exclude all games of chance and
card games where players do not reveal their hands, since these are not games
with perfect information. A board game will have a configuration (a state or a
state of play) and it must be made precise to include all information about any
situation in the game. The configuration describes the current state or stand-
ing of the game. Of significant importance, the configuration will dictate which
player is to move next. Hence, in board games, the play moves go from one player
to the other. A board game such as *tic-tac-toe* has only a very small number of
configurations. Here we can easily compute (via computer search techniques) all
the configurations and hence this game is not very interesting. However, and on

This research was supported by the following grant: PSC-CUNY Research Award:
TRADA-47-445.

Y. D. Sergeyev and D. E. Kvasov (Eds.): NUMTA 2019, LNCS 11974, pp. 346–353, 2020.
https://doi.org/10.1007/978-3-030-40616-5_29

the other hand, games of *checkers*, *chess* and *go* have an extremely large number of configurations and command a lot of attention from computer scientists and mathematicians.

Finite board games that are played to infinity may sound like science or mathematical fiction. Indeed, following the traditional Turing machine model, a computation is complete when it **halts** and produces some type of result. However when a game is played to infinity, it is implied that the game continues for an indefinite period (play continues without bound). For instance, a typical application that can be considered an infinite game is the operating system of a computer (a multiprogramming machine). The operating system has to manage multiple processes (or users on a server) without termination. When one process (or user) is satisfied, there are others waiting for system resources to be processed. Hence process-oriented theory is an application of infinite games to computer science (see [1]).

2   The Infinite Unit Axiom and Grossone

Applying the following new paradigm facilitates us to better understand the notion of infinite games on graphs. The problem of better understanding the notion of computing with infinity was approached beginning in 2003 by Yaroslav Sergeyev (see [15–18]). In these works, a new unit of measure on the set of natural numbers, $\mathbb{N}$ is defined. Thus, the following axiom evolves the idea of the infinite unit.

Axiom 1. *Infinite Unit Axiom. The number of elements in the set $\mathbb{N}$ of natural numbers is equal to the infinite unit denoted as ① and called grossone.*

The following properties are part of the *Infinite Unit Axiom*:

1. Infinity: For any finite natural number n, $n < ①$.
2. Identity: The following relationships hold and are extended from the usual identity relationships of the natural numbers:

$$0 \cdot ① = ① \cdot 0 = 0 \qquad\qquad ① - ① = 0$$

$$\frac{①}{①} = 1 \qquad ①^0 = 1 \qquad 1^① = 1$$

3. Divisibility: For any finite natural number n, the numbers

$$①, \frac{①}{2}, \frac{①}{3}, \frac{①}{4}, ..., \frac{①}{n}, ...$$

are defined as the number of elements in the nth part of $\mathbb{N}^1$

---

[1] In [15], Sergeyev formally presents the divisibility axiom as saying for any finite natural number n sets $\mathbb{N}_{k,n}$, $1 \le k \le n$, being the nth parts of the set $\mathbb{N}$, have the same number of elements indicated by the numeral $\frac{①}{n}$ where

$$\mathbb{N}_{k,n} = \{k, k+n, k+2n, k+3n, ...\}, \ 1 \le k \le n, \ \bigcup_{k=1}^{n} \mathbb{N}_{k,n} = \mathbb{N}.$$

The divisibility property will be of significant importance in determining a winner of an infinite game. Indeed, determining a winner will result by counting the number of elements in a sequence. It is important to mention, with the introduction of the Infinite Unit Axiom and grossone, ①, we list the natural numbers as

$$\mathbb{N} = \{1, 2, 3, 4, ..., ① - 2, ① - 1, ①\}$$

and as a consequence of this new paradigm, we have the following important theorem.

Theorem 1. *The number of elements of any infinite sequence is less or equal to* ①.

Proof. See [16] or [20]. ∎

Recently there has been a large amount of research activity on the logical theory and applications of grossone. To name a few, see [2–6, 8, 10–14, 20, 21]. This next section will describe a new application of grossone to infinite games.

3   Infinite Games

Formally, an *infinite graph game* is defined on a finite bipartite directed graph whose set, Q, of vertices are partitioned into two sets: **R**, the set of vertices from which player Red moves, and **B**, the set of vertices from which player Blue moves. The game has a place marker which is moved from vertex to vertex along the directed edges. The place marker signifies the progress of the play. When the marker is on a vertex of **R**, it is Red's move to move to a vertex in **B**. When the marker is on a vertex of **B**, it is Blue's turn to move to a vertex of set **R** and the play continues in this fashion.

Definition 1. *An infinite game, G, is a 6-tuple*

$$G = (Q, B, R, E, W(B), W(R))$$

where,

1. *Q is the finite set of positions (vertices).*
2. *B and R are subsets of Q, such that $B \cup R = Q$ and $B \cap R = \emptyset$*
3. *E is a set of directed edges between B and R such that:*
   (a) for each $b \in B$ there exists $r \in R$ such that $(b, r) \in E$.
   (b) for each $r \in R$ there exists $b \in B$ such that $(r, b) \in E$.
4. *W(B) is called the winning set for Blue.*
5. *W(R) is called the winning set for Red.*
6. *$W(B) \cap W(R) = \emptyset$.*

At this time it should be noted that the winning sets for each player are not limited to vertices of the player's color.

Definition 2. *A play that begins from position q is a complete[2] infinite sequence* $p = q_1q_2q_3q_4...q_{①-1}q_①$ *such that* $q = q_1$ *and* $(q_i, q_{i+1}) \in E$, $\forall i \in \mathbb{N}$, E *is the edge relation.*

Hence a play is a sequence of states of the game. That is,

$$p : \mathbb{N} \to Q$$

To determine how a player can win, let p be a play and consider the set of all vertices that occur infinitely often. We now have the following definition.

Definition 3. $In(p)$ *is the set of vertices, in play p, that occur infinitely often, called the* **infinity set** *of p.*

We now have the following cases to determine a win:

1. $W(B) \subset In(p)$ and $W(R) \not\subset In(p)$, then Blue wins.
2. $W(B) \not\subset In(p)$ and $W(R) \subset In(p)$, then Red wins.
3. $W(B) \not\subset In(p)$ and $W(R) \not\subset In(p)$, then Draw.
4. $W(B) \subset In(p)$ and $W(R) \subset In(p)$, then the frequencies of occurrence of the elements in each set must be considered; the player with the higher frequency wins.

Cases 1 and 2 above are the result that whatever winning set a player chooses, all vertices must occur infinitely often for a player to have a chance of winning (this concept is consistent with the ideology presented in [19]). All vertices must occur infinitely often also prevents a player from choosing too many vertices for their winning set[3]. Next we look at a simple example to analyze the situation when a player chooses the empty set.

Example 1. *Suppose Blue chooses $\emptyset$ as their winning set (this is consistent with the premise that no choice is also a choice). That is, $W(B) = \emptyset$. The reason for Blue's choice is clear. $\emptyset \subset In(p)$, hence Blue is hoping that $W(R) \not\subset In(p)$ and Blue wins the game (the same can be true for Red, if Red chooses the empty set). Of course the situation can arise if both players choose $\emptyset$. In that case, the game will result in a draw. However, to show this we first need to define more machinery.*

It is necessary to define a frequency function to count the number of occurrences of a given vertex in a play sequence. This gives rise to the next two definitions.

Definition 4. *Given $Q = \{q_1, q_2, ..., q_n\}$ is the finite set of states and let D be a subset of Q. Let p be an infinite sequence of states, from a play, define a new sequence by the function*

$$\psi_{D,p} : \mathbb{N} \to \{0, 1\}$$

---

[2] Here we use the notion of complete taken from [15], that is the sequence containing ① elements is complete.
[3] It is noted here that, as is usual, the $\subset$ symbol can also imply equality.

where,

$$\psi_{D,p}(i) = \begin{cases} 1 & if \ p(i) \in D \\ 0 & otherwise \end{cases}$$

Definition 5. *Define the frequency function, $freq_p$, as*

$$freq_p(D) = \sum_{i=1}^{①} \psi_{D,p}(i).$$

These definitions are in general, however here they are applied to the winning sets for Blue and Red, respectively $W(B)$ and $W(R)$.

   With the previous definitions, if both winning sets are subsets of the infinity set (the elements of both player's winning sets occur infinitely often) a winner can be determined. If the frequency of the elements in $W(B)$ is greater than the frequency of the elements in $W(R)$, then Blue is the winner. If the frequency of the elements in $W(R)$ is greater than the frequency of the elements in $W(B)$, then Red is the winner. If the frequencies are equal, then a draw results. This is a key advancement as a result of the grossone theory. As an immediate consequence from the above definitions, the following propositions are true.

Proposition 1. *For any sequence p, $freq_p(\emptyset) = 0$.*

Proof. $p(i) \notin \emptyset \ \forall i \in \mathbb{N}$. Hence $\psi_{\emptyset,p}(i) = 0 \ \forall i \in \mathbb{N}$ and $freq_p(\emptyset) = 0$.

Proposition 2. *If both players choose the empty set as their winning set, then the game is a draw.*

Proof. By Proposition 1, $freq_p(W(B)) = freq_p(W(R)) = freq_p(\emptyset) = 0$.

4   Examples and Results

Example 2. *Referring to the game in Fig. 1. Assume that $W(B) = \{b1\}$ and $W(R) = \{r1\}$. Then Blue is always the winner, no matter where the game begins. If $W(B) = \{b1\}$ and if $W(R) = \{r1, r2\}$, then Blue's winning strategy would be to move to either r1 or r2 finitely many times and the other infinitely times. Therefore $W(R) \not\subset In(p)$.*

   For instance, if the following sequence is played

$$p = r1, b1, r2, b1, r1, b1, r2, b1, r1, b1, r1, b1, r1, b1, r1, \dots$$

then $In(p) = \{b1, r1\}$ and $W(R) \not\subset In(p)$, however $W(B) \subset In(p)$, which implies Blue wins the game.

   The following theorem and corollaries provide a better understanding of the frequency function.

Theorem 2. *For any set A and play p, $freq_p(A) \leq ①$.*

Proof. This follows directly from the properties of ① and Theorem 1.

Corollary 1. *For any game, the frequency of occurrence of any single vertex is* $\leq$ ①$/2$.

Proof. Follows from Theorem 2 and the definition of a game, since there are two players.

Corollary 2. *For any game where Q is the set of vertices,* $freq_p(Q) =$ ①.

Proof. Using the premise of a complete sequence, the corollary directly follows from Theorems 1 and 2.

Example 3. *Again, referring to Fig. 1, if $W(B) = \{r1\}$ and $W(R) = \{r2\}$ (as mentioned previously, a player does not have to choose their color as their winning set) then Blue wins the game. The winning strategy for Blue consists of moving to r2 finitely many times. Actually Blue can move to r2 infinitely many times, however it must be less than* ①$/4$ *times.*

This next example will illustrate this new application of the grossone paradigm to infinite games.

Example 4. *Referring to the game in Fig. 1, suppose the play goes as follows:*

$$r2\ skip$$
$$p = r2, b1, r1, b1, r2, b1, r1, b1, \overbrace{r1}, b1, r1, b1, r2, b1, r1, b1, r2, b1, r1, b1, \ldots$$

Here the $In(p) = \{b1, r1, r2\}$. Hence, the frequency of occurrence for each vertex in the $In(p)$ is:

$$freq(\{b1\}) = ①/2 \quad freq(\{r1\}) = ①/4 + 1 \quad freq(\{r2\}) = ①/4 - 1$$

Using the same winning sets for Red and Blue as in Example 3, namely $W(B) = \{r1\}$ and $W(R) = \{r2\}$, Blue wins the game.

Fig. 1. A game (Color figure online)

Example 5. *In Fig. 2, if Blue chooses b4, that is $W(B) = \{b4\}$, a strategy for Red would be to choose $\emptyset$. Then from r3, Red can always move to b3 an infinite number of times or move to b4 a finite number of times.*

Example 6. *Referring again to Fig. 2, if each node is visited once in the 6 node outside cycle, that is via edges $(r1, b3)$, $(b3, r3)$, $(r3, b4)$, $(b4, r2)$, $(r2, b1)$, $(b1, r1)$, then the frequency of each vertex occurrence is* ①$/6$. *The sequence that will ensure this is:*

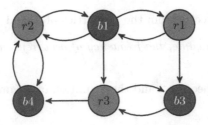

Fig. 2. A more complex game (Color figure online)

$$p = r1, b3, r3, b4, r2, b1, r1, b3, r3, b4, r2, b1, r1, b3...$$

If player Blue chooses their winning sets $W(B) = \{b1, b3\}$, then Red can choose $W(R) = \{r2, r3\}$ and Red has a winning strategy. When Blue lands on vertex b1, Blue must move to r1 to get to b3 (part of Blue's winning set). The play continues and can follow the outside cycle. However, at some point, Red moves from r2 back to b4 a finite number of times. For instance, a play can follow:

$$p = r1, b3, r3, b4, r2, b4, r2, b4, r2, b4, r2, b1, r1, b3, r3, b4, r2, b1, r1, ...$$

hence

$$freq_p(\{b1, b3\}) = \tfrac{①}{3} - 2 \quad freq_p(\{r2, r3\}) = \tfrac{①}{3} + 1$$

and Red wins the game.

5   Conclusion

This paper has presented a new model of infinite games played on finite graphs by applying the theory of grossone and the Infinite Unit Axiom. In his original work, McNaughton (see [1]) presented and developed a model of infinite games played on finite graphs using traditional methods of dealing with infinity. This paper has extended that work to count the number of times vertices in a board game are visited, although vertices can5 be visited an infinite number of times. Indeed, two players choose their winning sets and the player whose winning set is visited more frequently wins the game. With this new paradigm, as is common in the usual finite duration board games (chess, checkers, go), a draw can result. This was not the case in McNaughton's original work. Hence a more finer decision process is used in determining the winner or draw.

References

1. McNaughton, R.: Infinite games played on finite graphs. Ann. Pure Appl. Logic **65**, 149–184 (1993)
2. Caldarola, F.: The exact measures of the Sierpiński d-dimensional tetrahedron in connection with a Diophantine nonlinear system. Commun. Nonlinear Sci. Numer. Simul. **63**, 228–238 (2018)

3. De Leone, R., Fasano, G., Sergeyev, Ya.D.: Planar methods and grossone for the Conjugate Gradient breakdown in nonlinear programming. Comput. Optim. Appl. **71**(1), 73–93 (2018)
4. De Leone, R.: Nonlinear programming and grossone: quadratic programming and the role of constraint qualifications. Appl. Math. Comput. **318**, 290–297 (2018)
5. D'Alotto, L.: A classification of one-dimensional cellular automata using infinite computations. Appl. Math. Comput. **255**, 15–24 (2015)
6. Fiaschi, L., Cococcioni, M.: Numerical asymptotic results in Game Theory using Sergeyev's Infinity Computing. Int. J. Unconv. Comput. **14**(1), 1–25 (2018)
7. Khoussainov, B., Nerode, A.: Automata Theory and Its Applications. Birkhauser, Basel (2001)
8. Iudin, D.I., Sergeyev, Ya.D., Hayakawa, M.: Infinity computations in cellular automaton forest-fire model. Commun. Nonlinear Sci. Numer. Simul. **20**(3), 861–870 (2015)
9. Lolli, G.: Infinitesimals and infinites in the history of mathematics: a brief survey. Appl. Math. Comput. **218**(16), 7979–7988 (2012)
10. Lolli, G.: Metamathematical investigations on the theory of Grossone. Appl. Math. Comput. **255**, 3–14 (2015)
11. Margenstern, M.: Using Grossone to count the number of elements of infinite sets and the connection with bijections. p-Adic Numbers Ultrametric Anal. Appl. **3**(3), 196–204 (2011)
12. Montagna, F., Simi, G., Sorbi, A.: Taking the Pirahá seriously. Commun. Nonlinear Sci. Numer. Simul. **21**(1–3), 52–69 (2015)
13. Rizza, D.: Numerical methods for infinite decision-making processes. Int. J. Unconv. Comput. **14**(2), 139–158 (2019)
14. Rizza, D.: How to make an infinite decision. Bull. Symb. Logic **24**(2), 227 (2018)
15. Sergeyev, Y.D.: Arithmetic of Infinity. Edizioni Orizzonti Meridionali, Cosenza (2003)
16. Sergeyev, Y.D.: A new applied approach for executing computations with infinite and infinitesimal quantities. Informatica **19**(4), 567–596 (2008)
17. Cococcioni, M., Pappalardo, M., Sergeyev, Y.D.: Lexicographic multiobjective linear programming using grossone methodology: theory and algorithm. Appl. Math. Comput. **318**, 298–311 (2018)
18. Sergeyev, Y.D.: Computations with grossone-based infinities. In: Calude, C.S., Dinneen, M.J. (eds.) UCNC 2015. LNCS, vol. 9252, pp. 89–106. Springer, Cham (2015). https://doi.org/10.1007/978-3-319-21819-9_6
19. Sergeyev, Ya.D.: The Olympic medals ranks, lexicographic ordering and numerical infinities. Math. Intelligencer **37**(2), 4–8 (2015)
20. Sergeyev, Ya.D.: Numerical infinities and infinitesimals: methodology, applications, and repercussions on two Hilbert problems. EMS Surv. Math. Sci. **4**(2), 219–320 (2017)
21. Sergeyev, Ya.D., Garro, A.: Observability of Turing Machines: a refinement of the theory of computation. Informatica **21**(3), 425–454 (2010)
22. Zhigljavsky, A.: Computing sums of conditionally convergent and divergent series using the concept of grossone. Appl. Math. Comput. **218**, 8064–8076 (2012)

A Novel Geometric Approach to the Problem of Multidimensional Scaling

Gintautas Dzemyda[(✉)](ID) and Martynas Sabaliauskas(ID)

Vilnius University Institute of Data Science and Digital Technologies,
Akademijos Street 4, 08412 Vilnius, Lithuania
{gintautas.dzemyda,martynas.sabaliauskas}@mii.vu.lt

Abstract. Multidimensional scaling (MDS) is one of the most popular methods for a visual representation of multidimensional data. A novel geometric interpretation of the stress function and multidimensional scaling in general (Geometric MDS) has been proposed. Following this interpretation, the step size and direction forward the minimum of the stress function are found analytically for a separate point without reference to the analytical expression of the stress function, numerical evaluation of its derivatives and the linear search. It is proved theoretically that the direction coincides with the steepest descent direction, and the analytically found step size guarantees the decrease of stress in this direction. A strategy of application of the discovered option to minimize the stress function is presented and examined. It is compared with SMACOF version of MDS. The novel geometric approach will allow developing a new class of algorithms to minimize MDS stress, including global optimization and high-performance computing.

Keywords: Multidimensional scaling · Geometric approach · Minimization · Analytical derivatives · Analytical step size · Geometric MDS

1 Introduction

Recent approaches to minimize the stress in multidimensional scaling (MDS) suggest wide possibilities for dimensionality reduction [1,2]. Recently, it finds applications of various nature: face recognition [3], analysis of regional economic development [4], image graininess characterization [5].

Suppose, we have a set $X = \{X_i = (x_{i1}, \ldots, x_{in}), \ i = 1, \ldots, m\}$ of n-dimensional data points (observations) $X_i \in \mathbb{R}^n$, $n \geqslant 3$.

Dimensionality reduction and visualization requires estimating the coordinates of new points $Y_i = (y_{i1}, \ldots, y_{id})$, $i = 1, \ldots, m$, in a lower-dimensional space $(d < n)$ by holding proximities δ_{ij} between multidimensional points X_i and X_j, $i, j = 1, \ldots, m$, as much as possible. Proximity δ_{ij} can be measured e.g. by the distance between X_i and X_j.

© Springer Nature Switzerland AG 2020
Y. D. Sergeyev and D. E. Kvasov (Eds.): NUMTA 2019, LNCS 11974, pp. 354–361, 2020.
https://doi.org/10.1007/978-3-030-40616-5_30

The input data for MDS consists of the symmetric $m \times m$ matrix $\mathbf{D} = \{d_{ij}, i, j = 1, \ldots, m\}$ of proximities between pairs of points X_i and X_j. If the Minkowski distance is used as the proximity, then

$$d_{ij} = \left(\sum_{k=1}^{n} |x_{ik} - x_{jk}|^q \right)^{\frac{1}{q}}, \quad 1 \leqslant i, j \leqslant m. \tag{1}$$

If $q = 1$, then (1) defines the city-block or Manhattan distance. If $q = 2$, (1) becomes the Euclidean distance.

MDS finds the coordinates of new points Y_i representing X_i in a lower-dimensional space $\mathbb{R}^d$ by minimizing the multimodal stress function. Consider the raw stress function [6]:

$$S(Y_1, \ldots, Y_m) = \sum_{i=1}^{m} \sum_{j=i+1}^{m} (d_{ij} - d_{ij}^*)^2, \tag{2}$$

where d_{ij}^* is the Euclidean distance between points Y_i and Y_j in a lower dimensional space. In (2), other proximities may be used as well. The MDS-based dimensionality reduction optimization problem may be formulated as follows:

$$\min_{Y_1, \ldots, Y_m \in \mathbb{R}^d} S(Y_1, \ldots, Y_m). \tag{3}$$

In case $1 \leqslant d < n$, the stress function has many local minima, often. The optimization problem (3) can be solved using well-known descent methods, e.g. Quasi-Newton or conjugate gradient methods [7]. However, these algorithms cannot guarantee to find a global minimum.

Various attempts to find the global minimum are suggested. However, they are computational expensive and do not guarantee to find the global minimum, too. This lead to the conclusion that the classical approaches [8–10] to minimize the stress reached their limits in this sense. New viewpoint to the problem is necessary, including its formulation and ways of solving.

In this paper, a novel geometric interpretation of the stress function and multidimensional scaling has been proposed. It will allow developing a new class of algorithms to minimize MDS stress, including global optimization and high-performance computing. Denote this approach by Geometric MDS.

2   The Geometric Approach – Geometric MDS

A new approach, Geometric MDS, has been developed to minimize the stress function (2). Suppose, we have $m \times m$ matrix $\mathbf{D} = \{d_{ij}, i, j = 1, \ldots, m\}$ of proximities (e.g. distances) between n-dimensional points $X_i = (x_{i1}, \ldots, x_{in})$, $i = 1, \ldots, m$. We aim to find two-dimensional points $Y_i = (y_{i1}, \ldots, y_{id})$, $i = 1, \ldots, m$ by solving (3).

At first, let's have some initial configuration of points $Y_1, \ldots, Y_m$. Then, let's optimize the position of the particular point Y_j when the position of remaining

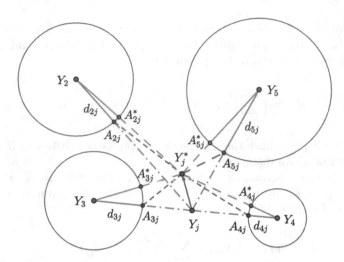

Fig. 1. An example of a single step of geometric method.

points $Y_1, \ldots, Y_{j-1}, Y_{j+1}, \ldots, Y_m$ is fixed. In this case, we tend to minimize $S(\cdot)$ in (3) by minimizing the so-called local stress function $S^*(\cdot)$ depending on Y_j, only:

$$S^*(Y_j) = \sum_{\substack{i=1 \\ i \neq j}}^{m} \left(d_{ij} - \sqrt{\sum_{k=1}^{d} (y_{ik} - y_{jk})^2} \right)^2 . \tag{4}$$

Figure 1 illustrates an example, where $m = 5$, $d = 2$. The location of points $Y_1, \ldots, Y_m$ and proximities d_{ij}, $i, j = 1, \ldots, m$ between points $X_1, \ldots, X_m$ are chosen such for better illustration of the idea. Position of point Y_1 is optimized. Y_1 is denoted by Y_j in Fig. 1 seeking for the better correspondence with notations in (4). In the centre of each circle, we have a corresponding point Y_i. Radius of the i-th circle is equal to the proximity d_{ij} between the points X_i and X_j in n-dimensional space. Point A_{ij} lies on the line between Y_i and Y_j, $i \neq j$, i.e. vectors $\overrightarrow{Y_i A_{ij}}$ and $\overrightarrow{A_{ij} Y_j}$ are collinear. Denote a new position of Y_j by Y_j^*. Let Y_j^* be chosen so that

(a) vectors $\overrightarrow{Y_i A_{ij}^*}$ and $\overrightarrow{A_{ij}^* Y_j^*}$ are collinear, $i \neq j$, \hfill (5)

$$\text{(b) } Y_j^* = \frac{1}{m-1} \sum_{\substack{i=1 \\ i \neq j}}^{m} A_{ij}. \tag{6}$$

We will analyse the value of the local stress function $S^*(Y_j^*)$ and compare it with the value $S^*(Y_j)$. According to (6), Y_j^* is an average point of the points A_{ij} over $i = 1 \ldots m, i \neq j$. According to (5), when we make a step from Y_j to Y_j^*, we get new intersection points A_{ij}^* on circles that correspond to Y_j, and these points are on the line between Y_i and Y_j^*.

Proposition 1. *The gradient of local stress function $S^*(\cdot)$ is as follows:*

$$\nabla S^*|_{Y_j} = \left(2 \sum_{\substack{i=1 \\ i \neq j}}^{m} \frac{d_{ij} - \sqrt{\sum_{l=1}^{d} (y_{il} - y_{jl})^2}}{\sqrt{\sum_{l=1}^{d} (y_{il} - y_{jl})^2}} (y_{ik} - y_{jk}), \ k = 1, \ldots, d \right).$$

The proof follows from (4) by differentiating $S^*(\cdot)$.

Proposition 2. *The step direction from Y_j to Y_j^* corresponds to the anti-gradient of the function $S^*(\cdot)$ at the point Y_j:*

$$Y_j^* = Y_j - \frac{1}{2(m-1)} \nabla S^*|_{Y_j}. \tag{7}$$

Proof

$$Y_j^* - Y_j = \left(\frac{1}{m-1} \sum_{\substack{i=1 \\ i \neq j}}^{m} \left(\frac{d_{ij}(y_{jk} - y_{ik})}{\sqrt{\sum_{l=1}^{d} (y_{il} - y_{jl})^2}} + y_{ik} - y_{jk} \right), \ k = 1, \ldots, d \right)$$

$$= \left(-\frac{1}{2(m-1)} 2 \sum_{\substack{i=1 \\ i \neq j}}^{m} \frac{d_{ij} - \sqrt{\sum_{l=1}^{d} (y_{il} - y_{jl})^2}}{\sqrt{\sum_{l=1}^{d} (y_{il} - y_{jl})^2}} (y_{ik} - y_{jk}), \ k = 1, \ldots, d \right)$$

$$= -\frac{\nabla S^*|_{Y_j}}{2(m-1)}. \quad \square$$

Proposition 3. *Size of a step from Y_j to Y_j^* is equal to*

$$\frac{\|\nabla S^*|_{Y_j}\|}{2(m-1)} = \frac{1}{m-1} \sqrt{\sum_{k=1}^{d} \left(\sum_{\substack{i=1 \\ i \neq j}}^{m} \frac{d_{ij} - \sqrt{\sum_{l=1}^{d} (y_{il} - y_{jl})^2}}{\sqrt{\sum_{l=1}^{d} (y_{il} - y_{jl})^2}} (y_{ik} - y_{jk}) \right)^2 }.$$

Proposition 4. *Let Y_j does not match to any local extreme point of the function $S^*(\cdot)$. If Y_j^* is chosen by (6), then a single step from Y_j to Y_j^* reduces a local stress $S^*(\cdot)$:*

$$S^*(Y_j^*) < S^*(Y_j).$$

Proof. Let's have following functions:

$$S^*(Y_j) = \sum_{\substack{i=1 \\ i \neq j}}^{m} d^2(A_{ij}, Y_j), \tag{8}$$

$$S_A^*(Y_j^*) = \sum_{\substack{i=1 \\ i \neq j}}^{m} d^2(A_{ij}, Y_j^*), \ S^*(Y_j^*) = \sum_{\substack{i=1 \\ i \neq j}}^{m} d^2(A_{ij}^*, Y_j^*). \tag{9}$$

where $d(\cdot, \cdot)$ is the Euclidean distance between two points.

Figure 1 illustrates a case, where position of point Y_j is optimized to Y_j^*. It is enough to show that

$$S^*(Y_j^*) < S_A^*(Y_j^*) < S^*(Y_j).$$

Firstly, we show that $S_A^*(Y_j^*) < S^*(Y_j)$. Define $A_{ij} = (a_{ij1}, \ldots, a_{ijd})$. From (8), it follows that the gradient of $S^*(Y_j)$ is equal to

$$\nabla S^*(Y_j) = \Big(\sum_{\substack{i=1 \\ i \neq j}}^{m} 2(a_{ijk} - y_{jk}), \ k = 1, \ldots, d \Big).$$

At the local minimum Y_j of function $S^*(Y_j)$, the condition $\nabla S^*(Y_j) = (0, \ldots, 0)$ is valid, and then we have a unique solution of Y_j:

$$(m-1)y_{jk} - \sum_{\substack{i=1 \\ i \neq j}}^{m} a_{ijk} = 0, \ k = 1, \ldots, d \implies (m-1)Y_j - \sum_{\substack{i=1 \\ i \neq j}}^{m} A_{ij} = 0.$$

We see that the solution is defined as Y_j^*, which is given in (6). Such Y_j^* corresponds to minimized local stress $S_A^*(Y_j^*)$. Therefore, $S_A^*(Y_j^*) < S^*(Y_j)$.
For the proof that $S^*(Y_j^*) < S_A^*(Y_j^*)$, it is enough to show that

$$d(Y_j^*, A_{ij}^*) < d(Y_j^*, A_{ij}), \quad i = 1, \ldots, m, \quad i \neq j.$$

Using the triangle inequality, we have a valid condition

$$d(Y_i^*, Y_j^*) = d(Y_i^*, A_{ij}^*) + d(A_{ij}^*, Y_j^*) < d(Y_i^*, A_{ij}) + d(A_{ij}, Y_j^*).$$

Since the radius of the i-th circle satisfies condition $d(Y_i^*, A_{ij}^*) = d(Y_i^*, A_{ij})$, then $d(Y_j^*, A_{ij}^*) < d(Y_j^*, A_{ij})$. $\quad\square$

Proposition 5. *The value of the local stress function $S^*(\cdot)$ (4) will converge to a local minimum when repeating steps (7) and $Y_j := Y_j^*$.*

Proposition 6. *Let Y_j does not match to any local extreme point of the function $S^*(\cdot)$. Movement of any projected point by the geometric method reduces the stress (2) of MDS: if Y_j^* is chosen by (6), then the stress function $S(\cdot)$, defined by (2), decreases:*

$$S(Y_1, \ldots, Y_{j-1}, Y_j^*, Y_{j+1}, \ldots, Y_m) < S(Y_1, \ldots, Y_{j-1}, Y_j, Y_{j+1}, \ldots, Y_m).$$

Proof. Before the step from Y_j to Y_j^*, we have following stress function

$$S(Y_1, \ldots, Y_{j-1}, Y_j, Y_{j+1}, \ldots, Y_m) = S^*(Y_j) + \sum_{\substack{i=1 \\ i \neq j}}^{m} \sum_{\substack{k=i+1 \\ k \neq j}}^{m} (d_{ik} - d_{ik}^*)^2.$$

Since $S^*(Y_j^*) < S^*(Y_j)$ and $\sum_{\substack{i=1 \\ i \neq j}}^{m} \sum_{\substack{k=i+1 \\ k \neq j}}^{m} (d_{ik} - d_{ik}^*)^2$ remain constant after the step, the stress function $S(\cdot)$ is reduced after the step. $\quad\square$

3 Multimodality of the Local Stress Function of Geometric MDS

Proposition 7. *Function* $f(\delta) = S^* \left(Y_j - \delta \frac{\nabla S^*|_{Y_j}}{\|\nabla S^*|_{Y_j}\|} \right)$ *is not unimodal, where* δ *is a step size.*

Proof. Consider a dataset X of six five-dimensional points and their Euclidean distances as proximities:

$X_1 = (3.142, 2.718, 1.618, 1.202, 0.2078)$, $X_2 = (16.462, 2.718, 1.618, 1.202, 0.2078)$, $X_3 = (3.142, 7.648, 1.618, 1.202, 0.2078)$, $X_4 = (3.142, 2.718, 4.818, 1.202, 0.2078)$, $X_5 = (3.142, 2.718, 1.618, 4.952, 0.2078)$, $X_6 = (3.142, 2.718, 1.618, 1.202, 4.0278)$.

Let the values of Y $(d = 2)$ be such:

$Y_1 = (18.723, -1.880)$, $Y_2 = (19.025, 6.247)$, $Y_3 = (12.147, 11.208)$, $Y_4 = (11.338, 2.585)$, $Y_5 = (3.909, 3.546)$, $Y_6 = (10.560, -4.654)$. Consider point Y_1 for its moving to a new position Y_1^* according to the anti-gradient direction by (7). See Fig. 2 for details. The local stress function reaches its two different local minima depending on the step δ. □

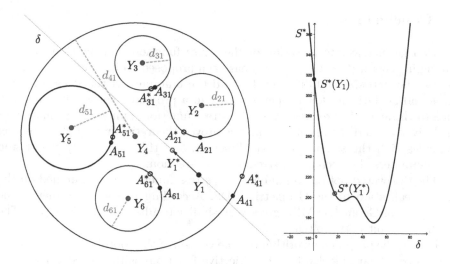

Fig. 2. Example of the anti-gradient search

4 Experiments with Geometric MDS

Simple realizations of Geometric MDS are based on fixing some initial positions of points $Y_i = (y_{i1}, \ldots, y_{id})$, $i = 1, \ldots, m$ (at random, using principal component analysis, etc.), and further changing the positions of Y_j (once by (7) or multistep descent by several steps using (7)) in consecutive order from $j = 1$ to $j = m$

many times till some stop condition is met: e.g. number of runs from $j = 1$ to $j = m$ reaches some limit or the decrease of stress function $S(\cdot)$ becomes less than some small constant after two consecutive runs.

In the experiments, minimization of stress $S(\cdot)$ was performed by consecutive one-step (not multistep) changing of positions of points $Y_1, \ldots, Y_m$ many times. 1000 random sets X of 30 points ($m = 30$) were generated inside the 4-dimensional unit hypercube ($n = 4$) and represented in $d = 2$ and $d = 3$ spaces. For comparison, the same data sets were analysed by multidimensional scaling based on stress $S(\cdot)$ minimization using majorization (SMACOF) that is realized in R [11,12]. Both Geometric MDS and SMACOF used the same initial values of points $Y_1, \ldots, Y_m$ obtained by Torgerson Scaling [13] realized in R [14].

When $d = 2$, Geometric MDS and SMACOF gave the same results (stress values) in 997 cases, however the average value of $S(\cdot)$ is obtained a bit better by Geometric MDS and equals 13.7570 as compared with 13.7613 by SMACOF. When $d = 3$, Geometric MDS gave the same results in 922 cases. Average values of $S(\cdot)$ are almost the same: 2.9789 (Geometric MDS) and 2.9787 (SMACOF). These preliminary results are very promising, because the evaluated efficiency of the Geometric MDS and the SMACOF is the same, however Geometric MDS is much easier realizable and interpreted.

5   Conclusions

A novel geometric interpretation of the stress function and multidimensional scaling in general (Geometric MDS) has been proposed. Following this interpretation, the step size and direction forward the minimum of the stress function are found analytically for a separate point in a projected space without reference to the analytical expression of the stress function, numerical evaluation of its derivatives and the linear search. It is proved theoretically that the direction coincides with the steepest descent direction, and the analytically found step size guarantees the decrease of stress in this direction.

The discovered option to minimize the stress function was examined on the simple realization of the Geometric MDS. According to the experiments, the realization of Geometric MDS gives very similar results as SMACOF [11]. The results are a bit better often.

In fact, the proposed algorithm is some version of the coordinate-wise descent using d-coordinate blocks. For the objective functions with curved valleys, the convergence of those algorithms normally is slow. However, the geometric approach guarantees the decrease of stress in every step, where the direction and size of the step is determined analytically. In the realisation of Geometric MDS, one step of descent is done only for a separate block taking into account that the most decrease in stress is in the first steps, usually. Despite the fact that the Geometric MDS uses the simplest stress function, there is no need for its normalization depending on the number m of data points and the scale of proximities d_{ij}. These are the reasons that a good performance of the proposed algorithm can be expected as compared with other (e.g. majorization) algorithms. Moreover, more sophisticated realizations of ideas presented in this paper should be developed.

Acknowledgements. This research is funded by Vilnius University, grant No. MSF-LMT-4. The authors are grateful to the reviewers for their comments that made the results of this paper more valuable.

References

1. Dzemyda, G., Kurasova, O., Žilinskas, J.: Multidimensional Data Visualization: Methods and Applications. Springer Optimization and its Applications, vol. 75. Springer, New York (2013). https://doi.org/10.1007/978-1-4419-0236-8
2. Borg, I., Groenen, P.J.F., Mair, P.: Applied Multidimensional Scaling and Unfolding, 2nd edn. Springer, Heidelberg (2018)
3. Li, F., Jiang, M.: Low-resolution face recognition and feature selection based on multidimensional scaling joint L-2, L-1-norm regularisation. IET Biom. **8**(3), 198–205 (2019)
4. Dzemyda, G., Kurasova, O., Medvedev, V., Dzemydaitė, G.: Visualization of data: methods, software, and applications. In: Singh, V.K., Gao, D., Fischer, A. (eds.) Advances in Mathematical Methods and High Performance Computing. AMM, vol. 41, pp. 295–307. Springer, Cham (2019). https://doi.org/10.1007/978-3-030-02487-1_18
5. Perales, E., Burgos, F.J., Vilaseca, M., Viqueira, V., Martinez-Verdu, F.M.: Graininess characterization by multidimensional scaling. J. Mod. Opt. **66**(9), 929–938 (2019)
6. Kruskal, J.B.: Multidimensional scaling by optimizing goodness of fit to a nonmetric hypothesis. Psychometrika **29**(1), 1–27 (1964)
7. Žilinskas, A.: A quadratically converging algorithm of multidimensional scaling. Informatica **7**(2), 268–274 (1996)
8. Orts Gomez, F.J., Ortega Lopez, G., Filatovas, E., Kurasova, O., Garzon, G.E.M.: Hyperspectral image classification using Isomap with SMACOF. Informatica **30**(2), 349–365 (2019)
9. Groenen, P., Mathar, R., Trejos, J.: Global optimization methods for multidimensional scaling applied to mobile communication. In: Gaul, W., Opitz, O., Schander, M. (eds.) Data Analysis: Scientific Modeling and Practical Applications, pp. 459–475. Springer, Heidelberg (2000). https://doi.org/10.1007/978-3-642-58250-9_37
10. Orts, F., et al.: Improving the energy efficiency of SMACOF for multidimensional scaling on modern architectures. J. Supercomput. **75**(3), 1038–1050 (2018)
11. De Leeuw, J., Mair, P.: Multidimensional scaling using majorization: SMACOF in R. J. Stat. Softw. **31**(3), 1–30 (2009)
12. Symmetric Smacof. https://www.rdocumentation.org/packages/smacof/versions/2.0-0/topics/smacofSym
13. Borg, I., Groenen, P.J.F.: Modern Multidimensional Scaling, 2nd edn. Springer, New York (2005). https://doi.org/10.1007/0-387-28981-X
14. Torgerson Scaling. https://www.rdocumentation.org/packages/smacof/versions/2.0-0/topics/torgerson

A Simulink-Based Infinity Computer Simulator and Some Applications

Alberto Falcone[✉][iD], Alfredo Garro[iD], Marat S. Mukhametzhanov[iD], and Yaroslav D. Sergeyev[iD]

Department of Informatics, Modeling, Electronics and Systems Engineering (DIMES), University of Calabria, 87036 Rende, CS, Italy
{a.falcone,alfredo.garro,m.mukhametzhanov,yaro}@dimes.unical.it

Abstract. This paper is dedicated to the Infinity Computer – a new type of a supercomputer allowing one to work *numerically* with finite, infinite, and infinitesimal numbers in one general framework. The existent software simulators of the Infinity Computer are used already for solving important real-world problems in applied mathematics. However, they are not efficient for solving difficult problems in control theory and dynamics, where visual programming tools like Simulink are used frequently. For this purpose, the main aim of this paper is to introduce a new Simulink-based solution of the Infinity Computer.

Keywords: Infinity computer · Scientific computing · Numerical differentiation

1 Introduction

The Infinity Computer is a computational system based on the Infinity Computing framework and allowing one to work numerically with finite, infinite and infinitesimal numbers in one general framework. It can be applied for computations with infinite and infinitesimal quantities in any field of mathematics and physics, where it is required. The existent simulator of the Infinity Computer if written in C++ and is already used in practice (see, e.g., [5,13,14,26]). However, it is not optimized for solving difficult real-life problems, e.g., in control theory or in dynamics due to the difficulties in extending the C++ code of the arithmetical operations and elementary functions in the external environments like Simulink. The main scope of this paper is to introduce such an extension and to present a new Simulink-based solution of the Infinity Computer.

Simulink is a graphical programming environment, developed by MathWorks, for modeling, simulating, and analyzing dynamical systems. It offers a graphical block diagramming tool and a customizable set of blocks that are tightly integrated with the MATLAB environment. Simulink is widely used in the modeling

---

The work of M.S. Mukhametzhanov was supported by the INdAM-GNCS funding "Giovani Ricercatori 2018-2019".

Y. D. Sergeyev and D. E. Kvasov (Eds.): NUMTA 2019, LNCS 11974, pp. 362–369, 2020.
https://doi.org/10.1007/978-3-030-40616-5_31

and simulation domain, including distributed simulation and model-based design [8,9,11,12,17].

The rest of the paper is organized as follows. Section 2 describes briefly the Infinity Computing methodology. The new Simulink-based solution for it is described in details in Sect. 3. The Simulink-based solution is exemplified in Sect. 4 by considering a higher order differentiation equation. Finally, some concluding remarks are presented in Sect. 5.

2   An Overview on the Infinity Computing

The Infinity Computing is a novel methodology allowing one to work numerically with different finite, infinite and infinitesimal numbers in one general framework (see, e.g., [20,27]). It is based on the positional numeral system with the infinite radix ① called *grossone* and introduced as the number of elements of the set of natural numbers $\mathbb{N}$ (see, e.g., a recent survey of this methodology in [27]).

A number C in this numeral system can be represented as follows:

$$C = d_n ①^{p_n} d_{n-1} ①^{p_{n-1}} ... d_0 ①^{p_0} ... d_{-k} ①^{p_{-k}}, \tag{1}$$

where d_i, $i = n, ..., -k$, are positive or negative finite numbers expressed in the traditional computational framework and called *grossdigits*; p_i, $i = n, ..., -k$, are called *grosspowers*, are written in the decreasing order with $p_0 = 0$ and can be finite, infinite and infinitesimal[1] of the form (1). One can see that the numeral system based on the form (1) allows one to work with finite, infinite and infinitesimal numbers in the same way: for instance, numbers containing at least one term $d_i ①^{p_i}$ with positive finite or infinite grosspower p_i are *infinite*, numbers containing the term $d_0 ①^{p_0}$ and no infinite terms are *finite*, while numbers containing only the term $d_0 ①^{p_0}$ are *purely finite*. Finally, numbers containing only negative finite or infinite grosspowers are *infinitesimal*. For example, the following numbers written in the form (1) are infinite: $① = 1①^1$, $1.8①^{1.2} - 0.31①^0 5.2①^{-1.4}$, $-1.3①^{3.2} 0.1①^{-1}$, etc. The simplest infinitesimal representable in the form (1) is $①^{-1} = \frac{1}{①}$, which is positive, since it is the result of division of two positive numbers 1 and ①.

The main advantage of this system is that arithmetical operations with ① are similar to the operations with traditional finite numbers, for instance:

$$①^0 = \frac{①}{①} = ① \cdot ①^{-1} = 1, \quad ① - ① = 0 = 0①,$$

$$(2.6①^{1.2} - 1.3①^0 1.8①^{-3.1}) - (3.5①^{1.2} - 1.3①^0 - 2.7①^{-5.2})$$

$$= -0.9①^{1.2} 1.8①^{-3.1} 2.7①^{-5.2}.$$

The software simulator of the Infinity Computer (see, e.g., patents [19]) has been already successfully used for solving different problems in the following fields of applied mathematics: optimization (see, e.g., [5–7,13]), numerical

---

[1] In this paper, only finite grosspowers are implemented for the simplicity.

solution to ODEs (see, e.g., [1, 14, 16, 26]), numerical differentiation (see, e.g., [21, 22]), handling ill-conditioning (see, e.g., [13, 15, 25]), probability theory (see, e.g., [4, 18]), Turing machines (see, e.g., [23, 24]), etc.

3   A Simulink-Based Solution

In this section, with regards to the Infinity Computing concepts delineated in Sect. 2, a Simulink-based solution is described.

The proposed Simulink-based solution is general-purpose and domain-independent; this means that it can be used in all industrial and scientific domains where a high level of accuracy in the calculations represents a key factor (e.g., Cyber-Physical Systems, Robotics and Automation (see [2, 10])). This solution allows engineers to focus on the specific aspects of their system, without dealing with low level APIs as well as complex procedures of the emulator of the Infinity Computer Arithmetic C++ library (*ICA-lib*) that can distract them from their high level design. The Simulink-based solution abstracts the design of a system while preserving the flexibility and performance of the C++ prototype.

Our design and implementation of the Simulink-based solution for the Infinity Computer have been centered on typical software engineering methods and, in particular, on the *Agile* software development process. Furthermore, it has been developed through the use of standard Simulink Blocks and S-Functions. In this way, engineers can jointly exploit the benefit coming from the Simulink-based solution and the standard Matlab/Simulink functionalities. Figure 1 presents an overview of the Simulink-based solution and its integration with the Matlab/Simulink core components.

Figure 1 delineates three main parts: (i) *Application Layer*, which represents the Simulink graphical programming environment that researchers use for modeling, analyzing and simulating dynamic systems through blocks according to the model-based paradigm (see [3]); (ii) *Simulink Environment*, which contains all the standard blocks provided by Simulink along with the ones designed and developed to provide Infinity Computer operations; and (iii) *Matlab Environment*, which represents the Matlab environment in which the Infinity Computer arithmetic C++ library has been integrated for managing arithmetic operations based on infinity computations.

Four Simulink blocks have been created, in the *Infinity Computer Arithmetic Blocks* section, to manage elementary computations on infinite, finite, and infinitesimal quantities that are: *Sum*, *Subtraction*, *Multiplication* and *Division*. Each block takes as input infinite, finite, and infinitesimal quantities that are forwarded to the associated S-Function to perform the computation by interacting with *ICA-lib*.

The *ICA-lib* provides a set of domain-independent services. Each service defines some C++ classes and interfaces that implement specific functionalities. The architecture is shown in Fig. 2. The *Infinity Computer Arithmetic Services* layer represents the kernel of the library and provides a set of low level services to manage infinite, finite, and infinitesimal computations. It is composed of the following five services.

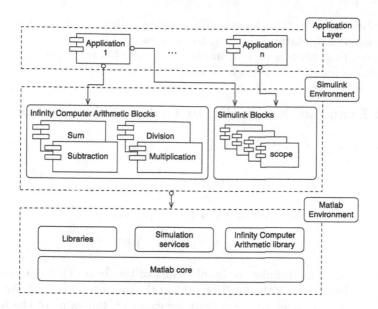

Fig. 1. Overview of the Simulink-based solution for the Infinity Computer and its integration in the Matlab/Simulink environment

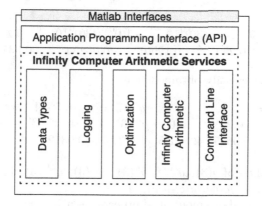

Fig. 2. The architecture of the infinity computer arithmetic library

The *Data Types* service provides data types for managing infinite, finite, and infinitesimal quantities. The *Logging* service stores into a log file data related to the activities carried out by *ICA-lib*. It represents a useful service for finding out problems/errors occurred during a computation and for understanding how the *ICA-lib* services work. The *Optimization* service delineates functionalities to separate the variable part of a class from the one that can be shared among instances. The provided functionalities allow to minimize the memory usage by sharing as much information among instances in order to speed up computations. The *Infinity Computer Arithmetic* service offers low-level infinite, finite, and infinitesimal arithmetic functionalities along with a set of utility classes for high precision computations. Finally, the *Command Line Interface* service provides classes that allow a user to interact with the library through input text commands via an alphanumeric keyboard.

4   An Example: Exact Higher Order Differentiation

Let us consider the following function

$$f(x) = \frac{x+1}{x-1} \tag{2}$$

implemented on the Infinity Computer by the procedure

$$g(x) = (x + 1①^0)/(x - 1①^0), \tag{3}$$

where x can be finite, infinite, or infinitesimal of the form (1). Let us calculate the first five derivatives of the function $g(x)$ at the finite point $x^* = 3$. According to [22], in order to calculate the derivatives exactly[2], the value of the function $g(x)$ at the point $y = x^* + ①^{-1} = 3 + ①^{-1} = 3①^0 1①^{-1}$ should be calculated (see the output of the block $g(x)$ in Figs. 3–4):

$$g(3①^0 1①^{-1}) = (3①^0 1①^{-1} + 1①^0)/(3①^0 1①^{-1} - 1①^0) \tag{4}$$
$$= 2①^0 - 0.5①^{-1} + 0.25①^{-2} - 0.125①^{-3} + 0.0625①^{-4} - 0.03125①^{-5} + ...,$$

from where, it can be easily obtained that

$$\begin{array}{llll}
g(3) & = & 2, & g'(3) = & -0.5 \cdot 1! & = -0.5, \\
g''(3) & = & 0.25 \cdot 2! = 0.5, & g'''(3) = & -0.125 \cdot 3! & = -0.75, \\
g^{(4)}(3) & = 0.0625 \cdot 4! = 1.5, & g^{(5)}(3) = & -0.03125 \cdot 5! = -3.75,
\end{array} \tag{5}$$

being the exact values of $g(3)$, $g'(3)$, $g''(3)$, $g'''(3)$, $g^{(4)}(3)$, and $g^{(5)}(3)$ (see the output of the block *differentiate* in Fig. 4).

The scheme of the above mentioned procedures using the presented Simulink-based solution is presented in Fig. 4. In this solution, each number is represented by the variable sized matrix, where the first column contains the grossdigits and the second column contains the grosspowers of a number written in the form (1). E.g., the number 1 is represented by the matrix $\begin{bmatrix} 1 & 0 \end{bmatrix}$, while the number $3①^0 1①^{-1}$ is represented by the matrix $\begin{bmatrix} 3 & 0 \\ 1 & -1 \end{bmatrix}$. The block *differentiate* executes only the extraction of the coefficients from the output matrix from the block $g(x)$ and its code is written in Matlab, since both the input and the output of this block are finite floating-point numbers.

---

[2] The word "exactly" means with the machine precision, since all the computations are numerical.

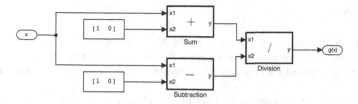

Fig. 3. Simulink block of the function $g(x) = (x + 1①^0)/(x - 1①^0)$ from (3) using the Infinity Computer Arithmetic blocks of the elementary operations $+$, $-$ and $/$ (see Fig. 1).

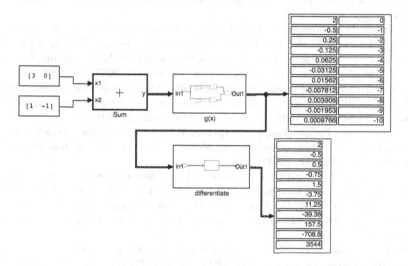

Fig. 4. Simulink diagram of the exact computation of the derivatives of the procedure $g(x)$. The input consists of the number $x^* = 3$ represented by the matrix $[3 \ 0]$ and the infinitesimal step $h = ①^{-1}$ represented by the matrix $[1 \quad -1]$. First, the value $y = x^* + ①^{-1} = x^*①^0 1①^{-1}$ and the value of the function $g(y)$ are calculated by the blocks *Sum* and $g(x)$, respectively. Then, the values of the derivatives are extracted from the result $g(y)$ following (4)–(5) by the block *differentiate*.

5   Conclusion

A new Simulink-based solution of the Infinity Computer Arithmetic library has been developed. The software solution is simple and powerful and allows one to apply the Infinity Computer Arithmetic library (ICA-lib) for solving real-life problems without the necessity to refer to its low-level implementation. This fact is very important from the practical point of view. The presented solution has been implemented and successfully applied for the exact higher order numerical differentiation.

References

1. Amodio, P., Iavernaro, F., Mazzia, F., Mukhametzhanov, M.S., Sergeyev, Y.D.: A generalized Taylor method of order three for the solution of initial value problems in standard and infinity floating-point arithmetic. Math. Comput. Simul. **141**, 24–39 (2016)
2. Bocciarelli, P., D'Ambrogio, A., Falcone, A., Garro, A., Giglio, A.: A model-driven approach to enable the distributed simulation of complex systems. In: Complex Systems Design & Management, Proceedings of the Sixth International Conference on Complex Systems Design & Management, CSD&M 2015, Paris, France, 23–25 November 2015, pp. 171–183 (2015). https://doi.org/10.1007/978-3-319-26109-6_13
3. Bocciarelli, P., D'Ambrogio, A., Falcone, A., Garro, A., Giglio, A.: A model-driven approach to enable the simulation of complex systems on distributed architectures. In: SIMULATION: Transactions of the Society for Modeling and Simulation International (2018, in Press). https://doi.org/10.1177/0037549719829828
4. Calude, C., Dumitrescu, M.: Infinitesimal probabilities based on grossone. Report CDMTCS-536, Centre for Discrete Mathematics and Theoretical Computer Science, University of Auckland, Auckland, New Zealand (2019)
5. Cococcioni, M., Pappalardo, M., Sergeyev, Y.D.: Lexicographic multiobjective linear programming using grossone methodology: theory and algorithm. Appl. Math. Comput. **318**, 298–311 (2018)
6. De Leone, R.: Nonlinear programming and grossone: quadratic programming and the role of constraint qualifications. Appl. Math. Comput. **318**, 290–297 (2018)
7. De Leone, R., Fasano, G., Sergeyev, Y.D.: Planar methods and grossone for the conjugate gradient breakdown in nonlinear programming. Comput. Optim. Appl. **71**(1), 73–93 (2018)
8. Falcone, A., Garro, A.: Using the HLA standard in the context of an international simulation project: the experience of the "SMASHTeam". In: 15th International Conference on Modeling and Applied Simulation, MAS 2016, Held at the International Multidisciplinary Modeling and Simulation Multiconference, I3M 2016, Larnaca, Cyprus, 26–28 September 2016, pp. 121–129. Dime University of Genoa (2016)
9. Falcone, A., Garro, A., Anagnostou, A., Taylor, S.J.E.: An introduction to developing federations with the high level architecture (HLA). In: 2017 Winter Simulation Conference, WSC 2017, Las Vegas, NV, USA, 3–6 December 2017, pp. 617–631 (2017)
10. Falcone, A., Garro, A., D'Ambrogio, A., Giglio, A.: Engineering systems by combining BPMN and HLA-based distributed simulation. In: 2017 IEEE International Conference on Systems Engineering Symposium, ISSE 2017, Vienna, Austria, 11–13 October 2017, pp. 1–6. Institute of Electrical and Electronics Engineers Inc. (2017). https://doi.org/10.1109/SysEng.2017.8088302
11. Falcone, A., Garro, A., Taylor, S.J.E., Anagnostou, A.: Simplifying the development of hla-based distributed simulations with the HLA development kit software framework (DKF). In: 21st IEEE/ACM International Symposium on Distributed Simulation and Real Time Applications, DS-RT 2017, Rome, Italy, 18–20 October 2017, pp. 216–217 (2017). https://doi.org/10.1109/DISTRA.2017.8167691

12. Falcone, A., Garro, A., Tundis, A.: Modeling and simulation for the performance evaluation of the on-board communication system of a metro train. In: 13th International Conference on Modeling and Applied Simulation, MAS 2014, Held at the International Multidisciplinary Modeling and Simulation Multiconference, I3M 2014, Bordeaux, France, 10–12 September 2014, pp. 20–29. Dime University of Genoa (2014)

13. Gaudioso, M., Giallombardo, G., Mukhametzhanov, M.S.: Numerical infinitesimals in a variable metric method for convex nonsmooth optimization. Appl. Math. Comput. **318**, 312–320 (2018)

14. Iavernaro, F., Mazzia, F., Mukhametzhanov, M.S., Sergeyev, Y.D.: Conjugate-symplecticity properties of Euler-Maclaurin methods and their implementation on the Infinity Computer. Preprint submitted in Archiv: https://arxiv.org/abs/1807. 10952 (2018)

15. Kvasov, D.E., Mukhametzhanov, M.S., Sergeyev, Y.D.: Ill-conditioning provoked by scaling in univariate global optimization and its handling on the Infinity Computer. In: AIP Conference Proceedings, vol. 2070, p. 20011 (2019)

16. Mazzia, F., Sergeyev, Y.D., Iavernaro, F., Amodio, P., Mukhametzhanov, M.S.: Numerical methods for solving ODEs on the Infinity Computer. In: AIP Conference Proceedings, New York, vol. 1776, p. 090033 (2016)

17. Möller, B., Garro, A., Falcone, A., Crues, E.Z., Dexter, D.E.: On the execution control of HLA federations using the SISO space reference FOM. In: 21st IEEE/ACM International Symposium on Distributed Simulation and Real Time Applications, DS-RT 2017, Rome, Italy, 18–20 October 2017, pp. 75–82. Institute of Electrical and Electronics Engineers Inc. (2017). https://doi.org/10.1109/DISTRA.2017. 8167669

18. Rizza, D.: A study of mathematical determination through Bertrand's paradox. Comput. Optim. Appl. **26**(3), 375–395 (2018)

19. Sergeyev, Y.D.: Computer system for storing infinite, infinitesimal, and finite quantities and executing arithmetical operations with them. USA patent 7,860,914 (2010), EU patent 1728149 (2009), RF patent 2395111 (2010)

20. Sergeyev, Y.D.: Arithmetic of Infinity. Edizioni Orizzonti Meridionali, CS (2003). 2nd edn. (2013)

21. Sergeyev, Y.D.: Numerical point of view on Calculus for functions assuming finite, infinite, and infinitesimal values over finite, infinite, and infinitesimal domains. Nonlinear Anal. Ser. A: Theory Methods Appl. **71**(12), e1688–e1707 (2009)

22. Sergeyev, Y.D.: Higher order numerical differentiation on the Infinity Computer. Optim. Lett. **5**(4), 575–585 (2011)

23. Sergeyev, Y.D., Garro, A.: Observability of turing machines: a refinement of the theory of computation. Informatica **21**(3), 425–454 (2010)

24. Sergeyev, Y.D., Garro, A.: Single-tape and multi-tape Turing machines through the lens of the grossone methodology. J. Supercomput. **65**(2), 645–663 (2013)

25. Sergeyev, Y.D., Kvasov, D.E., Mukhametzhanov, M.S.: On strong homogeneity of a class of global optimization algorithms working with infinite and infinitesimal scales. Commun. Nonlinear Sci. Numer. Simul. **59**, 319–330 (2018)

26. Sergeyev, Y.D., Mukhametzhanov, M.S., Mazzia, F., Iavernaro, F., Amodio, P.: Numerical methods for solving initial value problems on the Infinity Computer. Int. J. Unconv. Comput. **12**(1), 3–23 (2016)

27. Sergeyev, Y.D.: Numerical infinities and infinitesimals: methodology, applications, and repercussions on two Hilbert problems. EMS Surv. Math. Sci. **4**, 219–320 (2017)

Generalizing Pure and Impure Iterated Prisoner's Dilemmas to the Case of Infinite and Infinitesimal Quantities

Lorenzo Fiaschi[ID] and Marco Cococcioni[✉][ID]

University of Pisa, Largo Lucio Lazzarino, 1, 56122 Pisa, Italy
lorenzo.fiaschi@gmail.com, marco.cococcioni@unipi.it

Abstract. In this work, a generalization of both Pure and Impure iterated Prisoner's Dilemmas is presented. More precisely, the generalization concerns the use of non-Archimedean quantities, i.e., payoffs that can be infinite, finite or infinitesimal and probabilities that can be finite or infinitesimal. This new approach allows to model situations that cannot be adequately addressed using iterated games with purely finite quantities. This novel class of models contains, as a special case, the classical known ones. This is an important feature of the proposed methodology, which assures that we are proposing a generalization of the already known games. The properties of the generalized models have also been validated numerically, by using a Matlab simulator of Sergeyev's Infinity Computer.

Keywords: Game Theory · Prisoner's Dilemma · Non-archimedean payoffs and probabilities · Infinity computer · Grossone Methodology

1 Introduction

Game Theory (GT) is a widely and deeply studied mathematical branch which models the strategic behavior of rational individuals in competitive scenarios. For its intrinsic nature, GT found (and still nowadays finds) a lot of interests and applications in the economic and engineering fields. In this work we focused on two of the most known and simplest game theoretic models: the so called Prisoner's Dilemma (PD) and its iterated version, the Iterated Prisoner's Dilemma (IPD). In particular, we aim at showing a numerical extension of the IPD model to cases in which non-Archimedean (i.e., incommensurable) quantities are involved. For example, this happens when payoffs may assume finite, infinite and/or infinitesimal values, or when game probabilities may be finite or infinitesimal. Such kind of scenarios are not so rare nor improbable, e.g., loosing a finite amount of money with respect to the loss of a relative's life, or the necessity to distinguish between probable events and extremely rare, but still possible, ones. The classical game theoretic machinery used to model such environments adopting values several orders of magnitude far from each other.

© Springer Nature Switzerland AG 2020
Y. D. Sergeyev and D. E. Kvasov (Eds.): NUMTA 2019, LNCS 11974, pp. 370–377, 2020.
https://doi.org/10.1007/978-3-030-40616-5_32

However, in iterated contexts where the number of iterations increases arbitrarily, the repeated loss of an amount of money may be perceived by the model worse than the loss of a life, for instance. In order to avoid this kind of misleading behaviours, we employed Sergeyev's Grossone Methodology (GM), for a more proper representation of such non-Archimedean quantities. We also used Sergeyev's Infinity Computer (simulated in Matlab) to perform the numerical simulations of the resulting models. Thanks to this approach as a whole, we have been able to compare our results to the ones of the purely finite cases. Finally, please observe that, even though the non-Archimedean quantities could have been modeled using other methods like Non-Standard Analysis, the possibility to execute numerical computations is a peculiarity of the Infinity Computer.

2 Grossone Methodology

Grossone Methodology (GM), also known as Arithmetic of Infinity (AoI), is a novel way to *numerically* deal with infinite and infinitesimal quantities, originally proposed by Sergeyev in 2003 (see [15] and the references therein). GM has found application in a wide range of research fields, such as optimization [2–4,10], Bertrand's Paradox and mathematical determination [13] and numerical solution of ordinary differential equations [17], to cite a few. Also in GT we can find a case of AoI usage: in [5] the same authors exploited GM to deal with deterministic and stochastic tournaments involving finite, infinite and infinitesimal quantities.

The key pillar of the AoI is an infinite number called *Grossone*, indicated with ① and defined as the cardinality of the natural numbers set $\mathbb{N}$. It has been introduced by means of three methodological postulates and the Infinite Unit Axiom [15]. In particular, *Grossone* is used as the base of a new numeral system whose its generic element c (called *Gross*-scalar or *Gross*-number) is represented as follows:

$$c = c_{p_m} ①^{p_m} + ... + c_{p_1} ①^{p_1} + c_{p_0} ①^{p_0} + c_{p_{-1}} ①^{p_{-1}} + ... + c_{p_{-k}} ①^{p_{-k}},$$

where $m, k \in \mathbb{N}$, the exponents p_i are called *Gross-powers* and the finite real coefficients $c_{p_i} \neq 0$ *Gross-digits*, $i = m, ..., 1, 0, -1, ..., -k$. Two relevant features of GM make it so attractive. Firs of all, it offers the possibility to tackle the problem of the mathematical determination of uneasy concepts like infinitesimal probabilities [1,13] in a handy way. Secondly, the numerical attitude of GM, which sets it apart from Non-Standard Analysis and similar methodologies (see [16] for a discussion on the differences between the former and the latter).

3 A Glimpse on the PD Classical Formulation

In this section we introduce the class of game theoretic models referred as PDs. First of all, their theoretic formulation is provided, then a more complex class of games built upon them is shown. Such kind of games is known as Iterated Prisoner's Dilemmas and consists in playing repeatedly (potentially an infinite, but numerable, number of times) a PD.

3.1 Prisoner's Dilemma

A PD is a game theoretic model which involves two players ($\wp_1$ and $\wp_2$) who have to independently choose whether to cooperate or to defect [7,18]. After the simultaneous choice, each player obtain a reward depending on the outcome of the decision process. Such prizes are called payoffs and are indicated with the symbols $T_i, R_i, P_i, S_i, i = 1, 2$. For the purposes of this work, we considered only symmetric games, i.e., games where $T_1 = T_2, R_1 = R_2$, and so on. The policy under which the payoffs are assigned is the following: if both cooperate they receive R, if $\wp_i$ cooperates and $\wp_j$ defects ($i \neq j$, $i, j = 1, 2$) the first receive S and the second receive T, while if both defect they receive P. Furthermore, in any PD the relation on the payoffs $T > R > P > S$ (henceforth called *fundamental law*) must hold. Finally, a PD can be also classified in one of two disjoint classes, known as Pure and Impure PDs. In particular, a symmetric PD is defined as Pure if the payoffs property $2R > T + S$ is satisfied. On the other hand, it is called Impure if the payoffs inequality $2R < T + S$ holds.

3.2 Iterated Prisoner's Dilemma

When a PD is played repeatedly by the two players, the resulting game is called an Iterated PD (IPD) and every basic PD is referred to as stage or one-shot game [18]. Iterated models are interesting and still nowadays deeply studied [9,20] because they happen to be notably richer than their one-shot counterparts, thanks to the temporal component of such models. Indeed, the latter is a fundamental components for the birth of cooperation in the societies and collusion in competitive environments. We refer the reader to [11,12] for some applications of the IPD for collusion analysis in industrial economy scenarios. Such richness of the iterated models comes from the fact that the number of stages in an IPD is not known a priori or assumed (numerable) infinite. Otherwise, applying the so called backward induction, each stage's outcome becomes immediately predetermined, preventing any impact of the temporal component on the game. Thus it happens that the most interesting quantity to analyze in an IPD is the asymptotic per-stage expected reward of each player [8]. Henceforth we will refer to this reward with the symbol ρ_i for the player $\wp_i$, $i = 1, 2$.

There are still three relevant information about the IPD to point out before concluding this section. Firstly, also the IPD can be classified as Pure or Impure. The property depends on the nature of the one-shot PD played at each stage of the game. Namely, if it is Pure the IPD is labeled as Pure, vice versa if it is Impure. Hereinafter we will refer to them as Pure IPD (or simply PIPD) and Impure IPD (IIPD). Secondly, the IPD configuration we have dealt with in this work is the simplest one, where each player is modeled with an a priori fixed probability of cooperation, p_1 for $\wp_1$ and p_2 for $\wp_2$. It is worth to stress that such probability choices must be considered made up by the two players regardless (i.e., independently) one from the other. Note that this fact has a relevant impact on the set of the feasible couples of the form $\rho = (\rho_1, \rho_2)$ the game can output. In particular such set can be notably smaller than the one with players collusion (i.e., when the players can agree on the strategies to choose). Finally, one of the most common ways for graphically represent an IPD is by means of the just mentioned couple ρ, i.e., identifying the game outcome as a point of the Cartesian plane where on the X-axis is reported ρ_1's values and on the Y-axis ρ_2's. In Fig. 1 is reported the typical graphic resulting from several simulations of the same IPD. Here the simulations are realized fixing the payoffs values and letting the couple of probabilities (p_1, p_2) free to vary (uniformly) in $[0, 1]^2$, while the graphic plotting each game outcome ρ on the same plane. The same result can be found in [18] along with a more detailed introduction to PDs and IPDs.

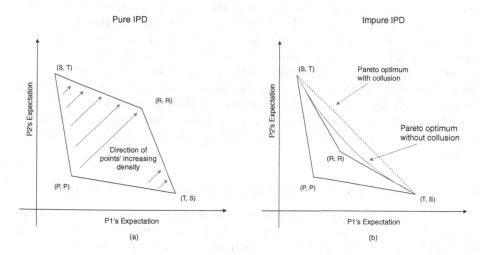

Fig. 1. The well-known finite case of Pure (a) and Impure (b) IPDs graphics.

4   PD Generalization by Means of Grossone Methodology

The goal of this section is to numerically extend the Pure and Impure IPDs to cases where also infinite and infinitesimal quantities are involved in terms of payoffs and probabilities. The idea is to compute several simulations of both a PIPD and IIPD and compare their outcomes with the graphics of Fig. 1. The remainder of the section proceeds in two steps: firstly we discuss the expedients we resorted to for realizing the simulations and the results comparisons, then we show the simulations setups and discuss the graphical outcomes.

4.1   How to Tackle Non-archimedean Simulations

In order to deal with probabilities and payoffs made up by infinite, finite and infinitesimal quantities, we make use of the GM, therefore representing them as Gross-scalars. For the simulation part, we decided to execute the computations exploiting an Infinity Computer simulator implemented in Matlab. Concerning the simulation graphics drawing, we used a Gross-variant of the Cartesian plane, i.e., a bi-dimensional space where each point is represented by coordinates expressed with Gross-numbers. Hereinafter we refer to such Cartesian plane as *Gross-Cartesian plane*. For the sake of simplicity but without loss of generality, we have imposed constraints on the possible inputs of the tool. More precisely, payoffs and probabilities are represented with positive Gross-numbers of the form $a①^1 b①^0 c①^{-1}$ and $d①^0 e①^{-1} f①^{-2}$, with $a, b, c, d, e, f \in \mathbb{R}$, respectively. Note that the algorithm for ρ's value computation has been maintained unaltered, with the exception that now the input parameters are Gross-number not necessarily finite. Before starting discussing the simulations setup and the resulting graphics, with reference to Fig. 2 let us spend few lines for the structure of the Matlab tool's output. The nine planes which can be seen denote the nine regions in which we can appreciate the Gross-Cartesian plane, once constrained by the structure of both the payoffs and the probabilities we have just assumed. Such division comes from the point's magnitude. From left to right and from bottom to top, the point's components increase their magnitude from $\mathcal{O}(①^{-1})$ to $\mathcal{O}(①^1)$, region by region. Thus, if a point falls in the top left region it has $\rho_1 \in \mathcal{O}(①^{-1})$ and $\rho_2 \in \mathcal{O}(①^1)$, while if it falls in the middle one $\rho_1, \rho_2 \in \mathcal{O}(①^0)$, and so on.

4.2   Simulations and Critical Analysis of the Obtained Results

In Figs. 2 and 3 are shown the simulations' outcomes when both the payoffs and the probabilities are non-Archimedean. To be specific, Fig. 2 refers to an IIPD, while Fig. 3 to a PIPD. The payoffs used are $T = 3①$, $R = 10$, $P = 7①^{-1}$, $S = 2①^{-1}$ and $T = 3①$, $R = 2① + 10$, $P = 7①^{-1}$, $S = 2①^{-1}$ respectively, while both the probabilities have been left free to vary within $(0, 1)$ contemplating the possibility to assume also infinitesimal values. The graphics have been obtained displaying a red point in correspondence of the game outcome ρ associated to each couple of sampled probabilities (p_1, p_2). Moreover, the black lines define the game diagram and, focusing on the IIPD, the dashed line identifies the Pareto optimal in case of collusion between the players.

   After an initial and shallow comparison, the results obtained numerically extending the IPDs to scenarios involving infinite and infinitesimal quantities already appear to have strong analogy with the finite case ones that are known in literature (see, for example, chapter 6 of [18], or Fig. 1 in this work). For example, we can point out the pseudo-quadrilateral shape of the game diagram (the black lines), the quadratic behavior of ρ's Pareto optimum in the IIPD (see the top-right plane of Fig. 2), the Pareto optimum (R, R) for the PIPD, the similar distributional behavior of ρ's points (the density increases from bottom-left to top-right). But this is not the only analogy. After a deeper analysis of the tool's outcomes, we can find even more correlations between the finite case and its non-Archimedean extension. More precisely, it can be shown that: (i) the parallelism of the diagram edges to the plane's axes is only apparent (otherwise the fundamental law would have been broken); (ii) the diagram edges

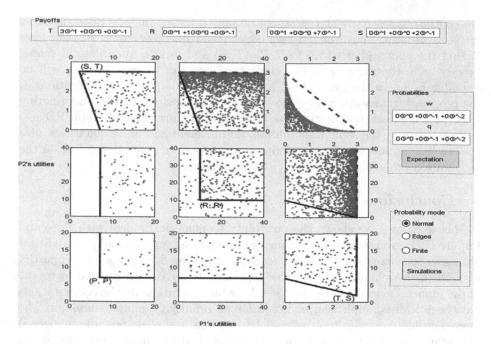

Fig. 2. The Gross-IIPD graphic, after 10,000 simulations performed by the Infinity Computer simulator. It is the generalization of the *impure* IPD shown in Fig. 1(b).

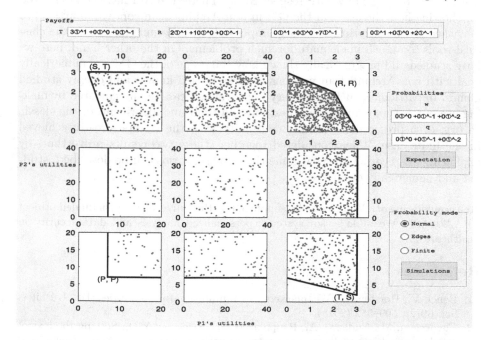

Fig. 3. The Gross-PIPD graphic, after 10,000 simulations performed by the Infinity Computer simulator. It is the generalization of the *pure* IPD shown in Fig. 1(a).

are continuous (in particular *Scott*-continuous [14, 19]); (iii) the diagram edges are straight lines. Thus, the diagram edges result into continuous straight lines linking the points $(P, P), (S, T), (R, R), (T, S)$ and form a quadrilateral shape, exactly as in the finite case. The theoretical proof of these properties will be available in [6]. In the same place, a discussion about practical applications of the proposed framework will be provided. Before concluding, it is worth noting that the numerical computations we have realized show that when an IPD scenario involves at the same time infinite, infinitesimal and finite quantities, also the rewards ρ_1 and ρ_2 may be infinite, infinitesimal or finite, depending on the cooperation probabilities chosen by the two players.

5   Conclusions

In this work we have shown the possibility to numerically deal with Pure and Impure IPDs even when non-Archimedean quantities (payoffs and probabilities) are used. Such capability has been reached exploiting Sergeyev's AoI. In particular, the non-Archimedean quantities representation has been realized by means of the numeral system offered by the GM, while operationally the computations have been realized implementing in Matlab an Infinity Computer simulator. The more sensible modeling tools offered by the GM and their possible numerical exploitation open the doors to finer studies of all the scenarios involving at the same time infinite, finite and infinitesimal quantities. Concerning the GT, this work moves orthogonally with respect to [5]. There, the AoI has been exploited for showing how the usage of the GM in non-Archimedean contexts may lead to new theoretical scenarios previously impossible even to imagine, being the classical tools too much inadequate for such problems. On the other hand, here we have studied and pragmatically shown how the usage of the GM can numerically deal with non-Archimedean versions of already well known and deeply studied games. In particular, such machinery improvement can be implemented by making use of the very same and well established tools and algorithms of the classic theory, even preserving a perfect continuity with the results already achieved in the finite case. When considered together, these two results prove that, by exploiting Sergeyev's GM, new game theoretic scenarios can be described, studied and numerically evaluated.

Acknowledgment. Work partially supported by the University of Pisa funded project PRA_2018_81 "Wearable sensor systems: personalized analysis and data security in healthcare".

References

1. Benci, V., Horsten, L., Wenmackers, S.: Infinitesimal probabilities. Br. J. Philos. Sci. **69**(2), 509–552 (2018)
2. Cococcioni, M., Cudazzo, A., Pappalardo, M., Sergeyev, Y.D.: Solving the lexicographic mixed-integer linear programming problem using Branch-and-Bound and Grossone methodology. Commun. Nonlinear Sci. Numer. Simul. **84**, 105177 (2020). https://doi.org/10.1016/j.cnsns.2020.105177

3. Cococcioni, M., Pappalardo, M., Sergeyev, Y.D.: Towards lexicographic multi-objective linear programming using grossone methodology. In: Sergeyev, Y.D., Kvasov, D., Dell'Accio, F., Mukhametzhanov, M. (eds.) Proceedings of the 2nd International Conference "Numerical Computations: Theory and Algorithms", vol. 1776, p. 090040. AIP Publishing, New York (2016)
4. Cococcioni, M., Pappalardo, M., Sergeyev, Y.D.: Lexicographic multi-objective linear programming using grossone methodology: theory and algorithm. Appl. Math. Comput. **318**, 298–311 (2018)
5. Fiaschi, L., Cococcioni, M.: Numerical asymptotic results in game theory using sergeyev's infinity computing. Int. J. Unconv. Comput. **14**(1), 1–25 (2018)
6. Fiaschi, L., Cococcioni, M.: Non-archimedean game theory: a numerical approach. Appl. Math. Comput. (submitted)
7. Fudenberg, D., Tirole, J.: Game Theory. MIT Press, Cambridge (1991)
8. Gale, D., Stewart, F.M.: Infinite Games with Perfect Information, vol. 2. Princeton University Press, Princeton (1953)
9. Hilbe, C., Traulsen, A., Sigmund, K.: Partners or rivals? Strategies for the iterated prisoner's dilemma. Games Econ. Behav. **92**(Suppl. C), 41–52 (2015)
10. Lai, L., Fiaschi, L., Cococcioni, M.: Solving mixed pareto-lexicographic multi-objective optimization problems: the case of priority chains (submitted)
11. Lambertini, L.: Prisoners' dilemma in duopoly (super) games. J. Econ. Theory **77**, 181–191 (1997)
12. Lambertini, L., Sasaki, D.: Optimal punishments in linear duopoly supergames with product differentiation. J. Econ. **69**(2), 173–188 (1999)
13. Rizza, D.: A study of mathematical determination through bertrand's paradox. Philos. Math. **26**(3), 375–395 (2017)
14. Scott, D.: Continuous lattices. In: Lawvere, F.W. (ed.) Toposes, Algebraic Geometry and Logic. LNM, vol. 274, pp. 97–136. Springer, Heidelberg (1972). https://doi.org/10.1007/BFb0073967
15. Sergeyev, Y.D.: Numerical infinities and infinitesimals: methodology, applications, and repercussions on two Hilbert problems. EMS Surv. Math. Sci. **4**, 219–320 (2017)
16. Sergeyev, Y.D.: Independence of the grossone-based infinity methodology from non-standard analysis and comments upon logical fallacies in some texts asserting the opposite. Found. Sci. **24**, 153–170 (2019)
17. Sergeyev, Y.D., Mukhametzhanov, M., Mazzia, F., Iavernaro, F., Amodio, P.: Numerical methods for solving initial value problems on the infinity computer. Int. J. Unconv. Comput. **12**(1), 3–23 (2016)
18. Steven, K.: Prisoner's Dilemma. The Stanford Encyclopedia of Philosophy, Spring 2017. https://plato.stanford.edu/archives/spr2017/entries/prisoner-dilemma/
19. Vickers, S.: Topology via Logic. Cambridge Tracts in Theoretical Computer Science, vol. 5. Cambridge University Press, Cambridge (1989)
20. Zeng, W., Li, M., Chen, F., Nan, G.: Risk consideration and cooperation in the iterated prisoner's dilemma. Soft. Comput. **20**(2), 567–587 (2016)

Multidimensional Global Search Using Numerical Estimations of Minimized Function Derivatives and Adaptive Nested Optimization Scheme

Victor Gergel$^{(\boxtimes)}$ ⓘ and Alexey Goryachikh ⓘ

Department of Software and Supercomputing, Lobachevsky State University,
Gagarin Avenue 23, Nizhni Novgorod 603950, Russia
gergel@unn.ru, a_goryachih@mail.ru

Abstract. This paper proposes a novel approach to the solution of time-consuming multivariate multiextremal optimization problems. This approach is based on integrating the global search method using derivatives of minimized functions and the nested scheme for dimensionality reduction. In contrast with related works novelty is that derivative values are calculated numerically and the dimensionality reduction scheme is generalized for adaptive use of the search information. The obtained global optimization method demonstrates a good performance, which has been confirmed by numerical experiments.

Keywords: Multiextremal optimization · Global search algorithms · Lipschitz condition · Numerical estimations of derivative values · Dimensionality reduction · Numerical experiments

1 Introduction

Global optimization problems are frequently encountered in various applications in the field of science and technology – see, for example, [1–6].

The global unconstrained optimization problem can be stated as follows

$$\min_{y \in D} \varphi(y), D = \{y \in \mathbb{R}^N : a_i \le y_i \le b_i, 1 \le i \le N\}, \tag{1}$$

where $\varphi(y)$ is the minimized function, $y = (y_1, y_2, \ldots, y_N)$ is an optimization parameter vector, N is the problem dimensionality, and the boundary vectors a and b define the search domain D.

To construct reliable estimates of the minimized function behavior, it is assumed that the function $\varphi(y)$ and its first derivatives $\varphi_i'(y)$, $1 \le i \le N$ satisfy the Lipschitz condition

$$|\varphi(y_2) - \varphi(y_1)| \le L \, \| y_2 - y_1 \|, y_1, y_2 \in D,$$
$$\left| \varphi_i'(y_2) - \varphi_i'(y_1) \right| \le L_i \, \| y_2 - y_1 \|, y_1, y_2 \in D, 1 \le i \le N, \tag{2}$$

© Springer Nature Switzerland AG 2020
Y. D. Sergeyev and D. E. Kvasov (Eds.): NUMTA 2019, LNCS 11974, pp. 378–385, 2020.
https://doi.org/10.1007/978-3-030-40616-5_33

where $L > 0$ and $L_i > 0$, $1 \leq i \leq N$ are the Lipschitz constants and $\|\cdot\|$ denotes the Euclidean norm in $\mathbb{R}^N$.

The proposed approach is described as follows. Firstly, in Sect. 2, the core univariate global optimization method is presented. Next, in Sect. 3, the adaptive nested reduction scheme providing for the use of the univariate method for solving the multidimensional problems is considered. Finally, in Sect. 4, we present the numerical results of our experiments.

2   One-Dimensional Global Search Method Using Numerical Derivatives

In the one-dimensional case, the global optimization problem can be stated in a simpler form

$$\min_{x \in [a,b]} \varphi(x). \tag{3}$$

To solve the problem (3), global optimization methods usually construct some estimates of the minimized function – mostly in the form of function minorants based on the condition (2). First such methods are given in [7–9], these methods use only the values of the optimized function. Later, some methods using the values of the function derivatives were proposed [10,11]. The latter methods are more efficient than the former ones but they use the derivatives which are not always available and computing the derivative values requires some additional computations. As a result, more recently some methods were developed in which the derivative values of the optimized function are computed numerically [12] – just this type of methods are used in the framework of the proposed approach.

The Adaptive Global Method using Numerical Derivatives (AGMND) being applied is given in [12,23]. For the sake of brevity, AGMND is described in the most general way.

The first two iterations of the method are performed at the boundary points a and b of the search domain $[a, b]$ (computing the optimized function value at each iteration of the search will be hereinafter called the trial). Suppose further k, $k > 2$, iterations of the global search have been performed. The choice of trial points x^{k+1} for the next iteration is determined by the following rules.

Rule 1. Renumber the points from the set of trial points with the subscripts in the ascending order of coordinate values

$$a = x_0 < x_1 < \cdots < x_i < \cdots < x_k = b. \tag{4}$$

Rule 2. For each interval (x_{t-1}, x_t), $1 \leq i \leq k$ compute the value of $R(i)$, hereinafter referred to as the interval characteristic.
Rule 3. Determine the interval with the maximum characteristic

$$R(t) = \max_{1 \leq i \leq k} R(i). \tag{5}$$

Rule 4. Perform a new trial (computing the value of the function $\varphi(y)$) at a point x^{k+1} located in the interval with the maximum characteristic from (5).

The stopping condition according to which the trials are terminated is determined by the condition

$$(x_t - x_{t-1}) \leq \varepsilon, \tag{6}$$

for the interval t, $1 \leq t \leq k$, with the maximum characteristic from (5), and $\varepsilon > 0$ is the specified solution accuracy for the problem. If the stopping condition is not satisfied, then the iteration number k is incremented by one, and a new global search iteration is performed.

As an estimate of the globally optimal solution to the problem (3), we take the point x_k^* at which the smallest value of the optimized function $\varphi(y)$ is calculated, i.e.

$$\varphi^*{}_k = \min_{0 \leq i \leq k} \varphi(x^i), x_k^* = \arg \min_{0 \leq i \leq k} \varphi(x^i). \tag{7}$$

It should be noted that many one-dimensional global search algorithms can be represented in accordance with this general scheme (4)–(7). The properties and effectiveness of each particular method are determined by the expression for calculating the characteristics of the intervals $R(i)$ from (5). To obtain the values of the characteristics, AGMND uses the computed values of the function $\varphi(y)$ being optimized and numerically calculated estimates of the Lipschitz constant as well as the values of the derivative $\varphi'(y)$ - for more details see [12,23].

3   Applying One-Dimensional Algorithms for Solving Multidimensional Optimization Problems

Effective algorithms used to solve multidimensional global optimization problems (1), as a rule, generate coverage of the search domain D, which is denser only in the vicinity of the sought-for global extrema. The construction of such non-uniform coverages can only be ensured by performing adaptive calculations when all the search information (previous trial points and the minimized function values at these points) obtained during the calculation is used when choosing the next trial points. The use of search information requires complex computational analysis of a large amount of multidimensional search information. Thus, many optimization algorithms use, to some extent, dimensional reduction methods (see, for example, [1,13–18,24]). Moreover, this approach allows the use of one-dimensional global search algorithms to solve multidimensional optimization problems.

One of the widely accepted methods for reducing the dimensionality is to use Peano *curves* or Peano *evolvents* that uniquely map the interval $[0,1]$ onto the N-dimensional search domain D [1,13,14,17]. As a result of such a reduction, the initial multidimensional global optimization problem (1) reduces to a one-dimensional problem:

$$\varphi(y(x^*)) = \min(\varphi(y(x)) : x \in [0,1]). \tag{8}$$

In the proposed approach, another commonly accepted method of dimension-ality reduction is used - the nested dimensionality reduction scheme [1, 18–20, 23]. This scheme allows the solution of problem (1) to be reduced to solving a series of nested one-dimensional problems:

According to this scheme, the solution of the multidimensional optimization problem (1) can be obtained by solving a series of nested one-dimensional global optimization problems:

$$\min \varphi(y) : y \in D = \min_{a_1 \leq y_1 \leq b_1} \Phi_1(y_1),$$

$$\downarrow$$

$$\Phi_1(\overline{y}_1) = \min_{a_2 \leq y_2 \leq b_2} \Phi_2(\overline{y}_1, y_2),$$

$$\downarrow$$

$$\cdots$$

$$\Phi_i(\overline{y}_1, \ldots, \overline{y}_i) = \min_{a_{i+1} \leq y_{i+1} \leq b_{i+1}} \Phi_{i+1}(\overline{y}_1, \ldots, \overline{y}_i, y_{i+1}),$$

$$\downarrow$$

$$\cdots$$

$$\Phi_N(\overline{y}_1, \ldots, \overline{y}_{N-1}) = \min_{a_N \leq y_N \leq b_N} \varphi(\overline{y}_1, \ldots, \overline{y}_{N-1}, y_N). \tag{9}$$

The problems of family (9) are handled in strictly sequential order. To compute the value of the function $\Phi_1(y_1)$ at some point y_1, it is necessary to minimize the function $\Phi_2(\overline{y}_1, y_2)$ for a fixed value of the variable $y_1 = \overline{y}_1$, etc. The values of the minimized function $\varphi(y)$ are computed only at the very last level of dimensionality reduction.

By applying the nested dimensionality reduction scheme, one can use many algorithms of one-dimensional global optimization for solving problem (1). At the same time, the efficiency of the global search with the use of this scheme is usually significantly lower compared to dimensionality reduction with the use of Peano curves.

Within the framework of the proposed approach, a new improved version of the computational procedure (9) is used - the adaptive nested dimensionality reduction scheme [18]. In the adaptive version of the scheme, the requirement of a sequential order for solving problems of family (9) is no longer relevant. All one-dimensional optimization problems

$$F_L(x) = \{f_l(x) : 1 \leq l \leq L\}, \tag{10}$$

generated in the course of calculations in accordance with (9) are solved simulta-neously. For this purpose, during the next iteration of the global search, the char-acteristics of the intervals $R(i)$ from (5) are calculated for each problem in the set $F_L(x)$ from [10] and a point of the new trial is chosen for the problem, for which the value of the characteristic $R(i)$ is the largest.

4   Numerical Results

To evaluate the efficiency of the proposed approach, computational experi-ments were performed in which for test global optimization problems, we used well-known two-dimensional multi-extremal functions determined by the relations [1]

$$
\varphi(y_1, y_2) = - \left\{ \left(\sum_{i=1}^{7} \sum_{j=1}^{7} [A_{ij} a_{ij}(y_1, y_2) + B_{ij} b_{ij}(y_1, y_2)] \right)^2 \right.
$$
$$
\left. + \left(\sum_{i=1}^{7} \sum_{j=1}^{7} [C_{ij} a_{ij}(y_1, y_2) + D_{ij} b_{ij}(y_1, y_2)] \right)^2 \right\}^{\frac{1}{2}}
$$

(11)

where

$$
a_{ij}(y_1, y_2) = \sin(\pi i y_1) \sin(\pi j y_2), \, b_{ij}(y_1, y_2) = \cos(\pi i y_1) \cos(\pi j y_2)
$$

are defined in the domain $0 \le y_1, y_2 \le 1$, and $-1 \le A_{ij}, Bij, C_{ij}, D_{ij} \le 1$ are independent random uniformly distributed values.

To achieve reliable conclusions, 100 problems of family (11) were solved in each series of experiments. The location of the global minimum points in these problems was known in advance, which made it possible to evaluate the degree of success in solving the problems.

When evaluating the productivity of the AGMND algorithm, the well-known Strongin Algorithm (GSA) was also considered, whose effectiveness has been shown in many publications [1,14,15,21,22]. Various dimensionality reduction techniques were considered for applying these one-dimensional algorithms AGMND and GSA to solve multidimensional global optimization problems. GSA was combined with the Peano curve reduction approach (GSA-P, see (8)) and with the standard (GSA-S, see (9)) and adaptive (GSA-A) nested reduction schemes. AGMND was applied with the standard (AGMND-S) and adaptive (AGMND-A) nested reduction schemes. The solution accuracy for the optimization problems was $\varepsilon = 0,001$, where ε is from (6).

As an efficiency indicator, we used the number of trials (computing the value of the function being optimized) performed by the optimization method to solve the problem with a given accuracy. Since the location of the global minimum points is known for test problems of family (11), in addition to the stopping condition (6), the optimization problem solving was also completed when the next trial point fell within the ε-neighborhood of the global minimum of the problem.

For a more integrated assessment of the efficiency of the developed approach, the frequently used operational characteristics (or the performance profile, P) of the compared methods were constructed using the results of the experiments performed (see [1,18,21,22]). The operational characteristic of the method is a set of pairs, each of them showing the dependence of the number of solved problems (ordinate axis) from a certain class on the number of trials (abscissa axis), i.e.:

$$
P = \{<k, p(k)>\}
$$

(12)

where k is the number of the global search iterations, and $p(k)$ is the proportion of successfully solved problems of the test class for the specified number of iterations. Such indicators can be calculated from the results of experiments and can be shown graphically in the form of a piecewise-broken line graph. In general, one may consider that the operational characteristic shows the probabilities of finding a global minimum with the required accuracy depending on the number of iterations performed by the method.

In the experiments performed, all 100 test problems were successfully solved by all the methods under comparison, exception for the AGMND-A algorithm. Using this algorithm, only 99 problems were solved. The operational characteristics of the methods under comparison are presented in Fig. 1. These results demonstrate that the AGMND method applied with the adaptive scheme (AGMND-A) shows the best performance (the smallest number of trials) in up to 84% of solved problems. With the score ranging from 84% to 100% of solved problems, the results of AGMND-A and GSA-A are very similar. The number of executed iterations averaged for all successfully resolved problems for all compared methods is given in Table 1.

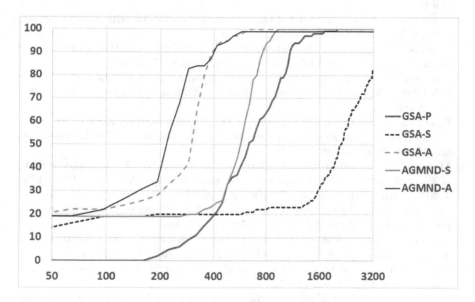

Fig. 1. Operational characteristics of the compared optimization methods. The vertical axis is the percentage of problems solved with the required accuracy, the horizontal axis is the number of executed trials in logarithmic scale

In conclusion, it should be noted that the results of numerical experiments confirm the potential of the proposed approach. Further research should include computational experiments for global optimization problems of higher dimensionality. Theoretical studies for investigating the convergence conditions of the

Table 1. Average number of executed iterations when solving 100 test problems

Methods and the dimensionality reduction schemes				
GSA-P	GSA-S	GSA-A	AGMND-S	AGMND-A
696.69	1974.75	252.59	494.74	206.08

AGMND method integrated with the adaptive nested dimensionality reduction scheme should also be performed.

Acknowledgements. The reported study was funded by the RFBR under research project No. 19-07-00242.

References

1. Strongin, R., Sergeyev, Y.D.: Global Optimization with Non-convex Constraints: Sequential and Parallel Algorithms, 3rd edn. Kluwer Academic Publishers, Dordrecht (2014)
2. Locatelli, M., Schoen, F.: Global Optimization: Theory, Algorithms, and Applications. SIAM, Philadelphia (2013)
3. Floudas, C., Pardalos, M.: Recent Advances in Global Optimization. Princeton University Press, Princeton (2016). https://doi.org/10.2307/2153139
4. Pardalos, M., Zhigljavsky, A., Žilinskas, J.: Advances in Stochastic and Deterministic Global Optimization. Springer, Cham (2016). https://doi.org/10.1007/978-3-319-29975-4
5. Famularo, D., Pugliese, P., Sergeyev, Y.D.: A global optimization technique for checking parametric robustness. Automatica **35**, 1605–1611 (1999). https://doi.org/10.1016/S0005-1098(99)00058-8
6. Modorskii, V., Gaynutdinova, D., Gergel, V., Barkalov, K.: Optimization in design of scientific products for purposes of cavitation problems. AIP Conf. Proc. **1738**, 400013 (2016). https://doi.org/10.1063/1.4952201
7. Piyavskij, S.: An algorithm for finding the absolute extremum of a function. Comput. Math. Math. Phys. **12**, 57–67 (1972). (in Russian)
8. Shubert, B.: A sequential method seeking the global maximum of a function. SIAM J. Numer. Anal. **9**, 379–388 (1972). https://doi.org/10.1137/0709036
9. Strongin, R.: On the convergence of an algorithm for finding a global extremum. Eng. Cybern. **11**, 549–555 (1973)
10. Gergel, V.: A method of using derivatives in the minimization of multiextremum functions. Comput. Math. Math. Phys. **36**, 729–742 (1996). (In Russian)
11. Sergeyev, Y.D.: Global one-dimensional optimization using smooth auxiliary functions. Math. Program. **81**, 127–146 (1998). https://doi.org/10.1007/bf01584848
12. Gergel, V., Goryachih, A.: Global optimization using numerical approximations of derivatives. In: Battiti, R., Kvasov, D.E., Sergeyev, Y.D. (eds.) LION 2017. LNCS, vol. 10556, pp. 320–325. Springer, Cham (2017). https://doi.org/10.1007/978-3-319-69404-7_25
13. Sergeyev, Y.D., Strongin, R., Lera, D.: Introduction to Global Optimization Exploiting Space-Filling Curves. Springer, New York (2013). https://doi.org/10.1007/978-1-4614-8042-6

14. Gergel, V.P., Strongin, R.G.: Parallel computing for globally optimal decision making. In: Malyshkin, V.E. (ed.) PaCT 2003. LNCS, vol. 2763, pp. 76–88. Springer, Heidelberg (2003). https://doi.org/10.1007/978-3-540-45145-7_7
15. Barkalov, K., Gergel, V., Lebedev, I.: Solving global optimization problems on GPU cluster. AIP Conf. Proc. **1738**, 400006 (2016). https://doi.org/10.1063/1.4952194
16. Sergeyev, Y.D., Kvasov, D.E.: A deterministic global optimization using smooth diagonal auxiliary functions. Commun. Nonlinear Sci. Numer. Simul. **21**, 99–111 (2015). https://doi.org/10.1016/j.cnsns.2014.08.026
17. Lera, D., Sergeyev, Y.D.: Deterministic global optimization using space-filling curves and multiple estimates of Lipschitz and Holder constants. Commun. Nonlinear Sci. Numer. Simul. **23**, 328–342 (2015). https://doi.org/10.1016/j.cnsns.2014.11.015
18. Gergel, V., Grishagin, V., Gergel, A.: Adaptive nested optimization scheme for multidimensional global search. J. Glob. Optim. **6**, 35–51 (2015). https://doi.org/10.1007/s10898-015-0355-7
19. Gergel, V., Grishagin, V., Israfilov, R.: Local tuning in nested scheme of global optimization. Procedia Comput. Sci. **51**, 865–874 (2015). https://doi.org/10.1016/j.procs.2015.05.216
20. Grishagin, V., Israfilov, R., Sergeyev, Y.D.: Convergence conditions and numerical comparison of global optimization methods based on dimensionality reduction scheme. Appl. Math. Comput. **318**, 270–280 (2018). https://doi.org/10.1016/j.amc.2017.06.036
21. Sergeyev, Y.D., Mukhametzhanov, M.S., Kvasov, D.E.: Operational zones for comparing metaheuristic and deterministic one-dimensional global optimization algorithms. Math. Comput. Simul. **141**, 96–109 (2017). https://doi.org/10.1016/j.matcom.2016.05.006
22. Grishagin, V., Israfilov, R., Sergeyev, Y.D.: Comparative efficiency of dimensionality reduction schemes in global optimization. AIP Conf. Proc. **1776**, 060011-1–060011-4 (2016). https://doi.org/10.1063/1.4965345
23. Gergel, V., Goryachih, A.: Multidimensional global optimization using numerical estimations of minimized function derivatives. Optim. Methods Softw. (2019). https://doi.org/10.1080/10556788.2019.1630624
24. Sergeyev, Y.D., Kvasov, D.E.: Deterministic Global Optimization: An Introduction to the Diagonal Approach. Springer, New York (2017). https://doi.org/10.1007/978-1-4939-7199-2

2-Approximation Polynomial-Time Algorithm for a Cardinality-Weighted 2-Partitioning Problem of a Sequence

Alexander Kel'manov[1,2] , Sergey Khamidullin[1] ,
and Anna Panasenko[1,2](✉)

[1] Sobolev Institute of Mathematics, 4 Koptyug Avenue, 630090 Novosibirsk, Russia
{kelm,kham,a.v.panasenko}@math.nsc.ru
[2] Novosibirsk State University, 2 Pirogova Street, 630090 Novosibirsk, Russia

Abstract. We consider a problem of 2-partitioning a finite sequence of points in Euclidean space into clusters of the given sizes with some constraints. The solution criterion is the minimum of the sum of weighted intracluster sums of squared distances between the elements of each cluster and its center. The weight of the intracluster sum is equal to the cluster size. The center of one cluster is given as input (is the origin without loss of generality), while the center of the other one is unknown and is determined as a geometric center. The following constraints hold: the difference between the indices of two subsequent points included in the first cluster is bounded from above and below by some given constants. In this paper, we have shown that the considered problem is the strongly NP-hard one and propose a polynomial-time 2-approximation algorithm for solving the problem.

Keywords: Euclidean space · Sequence of points · Weighted 2-partition · Quadratic variation · NP-hard problem · Approximation algorithm · Polynomial time

1 Introduction

The subject of this study is one quadratic cardinality-weighted problem of partitioning a sequence of points in Euclidean space into two subsequences of the given sizes with some additional constraints. The goal of our study is to analyze the computational complexity of the problem and to substantiate a polynomial-time approximation algorithm for this problem. The motivation of our study is the relevance of the considered problem, for example, for data mining and data clustering, when the data having in the hands is a time series.

The paper is organized as follows. In Sect. 2 the problem formulation and its interpretation are presented. Also, we present one closely related problem and some known results for it there. In addition, in the same section, we analyze the problem complexity. In Sect. 3 we formulate an auxiliary problem and some statements which underlie quality estimates for the proposed algorithm. Section 4

© Springer Nature Switzerland AG 2020
Y. D. Sergeyev and D. E. Kvasov (Eds.): NUMTA 2019, LNCS 11974, pp. 386–393, 2020.
https://doi.org/10.1007/978-3-030-40616-5_34

contains the approximation algorithm for the solution to the problem considered. The analysis of the algorithm properties is also in this section.

2   Problem Formulation, Related Problem, and Complexity

Everywhere below $\mathbb{R}$ denotes the set of real numbers, $\|\cdot\|$ denotes the Euclidean norm, and $\langle\cdot,\cdot\rangle$ denotes the scalar product.

We consider the following

Problem 1. Given a sequence $\mathcal{Y} = (y_1,\ldots,y_N)$ of points in $\mathbb{R}^d$ and some positive integers $T_{\min}$, $T_{\max}$, and $M > 1$. *Find* a subset $\mathcal{M} = \{n_1, n_2, \ldots\} \subset \mathcal{N} = \{1,\ldots,N\}$ of indices in $\mathcal{Y}$ such that

$$F(\mathcal{M}) = |\mathcal{M}| \sum_{j \in \mathcal{M}} \|y_j - \bar{y}(\{y_n | n \in \mathcal{M}\})\|^2 + |\mathcal{N} \setminus \mathcal{M}| \sum_{i \in \mathcal{N} \setminus \mathcal{M}} \|y_i\|^2 \longrightarrow \min, \quad (1)$$

where $\bar{y}(\{y_n | n \in \mathcal{M}\}) = \frac{1}{|\mathcal{M}|} \sum_{i \in \mathcal{M}} y_i$ is the centroid of $\{y_n | n \in \mathcal{M}\}$ with the following constraints

$$1 \leq T_{\min} \leq n_m - n_{m-1} \leq T_{\max} \leq N, \quad m = 2,\ldots,|\mathcal{M}|, \quad (2)$$

and $|\mathcal{M}| = M$.

The following applied interpretation can be proposed for Problem 1. As the input, we have a sequence $\mathcal{Y}$ of N measurement results. This sequence is the time-ordered one (i.e., the input is time series or discrete signal). The measurements are taken for d characteristics of some object. There are two different possible states for this object (active and passive, for example). We know that exactly M times the object was in the active state (or the probability of the active state is $\frac{M}{N}$). And we also know that there is an error for each measurement. The correspondence between the elements of the input sequence and the states is unknown. But the time interval between every two consecutive active states is bounded from below and above by some constants $T_{\min}$ and $T_{\max}$. It requires to find 2-partition of the input sequence and evaluate the object characteristics (i.e., $\bar{y}(\{y_n | n \in \mathcal{M}\})$) in accordance with (1)).

This application problem is very typical for processing time-series or discrete signals, for example, in distant object monitoring and in geophysics, in technical and medical diagnostics, etc. (see, for example, [1–5]).

The considered problem is closely related to

Problem 2. (Cardinality-weighted variance-based 2-clustering with given center). Given an N-element set $\mathcal{Y}$ of points in $\mathbb{R}^d$, and positive integer number M. *Find* a partition of $\mathcal{Y}$ into two non-empty clusters $\mathcal{C}$ and $\mathcal{Y} \setminus \mathcal{C}$ such that

$$|\mathcal{C}| \sum_{y \in \mathcal{C}} \|y - \bar{y}(\mathcal{C})\|^2 + |\mathcal{Y} \setminus \mathcal{C}| \sum_{y \in \mathcal{Y} \setminus \mathcal{C}} \|y\|^2 \rightarrow \min,$$

where $\bar{y}(\mathcal{C}) = \frac{1}{|\mathcal{C}|} \sum_{y \in \mathcal{C}} y$ is the centroid of $\mathcal{C}$, subject to constrain $|\mathcal{C}| = M$.

The strong NP-hardness of Problem 2 was established in [6]. The strong NP-hardness of Problem 1 follows from this result, as Problem 2 is the special case of Problem 1 when $T_{\min} = 1$ and $T_{\max} = N$.

Problem 2 has been studied in algorithmic direction in [7–11].

In [7], an exact pseudo-polynomial algorithm was constructed for the case of integer components of the input points and fixed dimension d of the space. The running time of this algorithm is $\mathcal{O}(N(MD)^d)$, where D is the maximum absolute value of coordinates of the input points.

In [8], an approximation scheme that allows one to find $(1 + \varepsilon)$-approximate solution in $\mathcal{O}\left(dN^2\left(\sqrt{\frac{2d}{\varepsilon}} + 2\right)^d\right)$ time was proposed. It implements an FPTAS in the case of the fixed space dimension.

Moreover, in [9], the modification of this algorithm with improved time complexity: $\mathcal{O}\left(\sqrt{d}N^2\left(\frac{\pi e}{2}\right)^{d/2}\left(\sqrt{\frac{2}{\varepsilon}} + 2\right)^d\right)$, was proposed. The algorithm implements an FPTAS in the case of fixed space dimension and remains polynomial for instances of dimension $\mathcal{O}(\log n)$. In this case, it implements a PTAS with $\mathcal{O}\left(N^{C\,(1.05 + \log(2 + \sqrt{\frac{2}{\varepsilon}}))}\right)$ time, where C is a positive constant.

In [10], an approximation algorithm that allows one to find a 2-approximate solution to the problem in $\mathcal{O}\left(dN^2\right)$ time was constructed.

In [11], a randomized algorithm was constructed. It allows one to find $(1+\varepsilon)$-approximate solution with probability not less than $1 - \gamma$ in $\mathcal{O}(dN)$ time for an established parameter value, a given relative error ε and fixed γ. The conditions are found under which the algorithm is asymptotically exact and runs in $\mathcal{O}(dN^2)$ time.

In this paper, we present the first result for strongly NP-hard Problem 1. Namely, we present a 2-approximation algorithm. This algorithm is based on the approaches and results presented in [10,12–14]. The running time of this algorithm is $\mathcal{O}(N^2(M(T_{\max} - T_{\min} + 1) + d))$.

3   Foundations of the Algorithm

In this section, we formulate some statements (e.g. about indices of $\mathcal{M}$) and formulate one more auxiliary problem which can be solved in polynomial time. All these statements and auxiliary problem are necessary for substantiation of our algorithms.

The following lemma is well known (see, for example, [15,16]).

Lemma 1. *For a finite set $\mathcal{Z} \subset \mathbb{R}^d$, if a point $t \in \mathbb{R}^d$ is closer (in terms of distance) to the centroid $\overline{z}$ of $\mathcal{Z}$ than any point in $\mathcal{Z}$, then*

$$\sum_{z \in \mathcal{Z}} \|z - t\|^2 \le 2 \sum_{z \in \mathcal{Z}} \|z - \overline{z}\|^2.$$

Two following lemmas were proved in [12,13].

Lemma 2. *If the elements of $\mathcal{M} = \{n_1, \ldots, n_M\}$ belong to $\mathcal{N} = \{1, \ldots, N\}$ and satisfy the system of constraints (2), then for every fixed $M \in \{2, \ldots, N\}$ we have:*

(1) the parameters of this system are related by inequality

$$(M - 1)T_{\min} \leq N - 1, \tag{3}$$

(2) the element n_m in $\{n_1, \ldots, n_m, \ldots, n_M\}$ belongs to the set

$$\omega_m = \{n | 1 + (m - 1)T_{\min} \leq n \leq N - (M - m)T_{\min}\}, \quad m = 1, \ldots, M, \tag{4}$$

(3) the feasibility domain of components n_{m-1} from this set under condition $n_m = n$ is defined by formula

$$\overline{\gamma_{m-1}}(n) = \{j | \max\{1 + (m - 2)T_{\min}, n - T_{\max}\} \leq j \leq n - T_{\min}\}, \tag{5}$$

where $n \in \omega_m$, $m = 2, \ldots, M$.

Lemma 3. *For every $M \in \{2, \ldots, N\}$ the system of constraints (2) is feasible if and only if inequality (3) holds.*

Consider the following function:

$$S(\mathcal{M}, b) = M \sum_{n \in \mathcal{M}} \|y_n - b\|^2 + (N - M) \sum_{n \in \mathcal{N} \setminus \mathcal{M}} \|y_n\|^2, \quad b \in \mathbb{R}^d, \quad \mathcal{M} \subset \mathcal{N}.$$

It is similar to the objective function of Problem 1. The only difference is the point b instead of the centroid $\overline{y}(\{y_n | n \in \mathcal{M}\})$. This function can be rewritten as follows:

$$S(\mathcal{M}, b) = (N - M) \sum_{n \in \mathcal{N}} \|y_n\|^2 - \sum_{n \in \mathcal{M}} \left(2M \langle y_n, b \rangle - (2M - N)\|y_n\|^2 - M\|b\|^2\right). \tag{6}$$

Note that the first summand is a constant if M and b are the fixed values. Hence the minimum of $S(\mathcal{M}, b)$ is reached on the subsequence that maximizes the second summand. This expression motivates us to formulate auxiliary:

Problem 3. Given a sequence $\mathcal{Y} = (y_1, \ldots, y_N)$ of points in $\mathbb{R}^d$, a point $b \in \mathbb{R}^d$, and some positive integers $T_{\min}$, $T_{\max}$ and M. Find a subset $\mathcal{M} = \{n_1, \ldots, n_M\} \subset \mathcal{N} = \{1, \ldots, N\}$ of indices in the sequence $\mathcal{Y}$ such that

$$G^b(\mathcal{M}) = \sum_{n \in \mathcal{M}} g^b(n) \longrightarrow \max, \tag{7}$$

where

$$g^b(n) = 2M \langle y_n, b \rangle - (2M - N)\|y_n\|^2 - M\|b\|^2, \quad n \in \mathcal{N}, \tag{8}$$

with additional constraints (2) on the elements of $\mathcal{M}$, if $M \neq 1$.

Let us define the set Ψ_M of subsets of admissible index tuples in the auxiliary problem:

$$\Psi_M = \begin{cases} \{(n_1)|n_1 \in \mathcal{N}\}, & \text{if } M = 1; \\ \{(n_1, \ldots, n_M)| \, n_i \in \mathcal{N}, \, i = 1, \ldots, M; \\ \quad 1 \le T_{\min} \le n_m - n_{m-1} \le T_{\max} \le N, \\ \quad m = 2, \ldots, M\}, & \text{if } 1 < M \le N. \end{cases} \quad (9)$$

For $M = 1$ the set Ψ_M is not empty for any parameters $T_{\min}$ and $T_{\max}$ by definition (9). For other feasible values of M we have [12,13]:

Lemma 4. *If $M \ge 2$, then the set Ψ_M is not empty if and only if an inequality (3) holds.*

Proofs of the following lemma and its corollary in [12,13] do not use (8) and so they hold for our case too.

Lemma 5. *Let $\Psi_M \ne \emptyset$ for some $M \ge 1$. Then for this M, the optimal value $G^b_{\max} = \max\limits_{\mathcal{M}} G^b(\mathcal{M})$ of objective function (7) can be found by formula*

$$G^b_{\max} = \max_{n \in \omega_M} G^b_M(n). \quad (10)$$

The values $G^b_M(n)$, $n \in \omega_M$, can be calculated by the following recurrent formulae:

$$G^b_m(n) = \begin{cases} g^b(n), & \text{if } n \in \omega_1, m = 1, \\ g^b(n) + \max\limits_{j \in \gamma^-_{m-1}(n)} G^b_{m-1}(j), & \text{if } n \in \omega_m, m = 2, \ldots, M, \end{cases} \quad (11)$$

where sets ω_m and $\gamma^-_{m-1}(n)$ are defined by formulae (4) and (5).

Corollary 1. *The elements $n^b_1, \ldots, n^b_M$ of the optimal set $\mathcal{M}^b = \arg\max\limits_{\mathcal{M}} G^b(\mathcal{M})$ can be found by the following recurrent formulae:*

$$n^b_M = \arg \max_{n \in \omega_M} G^b_M(n), \quad (12)$$

$$n^b_{m-1} = \arg \max_{n \in \gamma^-_m(n^b_m)} G^b_m(n), \quad m = M, M-1, \ldots, 2. \quad (13)$$

The following algorithm finds an optimal solution for auxiliary Problem 3. The step-by-step description of the algorithm looks like as follows.

Algorithm $\mathcal{A}_1$.
Input: a sequence $\mathcal{Y}$, a point b, some positive integer $T_{\min}$, $T_{\max}$, M.
Step 1. Compute $g^b(n)$, $n \in \mathcal{N}$, using formula (8).
Step 2. Using recurrent formulae (11), compute $G^b_m(n)$ for each $n \in \omega_n$ and $m = 1, \ldots, M$.

Step 3. Find the maximal value $G_{\max}^b$ of the objective function G^b using formula (10) and the optimal set $\mathcal{M}^b = \{n_1^b, \ldots, n_M^b\}$ by (12) and (13) from Corollary 1.
Output: the value of $G_{\max}^b$, the set $\mathcal{M}^b$.

The following theorem has been established in [13].

Theorem 1. *Algorithm $\mathcal{A}_1$ finds an optimal solution of Problem 3 in $\mathcal{O}(N(M(T_{\max} - T_{\min} + 1) + d))$ time.*

Remark 1. The value $(T_{\max} - T_{\min} + 1)$ is bounded by N. So the algorithm has a complexity at most $\mathcal{O}(N(MN + d))$.

Remark 2. If the values $T_{\max}$ and $T_{\min}$ are fixed then the algorithm has a complexity at most $\mathcal{O}(N(M + d))$.

4   Approximation Algorithm

The idea of the proposed algorithm is that the solution of the original NP-hard Problem 1 is replaced by an efficient algorithmic exact solution of auxiliary Problem 3. The solutions are found by Algorithm $\mathcal{A}_1$ presented in the previous section for each point $b \in \mathcal{Y}$ and then the best one is to be chosen.

Algorithm $\mathcal{A}$.
Input: a sequence $\mathcal{Y}$, some positive integers $T_{\min}$, $T_{\max}$, M.
Step 1. $i := 0$, $\mathcal{M}_{\mathcal{A}} := \emptyset$, $N := -\infty$.
Step 2. $i := i + 1$, $b := y_i$.
Step 3. Find the optimal solution $\mathcal{M}^b$ and the maximal value $G_{\max}^b$ of objective function (7) using algorithm $\mathcal{A}_1$.
Step 4. If $H < G_{\max}^b$ then $b_{\mathcal{A}} := b$, $H := G_{\max}^b$, $\mathcal{M}_{\mathcal{A}} = \mathcal{M}^b$.
Step 5. If $i < N$ then go to Step 2, else go to Step 6.
Step 6. Find the centroid $\bar{y}(\{y_n | n \in \mathcal{M}_{\mathcal{A}}\})$ and the value of the objective function $F(\mathcal{M}_{\mathcal{A}})$ by (1).
Output: the set $\mathcal{M}_{\mathcal{A}}$, the value $F(\mathcal{M}_{\mathcal{A}})$, the points $\bar{y}(\{y_n | n \in \mathcal{M}_{\mathcal{A}}\})$ and $b_{\mathcal{A}}$.

Theorem 2. *Algorithm $\mathcal{A}$ finds a 2-approximate solution of Problem 1 in $\mathcal{O}(N^2(M(T_{\max} - T_{\min} + 1) + d))$ time.*

Proof. Let $\mathcal{M}^*$ be an optimal solution of Problem 1, $y^* = \bar{y}(\{y_n | n \in \mathcal{M}^*\})$ be the centroid of this optimal solution, and $\mathcal{M}_{\mathcal{A}}$, $F(\mathcal{M}_{\mathcal{A}})$, $\bar{y}(\{y_n | n \in \mathcal{M}_{\mathcal{A}}\})$, $b_{\mathcal{A}}$ be the output of Algorithm $\mathcal{A}$. Let $t = \arg \min_{i \in \mathcal{M}^*} ||y_i - y^*||^2$ be a point from the optimal subsequence that is the closest to y^*.

Let us substantiate the accuracy of the algorithm. The following chain of equalities and inequalities holds:

$$F(\mathcal{M}_{\mathcal{A}}) = S(\mathcal{M}_{\mathcal{A}}, \overline{y}(\{y_n | n \in \mathcal{M}_{\mathcal{A}}\})) \leq^{(1)} S(\mathcal{M}_{\mathcal{A}}, b_{\mathcal{A}})$$

$$= (N - M) \sum_{n \in \mathcal{N}} \|y_n\|^2 - G^{b_{\mathcal{A}}}(\mathcal{M}_{\mathcal{A}}) =^{(2)} (N - M) \sum_{n \in \mathcal{N}} \|y_n\|^2 - \max_{b \in \mathcal{Y}} \max_{\mathcal{M} \in \Psi_M} G^b(\mathcal{M})$$

$$= \min_{b \in \mathcal{Y}} \min_{\mathcal{M} \in \Psi_M} ((N - M) \sum_{n \in \mathcal{N}} \|y_n\|^2 - G^b(\mathcal{M})) =^{(3)} \min_{b \in \mathcal{Y}} \min_{\mathcal{M} \in \Psi_M} (M \sum_{n \in \mathcal{M}} \|y_n - b\|^2$$

$$+ (N - M) \sum_{n \in \mathcal{N} \setminus \mathcal{M}} \|y_n\|^2) \leq^{(4)} M \sum_{n \in \mathcal{M}^*} \|y_n - t\|^2 + (N - M) \sum_{n \in \mathcal{N} \setminus \mathcal{M}^*} \|y_n\|^2$$

$$\leq^{(5)} 2M \sum_{n \in \mathcal{M}^*} \|y_n - y^*\|^2 + (N - M) \sum_{n \in \mathcal{N} \setminus \mathcal{M}^*} \|y_n\|^2 \leq 2(M \sum_{n \in \mathcal{M}^*} \|y_n - y^*\|^2$$

$$+ (N - M) \sum_{n \in \mathcal{N} \setminus \mathcal{M}^*} \|y_n\|^2) = 2F(\mathcal{M}^*).$$

Let us explain the numbered signs (we have chosen those that are not obvious). One can check by the differentiation that the minimum of $S(\mathcal{M}_{\mathcal{A}}, \cdot)$ (for the fixed $\mathcal{M}_{\mathcal{A}}$) is attained at $\overline{y}(\{y_n | n \in \mathcal{M}_{\mathcal{A}}\})$, so inequality (1) holds. Equality (2) follows from Theorem 1 and the description of Algorithm $\mathcal{A}_1$. Equality (3) is obvious by (6). Inequality (4) holds since $\mathcal{M}^* \in \Psi_M$ and $t \in \mathcal{Y}$ and inequality (5) holds by Lemma 1.

Hence $F(\mathcal{M}_{\mathcal{A}})/F(\mathcal{M}^*) \leq 2$ and Algorithm $\mathcal{A}$ finds a 2-approximate solution of Problem 1.

Let us estimate the time complexity. At Step 1 and Step 2 we need $\mathcal{O}(d)$ operations. At Step 3 we solve the auxiliary problem and so we need $\mathcal{O}(N(M(T_{\max} - T_{\min} + 1) + d))$ time. At Step 4 and Step 5 we need at most $\mathcal{O}(dN)$ operations. Step 2 – Step 5 are executed for N times. Step 6 requires $\mathcal{O}(dN)$ operations. Thus, the total time complexity of the algorithm is $\mathcal{O}(N^2(M(T_{\max} - T_{\min} + 1) + d))$. □

Remark 3. The value $(T_{\max} - T_{\min} + 1)$ is bounded by N. So the algorithm has a complexity at most $\mathcal{O}(N^2(MN + d))$.

Remark 4. If the values $T_{\max}$ and $T_{\min}$ are fixed then the algorithm has a complexity at most $\mathcal{O}(N^2(M + d))$.

5   Conclusion

In this paper, we have shown the strong NP-hardness of one cardinality-weighted quadratic partitioning problem of a sequence of points in Euclidean space into two clusters when the center of one of the desired clusters is fixed and the center of the other one is unknown and is determined as a geometric center. We also presented the first algorithmic result for the considered problem. This result is the polynomial-time 2-approximation algorithm. It seems important to continue studying the questions on the algorithmic approximability of the problem since the considered problem is poorly studied in the algorithmic direction.

Acknowledgments. The study presented in Sects. 3 and 4 was supported by the Russian Foundation for Basic Research, projects 19-07-00397, 19-01-00308 and 18-31-00398. The study presented in the other sections was supported by the Russian Academy of Science (the Program of basic research), project 0314-2019-0015, and by the Russian Ministry of Science and Education under the 5-100 Excellence Programme.

References

1. Fu, T.: A review on time series data mining. Eng. Appl. Artif. Intell. **24**(1), 164–181 (2011)
2. Kuenzer, C., Dech, S., Wagner, W.: Remote Sensing Time Series. RDIP, vol. 22. Springer, Cham (2015). https://doi.org/10.1007/978-3-319-15967-6
3. Liao, T.W.: Clustering of time series data – a survey. Pattern Recogn. **38**(11), 1857–1874 (2005)
4. Kel'manov, A.V., Jeon, B.: A posteriori joint detection and discrimination of pulses in a quasiperiodic pulse train. IEEE Trans. Signal Process. **52**(3), 645–656 (2004)
5. Carter, J.A., Agol, E., et al.: Kepler-36: a pair of planets with neighboring orbits and dissimilar densities. Science **337**(6094), 556–559 (2012)
6. Kel'manov, A.V., Pyatkin, A.V.: NP-hardness of some quadratic Euclidean 2-clustering problems. Dokl. Math. **92**(2), 634–637 (2015)
7. Kel'manov, A.V., Motkova, A.V.: Exact Pseudopolynomial Algorithms for a Balanced 2-Clustering Problem. J. Appl. Ind. Math. **10**(3), 349–355 (2016)
8. Kel'manov, A., Motkova, A.: A fully polynomial-time approximation scheme for a special case of a balanced 2-clustering problem. In: Kochetov, Y., Khachay, M., Beresnev, V., Nurminski, E., Pardalos, P. (eds.) DOOR 2016. LNCS, vol. 9869, pp. 182–192. Springer, Cham (2016). https://doi.org/10.1007/978-3-319-44914-2_15
9. Kel'manov, A., Motkova, A., Shenmaier, V.: An approximation scheme for a weighted two-cluster partition problem. In: van der Aalst, W., et al. (eds.) AIST 2017. LNCS, vol. 10716, pp. 323–333. Springer, Cham (2018). https://doi.org/10.1007/978-3-319-73013-4_30
10. Kel'manov, A.V., Motkova, A.V.: Polynomial-time approximation algorithm for the problem of cardinality-weighted variance-based 2-clustering with a given center. Comput. Math. Math. Phys. **58**(1), 130–136 (2018)
11. Kel'manov, A., Khandeev, V., Panasenko, A.: Randomized algorithms for some clustering problems. In: Eremeev, A., Khachay, M., Kochetov, Y., Pardalos, P. (eds.) OPTA 2018. CCIS, vol. 871, pp. 109–119. Springer, Cham (2018). https://doi.org/10.1007/978-3-319-93800-4_9
12. Kel'manov, A.V., Khamidullin, S.A.: Posterion detection of a given number of identical subsequences in a quasi-periodic sequence. Comp. Math. Math. Phys. **41**(5), 762–774 (2001)
13. Kel'manov, A.V., Khamidullin, S.A.: An approximation polynomial algorithm for a sequence partitioning problem. J. Appl. Ind. Math. 8(2), 236–244 (2014)
14. Kel'manov, A.V., Khamidullin, S.A., Khandeev, V.I.: Exact pseudopolynomial algorithm for one sequence partitioning problem. Autom. Remote Control **78**(1), 67–74 (2017)
15. Kel'manov, A.V., Romanchenko, S.M.: An FPTAS for a vector subset search problem. J. Appl. Ind. Math. 8(3), 329–336 (2014)
16. Kel'manov, A.V., Romanchenko, S.M.: An approximation algorithm for solving a problem of search for a vector subset. J. Appl. Ind. Math. 6(1), 90–96 (2012)

Exact Linear-Time Algorithm
for Parameterized K-Means Problem
with Optimized Number of Clusters
in the 1D Case

Alexander Kel'manov[1,2] and Vladimir Khandeev[1,2](✉)

[1] Sobolev Institute of Mathematics, 4 Koptyug Ave., 630090 Novosibirsk, Russia
{kelm,khan}@math.nsc.ru
[2] Novosibirsk State University, 2 Pirogova St., 630090 Novosibirsk, Russia

Abstract. We consider a well-known strongly NP-hard K-Means problem. In this problem, one needs to partition a finite set of N points in Euclidean space into K non-empty clusters minimizing the sum over all clusters of the intracluster sums of the squared distances between the elements of each cluster and its centers. The cluster's center is defined as the centroid (geometrical center). We analyze the polynomial-solvable one-dimensional case of the problem and propose a novel parameterized approach to this case. Within the framework of this approach, we, firstly, introduce a new parameterized formulation of the problem for this case and, secondly, we show that our approach and proposed algorithm allows one to find an optimal input data partition and, contrary to existing approaches and algorithms, simultaneously find an optimal clusters number in $\mathcal{O}(N)$ time.

Keywords: K-Means · One-dimensional case · Parameterized approach · Linear-time algorithm

1 Introduction

The subject of this study is the polynomial-solvable one-dimensional case of well-known strongly NP-hard K-Means problem. Our goal includes (1) substantiation of a novel parameterized approach to this case of the problem and (2) justification of new fast and exact algorithm for the problem-solving.

As is known, K-Means problem is important for Data mining, Pattern recognition, Computational geometry. Even in the one-dimensional case, fast algorithms are of great interest to mentioned applications. Some additional theoretical motivations we present and explain below.

The paper has the following structure. In Sect. 2, the parameterized problem formulation is given. We discuss our approach and existing ones there. The next Sect. presents auxiliary statements. These statements allow us to prove the main result. In Sect. 4, our main result, i.e., a new linear-time exact algorithm is presented.

© Springer Nature Switzerland AG 2020
Y. D. Sergeyev and D. E. Kvasov (Eds.): NUMTA 2019, LNCS 11974, pp. 394–399, 2020.
https://doi.org/10.1007/978-3-030-40616-5_35

2 Problem Formulation and Related Problems

Recall that in the well-known clustering K-Means problem, a set $\mathcal{Y} = \{y_1, \ldots, y_N\}$ of points in d-dimension Euclidean space and a positive integer K are given. It is required to find a partition of $\mathcal{Y}$ into non-empty clusters $\mathcal{C}_1, \ldots, \mathcal{C}_K$ minimizing the sum

$$\sum_{k=1}^{K} \sum_{y \in \mathcal{C}_k} \|y - \overline{y}(\mathcal{C}_k)\|^2, \tag{1}$$

where $\overline{y}(\mathcal{C}_k) = \frac{1}{|\mathcal{C}_k|} \sum_{y \in \mathcal{C}_k} y$ is the centroid (geometrical center) of the k-th cluster.

K-Means problem is known from the last century and is associated with Fisher (see, for example, [1,2]). In [3], it was shown that the problem can be solved in exponential $\mathcal{O}(N^{dK+1})$ time. Later in [4] the strong NP-hardness in the general case (when the space dimension is a part of the input) was proved, and in [5] it was proved that the problem is NP-hard even on a plane (when $d = 2$). But in the one-dimensional case (when $d = 1$), this problem is polynomially solvable [6].

The running time of the exact algorithm presented in [6] for the one-dimensional case is linear on K and quadratic on N. Recall that this algorithm relies an exact polynomial algorithm for solving the well-known *Nearest neighbor search* problem [7].

In recent years, for the one-dimensional case of K-Means problem, some exact algorithms with improved running time have been constructed. An overview of these algorithms and their properties can be found in [8]. In particularly, in [9], it was shown that if the input points are ordered, then the problem is solvable in a time that is linear both on K and N. But we have to note that the same result follows directly from the fundamental mathematical results obtained earlier back in 1965 [10] and 1969 [11] (even before [6]) for *Nearest neighbor search* problem. Later, these results were developed and applied to solve some other important mathematical and applied problems [12,13].

Below we recall some properties of problem K-means objective function. First, directly from (1) follows that if one needs to optimize the number K of clusters, then the solution is trivial. Indeed, in this case, the optimal value of the nonnegative objective function (1) is equal to zero when the number K of clusters is equal to the number N of the input points. It means that in the family of problem K-means solutions for all $K = 1, \ldots, N$, the best value of K is always N. Here we have to recall that, in Data analysis and Data mining, finding an optimal number of clusters is a very important issue.

Second, recall that if in the one-dimensional case of K-means problem the input points are ordered, i.e., $y_1 < \ldots < y_N$, then the optimal partition of $\mathcal{Y}$ into clusters corresponds to a partition of the index sequence $(1, \ldots, N)$ into disjoint segments [6].

Taking in attention above mentioned properties, it seems interesting to obtain an algorithm that allows finding an optimal partition of the one-dimensional input set of points and optimal clusters number simultaneously and in linear time. Below we present a new approach to find such partitioning and propose an algorithm. On this way, we consider some parameterized modification of K-Means problem.

Namely, we consider the following

Problem 1. Given a set $\mathcal{Y} = \{y_1 < \ldots < y_N\}$ of points in $\mathbb{R}$ and some positive integer parameters $T_{\min}$ and $T_{\max}$. *Find* a partition of $\mathcal{Y}$ into clusters $\mathcal{C}_1, \ldots, \mathcal{C}_K$ such that

$$F = \sum_{k=1}^{K} \sum_{y \in \mathcal{C}_k} (y - \overline{y}(\mathcal{C}_k))^2 \longrightarrow \min_{K, \{\mathcal{C}_1, \ldots, \mathcal{C}_K\}},$$

where $\overline{y}(\mathcal{C}_k) = \frac{1}{|\mathcal{C}_k|} \sum_{y \in \mathcal{C}_k} y$ is the centroid of $\mathcal{C}_k = \{y_{s_k}, y_{s_k+1}, \ldots, y_{n_k}\}$ with the following constraints

$$1 \leq T_{\min} \leq n_k - n_{k-1} \leq T_{\max} \leq N - 1, \quad k = 2, \ldots, K. \tag{2}$$

3   Auxiliary Statements

For constructing an algorithm for Problem 1, we formulate the following property of elements of the tuple $(n_1, \ldots, n_K)$.

Lemma 1. *Suppose that the elements of the tuple $(n_1, \ldots, n_K)$, belong to the set $\{1, \ldots, N\}$ and satisfy the system of constraints (2). Then the following hold:*

(i) $K \leq K_{\max}$, *where*

$$K_{\max} = \lfloor (N-1)/T_{\min} \rfloor \leq N - 1; \tag{3}$$

(ii) for each fixed $K \in \{1, \ldots, K_{\max}\}$ and $k \in \{1, \ldots, K\}$, the element n_k of the tuple $(n_1, \ldots, n_K)$ belongs to the set

$$\omega_k(K) = \{n \mid 1 + (k-1)T_{\min} \leq n \leq N - (K-k)T_{\min}\};$$

(iii) for each fixed $K \in \{2, \ldots, K_{\max}\}$ and $k \in \{2, \ldots, K\}$, if the element n_k of $(n_1, \ldots, n_K)$ is such that $n_k = n$, where $n \in \omega_k(K)$, then the element n_{k-1} belongs to the set

$$\gamma_{k-1}(n) = \{j \mid \max\{1 + (k-2)T_{\min}, n - T_{\max}\} \leq j \leq n - T_{\min}\};$$

(iv) for all $K \in \{2, \ldots, K_{\max}\}$ and $k \in \{2, \ldots, K\}$, if an element n_k of $(n_1, \ldots, n_K)$ is such that $n_k = n$, then the element n_{k-1} belongs to the set

$$\gamma(n) = \{j \mid \max\{1, n - T_{\max}\} \leq j \leq n - T_{\min}\}. \tag{4}$$

The formula (3) is obvious. The validity of statement (ii) follows from inequalities (2). The validity of statement (iii) follows from statements (i) and (ii). Finally, the validity of statement (iv) follows from statements (i), (ii) and (iii).

4   Exact Polynomial-Time Algorithm for 1D Case of the Problem

Let $\mathcal{Y}_{s,n} = \{y_s, \ldots, y_n\}$, where $1 \leq s \leq n \leq N$, be a subset of $n - s + 1$ points of $\mathcal{Y}$ with indexes from s to n.

Let

$$f_{s,n} = \sum_{i=s}^{n} (y_i - \overline{y}(\mathcal{Y}_{s,n}))^2 \equiv \sum_{k=s}^{n} y_k^2 - \frac{1}{n-s+1} \left(\sum_{k=s}^{n} y_k \right)^2, \tag{5}$$

where

$$\overline{y}(\mathcal{Y}_{s,n}) = \frac{1}{n-s+1} \sum_{k=s}^{n} y_k$$

is the centroid of the subset $\mathcal{Y}_{s,n}$.

The following lemma holds:

Lemma 2. *The optimal value F^* of Problem 1 objective function is found by the following formula*

$$F^* = F_N, \tag{6}$$

and the values F_n are calculated by the following recurrent formulas

$$F_n = \min \left\{ f_{1,n}, \min_{j \in \gamma(n)} (F_j + f_{j+1,n}) \right\}, \quad n = 1, \ldots, N, \tag{7}$$

where $f_{s,n}$ are calculated by the formula (5) and the set $\gamma(n)$ is defined by the formula (4).

Formula (7) implements the forward running of the algorithm presented below and is based on Bellman's principle of optimality.

The following lemma implements the backward running of the algorithm.

Lemma 3. *Let*

$$j^*(n) = \begin{cases} 0, & \text{if } F_n = f_{1,n}, \\ \arg\min_{j \in \gamma(n)} (F_j + f_{j+1,n}), & \text{if } F_n = \min_{j \in \gamma(n)} (F_j + f_{j+1,n}), \end{cases}$$

where $n = 1, \ldots, N$. Then the optimal number K^* of clusters is found by the formula

$$K^* = K_N, \tag{8}$$

and the values K_n are calculated by the following recurrent formulas

$$K_n = \begin{cases} 1, & \text{if } F_n = f_{1,n}; \\ K_{j^*(n)} + 1, & \text{if } F_n = \min_{j \in \gamma(n)} (F_j + f_{j+1,n}). \end{cases} \tag{9}$$

The optimal clusters $\mathcal{C}_1, \ldots, \mathcal{C}_K$ are found by the following recurrent rule.

Step 0. Put $n = N$.
Step 1. Put

$$C_{K_n} = \{y_{j^*(n)+1}, y_{j^*(n)+2}, \ldots, y_n\}.$$

Step 2. If $K_n > 1$, then put $n = j^*(n)$ and go to Step 1. Otherwise—the end of calculations.

The validity of this lemma we have proved by induction.
Let us formulate the following

Algorithm $\mathcal{A}$.
Input: an N-element set $\mathcal{Y}$ of 1D points, positive integers $T_{\min}$ and $T_{\max}$.

Step 1 (forward running).
Calculate the values of F_n, $n = 1, \ldots, N$, using formula (7).
Find the optimal value F^* by formula (6).

Step 2 (backward running).
Calculate the values of K_n, $n = 1, \ldots, N$, using formula (9).
Find the optimal number K^* of clusters by formula (8).
Find the optimal clusters $\mathcal{C}_1^*, \ldots, \mathcal{C}_{K^*}^*$ using Steps 0–2 of the backward rule.

Output: the value F^*, the number K^* of clusters, clusters $\mathcal{C}_1^*, \ldots, \mathcal{C}_{K^*}^*$.

From Lemmas 2 and 3 follows our main result.

Theorem 1. *Algorithm $\mathcal{A}$ finds an optimal solution of Problem 1 in $\mathcal{O}((T_{\max} - T_{\min} + 1)N)$ time.*

Remark 1. If $T_{\min}$ and $T_{\max}$ are fixed parameters, then Algorithm $\mathcal{A}$ finds an optimal solution of Problem 1 in $\mathcal{O}(N)$ time.

5   Conclusion

In this paper, we present a new parameterized approach for the one-dimensional case of the well-known strongly NP-hard K-Means problem. In this case, our approach allows one to find an optimal partition of the input set of points and optimal clusters number simultaneously. To justify this approach, we present an exact algorithm and show that its running time is linear on the size of the input points set. In our opinion, this algorithm will be useful for Data analysis and Data mining application problems.

Acknowledgments. The study was supported by the Russian Foundation for Basic Research, projects 19-01-00308, 19-07-00397, and 18-31-00398, by the Russian Academy of Science (the Program of basic research), project 0314-2019-0015, and by the Russian Ministry of Science and Education under the 5-100 Excellence Programme.

References

1. Fisher, R.A.: Statistical Methods and Scientific Inference. Hafner, New York (1956)
2. MacQueen, J.B.: Some methods for classification and analysis of multivariate observations. In: Proceedings of the 5th Berkeley Symposium on Mathematical Statistics and Probability, vol. 1, pp. 281–297. University of California Press, Berkeley (1967)
3. Inaba, M., Katoh, N., Imai, H.: Applications of weighted Voronoidiagrams and randomization to variance-based clustering. In: Proceedings of Annual Symposium on Computational Geometry, pp. 332–339 (1994)
4. Aloise, D., Deshpande, A., Hansen, P., Popat, P.: NP-hardness of Euclidean sum-of-squares clustering. Mach. Learn. **75**(2), 245–248 (2009)
5. Mahajana, M., Nimbhorkar, P., Varadarajan, K.: The planar k-means problem is NP-hard. Theor. Comput. Sci. **442**, 13–21 (2012)
6. Rao, M.: Cluster analysis and mathematical programming. J. Am. Stat. Assoc. **66**, 622–626 (1971)
7. Bellman, R.: Dynamic Programming. Princeton University Press, Princeton (1957)
8. Grønlund, A., Larsen, K.G., Mathiasen, A., Nielsen, J.S., Schneider, S., Song, M.: Fast exact k-means, k-medians and Bregman divergence clustering in 1D. CoRR arXiv:1701.07204 (2017)
9. Xiaolin, W.: Optimal quantization by matrix searching. J. Algorithms **12**(4), 663–673 (1991)
10. Glebov, N.I.: On the convex sequences. Discrete Anal. **4**, 10–22 (1965). (In Russian)
11. Gimadutdinov, E.K.: On the properties of solutions of one location problem of points on a segment. Control. Syst. **2**, 77–91 (1969). (In Russian)
12. Gimadutdinov, E.K.: On one class of nonlinear programming problems. Control. Syst. **3**, 101–113 (1969). (In Russian)
13. Gimadutdinov, E.K.: Some standartization problems with arbitrary sign instances and coherent, quasi-convex and almost quasi-convex matrices. Control. Syst. **27**, 3–11 (1987). (In Russian)

Polynomial-Time Approximation Scheme for a Problem of Searching for the Largest Subset with the Constraint on Quadratic Variation

Vladimir Khandeev[1,2]([⊠]) [iD]

[1] Sobolev Institute of Mathematics, 4 Koptyug Ave., 630090 Novosibirsk, Russia
khandeev@math.nsc.ru
[2] Novosibirsk State University, 2 Pirogova St., 630090 Novosibirsk, Russia

Abstract. The paper is addressed to one strongly NP-hard problem of searching for the largest subset in the finite set of points in Euclidean space. A restriction is imposed on the searched subset: quadratic variation of its points with respect to the unknown centroid of this subset must not exceed a given value. We present the first polynomial-time approximation scheme for this problem.

Keywords: Euclidean space · Largest subset · Quadratic variation · NP-hard problem · Polynomial-time approximation scheme

1 Introduction

In the paper, we consider a strongly NP-hard problem of searching for the largest subset in a given finite set of points in Euclidean space. The sum of squared distances between the elements of this subset and its unknown centroid (geometrical center) must not exceed a given value, which is defined as the percentage of the sum of squared distances between the elements of the input set and its centroid. The essence of the problem considered is searching for the largest well-concentrated subset of points.

Our research is motivated by the relevance of the problem for some applications, in particular, for Data mining [1], Data cleaning [2], Data reduction [3], statistics [4]. As is well known, the problems of searching in a finite set of objects for similar elements are typical for marked applications. For example, the considered problem models a common Data mining problem of finding the largest subset of pairwise similar objects (see next section). The additional motivation of our research is the fact that the considered NP-hard problem is poorly studied in terms of algorithms with performance guarantees. Our goal is to substantiate a polynomial-time approximation scheme (PTAS) for the problem.

The paper has the following structure. In Sect. 2, the formulation of the problem is given, as well as the formulation of the closely related problem. In the

© Springer Nature Switzerland AG 2020
Y. D. Sergeyev and D. E. Kvasov (Eds.): NUMTA 2019, LNCS 11974, pp. 400–405, 2020.
https://doi.org/10.1007/978-3-030-40616-5_36

same section, existing results for these problems are presented. The next Section contains auxiliary statements required for the justification of the algorithm. In Sect. 4, a PTAS for the considered problem is presented.

2 Problem Formulation and Related Problems, Existing and Obtained Results

Throughout what follows, $\| \cdot \|$ is the Euclidean norm.

The problem under consideration is as follows.

Problem 1. Given an N-element set $\mathcal{Y}$ of points in $\mathbb{R}^d$ and a real number $\alpha \in (0, 1)$. *Find* a subset $\mathcal{C} \subset \mathcal{Y}$ of largest size such that

$$F(\mathcal{C}) = \sum_{y \in \mathcal{C}} \|y - \overline{y}(\mathcal{C})\|^2 \leq \alpha \sum_{y \in \mathcal{Y}} \|y - \overline{y}(\mathcal{Y})\|^2 , \tag{1}$$

where $\overline{y}(\mathcal{C}) = \frac{1}{|\mathcal{C}|} \sum_{y \in \mathcal{C}} y$ and $\overline{y}(\mathcal{Y}) = \frac{1}{|\mathcal{Y}|} \sum_{y \in \mathcal{Y}} y$ are the centroids of the subset $\mathcal{C}$ and the input set $\mathcal{Y}$, respectively.

Note that for an arbitrary subset $\mathcal{C} \subseteq \mathcal{Y}$, the well-known (see, e.g., [5]) equality holds:

$$\sum_{y \in \mathcal{C}} \|y - \overline{y}(\mathcal{C})\|^2 = \frac{1}{2|\mathcal{C}|} \sum_{y \in \mathcal{C}} \sum_{z \in \mathcal{C}} \|y - z\|^2. \tag{2}$$

It follows from the right-hand side of (2) that one can interpret Problem 1 as the search for the largest subset with pairwise similar objects. The maximum total dissimilarity between objects is defined by the right-hand side of (1) and can be adjusted by the parameter α.

The strong NP-hardness of the problem was proved in [6]. In the same paper, a polynomial-time 1/2-approximation algorithm with $\mathcal{O}(N^2(d + \log N))$ running time was presented for the problem.

In [7], an exact algorithm for the case of Problem 1 in which the input points have integer-valued coordinates was proposed. If the space dimension is bounded by some constant, the algorithm runs in pseudopolynomial time $\mathcal{O}(N^2(NB))$, where B is the maximum absolute coordinate value in the input set.

Moreover, in [7,8], generalization of Problem 1 for the case when the input is a sequence was considered. An exact algorithm for the case in which the input points have integer-valued coordinates was proposed in [7]. If the space dimension is bounded by some constant, the algorithm runs in pseudopolynomial time $\mathcal{O}(N^4(NB))$, where B is the maximum absolute coordinate value in the input sequence. In [8], a polynomial-time approximation algorithm was proposed. If there are no solutions to the problem, the algorithm detects it in $\mathcal{O}(N^3(N^2 + d))$ time. Otherwise, the algorithm outputs a 1/2-approximate solution if the length M^* of an optimal subsequence is even, or it outputs a $(M^* - 1)/2M^*$-approximate solution if M^* is odd.

The following problem is closely related to Problem 1.

Problem 2. Given an N-element set $\mathcal{Y}$ of points in Euclidean space of dimension d and a positive integer M. *Find* a subset $\mathcal{C} \subset \mathcal{Y}$ of cardinality M minimizing the value of $F(\mathcal{C})$.

Problem 2 is also known as M-Variance [9]. Strong NP-hardness of this problem is substantiated in [10]. In the same paper, it was shown that there does not exist a fully polynomial-time approximation scheme (FPTAS) for this problem unless P = NP.

Note that Problem 1 is not equivalent to Problem 2, although it has a similar statement. A list of known algorithms with guaranteed quality bounds for Problem 2 can be found in [14]. Here we will mention only a few of them.

Exact algorithms for Problem 2 were proposed in [9,11]. Both these algorithms run in $\mathcal{O}(dN^{d+1})$ time and are polynomial if the space dimension d is fixed; in this case, their running time is $\mathcal{O}(N^{d+1})$.

In [12], a 2-approximation polynomial algorithm with $\mathcal{O}(dN^2)$ running time was presented.

A PTAS for the general case of Problem 2 was proposed in [13]. The time complexity of the scheme is $\mathcal{O}(dN^{2/\varepsilon+1}(9/\varepsilon)^{3/\varepsilon})$, where $\varepsilon > 0$ is a relative error of the algorithm.

Since Problems 1 and 2 are not equivalent, the algorithms for Problem 2 cannot be directly applied to Problem 1. In particular, the possibility of constructing approximate algorithms for Problem 1 with improved guaranteed quality bounds remained an open issue. This issue is resolved in this paper: we improve the result of [6] for Problem 1 and propose a polynomial-time algorithm for this problem with relative performance guarantee greater than 1/2. Moreover, we prove that our algorithm implements a PTAS.

3   Algorithm Foundations

To construct an algorithm, we need several auxiliary statements.

Lemma 1. *Let a sequence $a_1 \leq \ldots \leq a_k$ and a number $\beta \leq 1$ be given. Then, $g(\lfloor k\beta \rfloor) \leq \beta g(k)$, where $g(i) = a_1 + \ldots + a_i$, $i = 1, \ldots, k$.*

The proof of this lemma can be found in [6].

Let $\mathcal{B}_i(x)$, where $i \in \{1, \ldots, N\}$, $x \in \mathbb{R}^d$, be the set of i points in $\mathcal{Y}$ closest to x. Also let

$$f(\mathcal{C}, x) = \sum_{y \in \mathcal{C}} \|y - x\|^2 \, ,$$

where $\mathcal{C} \subseteq \mathcal{Y}$, $x \in \mathbb{R}^d$.

Properties of this function are described by

Lemma 2. *(i) For any fixed nonempty subset $\mathcal{C} \subseteq \mathcal{Y}$, the minimum of the function $f(\mathcal{C}, x)$ with respect to $x \in \mathbb{R}^d$ is achieved at the point $\overline{y}(\mathcal{C}) = \frac{1}{|\mathcal{C}|} \sum\limits_{y \in \mathcal{C}} y$.*

(ii) If $|\mathcal{C}| = M = const$, then, for any fixed point $x \in \mathbb{R}^d$, the minimum of the function $f(\mathcal{C}, x)$ with respect to $\mathcal{C} \subseteq \mathcal{Y}$ is achieved at the set $\mathcal{B}_M(x)$.

The validity of statement (i) follows from the well-known equation

$$\sum_{y \in C} \|y - x\|^2 = \sum_{y \in C} \|y - \overline{y}(C)\|^2 + |C| \cdot \|x - \overline{y}(C)\| .$$

The statement (ii) is obvious.

The next required statement is

Lemma 3. *Let C_i^* be the optimal solution of Problem 2 for $M = i$, $i \in \{1, \ldots, N\}$. Then, for $a, b \in \mathbb{N}$, $1 \leq a \leq b \leq N$, the following holds:*

$$F(C_a^*) \leq \frac{a}{b} F(C_b^*) .$$

The validity of this lemma follows from Lemmas 1 and 2.

Finally, we need the following

Lemma 4. *Let $M \geq 1/\varepsilon$, where M is positive integer, $\varepsilon \in (0, 1)$. Then for $\delta = \frac{M}{\lceil (1-\varepsilon)M \rceil} - 1$ it is true that $1/\delta < 2/\varepsilon$.*

The idea of the proof is to apply the inequality $\lfloor x \rfloor > \frac{x}{2}$ to the value $x = \varepsilon M$ and then use the equality $\frac{\lfloor \varepsilon M \rfloor}{M - \lfloor \varepsilon M \rfloor} = \delta$.

Remark 1. Note that for positive integer M, real $\varepsilon \in (0, 1)$, and $\delta = \frac{M}{\lceil (1-\varepsilon)M \rceil} - 1$, the inequality $\delta > 0$ holds if and only if $M \geq 1/\varepsilon$.

4   Approximation Scheme

Let us formulate the approximate algorithm for Problem 1.

Algorithm $\mathcal{A}$.

Input: set $\mathcal{Y}$, real $\varepsilon > 0$, $\alpha \in (0, 1)$.

Step 1. Compute $A = \alpha \sum_{y \in \mathcal{Y}} \|y - \overline{y}(\mathcal{Y})\|^2$ (the right-hand side of (1)).

Step 2. For each positive integer $M < 1/\varepsilon$, find an exact solution C_M of Problem 2 by exhaustive search of sets of size M.

Step 3. For each positive integer M such that $1/\varepsilon \leq M < N$, find a $(1 + \delta)$-approximate solution C_K of size $K = \lceil (1-\varepsilon)M \rceil$ of Problem 2 for $\delta = \frac{M}{\lceil (1-\varepsilon)M \rceil} - 1$.

Step 4. In the family of solutions found at Steps 2 and 3, find subset C_A of the largest size for which $F(C_A) \leq A$.

Output: subset C_A.

Remark 2. At Step 3, PTAS from [13] can be used. In that case, for each K, subset C_K is found in $\mathcal{O}(dN^{2/\delta+1}(9/\delta)^{3/\delta})$ time.

The following theorem is the main result of the paper.

Theorem 1. *Algorithm $\mathcal{A}$ finds a $(1 - \varepsilon)$-approximate solution of Problem 1 in $\mathcal{O}(dN^{4/\varepsilon+2}(18/\varepsilon)^{6/\varepsilon})$ time.*

The proof of the algorithm accuracy bound is split into two cases depending on M^* and $1/\varepsilon$, where M^* is cardinality of the optimal solution $\mathcal{C}^*$ of Problem 1.

If $M^* < 1/\varepsilon$, then is it easy to show that the algorithm at Step 2 for $M = M^*$ finds an optimal solution to the problem.

If $M^* \geq 1/\varepsilon$, then the set $\mathcal{C}_K$, which is constructed at Step 3 for $K = \lceil (1 - \varepsilon)M^* \rceil$, is a $(1 - \varepsilon)$-approximate solution of Problem 1. To show this, it is enough to prove inequality $F(\mathcal{C}_K) \leq A$. This inequality is justified by the following chain of bounds:

$$F(\mathcal{C}_K) \leq (1 + \delta)F(\mathcal{C}_K^*) \leq (1 + \delta)\frac{K}{M^*}F(\mathcal{C}_{M^*}^*) \leq A,$$

where $\mathcal{C}_K^*$ and $\mathcal{C}_{M^*}^*$ are the optimal solutions of Problem 2 for $M = K$ and $M = M^*$, respectively. In this chain, the first bound follows from the definition of Step 3; the second bound follows from Lemma 3 for $a = K$ and $b = M^*$; the third bound follows from the definitions of optimal solutions of Problems 1 and 2.

The running time of the algorithm is defined by Steps 2 and 3.

At Step 2, $\mathcal{O}(1/\varepsilon)$ values of M are considered. For each M, we need to iterate over $\mathcal{O}(N^M)$ subsets of size M; for each subset, the value of Problem 2 objective function is calculated in $\mathcal{O}(dM)$ operations. Therefore, due to $M < \frac{1}{\varepsilon}$, Step 2 is performed in $\mathcal{O}(d(1/\varepsilon)^2 N^{1/\varepsilon})$ operations.

At Step 3, there are $\mathcal{O}(N)$ values of M; for each of them, an approximate solution of Problem 2 can be found in $\mathcal{O}(dN^{2/\delta+2}(9/\delta)^{3/\delta})$ time. Therefore, due to $1/\delta < 2/\varepsilon$, the running time of Step 3 (as well as the running time of Algorithm $\mathcal{A}$) is $\mathcal{O}(dN^{4/\varepsilon+2}(18/\varepsilon)^{6/\varepsilon})$.

Since Problem 1 is a maximization problem, it follows from Theorem 1 that Algorithm $\mathcal{A}$ implements a PTAS.

5   Conclusion

In this paper, we have considered a poorly studied strongly NP-hard problem of searching for the largest subset in a finite set of points in Euclidean space with the constraint on quadratic variation. We have proposed a polynomial-time approximation scheme for the problem. This is the first polynomial-time algorithm for this problem with a relative performance guarantee greater than 1/2. The construction of faster polynomial-time algorithms with guaranteed performance bounds seems to be the important direction of future studies.

Acknowledgments. The study was supported by the Russian Foundation for Basic Research, projects 19-01-00308 and 18-31-00398, by the Russian Academy of Science (the Program of basic research), project 0314-2019-0015, and by the Russian Ministry of Science and Education under the 5-100 Excellence Programme.

References

1. Aggarwal, C.C.: Data Mining: The Textbook. Springer, Heidelberg (2015)
2. Osborne, J.W.: Best Practices in Data Cleaning: A Complete Guide to Everything You Need to Do before and after Collecting Your Data. SAGE Publication, Los Angeles (2013)
3. Greco, L., Farcomeni, A.: Robust Methods for Data Reduction. Chapman and Hall/CRC, London/Boca Raton (2015)
4. James, G., Witten, D., Hastie, T., Tibshirani, R.: An Introduction to Statistical Learning. Springer, New York (2013)
5. Edwards, A.W.F., Cavalli-Sforza, L.L.: A method for cluster analysis. Biometrics **21**, 362–375 (1965)
6. Ageev, A.A., Kel'manov, A.V., Pyatkin, A.V., Khamidullin, S.A., Shenmaier, V.V.: Approximation polynomial algorithm for the data editing and data cleaning problem. Pattern Recogn. Image Anal. **17**(3), 365–370 (2017)
7. Kel'manov, A., Khamidullin, S., Khandeev, V., Pyatkin, A.: Exact algorithms for two quadratic euclidean problems of searching for the largest subset and longest subsequence. In: Battiti, R., Brunato, M., Kotsireas, I., Pardalos, P.M. (eds.) LION 12 2018. LNCS, vol. 11353, pp. 326–336. Springer, Cham (2019). https://doi.org/10.1007/978-3-030-05348-2_28
8. Kel'manov, A., Pyatkin, A., Khamidullin, S., Khandeev, V., Shamardin, Y., Shenmaier, V.: An approximation polynomial algorithm for a problem of searching for the longest subsequence in a finite sequence of points in euclidean space. Commun. Comput. Inf. Sci. **871**, 120–130 (2018)
9. Aggarwal, H., Imai, N., Katoh, N., Suri, S.: Finding k points with minimum diameter and related problems. J. Algorithms **12**(1), 38–56 (1991)
10. Kel'manov, A.V., Pyatkin, A.V.: NP-completeness of some problems of choosing a vector subset. J. Appl. Ind. Math. **5**(3), 352–357 (2011)
11. Shenmaier, V.V.: Solving some vector subset problems by voronoi diagrams. J. Appl. Ind. Math. **10**(4), 560–566 (2016)
12. Kel'manov, A.V., Romanchenko, S.M.: An approximation algorithm for solving a problem of search for a vector subset. J. Appl. Ind. Math. **6**(1), 90–96 (2012)
13. Shenmaier, V.V.: An approximation scheme for a problem of search for a vector subset. J. Appl. Ind. Math. **6**(3), 381–386 (2012)
14. Kel'manov, A., Khandeev, V., Panasenko, A.: Randomized algorithms for some clustering problems. In: Eremeev, A., Khachay, M., Kochetov, Y., Pardalos, P. (eds.) OPTA 2018. CCIS, vol. 871, pp. 109–119. Springer, Cham (2018). https://doi.org/10.1007/978-3-319-93800-4_9

On a Comparison of Several Numerical Integration Methods for Ordinary Systems of Differential Equations

Anatoliy G. Korotchenko⬚ and Valentina M. Smoryakova[(✉)]⬚

N.I. Lobachevsky State University, Gagarin Av.23, 603950 Nizhni Novgorod, Russia
koangr@yandex.ru, smorykov@mail.ru

Abstract. The paper considers the numerical integration methods for ordinary systems of differential equations in which the end of the integration interval is a priori undefined but is defined during the integration process instead. Moreover, the calculation of right hand sides of such systems is an expensive procedure. The paper describes a new integration strategy based on an implicit fourth order method. The proposed strategy employs the behavior of obtained solution to control the integration process. In addition, the number of integration nodes selected by the mentioned method is minimal at every fixed interval under the limitations defined by the local error which results from the approximation of system derivatives.

Keywords: Finite difference formulas · Integration strategies · Optimal strategy

1 Introduction

Difficulties of constructing effective algorithms of numerical integration of the Cauchy problem are of great importance especially if calculation complexity of the right hand sides of a differential equations system is rather high and the system solution is strongly non-linear at one of subintervals of the change of the independent variable and close to invariable at the other subinterval. Thereto, there are such systems where the ending moment of the integration process is not assumed to be determined a priori, but during the system integration process. For example, such condition can be based on the fact that every component of the system solution comes up to some present value.

Therefore, taking into consideration high complexity of calculation of the right hand sides, when constructing the integration formulas, their parameters should be chosen in order to minimize the calculations, and this can be formulated as an optimization problem.

It should be noted that the issues of constructing effective algorithms of numerical integration for ordinary differential equation systems were tackled upon in the works [1–3,5].

© Springer Nature Switzerland AG 2020
Y. D. Sergeyev and D. E. Kvasov (Eds.): NUMTA 2019, LNCS 11974, pp. 406–412, 2020.
https://doi.org/10.1007/978-3-030-40616-5_37

The article presents a new strategy of choosing integration steps for finite difference scheme. The strategy is based on the idea that at every fixed interval the number of integration nodes of the finite difference scheme is to be minimal under the limitations defined by the local mistake resulted from the approximation of initial system derivatives.

In order to evaluate the effectiveness, the proposed strategy was compared with ode45 and ode15s methods provided in MATLAB. The compared methods were evaluated using the following criteria. The first criterion is the number of integration nodes, the second one is the number of right hand side calculations of the system at the integration interval and the third criterion is the maximum of difference of exact and approximate solution of the test (model) problems for each solution component at every integration node.

2   Problem Definition

Finding the solution $Y = Y(x)$ of the Cauchy problem, it is assumed that the solution $Y(x)$ is the one and only and four times continuously differentiable within the interval $[x_0, t)$ for the system of ordinary differential equations:

$$Y' = F(x, Y), Y(x_0) = Y_0, \qquad (1)$$

$$Y = Y(x) = (y^1(x), \ldots, y^n(x)), x_0 \leq x < t,$$

where the value of variable t is unknown a priori and determined during the system integration process. In case the system (1) is linear $F(x, Y) = B(x)Y + b(x)$, where $B(x)$ is $(n \times n)$-matrix, $b(x)$ is n-dimensional vector.

To determine the numerical values $Y_i = (y_1^i, \ldots, y_n^i)$, $i = 1, 2 \ldots$, of the solution of system (1) we use the method which is based on applying finite difference formula.

Consider the following finite difference formula:

$$Y_i = a_{i-3}Y_{i-3} + a_{i-2}Y_{i-2} + a_{i-1}Y_{i-1} + h_i a_i q_i, \qquad (2)$$

where

$$q_i = F(x_i, Y_i), Y_i = (y_i^1, y_i^2, \ldots, y_i^n),$$

$$a_{i-3} = 1 - a_{i-2} - a_{i-1},$$

$$a_{i-2} = \frac{-h_i^2 \psi_i^2}{h_{i-2}h_{i-1}(h_{i-1}^2 + h_{i-1}h_{i-2} + 2h_i h_{i-2} + 3h_i^2 + 4h_{i-1}h_i)},$$

$$a_{i-1} = \frac{(h_i + h_{i-1})^2 \psi_i^2}{h_{i-1}(h_{i-2} + h_{i-1})(h_{i-1}^2 + h_{i-1}h_{i-2} + 2h_i h_{i-2} + 3h_i^2 + 4h_{i-1}h_i)},$$

$$a_i = \frac{(h_i^2 + h_{i-1}h_i)\psi_i}{h_i(h_{i-1}^2 + h_{i-1}h_{i-2} + 2h_i h_{i-2} + 3h_i^2 + 4h_{i-1}h_i)},$$

herewith $\psi_i = h_{i-2} + h_{i-1} + h_i$, $h_i = x_i - x_{i-1}$ –is an integration step.

Suppose that the system (1) meets the conditions in case the vector $Y_i = (y_1^i, \ldots, y_n^i)$ may be obtained from the solution of the following system of equations

$$Y_i = a_{i-3}Y_{i-3} + a_{i-2}Y_{i-2} + a_{i-1}Y_{i-1} + h_i a_i F(x_i, Y_i). \tag{3}$$

The system (3) may be solved by means of some solution procedure for nonlinear system of equations.

The local mistake occuring at the i-th step for every j-th component of the solution as the main formula characteristics of $Y_i = (y_i^1, \ldots, y_i^n)$ is

$$R_i^j = \frac{h_i^6 + 2(h_{i-2} + 2h_{i-1})h_i^5 + 2h_{i-1}(h_{i-2} + 2h_{i-1})h_i^4}{3h_i^2 + 2(h_{i-2} + 2h_{i-1})h_i + (h_{i-1}^2 + h_{i-1}h_{i-2})} \tag{4}$$

$$+ \frac{2h_{i-1}(h_{i-2} + h_{i-1})(h_{i-2} + 2h_{i-1})h_i^3 + h_{i-1}^2(h_{i-2} + h_{i-1})^2 h_i^2}{3h_i^2 + 2(h_{i-2} + 2h_{i-1})h_i + (h_{i-1}^2 + h_{i-1}h_{i-2})}(y^{IV}(\Theta_i))^j,$$

where $x_{i-2} - h_i \leq \Theta_i \leq x_{i-1} + h_i$ and $(y^{IV}(\Theta_i))^j$ is the value of the fourth derivative of the j-th component of the solution vector $Y(x)$ at point Θ_i. The local mistake (4) results from the finite difference approximation of the derivatives of the differential equation system.

Further, we assume that the following conditions are met during the process of system integration:

$$\max_{1 \leq j \leq n} |R_i^j| \leq \varepsilon_i,$$

where ε_i is the fixed accuracy at the i-th step of integration.

Now suppose that each component of the fourth derivative of the solution $Y(x)$ meets the following condition within the interval $[x_0, x_0 + z]$:

$$\max_{1 \leq j \leq n} |y^{IV}(x)^j| \leq K,$$

where $z > 0$, $K > 0$ are real constants.

Thus, the formula (2) limitation defined by the calculation accuracy at the i-th step of integration can be set as the following in equation:

$$\varphi_i(h_{i-2}, h_{i-1}, h_i) = h_i^6 + 2(h_{i-2} + 2h_{i-1})h_i^5 + 2h_{i-1}(h_{i-2} + 2h_{i-1})h_i^4 \tag{5}$$

$$+ 2h_{i-1}(h_{i-2} + h_{i-1})(h_{i-2} + 2h_{i-1})h_i^3 + h_{i-1}^2(h_{i-2} + h_{i-1})^2 h_i^2 - 3\Delta_i h_i^2$$

$$- 2(h_{i-2} + 2h_{i-1})\Delta_i h_i - (h_{i-1}^2 + h_{i-1}h_{i-2})\Delta_i \leq 0,$$

here $\Delta_i = 2\varepsilon_i K^{-1}$, $i = 1, 2, \ldots$.

Assign $h_{2i-2} = \tau_{i-1}$, $h_{2i-1} = \tau_i$, $h_{2i} = \tau_i$ in functions $\varphi_{2i}(h_{2i-2}, h_{2i-1}, h_{2i})$, defined in (5), and take into consideration the following functions:

$$f_i(\tau_{i-1}, \tau_i) = 14\tau_i^5 + 12\tau_{i-1}\tau_i^4 + 3\tau_{i-1}^2\tau_i^3 - 8\Delta_i\tau_i - 3\Delta_i\tau_{i-1}, \tag{6}$$

herewith $\varphi_{2i}(h_{2i-2}, h_{2i-1}, h_{2i}) = \tau_i f_i(\tau_{i-1}, \tau_i)$, $i = 1, 2, \ldots$.

Then the formula (5) is satisfied by solving the inequation

$$f_i(\tau_{i-1}, \tau_i) \leq 0 \tag{7}$$

for $\tau_i > 0$ and assumptions made for the integration steps.

3   Integration Strategy

Consider a new integration strategy based on the optimal strategy, which is mentioned in [4]. Suppose $h_0 = \tau_0$, where τ_0 is a fixed positive number. Then find solutions Y_{-1}, Y_{-2}, Y_{-3} of the system (1) in integration nodes $x_{-1} = x_0 - h_0$, $x_{-2} = x_{-1} - h_0$, $x_{-3} = x_{-2} - h_0$ using second-order Runge–Kutta method.

Let's z be a fixed positive number. We will estimate the maximum $K = K_0$ of the absolute value of the fourth derivatives for the solution $Y(x)$ at the segment $[x_0, x_0 + z]$ by interpolating the solution with Lagrange polynomial of fourth order in points (x_{-3}, Y_{-3}), (x_{-2}, Y_{-2}), (x_{-1}, Y_{-1}), (x_0, Y_0), $\overline{x}_0, \overline{Y}_0)$, where $\overline{x}_0 = x_0 + z$, the solution $\overline{Y}_0(\overline{x}_0)$ is found by second-order Runge-Kutta method.

Integration steps h_{2i-1}, h_{2i} are taken equal τ_i, where τ_i may be found as the unique positive root of

$$f_i(\tau_{i-1}, \tau_i) = 14\tau_i^5 + 12\tau_{i-1}\tau_i^4 + 3\tau_{i-1}^2\tau_i^3 - 8\Delta_i\tau_i - 3\Delta_i\tau_{i-1} = 0,$$

with τ_{i-1} obtained at previous step, $i = 1, 2, \ldots$.

Let $x_i = x_{i-1} + h_{i-1}$ where $i = 1, 2, \ldots$.

Let again estimate the value K_i of maximum of the absolute value of the fourth derivatives of the solution $Y(x)$ at the segment $[x_{i-1}, x_i]$ by interpolating the solution with Lagrange polynomial of fourth order in points (x_{i-4}, Y_{i-4}), (x_{i-3}, Y_{i-3}), (x_{i-2}, Y_{i-2}), (x_{i-1}, Y_{i-1}), (x_i, Y_i).

If $K_m \leq K$, where m is a positive number, $x_m \geq x_0 + z$, then assign $K = \frac{K + K_m}{2}$, otherwise take $x_0 = x_{m-1}$, $x_{-1} = x_{m-2}$, $x_{-2} = x_{m-3}$, $x_{-3} = x_{m-4}$ and start the strategy over in point x_0.

The method based on the strategy described above stops when termination condition is satisfied.

4   Numerical Illustration

The paper is devoted to the comparison of the designed method with ode45 and ode15s methods provided in MATLAB [6]. Below denote by α_1 the proposed method. Recall that methods are compared using three criteria. Suppose N is the number of integration nodes, M is the number of right hand side calculations of the system at the integration interval and δ is the maximum of difference of exact solution and approximate solution of test (model) problems for each solution component at every integration node. A variety of experiments was carried out on a range of model problems. Here, for instance, let us consider the system with initial conditions $y_1(0) = 1$, $y_2(0) = 2$, $y_3(0) = 3$ [1,3]:

$$y_1' = \left(5 - \frac{2}{(x+1)^2} + \frac{2}{(x+1)^3}\right) y_1 + \left(-3 + \frac{2}{(x+1)^3} + \frac{3}{(x+1)^4}\right) y_2$$
$$+ \left(-3 + \frac{1}{(x+1)^2} - \frac{3}{(x+1)^4}\right) y_3 + e^{\frac{1}{x+1}} + e^{\frac{1}{(x+1)}^2} + e^{\frac{1}{(x+1)^3}},$$

$$y_2' = \left(12 - \frac{2}{(x+1)^2} + \frac{4}{(x+1)^3}\right) y_1 + \left(-10 - \frac{4}{(x+1)^3} + \frac{3}{(x+1)^4}\right) y_2$$
$$+ \left(-3 + \frac{1}{(x+1)^2} - \frac{3}{(x+1)^4}\right) y_3 + e^{\frac{1}{x+1}} + 2e^{\frac{1}{(x+1)^2}} + e^{\frac{1}{(x+1)^3}},$$

$$y_3' = \left(12 - \frac{2}{(x+1)^2} + \frac{4}{(x+1)^3} y_1\right) + \left(-6 - \frac{4}{(x+1)^3} + \frac{6}{(x+1)^4}\right) y_2$$
$$+ \left(-7 + \frac{1}{(x+1)^2} - \frac{6}{(x+1)^4}\right) y_3 + e^{\frac{1}{x+1}} + 2e^{\frac{1}{(x+1)^2}} + 2e^{\frac{1}{(x+1)^3}}.$$

The initial conditions are set in such a way that the integration interval can be split into two subintervals. At the first of them each vector component of the solution is significantly non-linear, whereas in the second one it practically does not change. The first component of the solution of the considered system is shown in Fig. 1. The second and the third components exhibit the same behavior.

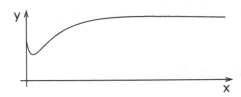

Fig. 1. The first component of the solution of the system

In the Figs. 2 and 3 the results are displayed on the diagrams. The results of `ode45`, `ode15s`, α_1 are presented here as triangle markers, circle markers and square markers respectively. The logarithmic scale is used to represent δ value. The initial step is fixed and equals 0.000001 for each of methods. The varied parameters are 'RelTol' for MATLAB methods, and ϵ for α_1 method. α_1 method needs more calculations of right hand sides of the system than MATLAB methods on high tolerance (ϵ), however on low tolerance α_1 yields lower maximal error (δ) and requires less calculations than MATLAB methods. On the basis of the results obtained we can conclude that α_1 is applicable when it is necessary to make a rough estimation of the solution behavior on a priory unknown segment.

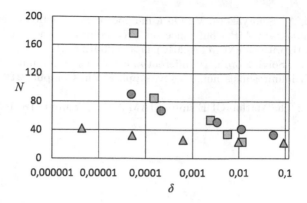

Fig. 2. The comparison of algorithms by the resulting number of nodes N and δ

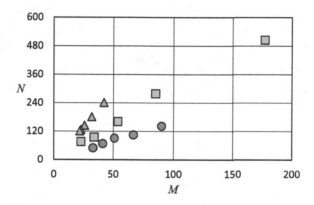

Fig. 3. The comparison of algorithms by the resulting number of nodes N and the number of system calculations M

Acknowledgments. The work is financially supported by the Federal Targeted Program for Research and Development in Priority Areas of Development of the Russian Scientific and Technological Complex for 2014–2020 under the contract No. 14.578.21.0246 (unique identifier RFMEFI57817X0246).

References

1. Effati, S., Roohparvar, H.: Iterative dynamic programming for solving linear and nonlinear differential equations. Appl. Math. Comput. **175**, 247–257 (2006)
2. Korotchenko, A.G., Lapin, A.V.: On an numerical integration algorithm with optimal choice of step. Vestn. Lobachevsky State Univ. Nizhni Novgorod **24**(2), 270–278 (2001)
3. Korotchenko, A.G., Lapin, A.V.: About construction of the approximately optimal algorithm of numerical integration. Vestn. Lobachevsky State Univ. Nizhni Novgorod **1**(26), 189–195 (2003)

4. Korotchenko, A., Smoryakova, V.: On a method of construction of numerical integration formulas. In: AIP Conference on Proceedings, Numerical Computations: Theory and Algorithms (NUMTA-2016) 1776, 090012 (2016)
5. Sergeyev, Y.D.: Solving ordinary differential equations on the infinity computer by working with infinitesimals numerically. Appl. Math. Comput. **219**, 10668–10681 (2013)
6. Shampine, L.: The Matlab ODE suite. SIAM J. Sci. Comput. **18**, 1–22 (1997)

On Acceleration of Derivative-Free Univariate Lipschitz Global Optimization Methods

Dmitri E. Kvasov[1,2]([✉]) [iD], Marat S. Mukhametzhanov[1,2] [iD],
Maria Chiara Nasso[1] [iD], and Yaroslav D. Sergeyev[1,2] [iD]

[1] University of Calabria, Rende, CS, Italy
{kvadim,m.mukhametzhanov,mc.nasso,yaro}@dimes.unical.it
[2] Lobachevsky State University, Nizhni Novgorod, Russia

Abstract. Univariate box-constrained global optimization problems are considered, where the objective function is supposed to be Lipschitz continuous and multiextremal. It is assumed that its analytical representation is unknown (the function is given as a "black-box") and even one its evaluation is a computationally expensive procedure. Geometric and information statistical frameworks for construction of global optimization algorithms are discussed. Several powerful acceleration techniques are described and a number of methods of both classes is constructed by mixing the introduced acceleration ideas. Numerical experiments executed on broad test classes taken from the literature show advantages of the presented techniques with respect to their direct competitors.

Keywords: Lipschitz global optimization · Univariate black-box functions · Geometric and information approaches · Local tuning

1 Introduction

It is well known that multiextremal optimization problems arise in many practical applications such as technological processes, engineering design, economic models, biology studies, etc. (see, e.g., [1,3–7,10,13,20,24,29]). Very often an analytical representation of the function to be optimized is unknown (i.e., it is given as a "black-box") and even one its evaluation is a computationally expensive procedure. In the literature (see, e.g., the references in [4,10,14,15,28,29]), there exist numerous optimization techniques proposed for different classes of problems. In particular, Lipschitz univariate global optimization problems that can be met frequently in electric and electronic engineering are studied actively (see, e.g., [3,8,10,15,20,22,28]). The fact that in these problems the objective functions are often non-differentiable explains the continuous interest of researches in derivative-free univariate Lipschitz global optimization methods (see, e.g., [2,8,18,19,21–23,26]).

Thus, in this work our attention is devoted to the global optimization problem

$$f^* = f(x^*) = \min f(x), \quad x \in D, \tag{1}$$

© Springer Nature Switzerland AG 2020
Y. D. Sergeyev and D. E. Kvasov (Eds.): NUMTA 2019, LNCS 11974, pp. 413–421, 2020.
https://doi.org/10.1007/978-3-030-40616-5_38

with the objective black-box function $f(x)$, $x \in D = [a, b]$, satisfying the Lipschitz condition with an unknown Lipschitz constant $0 < L < \infty$.

2 Univariate Global Optimization Techniques

Let us consider geometric and information statistical classes of algorithms which have their origins in the methods of Piyasvskij (see [16]) and Strongin (see [27]), respectively. These methods have been chosen for this study because they have shown their efficacy on several classes of problems (see, e.g., [15,18,19,22,26,28]) and it is also known that they can be improved with some powerful acceleration techniques (see [9,17,18,22]). These two classes of methods have a different nature: Piyasvskij's method requires the knowledge of an overestimate of the Lipschitz constant L and uses geometric ideas based on the Lipschitz property whereas in the Information approach, introduced by Strongin, an adaptive estimate of L calculated during the search is used in a statistical model.

The main idea of geometric algorithms is to construct a minorant $\varphi_i(x)$ for $f(x)$ over each subinterval $[x_{i-1}, x_i]$ of the search region D, where x_i, $i = 1, \ldots, k$, are so-called *trial* points, i.e., points where the values $z_i = f(x_i)$ have been evaluated. If we suppose that $\hat{L}$ is an overestimate of L, then it follows

$$f(x) \geq \varphi_i(x) = \max\{z_{i-1} - \hat{L}(x - x_{i-1}), z_i + \hat{L}(x - x_i)\}, \quad x \in [x_{i-1}, x_i],$$

and the minimal value of $\varphi_i(x)$, $x \in [x_{i-1}, x_i]$, denoted by R_i, is called *characteristic* of the interval $[x_{i-1}, x_i]$ (see, e.g., [21]),

$$R_i = 0.5(z_{i-1} + z_i - \hat{L}(x_i - x_{i-1})). \tag{2}$$

In contrast, the information approach uses the Bayesian ideas and considers the objective function from a stochastic point of view. The characteristic R_i of the Strongin information algorithm associated to each subinterval $[x_{i-1}, x_i]$ is

$$R_i = 2(z_i + z_{i-1}) - \overline{L}_k(x_i - x_{i-1}) - (z_i - z_{i-1})^2(\overline{L}_k(x_i - x_{i-1}))^{-1}, \tag{3}$$

where $\overline{L}_k$ is an adaptive estimate of the global (i.e., valid for the whole search region D) Lipschitz constant L during the search

$$\overline{L}_k = r \cdot \max\{H^k, \xi\}, \tag{4}$$

$$H^k = \max\{H_i : i = 2, \ldots, k\}, \quad H_i = |z_i - z_{i-1}|/(x_i - x_{i-1}), \tag{5}$$

and $\xi > 0$ is a small technical parameter allowing the correct work of the method.

It has been shown in [17–19,25] that the usage of global estimates of L can slow down the search significantly. However, for both the methodologies, geometric and information, the so-called *local tuning approach* introduced in [17,18] can be used to accelerate the global search. It allows one to tune the behavior of the algorithm according to behavior of the objective function at different subintervals using adaptive estimates of the local Lipschitz constants.

In fact, when subinterval $[x_{i-1}, x_i]$ is narrow, only the local information near trial points x_{i-1}, x_i has a decisive influence on the method. In contrast, when the subintervals is wide, the local information becomes less reliable.

In order to introduce the local tuning techniques let us denote as $\{x_i\}_1^k$ the ordered trial points and $k \geq 2$ the number of iterations of the algorithm (for $k = 2$: $x_1 = a$ and $x_2 = b$). Let $r > 1$ the reliability parameter. In the local tuning approach, we compute estimates l_i of local Lipschitz constants for each interval $[x_{i-1}, x_i]$, $i = 2, \ldots, k$, in one of the three following procedures:

1. "Maximum" Local Tuning

$$l_i = r \cdot \max\{\lambda_i, \gamma_i, \xi\}, \tag{6}$$

$$\lambda_i = \max\{H_{i-1}, H_i, H_{i+1}\}, \quad \gamma_i = H^k \frac{(x_i - x_{i-1})}{\max\{x_i - x_{i-1} : i = 2, \ldots, k\}},$$

where H_i, H^k are from (5) (when $i = 2$ and $i = k$ only H_2, H_3 and H_{k-1}, H_k should be considered, respectively) and ξ is the technical parameter.

2. "Maximum-Additive" Local Tuning

$$l_i = r \cdot \max\{H_i, 0.5(\lambda_i + \gamma_i), \xi\}, \tag{7}$$

where r, ξ, H_i, λ_i, and γ_i have the same meaning as above.

3. "Mixed" Local Tuning

$$l_i = 0.5 \left(r\eta + H_i^2(r\eta)^{-1}\right), \quad \eta = \max\{H^k, \xi\}. \tag{8}$$

Let us give an explanation of the last procedure. It has been observed in [28] that (3) can be rewritten in the form

$$R_i = 2(z_i + z_{i-1}) - (x_i - x_{i-1}) \left(\overline{L}_k + H_i^2 \cdot (\overline{L}_k)^{-1}\right),$$

so it can be interpreted as an auxiliary piecewise-linear function with local slopes

$$0.5 \left(\overline{L}_k + H_i^2 \cdot (\overline{L}_k)^{-1}\right).$$

Therefore, the stochastic model has a geometric interpretation and, as we shall see in the next section, the local estimates (8) can be very useful when used together with the characteristic (3).

The second acceleration technique used hereinafter is the *local improvement technique* (see [11, 12, 25]). We distinguish the "optimistic" and the "pessimistic" approaches. The optimistic method alternates local steps with the global ones until a local stopping rule is satisfied. In its turn, the pessimistic strategy continues the global search after the local search accuracy has been achieved and stops if the global stopping rule is satisfied (see [25] for details). In both cases the information obtained during the local searches is used in the global search, too.

By mixing the above procedures we obtained the following 19 methods:

- **GEOM-AL, GEOM-GL, GEOM-LTM, GEOM-LTMA** that are Geometric methods (in the sense that (2) is used) which use, respectively: A priori given Lipschitz constant; Global estimate of the Lipschitz constant; Maximum Local Tuning and Maximum-Additive Local Tuning. Each of these methods does not perform local improvement.
- **INF-AL, INF-GL, INF-LTM, INF-LTMA, INF-LTMI** that are Information methods ((3) is used) which use, respectively : A priori given Lipschitz constant; Global estimate of the Lipschitz constant; Maximum Local Tuning; Maximum-Additive Local Tuning and Mixed Local Tuning. Each of these methods does not perform local improvement.
- **GEOM-LTIMO, GEOM-LTIMAO** are Geometric methods which use Maximum Local Tuning and Maximum-Additive Local Tuning. Each of these methods uses the Optimistic strategy of the local improvement.
- **INF-LTIMO, INF-LTIMAO, INF-LTIMIO** are Information methods which use Maximum Local Tuning; Maximum-Additive Local Tuning and Mixed Local Tuning. Each of these methods uses the Optimistic strategy of the local improvement.
- **GEOM-LTIMP, GEOM-LTIMAP** are Geometric methods which use Maximum Local Tuning and Maximum-Additive Local Tuning. Each of these methods uses the Pessimistic strategy of the local improvement.
- **INF-LTIMP, INF-LTIMAP, INF-LTIMIP** are Information methods which use Maximum Local Tuning; Maximum-Additive Local Tuning and Mixed Local Tuning. Each of these methods uses the Pessimistic strategy of the local improvement.

3   Numerical Experiments and Discussion

All the methods presented in the previous section have been compared on two classes of functions: 100 Shekel (see [28]) type test functions (Class 1) and the opposite of 100 Shekel type test functions selected so that $x^* \neq a \wedge x^* \neq b$, where x^* is from (1). Notice that given a function $f : \mathbb{R} \to \mathbb{R}$, we denoted by "the opposite of f" the function $g : \mathbb{R} \to \mathbb{R}$ defined as $g(x) = -f(x)$. Functions from Class 1 were generated as follows

$$\varphi(x) = -\sum_{i=1}^{10} \left[k_i^2 (10x - a_i)^2 + c_i \right]^{-1}, \quad 0 \leq x \leq 1,$$

where $1 \leq k_i \leq 3$, $0.1 \leq c_i \leq 0.3$, $0 \leq a_i \leq 10$, $1 \leq i \leq 10$, and all the parameters are supposed to be the pseudorandom numbers in the corresponding ranges. For each method the technical parameter ξ from (6) was set to 10^{-8}.

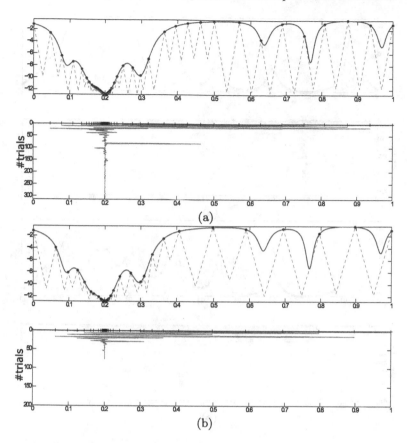

Fig. 1. Function 3 from Class 1 and trial points generated by GEOM-AL (a) and GEOM LTMA (b)

The initial values $r = 1.1$ and $r = 2$ were used respectively for the geometric and information methods without optimistic local improvement over all the classes of test functions and they were increased with step equal to 0.1 until all test problems were solved, i.e., the tested algorithm has generated a point x^k after k trials such that $|x^k - x^*| \leq \varepsilon$ with $\varepsilon = 10^{-5}$. Figures 1 and 2 show two examples of application of the methods respectively on Class 1 and on Class 2. In the same figures appears the auxiliary function and trial points generated by the methods. As the objective functions $f(x)$ are considered to be hard to evaluate, the number of trials was chosen as the comparison criterion. We reported in Table 1 the averages of trials for each method on both classes. The best results are shown in bold.

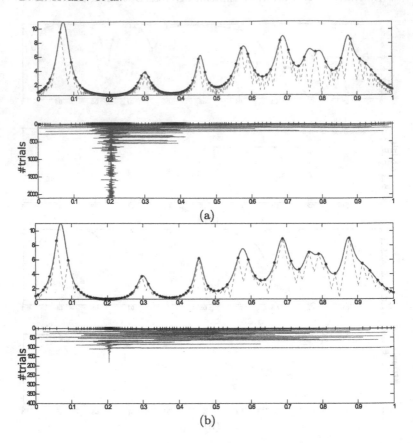

Fig. 2. Function 85 from Class 2 and trial points generated by GEOM AL (a) and GEOM LTMA (b)

Since the majority of test problems in both classes can be solved with smaller values of the parameter r w.r.t those reported in Table 1, the use of a common value of this parameter for the whole classes of test functions can increase considerably the number of trials. For this reason, for the geometric and information methods with the use of optimistic local improvement applied to Class 1 we have not increased r until all test problems were solved to further appreciate the speed of this approach. We have chosen to stop until at least 90 of the 100 problems were solved and for the remaining ones we obtained a local minimum. In Table 2 we reported the averages of trials using the optimistic local improvement on both classes and the percentage of problems solved.

Table 1. Results of numerical experiments without the optimistic local improvement

Method	Class 1		Class 2	
	r	Average	r	Average
GEOM-AL	1.1	190.28	1.1	3756.54
GEOM-GL	1.4	205.42	1.3	2905.17
GEOM-LTM	1.7	93.85	1.3	217.11
GEOM-LTMA	2.5	97.92	1.5	196.00
INF-AL	2.0	188.79	2.0	3709.51
INF-GL	2.8	211.64	2.0	2211.69
INF-LTM	3.7	103.00	2.1	158.12
INF-LTMA	4.0	**78.05**	2.3	130.71
INF-LTMI	5.4	208.09	3.4	1913.82
GEOM-LTIMP	1.7	98.04	1.1	153.07
GEOM-LTIMAP	2.3	93.56	1.5	202.15
INF-LTIMP	3.7	107.98	2.1	160.84
INF-LTIMAP	4.2	86.01	2.1	**114.76**
INF-LTMIP	5.6	215.10	3.4	1892.04

Table 2. Results of numerical experiments with the optimistic local improvement

Method	Class 1			Class 2		
	r	Average	Success	r	Average	Success
GEOM-LTIMO	1.7	55.16	92%	1.8	55.40	100%
GEOM-LTIMAO	1.6	**47.90**	91%	1.7	55.74	100%
INF-LTIMO	2.3	54.48	91%	3.1	51.70	100%
INF-LTIMAO	4.0	61.34	95%	3.1	56.78	100%
INF-LTMIO	5.5	51.78	90%	3.4	**50.66**	100%

It has been shown that the two acceleration techniques, described in this work, increase considerably the speed of geometric and information methods, especially when they are tested on Class 2. Mixing the considered procedures we obtained 19 different derivative-free Lipschitz Global Optimization Methods. In particular, the use of the geometric interpretation of the stochastic model led to the introduction of 3 new methods (INF-LTIMI, INF-LTIMO, INF-LTIMIP).

Acknowledgements. The work of M.S. Mukhametzhanov was supported by the INdAM-GNCS funding "Giovani Ricercatori 2018–2019".

References

1. Barkalov, K.A., Gergel, V.P.: Parallel global optimization on GPU. J. Glob. Optim. **66**(1), 3–20 (2016)
2. Calvin, J.M., Žilinskas, A.: One-dimensional global optimization for observations with noise. Comput. Math. Appl. **50**(1–2), 157–169 (2005)
3. Daponte, P., Grimaldi, D., Molinaro, A., Sergeyev, Y.D.: Fast detection of the first zero-crossing in a measurement signal set. Measurement **19**(1), 29–39 (1996)
4. Floudas, C.A., Pardalos, P.M.: State of the Art in Global Optimization. Kluwer Academic Publishers, Dordrecht (1996)
5. Gergel, V.P., Grishagin, V.A., Israfilov, R.A.: Local tuning in nested scheme of global optimization. Procedia Comput. Sci. **51**, 865–874 (2015)
6. Grishagin, V.A., Israfilov, R.A., Sergeyev, Y.D.: Comparative efficiency of dimensionality reduction schemes in global optimization. In: Proceedings of the 2nd International Conference on "Numerical Computations: Theory and Algorithms", vol. 1776, p. 060011. AIP Publishing, New York (2016). https://doi.org/10.1063/1.4965345
7. Grishagin, V.A., Israfilov, R.A., Sergeyev, Y.D.: Convergence conditions and numerical comparison of global optimization methods based on dimensionality reduction schemes. Appl. Math. Comput. **318**, 270–280 (2018)
8. Hansen, P., Jaumard, B., Lu, S.H.: Global optimization of univariate Lipschitz functions: II. New algorithms and computational comparison. Math. Program. **55**(1–3), 273–292 (1992)
9. Kvasov, D.E., Pizzuti, C., Sergeyev, Y.D.: Local tuning and partition strategies for diagonal GO methods. Numer. Math. **94**(1), 93–106 (2003)
10. Kvasov, D.E., Sergeyev, Y.D.: Deterministic approaches for solving practical blackbox global optimization problems. Adv. Eng. Softw. **80**, 58–66 (2015)
11. Lera, D., Sergeyev, Y.D.: An information global minimization algorithm using the local improvement technique. J. Glob. Optim. **48**(1), 99–112 (2010)
12. Lera, D., Sergeyev, Y.D.: Acceleration of univariate global optimization algorithms working with Lipschitz functions and Lipschitz first derivatives. SIAM J. Optim. **1**(23), 508–529 (2013)
13. Modorskii, V.Y., Gaynutdinova, D.F., Gergel, V.P., Barkalov, K.A.: Optimization in design of scientific products for purposes of cavitation problems. In: Proceedings of the International Conference of Numerical Analysis and Applied Mathematics (ICNAAM 2015), vol. 1738, p. 400013. AIP Publishing, New York (2016). https://doi.org/10.1063/1.4952201
14. Paulavičius, R., Sergeyev, Y.D., Kvasov, D.E., Žilinskas, J.: Globally-biased DISIMPL algorithm for expensive global optimization. J. Glob. Optim. **59**(2–3), 545–567 (2014)
15. Pintér, J.D.: Global Optimization in Action (Continuous and Lipschitz Optimization: Algorithms, Implementations and Applications). Kluwer, Dordrecht (1996)
16. Piyavskij, S.A.: An algorithm for finding the absolute extremum of a function. USSR Comput. Math. Math. Phys. **12**(4), 57–67 (1972)
17. Sergeyev, Y.D.: An information global optimization algorithm with local tuning. SIAM J. Optim. **5**(4), 858–870 (1995)
18. Sergeyev, Y.D.: A one-dimensional deterministic global minimization algorithm. Comput. Math. Math. Phys. **35**(5), 705–717 (1995)
19. Sergeyev, Y.D.: Global one-dimensional optimization using smooth auxiliary functions. Math. Program. **81**(1), 127–146 (1998)

20. Sergeyev, Y.D., Daponte, P., Grimaldi, D., Molinaro, A.: Two methods for solving optimization problems arising in electronic measurements and electrical engineering. SIAM J. Optim. **10**(1), 1–21 (1999)
21. Sergeyev, Y.D., Grishagin, V.A.: A parallel method for finding the global minimum of univariate functions. J. Optimiz. Theor. Appl. **80**(3), 513–536 (1994)
22. Sergeyev, Y.D., Kvasov, D.E.: Deterministic Global Optimization: An Introduction to the Diagonal Approach. Springer, New York (2017)
23. Sergeyev, Y.D., Kvasov, D.E., Mukhametzhanov, M.S.: On strong homogeneity of a class of global optimization algorithms working with infinite and infinitesimal scales. Commun. Nonlinear Sci. **59**, 319–330 (2018)
24. Sergeyev, Y.D., Kvasov, D.E., Mukhametzhanov, M.S.: On the efficiency of nature-inspired metaheuristics in expensive global optimization with limited budget. Nat. Sci. Rep. **8**, Article 453 (2018). https://doi.org/10.1038/s41598-017-18940-4
25. Sergeyev, Y.D., Mukhametzhanov, M.S., Kvasov, D.E., Lera, D.: Derivative-free local tuning and local improvement techniques embedded in the univariate global optimization. J. Optimiz. Theor. Appl. **171**(1), 319–330 (2016)
26. Sergeyev, Y.D., Strongin, R.G., Lera, D.: Introduction to Global Optimization Exploiting Space-Filling Curves. Springer, New York (2013)
27. Strongin, R.G.: On the convergence of an algorithm for finding a global extremum. Eng. Cybern. **11**, 549–555 (1973)
28. Strongin, R.G., Sergeyev, Y.D.: Global Optimization with Non-convex Constraints: Sequential and Parallel Algorithms. Kluwer Academic Publishers, Dordrecht (2000)
29. Zhigljavsky, A., Žilinskas, A.: Stochastic Global Optimization. Springer, New York (2008)

Numerical Simulation of Ski-Jump Hydraulic Behavior

Agostino Lauria$^{(\boxtimes)}$ (iD) and Giancarlo Alfonsi (iD)

Fluid Dynamics Laboratory, University of Calabria, Via Bucci 42b, 87036 Rende, Cosenza, Italy
{agostino.lauria,giancarlo.alfonsi}@unical.it

Abstract. The hydraulic behavior of ski jumps is investigated numerically using the OpenFOAM digital library. A number of ski-jump cases has been simulated by following the RANS approach (Reynolds Averaged Navier-Stokes equations), using the k-ω SST closure model, and the VoF technique (Volume of Fluid) for the tracking of the free surface. Particular attention is given to the pressure distributions in the zone of impact of the falling jet, and to the length of the jet itself, as defined as the distance along the x-direction between the point of maximum dynamic pressure head, and the origin of the reference frame. A chart is proposed reporting the correlation line (and correspondent formal expression) between the approach Froude numbers and the lengths of the jets, in the limit of other parameters tested. The chart may serve as a useful tool for the determination of the length of the jet taking off from the bucket, starting from the value of the approach Froude number.

Keywords: Ski-Jump · Reynolds averaged Navier-Stokes equations · Volume of fluid

1 Introduction

Ski jumps have been first introduced in the field of dam construction in the 1930s, and have been studied in the early times within the potential-flow approach. By looking at the literature of the more recent years, a number of works of different type and dealing with different aspects of the phenomena at hand, can be found. Among others, in the experimental field, Bathe et al. [1] reported on a case study related to the optimization of the design of a two-tier spillway. De Lara et al. [2] studied the spillway flow originally designed with a conventional ski jump. Deng et al. [3] proposed a leak-floor flip bucket in which the middle of the water is lifted into the atmosphere and deflects transversely. Gou et al. [4] studied the effect of sediment concentration on the hydraulic characteristics of energy dissipation in a falling turbulent jet from a ski-jump energy dissipator. Felder and Chanson [5] executed a number of experiments on a two-scaled stepped spillway to investigate the scale effects in terms of air-water properties. Xu et al. [6] performed an experimental study on pressure and aeration characteristics in stepped-chute flows. Li et al. [7] executed a theoretical and experimental study on flaring gate pier on the surface spillway in an arch dam. Wu et al. [8] studied the hydraulic characteristic of slit-type energy dissipaters. In the numerical field, Deng et al. [9] proposed a new type of design for a streamwise-lateral spillway. Chanson [10] performed a theoretical study on aeration

© Springer Nature Switzerland AG 2020
Y. D. Sergeyev and D. E. Kvasov (Eds.): NUMTA 2019, LNCS 11974, pp. 422–429, 2020.
https://doi.org/10.1007/978-3-030-40616-5_39

of a free jet above a spillway, based on dimensional analysis. Overall, the majority of the above-mentioned works report on specific case studies executed on reduced-scale laboratory models, as related to the design of actual dams to be built. The results are interesting but not susceptible to generalization.

Differently, Juon and Hager [11] and Heller et al. [12] performed a systematic experimental research about the hydraulic behavior of ski jumps. They considered a simplified ski-jump configuration that involved an horizontal approach channel and a tailwater channel, and executed experimental tests with different values of the characteristic parameters of the phenomenon at hand, with the aim of obtaining results of a more general effectiveness.

In the present work, the hydraulic behavior of ski jumps is investigated numerically using the OpenFOAM digital library. The numerical model has been first calibrated by comparing the numerical results with some of those obtained by Heller et al. [12], and then a number of ski-jump cases has been simulated. Particular attention is given to the pressure distributions in the zone of impact of the falling jet, and to the length of the jet itself. The results are reported in general form and may provide useful suggestions for those involved in dam construction and management.

2 Numerical Procedures

The flow cases at hand have been simulated by solving the three-dimensional Reynolds Averaged Navier-Stokes (RANS) equations (Alfonsi [13]) in conservative form (the fluid is incompressible and viscous, Einstein summation convention applies to repeated indices, $i, j = 1, 2, 3$):

$$\rho \frac{\partial \bar{u}_i}{\partial t} + \rho \bar{u}_j \frac{\partial \bar{u}_i}{\partial x_j} = -\frac{\partial \bar{p}}{\partial x_i} + \frac{\partial}{\partial x_j} \left(2\mu \bar{s}_{ij} - \rho \overline{u'_i u'_j} \right) \tag{1}$$

$$\frac{\partial \bar{u}_i}{\partial x_i} = 0 \tag{2}$$

where ρ is the fluid density, μ is the fluid dynamic viscosity, $\bar{p}$ is the mean fluid pressure, $\bar{u}_i$ are the fluid mean velocity components, $\bar{s}_{ij}$ is the mean strain-rate tensor, and $\tau_{ij} = -\rho \overline{u'_i u'_j}$ is the Reynolds stress tensor. The Reynolds averaging procedure introduces six new unknown quantities (the six independent components of τ_{ij}) without providing additional equations. To close the system, the Boussinesq approximation is introduced (Wilcox [14]) and the eddy-viscosity is expressed as a function of the turbulent kinetic energy (k) and the energy-dissipation rate (ω), leading to a two-equation turbulence model. Two-equations models provide one equation to compute k and another equation to calculate ω. In the present study in particular, the k-ω SST model developed by Menter et al. [15] has been used. In this model, the turbulent kinetic energy and the energy-dissipation rate are computed using the following expressions:

$$\rho \frac{\partial k}{\partial t} + \rho \bar{u}_j \frac{\partial k}{\partial x_j} = \tilde{P}_k - \beta^* \rho k \omega + \frac{\partial}{\partial x_j} \left[(\mu + \sigma_{k1} \mu_t) \frac{\partial k}{\partial x_j} \right] \tag{3}$$

$$\rho \frac{\partial \omega}{\partial t} + \rho \bar{u}_j \frac{\partial \omega}{\partial x_j} = \alpha \rho S^2 - \beta \rho \omega^2 + \frac{\partial}{\partial x_j} \left[(\mu + \sigma_{\omega 1} \mu_t) \frac{\partial \omega}{\partial x_j} \right]$$
$$+ 2(1 - F_1) \rho \sigma_{\omega 2} \frac{1}{\omega} \frac{\partial k}{\partial x_j} \frac{\partial \omega}{\partial x_j} \qquad (4)$$

where $\tilde{P}_k$ represents a production limiter used in the model to prevent the build-up of the turbulence in stagnation regions, F_1 represents a blending function defined as:

$$F_1 = \tanh \left\{ \left\{ \min \left[\max \left(\frac{\sqrt{k}}{\beta^* \omega y}, \frac{500 \upsilon}{y^2 \omega} \right), \frac{4 \rho \sigma_{\omega 2} k}{C D_{k\omega} y^2} \right] \right\}^4 \right\} \qquad (5)$$

with $C D_{k\omega} = \max \left(2 \rho \sigma_{\omega 2} \frac{1}{\omega} \frac{\partial k}{\partial x_j} \frac{\partial \omega}{\partial x_j}, 10^{-10} \right)$, and y represent the distance from the nearest wall. The turbulent eddy viscosity is $\mu_t = \frac{\rho a_1 k}{\max(a_1 \omega, SF_2)}$, where $a_1 = 0.31$, S the second invariant of the deviatoric stress tensor and F_2 is a second blending function:

$$F_2 = \tanh \left[\left[\max \left(\frac{2\sqrt{k}}{\beta^* \omega y}, \frac{500 \upsilon}{y^2 \omega} \right) \right]^2 \right] \qquad (6)$$

All the constants are predicted through a blend from the corresponding constants. The choice of the k-ω SST model is due to its generally-superior performance with respect, for example, to the more classical two-equation k-ε model in a variety of complex flow cases [15–17].

The system of the governing equations has been solved by means of the Inter-Foam solver, embedded in the OpenFoam C++ libraries. The InterFoam solver has been designed for incompressible, isothermal, immiscible fluids and incorporates the VoF (Volume of Fluid) technique for the capturing of the interfaces. The VoF method (Hirt and Nichols [18]) has been already used in several different flow cases always giving satisfactory results ([19–21] among others). The governing equations are discretized with the method of the Finite Volumes. As for the discretization of the solution domain, an unstructured multi-block mesh has been built where the dependent variables are stored at the cell center of each cell-space domain in a co-located arrangement. The PISO (Pressure Implicit with Split Operator) technique suggested by Issa [22, 23] has been employed to couple pressure and velocity in the transient computations. The PISO procedure adopts a segregated approach, and the system of the equations is solved sequentially.

Table 1. Fluid properties used in the simulations.

Air density	Water density	Air kinematic viscosity	Water kinematic viscosity
1.225 kg/m^3	1000 kg/m^3	1.48×10^{-5} m^2/s	1.0×10^{-6} m^2/s

The stability of the solution procedure has been ensured utilizing an adaptive time step with an initial value of 1×10^{-6} s, in conjunction with a mean Courant-Friedrichs-Lewy (CFL) number limit of 0.5. The fluid properties used in the simulations are reported in Table 1.

As for the numerical channel, a rectangular horizontal approach channel with the same characteristics of that of Heller et al. [12] has been considered (Fig. 1). The width of the channel is b (0.50 m), the height is 0.70 m, the total length is about 7 m, the bucket radius is R and the bucket angle is β. The water depth of the approaching flow is h_0, the flowrate is Q and the approaching-flow bulk velocity is $V_0 = Q/(bh_0)$. The approaching flow can be described in terms of the approach Froude number $F_0 = V_0/(gh_0)^{1/2}$, and the relative bucket radius h_0/R, while the values of the bucket height $w = R(1 - cos\beta)$ have been considered as only related to free-bucket flow (not chocked). The maxima values of the upper and lower jet trajectories are, respectively h_U and h_L while the take-off angles are α_U and α_L. The elevation difference $(s - w)$ between the approach and the tailwater channel has been kept constant $(s = 0.25$ m). The point of maximum dynamic pressure head in the zone in which the falling jet impacts with the bottom of the -initially empty - tailwater channel along its centerline is A, while the length of the jet, as defined as the distance of A from the origin of the reference frame along the x-direction, is L (Fig. 1). Boundary conditions of no-slip and zero wall-normal velocity at the solid walls have been imposed.

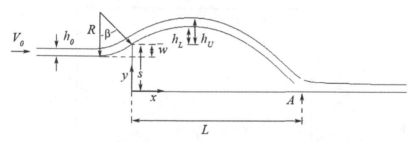

Fig. 1. Definition sketch of ski-jump flow

A CPU-based computing system has been used for the calculations. The system includes 3 worker nodes, each equipped with 4 E5-2640 CPU (a total of 96 cores/16 threads at 2.0 GHz), 128 GB RAM at 1899 MHz, and 1 TB disk space. The simulations have been executed using 16 processors through the public domain OpenMPI implementation of the standard Message Passing Interface (MPI) for parallel running.

Table 2. Characteristic parameters of the computational grid.

N_x	N_y	N_z	N_{tot}	Δx_min (m)	Δy_min (m)	Δz_min (m)
350	100	50	1,750,000	0.01	0.005	0.005

For parallel computing, the technique of the domain decomposition has been adopted to split the geometry and the associated fields into segments. In this study, the "simple geometric decomposition" technique has been used, in which the domain is broken into segments by direction. The elapsed computational time has been of about 5 h of CPU time for each simulation. As for grid refinement, the unstructured computational mesh has been refined through different steps, up to the point in which the comparisons of the computed results with those obtained by other authors became satisfactory. The final configuration of the computational grid is reported in Table 2.

3 Results

Seventeen ski-jump cases have been simulated numerically. In Fig. 2 some of the computed nondimensional dynamic pressure head distributions along the bucket centerline at the value of β equal to 40° are compared with reference Eq. (5) of the work of Heller er al. [12]. In Fig. 3 a visual comparisons of jets from ski jumps ($R = 0.10$ m, $\beta = 40°$, $h_0 = 0.05$ m) at approach Froude number $F_0 = 5$ is shown. In the figure the experimental jet of Heller et al. [12] is compared with the flow visualization of present work in terms of velocity fields, where reddish colors mirror the highest values of the fluid velocities, and bluish colors mirror the lowest values. Overall the comparison is rather satisfactory.

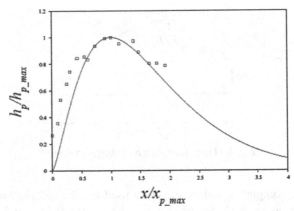

Fig. 2. Local pressure head distribution along bucket centerline: (—) measured values from Eq. (5) of Heller et al. [12], versus values from present work (symbols).

As mentioned before, we define here the length of the jet as the distance (L) along the x-direction, between the point of maximum dynamic pressure head in the zone of impact of the jet along the centerline of the tailwater channel, and the origin of the reference frame (Fig. 1). In Fig. 4 a chart is shown in which the correlation line between values of the approach Froude number (F_0) and the nondimensional lengths of the jet (L/s) as previously defined, and involving some of the simulated ski-jump cases with $R = 0.10$ m, are reported. The line in the chart of Fig. 4 mirror the expression reported in Table 3.

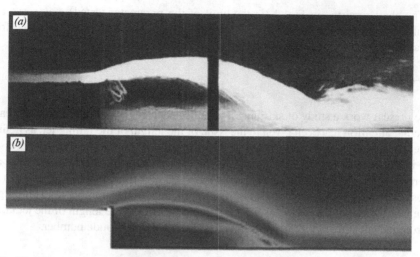

Fig. 3. Visual comparison (side view) between jets from ski jumps with $R = 0.10$ m, $\beta = 40°$, $h_0 = 0.05$ m, $F_0 = 5$: (a) picture from Heller et al. [12], (b) flow visualization of present work in terms of velocity field (reddish colors mirror highest fluid velocities, bluish colors mirror lowest fluid velocities). (Color figure online)

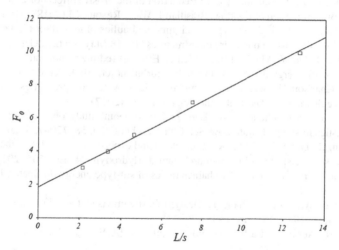

Fig. 4. Chart of correlation line between values of the approach Froude number (F_0) and nondimensional length of the jet (L/s) at $\beta = 40°$ involving some of the simulated ski-jump cases with $R = 0.10$ m.

Table 3. Expression of correlation line in chart of Fig. 4.

R (m)	β (°)	r^2	Expression	Expression number
0.10	40	0.994	$F_0 = 0.652$ (L/s) $-$ 1.832	(7)

The chart reported in Fig. 4 (in the limit of the parameter tested) may serve as a useful tool to determine the length of the jet taking off from the bucket, starting from the value of the approach Froude number.

4 Conclusions

In the present work a study of ski-jump hydraulics has been accomplished numerically. Particular attention has been given to the issue of the length of the falling jet, as defined as the distance along the x-direction between the point of maximum dynamic pressure head in the zone of impact of the jet along the centerline of the tailwater channel, and the origin of the reference frame. A chart is proposed reporting the correlation line (and corresponding formal expression) between the approach Froude numbers and the lengths of the jets. The chart may serve as a useful tool to determine the length of the jet taking off from the bucket, starting from the value of the approach Froude number.

References

1. Bathe, R.R., More, K.T., Bhajantri, M.R., Bhosekar, V.V.: Hydraulic model studies for optimizing the design of two tier spillway. ISH J. Hydraulic Eng. **25**, 28–37 (2019)
2. De Lara, R., Ota, J.J., Fabiani, A.L.T.: Reduction of the erosive effects of effluent jets from spillways by contraction of the flow. Brasilian J. Water Resour. **23** (2018). Art. E11
3. Deng, J., Wei, W., Tian, Z., Zhang, F.: Ski jump hydraulics of leak-floor flip bucket. In: 7th International Symposium on Hydraulic Structures, 15–18 May Aachen, Germany (2018)
4. Gou, W., Li, H., Du, Y., Yin, H., Liu, F., Lian, J.: Effect of sediment concentration on hydraulic characteristic of energy dissipation in a falling turbulent jet. Appl. Sci. **8** (2018). Art. 1672
5. Felder, S., Chanson, H.: Scale effects in microscopic air-water flow properties in high-velocity free-surface flows. Exp. Thermal Fluid Sci. **83**, 19–36 (2017)
6. Xu, W.L., Luo, S.J., Zheng, Q.W., Luo, J.: Experimental study on pressure and aeration characteristics in stepped chute flow. Sci. China Technol. Sci. **58**, 720–726 (2015)
7. Li, N., Liu, C., Deng, J., Zhang, X.: Theoretical and experimental studies of the flaring gate pier on the surface spillway in a high-arch dam. J. Hydrodyn. **24**, 496–505 (2012)
8. Wu, J., Ma, F., Yao, L.: Hydraulic characteristics of slit-type energy dissipater. J. Hydrodyn. **24**, 883–887 (2012)
9. Deng, J., Wei, W., Tian, Z., Zhang, F.: Design of a streamwise lateral ski-jump flow discharge spillway. Water **10** (2018). Art. 1585
10. Chanson, H.: Aeration of a free jet above a spillway. IAHR J. Hydraulic Res. **29**, 655–667 (1991)
11. Juon, R., Hager, W.H.: Flip bucket without and with deflectors. ASCE J. Hydraulic Eng. **11**, 837–845 (2000)
12. Heller, V., Hager, W.H., Minor, H.E.: Ski jump hydraulics. ASCE J. Hydraulic Eng. **5**, 347–355 (2005)
13. Alfonsi, G.: Reynolds averaged Navier-Stokes equations for turbulence modeling. Appl. Mech. Rev. **63** (2009). Art. 040802
14. Wilcox, D.C.: Turbulence modeling for CFD. DCW Industries, La Cañada, California, USA (1998)
15. Menter, F.R., Kunz, M., Langtry, R.: Ten years of industrial experience with the SST turbulence model. In: Proceedings 4th International Symposium on Turbulence, Heat and Mass Transfer, 12–17 October, Antalya, Turkey (2003)

16. Calomino, F., Alfonsi, G., Gaudio, R., D'Ippolito, A., Lauria, A., Tafarojnoruz, A., Artese, S.: Experimental and numerical study of free-surface flows in a corrugated pipe. Water **10** (2018). Art. 638
17. D'Ippolito, A., Lauria, A., Alfonsi, G., Calomino, F.: Investigation of flow resistance exerted by rigid emergent vegetation in open channel. Acta Geophysica Online (2019). https://doi.org/10.1007/s11600-019-00280-8
18. Hirt, C.W., Nichols, B.D.: Volume of Fluid (VoF) method for the dynamics of free boundaries. J. Comput. Phys. **39**, 201–225 (1981)
19. Alfonsi, G., Lauria, A., Primavera, L.: On evaluation of wave forces and runups on cylindrical obstacles. J. Flow Vis. Image Process. **20**, 269–291 (2013)
20. Alfonsi, G., Lauria, A., Primavera, L.: The field of flow structures generated by a wave of viscous fluid around vertical circular cylinder piercing the free surface. Procedia Eng. **116**, 103–110 (2015)
21. Alfonsi, G., Lauria, A., Primavera, L.: Recent results from analysis of flow structures and energy modes induced by viscous wave around a surface-piercing cylinder. Math. Probl. Eng. **2017** (2017). Art. 5875948
22. Issa, R.I.: Solution of the implicitly discretized fluid flow equations by operator splitting. J. Comput. Phys. **62**, 40–65 (1986)
23. Alfonsi, G., Lauria, A., Primavera, L.: A study of vortical structures past the lower portion of the Ahmed car model. J. Flow Vis. Image Process. **19**(1), 81–95 (2013)

On Linear Spline Wavelets
with Shifted Supports

Svetlana Makarova[1] and Anton Makarov[2]

[1] St. Petersburg State University of Aerospace Instrumentation,
67, Bolshaya Morskaya str., St. Petersburg 190000, Russia
[2] St. Petersburg State University,
7/9, Universitetskaya nab., St. Petersburg 199034, Russia
a.a.makarov@spbu.ru

Abstract. We examine Faber's type decompositions for spaces of linear minimal splines constructed on nonuniform grids on a segment. A characteristic feature of the Faber decomposition is that the basis wavelets are centered around the knots that do not belong to the coarse grid. The construction of the lazy wavelets begins with the use of the basis functions in refined spline space centered at the odd knots. We propose to use as wavelets the functions centered at the even knots under some conditions. In contrast to lazy wavelets, in this case the decomposition system of equations has a unique solution, which can be found by the sweep method with the guarantee of well-posedness and stability.

Keywords: Minimal splines · B-spline · Wavelets · Nonuniform grid

1 Introduction

One of the simplest and the most common processing schemes for digital data arrays is piecewise linear interpolation on uniform grids. Investigating the construction of a sequence of continuous piecewise linear functions that converge pointwise to a continuous nowhere differentiable function, Faber [1] introduced a hierarchical representation of functions by linear B-splines as a series based on piecewise linear interpolation on nested dyadic grids. It is well known if this scheme is used as a basis for multiresolution analysis, then it is often referred to as Faber decomposition. If a grid is uniform, one can apply classical wavelets using the powerful tools of harmonic analysis (in the space of functions $L^2(\mathbb{R}^1)$ and in the space of sequences l^2). However, many applications require considering bounded intervals and nonuniform grids. In this important case the methods of harmonic analysis are not easily applicable, so it is less well studied. Some examples and recommendations can be found in [2–8].

The construction of classical wavelets is based on the solution of refinement equations. Instead of this we examine the calibration relations such that each

The reported study was funded by a grant of the President of the Russian Federation (MD-2242.2019.9).

Y. D. Sergeyev and D. E. Kvasov (Eds.): NUMTA 2019, LNCS 11974, pp. 430–437, 2020.
https://doi.org/10.1007/978-3-030-40616-5_40

basis function on a coarse grid can be represented as a linear combination of the basis functions on a refine grid. This paper continues the studies initiated in [9–11]. We consider Faber's type decompositions for spaces of linear minimal splines constructed on nonuniform grids on a segment. The approximation relations are used as an initial structure for constructing the mentioned spaces. The method of approximation relations used for construction of different types of minimal splines has been initiated by Dem'yanovich [12]. The minimal splines of maximal smoothness [13] are nonpolynomial generalization of B-splines and as a special case include well known polynomial B-splines and share most properties of B-splines (smoothness, nonnegativity, etc.).

The construction of the lazy wavelets begins with the use of the basis functions in refined spline space centered at the odd knots. We propose to use as wavelets the functions centered at the even knots (on nonuniform grid) under the additional condition that the spline must vanish at the last knot on the considered segment. The approach is based on direct decomposition of space of linear minimal splines by using the calibration relations for two-fold refining of almost arbitrary initial nonuniform grid. Against to lazy wavelets, in this case the decomposition system of equations has a unique solution, which can be found by the sweep method with the guarantee of well-posedness and stability.

2   The Space of Coordinate Splines

On a closed interval $[a, b] \subset \mathbb{R}^1$ consider a partition X with two supplementary knots outside of $[a, b]$:

$$X : x_{-1} < a = x_0 < x_1 < \ldots < x_{n-1} < x_n = b < x_{n+1}. \tag{1}$$

Introduce the notation $J_{i,k} := \{i, i+1, \ldots, k\}, i, k \in \mathbb{Z}, i < k$. Let $\{\mathbf{a}_j\}$ be an ordered set of vectors $\mathbf{a}_j \in \mathbb{R}^2$, $j \in J_{-1,n-1}$. The vector components are denoted by square brackets and supplied with subscripts for convenience. For instance, $\mathbf{a}_j = ([\mathbf{a}_j]_0, [\mathbf{a}_j]_1)^T$, where T is the transpose operation.

We assume that the square matrices of the second order $(\mathbf{a}_{j-1}, \mathbf{a}_j)$, composed of the two column vectors $\mathbf{a}_{j-1}, \mathbf{a}_j$, are nonsingular:

$$\det(\mathbf{a}_{j-1}, \mathbf{a}_j) \neq 0 \qquad \forall j \in J_{-1,n-1}. \tag{2}$$

Let $M := \cup_{j \in J_{-1,n}}(x_j, x_{j+1})$ be the union of all elementary partition intervals and let $\mathbb{X}(M)$ be the linear space of real-valued functions given on the set M.

Consider a vector function $\boldsymbol{\varphi} : [a, b] \mapsto \mathbb{R}^2$ with components in the space $C^1[a, b]$ and nonzero Wronskian $|\det(\boldsymbol{\varphi}, \boldsymbol{\varphi}')(t)| \geq const > 0$ for all $t \in [a, b]$.

Suppose that functions $\omega_j \in \mathbb{X}(M)$, $j \in J_{-1,n-1}$, satisfy the approximation relations

$$\sum_{j'=k-1}^{k} \mathbf{a}_{j'} \omega_{j'}(t) \equiv \boldsymbol{\varphi}(t) \quad \forall t \in (x_k, x_{k+1}), \forall k \in J_{-1,n-1},$$

$$\omega_j(t) \equiv 0 \qquad \forall t \notin [x_j, x_{j+2}] \cap M. \tag{3}$$

For each fixed $t \in (x_k, x_{k+1})$ relations (3) can be regarded as a system of linear algebraic equations with respect to the unknowns $\omega_j(t)$. By virtue of assumption (2) the system (3) is uniquely solvable and $\operatorname{supp} \omega_j(t) \subset [x_j, x_{j+2}]$.

We introduce the notation $\varphi_j := \varphi(x_j)$ and consider the vectors $\mathbf{d}_j \in \mathbb{R}^2$, $j \in J_{-1,n+1}$, defined by the identity $\mathbf{d}_j^T \mathbf{z} \equiv \det(\varphi_j, \mathbf{z})$, where $\mathbf{z} \in \mathbb{R}^2$.

As is known [13], if the vectors $\mathbf{a}_j \in \mathbb{R}^2, j \in J_{-1,n-1}$, are defined by the formula $\mathbf{a}_j := \varphi_{j+1}$, then $\omega_j \in C[a,b]$ and the following formulas are valid

$$\omega_j(t) = \begin{cases} \dfrac{\mathbf{d}_j^T \varphi(t)}{\mathbf{d}_j^T \mathbf{a}_j}, & t \in [x_j, x_{j+1}), \\ \dfrac{\mathbf{d}_{j+2}^T \varphi(t)}{\mathbf{d}_{j+2}^T \mathbf{a}_j}, & t \in [x_{j+1}, x_{j+2}). \end{cases} \tag{4}$$

The space

$$\mathbb{S}(X) := \{u \; : \; u = \sum_{j=-1}^{n-1} c_j \, \omega_j \quad \forall c_j \in \mathbb{R}^1\}$$

is called the *space of minimal linear B_φ-splines (of the second order)* on the partition X. The splines themselves are called the *minimal splines of maximal smoothness*.

The vectors $\mathbf{d}_j$ and $\mathbf{a}_j$ in componentwise form have the representation as follows $\mathbf{d}_j = (-[\varphi_j]_1, [\varphi_j]_0)^T$, $\mathbf{a}_j = ([\varphi_{j+1}]_0, [\varphi_{j+1}]_1)^T$.

If $[\varphi(t)]_0 \equiv 1$, i. e. $\varphi(t) = (1, \rho(t))^T$, where $\rho \in C^1[a,b]$, then we have the identity

$$\sum_{j=-1}^{n-1} \omega_j(t) \equiv 1 \quad \forall t \in [a,b],$$

and if we let $\rho_j := \rho(x_j)$, then we obtain the formulas (4) in the form

$$\omega_j(t) = \begin{cases} \dfrac{\rho(t) - \rho_j}{\rho_{j+1} - \rho_j}, & t \in [x_j, x_{j+1}), \\ \dfrac{\rho_{j+2} - \rho(t)}{\rho_{j+2} - \rho_{j+1}}, & t \in [x_{j+1}, x_{j+2}). \end{cases} \tag{5}$$

Obviously, $\omega_j(x_i) = \delta_{j,i-1}$, where $\delta_{j,i}$ is the Kronecker symbol. Moreover, if a function $\rho(t)$ is strictly monotone on the set M, then the spline $\omega_j(t) > 0$ for all $t \in (x_j, x_{j+2})$.

For $\varphi(t) = (1, t)^T$ the functions ω_j coincide with the known polynomial B-splines of the first degree, i. e. with the one-dimensional Courant functions.

3   Calibration Relations

If we extend the original partition X by a new knot $\xi \in (x_k, x_{k+1})$, $k \in J_{0,n-1}$, then as a result we get a *refined partition* $\overline{X} := \{\overline{x}_j \; : \; j \in J_{-1,n+2}\}$ such that

$$\overline{x}_j := \begin{cases} x_j, & j \le k, \\ \xi, & j = k+1, \\ x_{j-1}, & j \ge k+2. \end{cases} \tag{6}$$

For convenience, we will explicitly indicate the partition on which a certain object is considered. For instance, the functions $w_j^{\overline{X}}(t)$, $j \in J_{-1,n}$, can be found using (5) by replacing the knots x_j with the knots $\overline{x}_j$, $j \in J_{-1,n+2}$.

It is easy to see that

$$\mathbf{d}_j^{\overline{X}} = \mathbf{d}_j^X, \ \ j \leq k, \qquad \mathbf{d}_j^{\overline{X}} = \mathbf{d}_{j-1}^X, \ \ j \geq k+2, \tag{7}$$

$$\mathbf{d}_{k+1}^{\overline{X}} \, ^T \mathbf{x} \equiv \det(\boldsymbol{\varphi}(\xi), \mathbf{x}), \ \ \mathbf{x} \in \mathbb{R}^2, \qquad \mathbf{a}_k^{\overline{X}} = \boldsymbol{\varphi}(\overline{x}_{k+1}) = \boldsymbol{\varphi}(\xi), \tag{8}$$

$$\mathbf{a}_j^{\overline{X}} = \mathbf{a}_j^X, \ \ j \leq k-1, \qquad \mathbf{a}_j^{\overline{X}} = \mathbf{a}_{j-1}^X, \ \ j \geq k+1. \tag{9}$$

Lemma 1. *For $k \in J_{0,n-1}$ and $t \in [a,b]$ the following calibration relations hold:*

$$w_i^X(t) = \sum_{j \in J_{-1,n}} \mathbf{p}_{i,j}^{\overline{X}} w_j^{\overline{X}}(t) \quad \forall i \in J_{-1,n-1}, \tag{10}$$

the values $\mathbf{p}_{i,j}^{\overline{X}} \in \mathbb{R}^1$ are given by the formulas

$$\mathbf{p}_{i,j}^{\overline{X}} = \begin{cases} \delta_{i,j}, & i \leq k-2, \ \forall j, \\ \delta_{k-1,j}, & i = k-1, j \neq k, \\ \delta_{k+1,j}, & i = k, j \neq k, \\ \delta_{i,j-1}, & i \geq k+1, \ \forall j, \end{cases} \tag{11}$$

and by the formulas

$$\mathbf{p}_{k-1,k}^{\overline{X}} = \frac{\mathbf{d}_{k+1}^{\overline{X}}\, ^T \mathbf{a}_{k+1}^{\overline{X}}}{\mathbf{d}_k^{\overline{X}}\, ^T \mathbf{a}_{k+1}^{\overline{X}}}, \tag{12}$$

$$\mathbf{p}_{k,k}^{\overline{X}} = \frac{\mathbf{d}_k^{\overline{X}}\, ^T \mathbf{a}_k^{\overline{X}}}{\mathbf{d}_k^{\overline{X}}\, ^T \mathbf{a}_{k+1}^{\overline{X}}}. \tag{13}$$

Proof. The validity of the assertions directly follows from the papers [9,11].

Below, we consider the refinement of the original partition X by insertion a single knot ξ_j into every elementary partition interval (x_j, x_{j+1}), $j \in J_{0,n-1}$.

The *two-fold refining* of the partition X is defined as the partition $Y := \{y_j : j \in J_{-1,2n+1}\}$ such that

$$y_j = \begin{cases} x_{-1}, & j = -1, \\ x_{j/2}, & j = 2l, l \in J_{0,n}, \\ \xi_{(j-1)/2}, & j = 2l-1, l \in J_{1,n}, \\ x_{n+1}, & j = 2n+1. \end{cases}$$

On the partition Y consider the splines w_j^Y, $j \in J_{-1,2n-1}$, which can be found using (5) by replacing the knots x_j with the knots y_j.

Theorem 1. *For* $t \in [a, b]$ *the following calibration relations hold:*

$$\omega_j^X(t) = \begin{cases} \omega_{-1}^Y(t) + p_{-1,2}^Y \omega_0^Y(t), & j = -1, \\ \sum_{i=0}^{2} p_{j,i}^Y \omega_{2j+i}^Y(t), & j \in J_{0,n-2}, \\ p_{n-1,0}^Y \omega_{2n-2}^Y(t) + \omega_{2n-1}^Y(t), & j = n-1, \end{cases} \tag{14}$$

the values $p_{j,i}^Y \in \mathbb{R}^1$, $i = 0, 1, 2$, *are given by the formulas*

$$\begin{aligned} p_{j,0}^Y &= \frac{\mathbf{d}_{2j}^{Y}{}^T \mathbf{a}_{2j}^Y}{\mathbf{d}_{2j}^{Y}{}^T \mathbf{a}_{2j+1}^Y}, & j \in J_{0,n-1}, \\ p_{j,1}^Y &= 1, & j \in J_{-1,n-1}, \\ p_{j,2}^Y &= \frac{\mathbf{d}_{2j+3}^{Y}{}^T \mathbf{a}_{2j+3}^Y}{\mathbf{d}_{2j+2}^{Y}{}^T \mathbf{a}_{2j+3}^Y}, & j \in J_{-1,n-2}. \end{aligned} \tag{15}$$

Proof. For the spline functions constructed on the partitions X and $\overline{X}$, in view of the calibration relations (10)–(13), we have

$$\omega_k^X(t) = \mathfrak{p}_{k,k}^{\overline{X}} \omega_k^{\overline{X}}(t) + \omega_{k+1}^{\overline{X}}(t). \tag{16}$$

Consider the partition $Z := \{z_j : z_j = \overline{x}_{j+2}\}$. In the same way as (6), construct the partition $\overline{Z} := \{\overline{z}_j\}$ by adding to the partition Z a new knot $\zeta \in (z_k, z_{k+1})$. Then, in view of the calibration relations (10)–(13), we have

$$\omega_{k-1}^Z(t) = \omega_{k-1}^{\overline{Z}}(t) + \mathfrak{p}_{k-1,k}^{\overline{Z}} \omega_k^{\overline{Z}}(t). \tag{17}$$

Using obvious equalities $\omega_k^{\overline{X}}(t) = \omega_{k-2}^Z(t) = \omega_{k-2}^{\overline{Z}}(t)$ and $\omega_{k+1}^{\overline{X}}(t) = \omega_{k-1}^Z(t)$, from the relations (16)–(17) we find

$$\omega_k^X(t) = \mathfrak{p}_{k,k}^{\overline{X}} \omega_{k-2}^{\overline{Z}}(t) + \omega_{k-1}^{\overline{Z}}(t) + \mathfrak{p}_{k-1,k}^{\overline{Z}} \omega_k^{\overline{Z}}(t). \tag{18}$$

By formulas (7)–(9), we have $\mathbf{d}_k^{\overline{X}} = \mathbf{d}_{k-2}^{\overline{Z}}$, $\mathbf{a}_k^{\overline{X}} = \mathbf{a}_{k-2}^{\overline{Z}}$, and $\mathbf{a}_{k+1}^{\overline{X}} = \mathbf{a}_{k-1}^{\overline{Z}}$, then, in accordance with (13), we find

$$\mathfrak{p}_{k,k}^{\overline{X}} = \frac{\mathbf{d}_{k-2}^{\overline{Z}}{}^T \mathbf{a}_{k-2}^{\overline{Z}}}{\mathbf{d}_{k-2}^{\overline{Z}}{}^T \mathbf{a}_{k-1}^{\overline{Z}}}. \tag{19}$$

Substituting (12) and (19) into (18), we have

$$\omega_k^X(t) = \frac{\mathbf{d}_{k-2}^{\overline{Z}}{}^T \mathbf{a}_{k-2}^{\overline{Z}}}{\mathbf{d}_{k-2}^{\overline{Z}}{}^T \mathbf{a}_{k-1}^{\overline{Z}}} \omega_{k-2}^{\overline{Z}}(t) + \omega_{k-1}^{\overline{Z}}(t) + \frac{\mathbf{d}_{k+1}^{\overline{Z}}{}^T \mathbf{a}_{k+1}^{\overline{Z}}}{\mathbf{d}_k^{\overline{Z}}{}^T \mathbf{a}_{k+1}^{\overline{Z}}} \omega_k^{\overline{Z}}(t). \tag{20}$$

Now we find the representation of the right-hand side of the obtained equality (20) on the partition Y. It is obvious that $\omega_{k-2}^{\overline{Z}}(t) = \omega_{2k}^Y(t)$, $\omega_{k-1}^{\overline{Z}}(t) = \omega_{2k+1}^Y(t)$, $\omega_k^{\overline{Z}}(t) = \omega_{2k+2}^Y(t)$, and $\mathbf{d}_k^{\overline{Z}} = \mathbf{d}_{2k+2}^Y$, $\mathbf{a}_k^{\overline{Z}} = \mathbf{a}_{2k+2}^Y$.

Substituting the obtained relations into (20), we have the provable statement (14). For boundary spline functions, the required equalities are found similarly.

4 Construction of Direct Decompositions

We denote by Δ^L a partition of the form (1), in which $n = 2^L m$, where $L, m \in \mathbb{Z}$, $L \geq 0, m \geq 1$. Against to [11] we will discuss construction of direct decompositions based on calibration relations for two-fold refining of almost arbitrary initial nonuniform grids. The splines can be indexed both by the left knot of the support (5) and also by the central knot. On the partition Δ^L we construct splines indexed by the central knots denoted by $\nu_j^L, j \in J_{0,n}$,

$$
\nu_j^L(t) = \begin{cases}
\dfrac{\rho(t) - \rho_{j-1}}{\rho_j - \rho_{j-1}}, & t \in [x_{j-1}, x_j), \\
\dfrac{\rho_{j+1} - \rho(t)}{\rho_{j+1} - \rho_j}, & t \in [x_j, x_{j+1}).
\end{cases}
$$

The space of such splines on the closed interval $[a, b]$ is denoted by

$$
V^L := \mathbb{S}(\Delta^L), \quad \dim V^L = 2^L m + 1.
$$

In view of the calibration relation (14)–(15) considered for splines with central indexing, it holds that $V^L \subset V^{L+1}$. Hence the following direct decomposition holds

$$
V^{L+1} = V^L \dot{+} W^L. \tag{21}
$$

There are two alternative possibilities for constructing the basis functions in the space W^L.

For instance, as the basis functions in the space W^L one can use the basis functions in V^{L+1} with centers at odd knots [14]. In this way, one obtains the so-called "lazy" wavelets, which require no additional computations being a subset of the scaling functions. Obviously, $\dim W^L = 2^L m$. Then the complementarity condition for the dimensions of the spaces under consideration holds

$$
\dim V^{L+1} = \dim V^L + \dim W^L.
$$

The second variant of choosing the basis functions in the space W^L consists in using the basis functions in V^{L+1} with centers at the even knots under the additional condition that the spline must vanish at the last knot on the closed interval $[a, b]$. In this case, we assume that the spline himself is vanish if its values are vanish at the ends of a single elementary partition interval. Hence the corresponding basis functions are removed from the bases of the spaces in question V^{L+1}, V^L, W^L. Supply the notation of the spaces considered with the index "0":

$$
V_0^L := V_0^L(\Delta^L) = \left\{ S^L : S^L(t) = \sum_{j=0}^{2^L m - 1} C_j^L \nu_j^L(t) \quad \forall C_j^L \in \mathbb{R}^1, t \in [a, b] \right\},
$$

$$
\dim V_0^L = 2^L m. \tag{22}
$$

Then $\dim W_0^L = 2^L m$, and the complementarity condition for the dimensions of the spaces in question holds:

$$\dim V_0^{L+1} = \dim V_0^L + \dim W_0^L.$$

Let $C^L := \left(C_0^L, C_1^L, \ldots, C_{2^L m - 1}^L \right)^T$, $N^L := \left(\nu_0^L, \nu_1^L, \ldots, \nu_{2^L m - 1}^L \right)$, then we can write (22) in vector form as

$$S^L(t) = N^L(t)\, C^L.$$

If the partition Δ^{L+1} is obtained by two-fold refining of the partition Δ^L, then there is a matrix of refining reconstruction of the scaling functions (or the subdivision matrix) $\mathfrak{P}^{L+1}$ of size $2^{L+1}m \times 2^L m$ such that

$$N^L = N^{L+1} \mathfrak{P}^{L+1},$$

where the columns are formed from the coefficients of the calibration relations (14)–(15), written for the splines with central indexing, taking into account that the basic functions in each space are one less:

$$\nu_j^L(t) = \begin{cases} \nu_0^{L+1}(t) + p_{-1,2}^{L+1} \nu_1^{L+1}(t), & j = 0, \\ \sum\limits_{i=0}^{2} p_{j,i}^{L+1} \nu_{2j+i-1}^{L+1}(t), & j \in J_{1,2^L m - 2}, \\ p_{2^L m - 1, 0}^{L+1} \nu_{2^{L+1} m - 1}^{L+1}(t) + \nu_{2^{L+1} m}^{L+1}(t), & j = 2^L m - 1. \end{cases}$$

Denote the basis wavelet functions by

$$\Psi_i^L(t) := \nu_{2i}^{L+1}(t), \quad i = 0, 1, \ldots, 2^L m - 1,$$

and introduce the notation $\boldsymbol{\Psi}^L := \left(\Psi_0^L, \Psi_1^L, \ldots, \Psi_{2^L m - 1}^L \right)$.

Denote the corresponding wavelet approximation coefficients by D_i^L, $i = 0, 1, \ldots, 2^L m - 1$, and introduce the vector $\boldsymbol{D}^L := \left(D_0^L, D_1^L, \ldots, D_{2^L m - 1}^L \right)^T$.

The corresponding reconstruction matrix $\mathfrak{Q}^{L+1}$ of size $2^{L+1}m \times 2^L m$ satisfies the equation

$$\boldsymbol{\Psi}^L = N^{L+1} \mathfrak{Q}^{L+1},$$

where all the entries in every column of the matrix $\mathfrak{Q}^{L+1}$ are zero, except for a unique unit entry.

In view of the direct decomposition (21) any function in V^{L+1} can be written as a sum of a certain function from V^L and a certain function from W^L, and the following string of equalities is valid:

$$S^{L+1}(t) = N^{L+1}(t)\, C^{L+1} = N^L(t)\, C^L + \boldsymbol{\Psi}^L(t)\, \boldsymbol{D}^L$$

$$= N^{L+1}(t)\, \mathfrak{P}^{L+1} C^L + N^{L+1}(t)\, \mathfrak{Q}^{L+1} \boldsymbol{D}^L.$$

Therefore, the coefficients C^{L+1} can be obtained from the coefficients C^L and D^L as follows:

$$C^{L+1} = \mathfrak{P}^{L+1} C^L + \mathfrak{Q}^{L+1} D^L,$$

or, in block notation,

$$C^{L+1} = [\mathfrak{P}^{L+1} \mid \mathfrak{Q}^{L+1}] \begin{bmatrix} C^L \\ \hline D^L \end{bmatrix}. \tag{23}$$

The reverse process of decomposing the coefficients of C^{L+1} into a coarser version C^L and the refining coefficients D^L consists in solving the sparse linear algebraic system (23). It is reasonable to split this system into strictly diagonally dominant systems for even and odd knots (for details in the case of B-splines see [7]). Such systems of equations can be solved, with guaranteed well-posedness and stability, by the sweep method.

References

1. Faber, G.: Über stetige functionen. Math. Annalen. **66**, 81–94 (1909)
2. Sweldens, W.: The lifting scheme: a custom-design construction of biorthogonal wavelets. Appl. Comput. Harmonic Analys. **3**(2), 186–200 (1996)
3. Stollnitz, E.J., DeRose, T.D., Salesin, D.H.: Wavelets for Computer Graphics: Theory and Applications. Morgan Kaufmann, San Francisco (1996)
4. Dem'yanovich, Y.K.: Smoothness of spline spaces and wavelet decompositions. Doklady Math. **71**(2), 220–223 (2005)
5. Lyche, T., Mørken, K., Pelosi, F.: Stable, linear spline wavelets on nonuniform knots with vanishing moments. Comput. Aided Geom. Design. **26**, 203–216 (2009)
6. Atkinson, B.W., Bruff, D.O., Geronimo, J.S., Hardin, D.P.: Wavelets centered on a knot sequence: theory, construction, and applications. J. Fourier. Anal. Appl. **21**(3), 509–553 (2015)
7. Shumilov, B.M.: Splitting algorithms for the wavelet transform of first-degree splines on nonuniform grids. Comput. Math. Math. Phys. **56**(7), 1209–1219 (2016)
8. Dem'yanovich, Y.K., Ponomarev, A.S.: Realization of the spline-wavelet decomposition of the first order. J. Math. Sci. **224**(6), 833–860 (2017)
9. Makarov, A.A.: On wavelet decomposition of spaces of first order splines. J. Math. Sci. **156**(4), 617–631 (2009)
10. Makarov, A.A.: Algorithms of wavelet compression of linear spline spaces. Vestnik St. Petersburg Univ.: Math. **45**(2), 82–92 (2012)
11. Makarov, A.A.: On two algorithms of wavelet decomposition for spaces of linear splines. J. Math. Sci. **232**(6), 926–937 (2018)
12. Dem'yanovich, Yu.K.: Local Approximation on a Manifold and Minimal Splines [in Russian], St. Petersburg State University (1994)
13. Makarov, A.A.: Construction of splines of maximal smoothness. J. Math. Sci. **178**(6), 589–604 (2011)
14. Makarov, A., Makarova, S.: On lazy Faber's type decomposition for linear splines. AIP Conf. Proc. **2164**, 110006 (2019)

An Online Learning Approach to a Multi-player N-armed Functional Bandit

Sam O'Neill[1]([✉])[iD], Ovidiu Bagdasar[1][iD], and Antonio Liotta[2][iD]

[1] University of Derby, Kedleston Road, Derby DE22 1GB, UK
{s.oneill,o.bagdasar}@derby.ac.uk
[2] Edinburgh Napier University, Sighthill Campus, Sighthill Court, Edinburgh EH11 4BN, UK
A.Liotta@napier.ac.uk

Abstract. Congestion games possess the property of emitting at least one pure Nash equilibrium and have a rich history of practical use in transport modelling. In this paper we approach the problem of modelling equilibrium within congestion games using a decentralised multi-player probabilistic approach via stochastic bandit feedback. Restricting the strategies available to players under the assumption of bounded rationality, we explore an online multiplayer exponential weights algorithm for unweighted atomic routing games and compare this with a ϵ-greedy algorithm.

Keywords: Congestion games · Online learning · Multi-armed bandit

1 Introduction

The multi-armed bandit (MAB) problem has received much attention in recent years within the online and machine learning community due to its appropriateness for demonstrating the fundamental trade-off between exploration and exploitation in online learning. The basic MAB problem is for an agent to maximise the cumulative reward received after playing a number of rounds (finite or infinite). In each round the agent is required to choose one of K bandits and subsequently receives an associated reward. For an agent to be successful it must employ a strategy which balances the trade-off between exploration and exploitation. Explore too little and the agent's preferred choice may remain suboptimal, explore too often and the agent fails to exploit the most optimal choices. Numerous algorithms have been studied for variants of the MAB problem and a popular measure of an algorithm's performance is the notion of expected regret, whereby the agent's received reward is compared with the expected reward that would have been received for the optimal choices [1].

In strategic repeated games, a natural approach towards equilibrium is to employ an online learning algorithm in which the expected regret of the player(s)

© Springer Nature Switzerland AG 2020
Y. D. Sergeyev and D. E. Kvasov (Eds.): NUMTA 2019, LNCS 11974, pp. 438–445, 2020.
https://doi.org/10.1007/978-3-030-40616-5_41

is minimised over the time horizon [2]. Whilst expected regret analysis and convergence of equilibrium are important and rich areas of research, they make some key assumptions that could, in certain modelling scenarios, be deemed too restrictive. First, when bounding the regret of an algorithm it is necessary that the utility received by a player is itself bounded, therefore restricting the types of utility function. Second, convergence to a state of equilibrium does not take into account the capricious nature of certain individuals and that a player's rationality is often bounded by both the intractability of the decision making process and the player's preference for exhaustive search [4]. Therefore the best one may be able to do is express a player's belief in the most preferable choices over a set of tractable strategies.

The above concepts are particularly inherent in routing games, a form of strategic repeated game in which multiple players (e.g. drivers of vehicles) simultaneously route flow across a network in an attempt to minimise their own cost. Routing games belong to the larger class of congestion games which possess the property of emitting at least one pure strategy Nash equilibrium [6] and have received much attention within the field of algorithmic game theory [7]. However, due to the underlying graph structure, the strategy set for these games suffers from the "curse of dimensionality" whereby the strategy set for a source sink pair (available paths) grows exponentially with the size of the underlying graph. Traditionally methods have employed a centralised approach in which full information of the costs associated with all strategies is known, and flow is shifted globally between paths so as to satisfy a set of constraints representing a state of equilibrium for the given problem [5]. Such approaches fail to consider both the decentralised nature of the decision making processes within the system and that individual players have a particularly myopic view of the system and, therefore, tend to make decisions on very little information.

Motivated by the concepts of bounded rationality and random/deliberate sub-optimal choices, the focus of this paper is to model unweighted atomic routing games under a restricted subset of strategies via noisy feedback, i.e. the utility may vary due to external factors. We investigate an exponential weights algorithm which at each time step (round) uses feedback as a mechanism for a player to update their personal beliefs (probability distribution) of the best course of action and an ϵ-greedy algorithm in which the best course of action is selected greedily with probability $p = 1 - \epsilon$. Variants of both algorithms are implemented for the semi-bandit and bandit feedback scenarios.

2 Preliminaries

2.1 Congestion Games

An N-player congestion game consists of a finite number of players $\mathcal{N} = \{1, \cdots, N\}$, a set of congestible elements $\mathcal{E}$ with associated cost (latency) functions $l_e : \mathbb{N} \mapsto \mathbb{R}$ for each element $e \in \mathcal{E}$ and a set of playable strategies $\mathcal{A}_i$ for each player i, where a given strategy $a_i \in \mathcal{A}_i$ is a set of congestible elements $a_i \subseteq \mathcal{E}$. The number of players choosing element e is $x_e = \sum_{i \in \mathcal{N}} \sum_{e_i \in a_i} \mathbb{1}(e_i = e)$, where

$\mathbb{1}$ is the indicator function. The associated cost to player i playing strategy a_i is $u_i(a_i; a_{-i}) = \sum_{e \in \mathcal{E}} \sum_{e_i \in a_i} \mathbb{1}(e_i = e) \cdot l_e(x_e)$.[1] That is each player picks a set of congestion elements and their associated costs are dependent not only on their own strategy, but on those played by the other players. The total cost U under strategy profile $\mathbf{a} = (a_i)_{i \in \mathcal{N}}$ is then,

$$U(\mathbf{a}) = \sum_{i \in \mathcal{N}} u_i(a_i; a_{-i}) = \sum_{i \in \mathcal{N}} \sum_{e \in \mathcal{E}} \sum_{e_i \in a_i} \mathbb{1}(e_i = e) \cdot l_e(x_e) = \sum_{e \in \mathcal{E}} x_e l_e(x_e).$$

Let $\mathcal{A} = \prod_i \mathcal{A}_i$ to be the set of all strategy profiles and $l = (l_e)_{e \in \mathcal{E}}$ the vector of cost functions associated with each e, then the congestion game is described by the tuple $(\mathcal{N}, \mathcal{E}, \mathcal{A}, l)$.

Rosenthal showed that a congestion game has at least one pure strategy Nash equilibrium found by minimising the potential function $\Phi = \sum_{e \in \mathcal{E}} \sum_{i=1}^{x_e} l_e(i)$ [6].

2.2 Unweighted Atomic Routing Game

For an unweighted atomic routing game, let the set of congestible elements $\mathcal{E}$ be the edges in the graph $G = (V, E)$ and for each player $i \in \mathcal{N}$ associate a source/sink pair (o_i, d_i) and traffic demand $k_i = 1$, i.e. players route themselves.[2] A player's strategy set $\mathcal{A}_i$ is the set of possible paths from source to sink, i.e. a strategy $a_i \in \mathcal{A}_i$ is a path consisting of edges $e \in E$ [7]. Therefore the cost to a given player choosing a particular path is dependent on the number of players choosing paths which share edges in the graph.

As a bandit problem, an unweighted atomic routing game consists of $\mathcal{N}$ players, a set E of functional bandit machines (edges), with corresponding congestion functions l. Each player $i \in \mathcal{N}$ then pulls a combination of bandit machines $a_i \subseteq E$ (path) from the strategy set $\mathcal{A}_i$ (set of available paths for (o_i, d_i) pair) and receives feedback given the strategy profile of played actions $\mathbf{a} = (a_i)_{i \in \mathcal{N}}$.

3   Learning Under Bandit Feedback

The following section introduces the exponential weights and ϵ-greedy algorithms for both semi-bandit and bandit feedback.

For each player i let $W_i^t = (W_{i a_i}^t)_{a_i \in \mathcal{A}_i}$ be a set of weights associated with the player's available strategies at a given round t. We denote the probability of a player selecting strategy a_i as,

$$\mathbf{P}_{i a_i}^t = \frac{W_{i a_i}^t}{\sum_{j=1}^{|\mathcal{A}_i|} W_{ij}^t},$$

and the probability distribution over all strategies $\mathcal{A}_i$ as $\mathbf{P}_i^t = (\mathbf{P}_{i a_i}^t)_{a_i \in \mathcal{A}_i}$.

---

[1] $(a_i; a_{-i})$ is commonly used to refer to player i's strategy given the strategy profile $\mathbf{a} = (a_1, \cdots, a_i, \cdots, a_N)$.

[2] In general an unweighted traffic rate routes the same quantity $k_i = k \quad \forall i \in \mathcal{N}$.

3.1 Semi-bandit Feedback

Under semi-bandit feedback, the player has access to the entire payoff vector of playable strategies. The noisy feedback for a given strategy a_i^t played by player i in round t is,

$$\hat{r}_{ia_i}(a_{-i}^t) = u_i(a_i^t; a_{-i}^t) + \xi_{ia_i}^t,$$

and the entire payoff vector for all strategies available to player i is then

$$\hat{\mathbf{r}}_i^t = (\hat{r}_{ia_i}(a_{-i}^t))_{a_i \in \mathcal{A}_i}$$

For each player i, the exponential weights algorithm (see Algorithm 1) maintains the probability distribution $\mathbf{P}_i^t = (\mathbf{P}_{ia_i}^t)_{a_i \in \mathcal{A}_i}$ reflecting the beliefs about player i's best strategy from the strategy set $\mathcal{A}_i$. At time t, player i samples an action $a_i^t \sim \mathbf{P}_i^t$ and updates the distribution $\mathbf{P}_i^{t+1}$ based on the semi-bandit feedback it receives [3]. Note that due to the interdependence of the congestion functions, all players actions must be selected and played before players receive their corresponding feedback.

Algorithm 1. Exponential weights with semi-bandit feedback [EW-SB]

Require: $\gamma_t = t^{-\frac{1}{\alpha}} \ \forall t \in [1, \ldots, T],\ W_i^0 \in \mathbf{1}^{|\mathcal{A}_i|} \ \forall i \in \mathcal{N}$
1: **for** $t = 1, \ldots, T$ **do**
2:      **for** each player i in $\mathcal{N}$ **do**
3:          $\mathbf{P}_i^t = \frac{W_i^t}{\sum_{j=1}^{|\mathcal{A}_i|} W_{ij}^t}$ ▷ Calculate probability distribution for strategies
4:          $a_i^t \sim \mathbf{P}_i^t$ ▷ Sample action from probability distribution
5:      **end for**
6:      **for** each player i in $\mathcal{N}$ **do**
7:          $\hat{\mathbf{r}}_i^t = (\hat{r}_{ia_i}(a_{-i}^t))_{a_i \in \mathcal{A}_i}$ ▷ Observe estimated reward for strategies
8:          $W_i^{t+1} = W_i^t \cdot \exp\left(\frac{\gamma_t \hat{\mathbf{r}}_i^t}{|\mathcal{A}_i|}\right)$ ▷ Update weights
9:      **end for**
10: **end for**

The ϵ-greedy algorithm (see Algorithm 2) updates the average reward for all player strategies via the feedback vector and greedily selects the best known strategy with probability $p = 1 - \epsilon$ and randomly selects an action with probability $p = \frac{\epsilon}{\mathcal{A}_i}$.

3.2 Bandit Feedback

Under bandit feedback the player only has access to feedback for the strategy played in round t and therefore a player must attempt to estimate the cost of strategies over time. The exponential Weights algorithm can be amended (see Algorithm 3) by utilising the importance sampling estimator.

Algorithm 2. ϵ-greedy with semi-bandit feedback [ϵG-SB]

Require: $W_i^0 \in 0^{|\mathcal{A}_i|} \ \forall i \in \mathcal{N}$
1: **for** $t = 1, \ldots, T$ **do**
2:     **for** each player i in $\mathcal{N}$ **do**
3:         **if** $\epsilon_t \sim unif(0,1) < \epsilon$ **then**
4:             $a_i^t \sim unif\{1, |\mathcal{A}_i|\}$ ▷ Choose at random with probability $p = \frac{1}{|\mathcal{A}_i|}$
5:         **else**
6:             $a_i^t = \underset{a_i \in \mathcal{A}_i}{\arg\max}(W_{ia_i}^t)$
7:         **end if**
8:     **end for**
9:     **for** each player i in $\mathcal{N}$ **do**
10:         $\hat{\mathbf{r}}_\mathbf{i}^\mathbf{t} = (\hat{r}_{ia_i}(a_{-i}^t))_{a_i \in \mathcal{A}_i}$ ▷ Observe estimated feedback for strategies
11:         $W_i^{t+1} = W_i^t + \frac{1}{t+1}[\hat{\mathbf{r}}_\mathbf{i}^\mathbf{t} - W_i^t]$ ▷ Update average feedback
12:     **end for**
13: **end for**

The feedback for strategy a_i^t received in round t is the individual cost incurred by the player,

$$\hat{u}_i^t = u_i(a_i^t; a_{-i}^t) + \xi_i^t$$

and the full feedback vector $\mathbf{r}_\mathbf{i}^\mathbf{t}(a_{-i}^t)$ can be estimated by $\hat{\mathbf{r}}_\mathbf{i}^\mathbf{t} = (\hat{r}_{ia_i}^t)_{a_i \in \mathcal{A}_i}$, where,

$$\hat{r}_{ia_i}^t = \begin{cases} \frac{\hat{u}_i^t}{\mathbf{P}_{ia_i}^t}, & \text{if } a_i = a_i^t. \\ 0, & \text{otherwise.} \end{cases} \quad \forall a_i \in \mathcal{A}_i.$$

It can be shown [3] that under certain probabilistic assumptions, $\hat{r}_{ia_i}^t$ results in an unbiased estimator of the feedback received by player i playing action a_i calculated over the joint probability of all other strategy profiles $a_{-i} \in \prod_{j \neq i} \mathcal{A}_j$,

$$\mathbf{P}_{-i}^t = (\mathbf{P}_{-ia_{-i}}^t)_{a_{-i} \in \mathcal{A}_{-i}},$$

namely,

$$\mathbb{E}_t[\hat{r}_{ia_i}^t] = u_i(a_i; \mathbf{P}_{-i}^t) = \sum_{a_{-i} \in \mathcal{A}_{-i}} \mathbf{P}_{-ia_{-i}}^t u_i(a_i^t; a_{-i}^t).$$

For the ϵ-greedy algorithm (see Algorithm 4) we amend the update of the average rewards W_i^{t+1} to only update the strategy that has been played at time t.

4   Preliminary Results

Algorithms 1–4 were tested on a bidirectional lattice network with 16 vertices and 48 edges. Given the stochastic nature of the algorithms, 10 randomly generated instances of the lattice network were generated and 250 players were routed between 4 origin destination pairs. The results were averaged over 10 episodes per

Algorithm 3. Exponential weights with bandit feedback [EW-B]

Replace lines $7 - 8$ in algorithm 1 with:

$$\hat{u}_i^t = u_i(a_i^t; a_{-i}^t) + \xi_i^t \qquad \qquad \triangleright \text{ Observe estimated feedback for played strategy}$$

$$\hat{r}_{ia_i}^t = \begin{cases} \frac{\hat{u}_i^t}{\mathbf{P}_{ia_i}^t}, & \text{if } a_i = a_i^t. \\ 0, & \text{otherwise.} \end{cases} \quad \forall a_i \in \mathcal{A}_i$$

$$\qquad \qquad \triangleright \text{ Estimate feedback vector } \hat{\mathbf{r}}_i^t$$

$$W_i^{t+1} = W_i^t \cdot \exp\left(\frac{\gamma_t \hat{\mathbf{r}}_i^t}{|\mathcal{A}_i|}\right) \qquad \qquad \triangleright \text{ Update weights}$$

Algorithm 4. ϵ-greedy with bandit feedback [ϵG-B]

Replace lines $10 - 11$ in algorithm 2 with:

$$\hat{u}_i^t = u_i(a_i^t; a_{-i}^t) + \xi_i^t \qquad \qquad \triangleright \text{ Observe estimated feedback for played strategy}$$

$$\hat{r}_{ia_i}^t = \begin{cases} \frac{1}{t+1}\left[\hat{u}_i^t - W_{ia_i}^t\right], & \text{if } a_i = a_i^t. \\ 0, & \text{otherwise.} \end{cases} \quad \forall a_i \in \mathcal{A}_i$$

$$\qquad \qquad \triangleright \text{ Estimate feedback vector } \hat{\mathbf{r}}_i^t$$

$$W_i^{t+1} = W_i^t + \hat{\mathbf{r}}_i^t \qquad \qquad \triangleright \text{ Update average rewards}$$

---

network - each episode consisting of a 100 rounds ($T = 100$).[3] Figure 1(a) plots the total cost U averaged over the data set and for comparison, the total cost U_Φ experienced at the equilibrium given by the potential function Φ. Figure 1(b) plots the regret of each algorithm defined to be the cumulative sum of the difference between the total cost of the played strategy profile $\mathbf{a}^t$ at time t and the equilibrium total cost U_Φ,

$$R_t = \sum_{i=t}^{t} \left[U(\mathbf{a^t}) - U_\Phi\right].$$

Finally Fig. 2 plots the individual costs for the players at the initial and the final (T) round. Clearly a more uniform cost has emerged at time T for the 4 origin/destination pairs and this compares well with the costs at equilibrium given by minimising Φ (indicated in red).

---

[3] The source code is available at https://github.com/samtoneill/congestionbandi tgames.

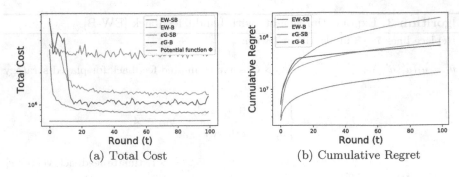

(a) Total Cost   (b) Cumulative Regret

Fig. 1. Log-lin plots of total cost and cumulative regret for the 4 algorithms averaged over all test data

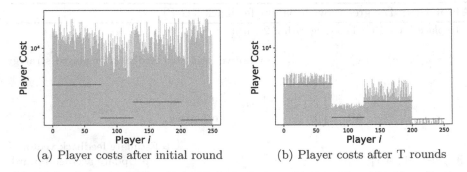

(a) Player costs after initial round   (b) Player costs after T rounds

Fig. 2. Log-lin plots illustrating the convergence of players costs for the 4 origin/destination pairs (Color figure online)

5 Concluding Remarks

On average, the exponential weights algorithm with semi-bandit feedback performs the best over the data set and compares reasonably well with the total cost associated with the Nash equilibrium given by minimising Φ. It is worth noting that the two bandit feedback algorithms which, arguably, in certain circumstances represent a more realistic model, e.g. a player would only experience or log their own travel time, perform comparably well. The poor performance of the ϵ-greedy algorithm with semi-bandit feedback is also of interest and, while it would require more investigation, a possible cause is that certain strategies $a_i \in \mathcal{A}_i$ experience an extreme cost under certain strategy profiles $(a_i; a_{-i})$ and therefore the averages maintained become unrepresentative of the more optimal choices.

Whilst it can be argued that these algorithms more realistically represent a player's decision making processes when taking into account human nature, they are not designed to be efficient in terms of computational complexity and therefore they may not be practical for use on larger networks. A future direction would be to employ similar techniques using a more scalable function approximation, such as a neural network, to keep track of a player's beliefs. There is also the possibility of using reinforcement learning techniques to employ autonomous agents whose primary role is to act altruistically for the benefit of the other agents within the network to reduce the overall congestion experienced [8].

References

1. Belmega, E.V., Mertikopoulos, P., Negrel, R., Sanguinetti, L.: Online convex optimization and no-regret learning: algorithms, guarantees and applications (2018). http://arxiv.org/abs/1804.04529
2. Cesa-Bianchi, N., Lugosi, G.: Prediction, Learning, and Games. Cambridge University Press, Cambridge (2006)
3. Cohen, J., Héliou, A., Mertikopoulos, P.: Learning with bandit feedback in potential games (2017). https://hal.archives-ouvertes.fr/hal-01643352
4. Gigerenzer, G., Selten, R.: Bounded Rationality: The Adaptive Toolbox. MIT Press, Cambridge (2001)
5. Patriksson, M.: The Traffic Assignment Problem: Models and Methods. Dover Publications, Mineola (1994)
6. Rosenthal, R.W.: A class of games possessing pure-strategy Nash equilibria. Int. J. Game Theory **2**(1), 65–67 (1973). https://doi.org/10.1007/BF01737559
7. Roughgarden, T.: Routing games. In: Nisan, N., Roughgarden, T., Tardos, E., Vazirani, V.V. (eds.) Algorithmic Game Theory, pp. 461–486. Cambridge University Press, Cambridge (2007). https://doi.org/10.1017/CBO9780511800481.020
8. Vinitsky, E., et al.: Benchmarks for reinforcement learning in mixed-autonomy traffic. In: Billard, A., Dragan, A., Peters, J., Morimoto, J. (eds.) Proceedings of the 2nd Conference on Robot Learning. Proceedings of Machine Learning Research, vol. 87, pp. 399–409. PMLR (2018). http://proceedings.mlr.press/v87/vinitsky18a.html

The Singular Value Decomposition of the Operators of the Dynamic Ray Transforms Acting on 2D Vector Fields

Anna P. Polyakova[1]([✉]) [iD], Ivan E. Svetov[1] [iD], and Bernadette N. Hahn[2] [iD]

[1] Sobolev Institute of Mathematics, Novosibirsk, Russia
anna.polyakova@ngs.ru, svetovie@math.nsc.ru
[2] University of Wuerzburg, Wuerzburg, Germany
bernadette.hahn@mathematik.uni-wuerzburg.de

Abstract. The problem of dynamic 2D vector tomography is considered. Object motion is a combination of rotation and shifting. Properties of the dynamic ray transform operators are investigated. Singular value decomposition of the operators is constructed with usage of classic orthogonal polynomials.

Keywords: Dynamic vector tomography · Longitudinal ray transform · Transverse ray transform · Singular value decomposition · Orthogonal polynomial

1 Introduction

Currently, directions in tomography, focused on the study of vector characteristics of mediums, are developing intensively. And above all, due to the fact that the areas of application of tomographic methods for studying nonscalar properties of objects are very wide.

We list the main mathematical tools on which numerical methods and algorithms for solving vector tomography problems are based. In the absence of the phenomenon of refraction, the inversion formulas [1–3] are very attractive from a mathematical point of view. Of particular interest are two very general methods: the least squares method and the method of singular value decomposition. In the numerical solution of the vector tomography problem, the least squares method was used with approximating sequences consisted of polynomials [4] and B-splines [5,6]. Note articles in which singular value decompositions of the operators of ray transforms of two-dimensional vector fields [7–9], the normal Radon transform operator of three-dimensional vector fields [10,11] were obtained, and numerical studies of algorithms based on the truncated singular decomposition method were performed. The method of approximate inverse is another powerful

This research was partially supported by RFBR and DFG according to the research project 19-51-12008.

Y. D. Sergeyev and D. E. Kvasov (Eds.): NUMTA 2019, LNCS 11974, pp. 446–453, 2020.
https://doi.org/10.1007/978-3-030-40616-5_42

approach used to solve Doppler tomography problems in $\mathbb{R}^3$ [12,13] and vector tomography problems in $\mathbb{R}^2$ [14,15].

All the above results were obtained under the assumption that the object of study is stationary. Often in practice, this assumption is not true. Such tasks are called dynamic tomography problems. There is a small number of works devoted to solving problems of dynamic tomography to restore the scalar characteristics of objects [16–18], while the vector case was not considered previously.

In this article we give definitions of dynamic longitudinal and transverse ray transforms of vector fields and formulate the dynamic vector tomography problem. It is necessary to recovery a vector field from its known values of the dynamic longitudinal and transverse ray transforms. We assume that the movement of the object under study is known and is a combination of rotation and shifting. The properties of the operators are investigated and their singular value decompositions are constructed.

2   Definitions

Let $\mathbf{x} = (x,y)$, $B = \{\mathbf{x} \in \mathbb{R}^2 \mid |\mathbf{x}| = \sqrt{x^2 + y^2} < 1\}$ be a unit disk with a center in the origin of Cartesian rectangular coordinate system, $\partial B = \{\mathbf{x} \in \mathbb{R}^2 \mid |\mathbf{x}| = 1\}$ be its boundary (unit circle), $Z = \{(s,\xi) \mid \xi \in \mathbb{R}^2, |\xi| = 1, s \in \mathbb{R}\}$ be a cylinder.

Functional space $L_2(B)$ consists of functions, which are square integrable in B. We also use functions in weight space $L_2(Z,\rho)$, $\rho > 0$. The inner product in the space $L_2(Z,\rho)$ is defined as

$$(f,g)_{L_2(Z,\rho)} = \int_Z f(z)g(z)\rho(z)dz.$$

The Radon transform $\mathcal{R}f : L_2(B) \to L_2(Z)$ of a function f is defined by formula

$$(\mathcal{R}f)(s,\xi) = \int_B f(\mathbf{x})\,\delta(\langle \xi, \mathbf{x}\rangle - s)\,d\mathbf{x},$$

where δ denotes the delta distribution, the unit vector $\xi^\perp = (-\sin\alpha, \cos\alpha)$ specifies the direction of integration, $\xi = (\cos\alpha, \sin\alpha)$ is the orthogonal direction. A set of parallel lines on the plane is defined by the fixed direction ξ and every line of the set is determined by distance s.

Let vector field $w = (w_1, w_2)$, $w : B \to \mathbb{R}^2$ be given. *The transverse ray transform*

$$\mathcal{P}^\perp : L_2(S^1(B)) \to L_2(Z)$$

acting on the vector field w is determined by the formula

$$(\mathcal{P}^\perp w)(s,\xi) = \int_B \langle w(\mathbf{x}), \xi\rangle\,\delta(\langle \xi, \mathbf{x}\rangle - s)\,d\mathbf{x}.$$

The longitudinal ray transform

$$\mathcal{P} : L_2(S^1(B)) \to L_2(Z)$$

of the vector field w is defined as

$$(\mathcal{P}w)(s,\xi) = \int\limits_B \langle w(\mathbf{x}), \xi^\perp \rangle \, \delta(\langle \xi, \mathbf{x} \rangle - s) \, d\mathbf{x}.$$

Remind that a vector field u is called *potential*, if there is a function φ, such that $u = \mathrm{d}\varphi = \left(\frac{\partial \varphi}{\partial x}, \frac{\partial \varphi}{\partial y} \right)$. A vector field v is called *solenoidal*, if its divergence is equal to 0, $\operatorname{div} v = \frac{\partial v_1}{\partial x} + \frac{\partial v_2}{\partial y} = 0$. In the other words, there is a function ψ, such that $v = \mathrm{d}^\perp \psi = \left(-\frac{\partial \psi}{\partial y}, \frac{\partial \psi}{\partial x} \right)$ (see, for example, [3]).

It is well known [1,5], that there is uniquely decomposition of arbitrary vector field w on a sum of potential and solenoidal parts

$$w = \mathrm{d}\varphi + \mathrm{d}^\perp \psi, \qquad \varphi, \psi|_{\partial B} = 0.$$

The operators of longitudinal and transverse ray transforms have nonzero kernels [5], namely

$$(\mathcal{P}\,\mathrm{d}\varphi)\,(s,\xi) = (\mathcal{P}^\perp \, \mathrm{d}^\perp \psi)(s,\xi) = 0, \qquad \varphi, \psi|_{\partial B} = 0.$$

Moreover, there are connections between the ray transforms and the Radon transform of the same potential [6]:

$$(\mathcal{P}\,\mathrm{d}^\perp \varphi)\,(s,\xi) = (\mathcal{P}^\perp \, \mathrm{d}\varphi)\,(s,\xi) = \frac{\partial(\mathcal{R}\varphi)}{\partial s}(s,\xi), \qquad \varphi|_{\partial B} = 0.$$

In the other words, if for the vector field $w = \mathrm{d}\varphi + \mathrm{d}^\perp \psi$ we know only values of the longitudinal ray transform $\mathcal{P}w$, we can reconstruct only solenoidal part $\mathrm{d}^\perp \psi$ of the field w, and we can reconstruct only potential part $\mathrm{d}\varphi$ of w by known values of the transverse ray transform $\mathcal{P}^\perp w$.

We can formulate the vector tomography problem in stationary case. Let the longitudinal ray transform $\mathcal{P}w$ and (or) the transverse ray transform $\mathcal{P}^\perp w$ of a vector field w be known for all $(s,\xi) \in Z$. It is required to determine the unknown vector field $w(\mathbf{x})$, $\mathbf{x} \in B$ from these data.

3   Singular Value Decomposition of the Ray Transform Operators

Need to remind a singular value decomposition of the operators of ray transforms [8]. Taking in consideration of the connection between the operators of ray transform, we can consider only the transverse ray transform operator, acting on a potential vector field $\mathcal{P}^\perp \, \mathrm{d}\phi$. Singular value decomposition for the longitudinal ray transform operator can be obtained analogously.

Lemma. *The singular value decomposition of the transverse ray transform operator*

$$\mathcal{P}^{\perp} : L_2(S^1(B)) \rightarrow L_2(Z, (1-s^2)^{-1/2})$$

is given by

$$\{(\sigma_{kn}, u_{kn}, v_{kn}) | \, k, n = 0, 1, ...\}$$

where

$$\sigma_{kn} = 2\sqrt{\frac{\pi}{k+2n+2}},$$

$$u_{kn}(\mathbf{x}) = \mathrm{d}\left(\sqrt{\frac{k+2n+2}{2\pi}} \frac{C_{n+k}^k}{n+1}(1-|\mathbf{x}|^2)|\mathbf{x}|^k Y_k(\mathbf{x}/|\mathbf{x}|) P_n^{(k+2,k+1)}(|\mathbf{x}|^2)\right),$$

$$v_{kn}(s, \alpha) = (-1)^{n+1}\frac{\sqrt{2}}{\pi}\sqrt{1-s^2}C_{k+2n+2}^{(1)}(s)Y_k(\alpha),$$

with the spherical harmonics Y_k, the Jakobi polynomials $P_n^{(p,q)}$ and the Gegenbauer polynomials $C_m^{(\mu)}$.

4   The Mathematical Model of the Dynamic Setting

Through the rotation of the x-ray source, the data collecting in computer tomography takes a definite amount of time. The position of the radiation source can be uniquely described via the following parametrization

$$\xi = (\cos(\phi t), \sin(\phi t))^T,$$

where t is a moment of time, ϕ is the source rotation angle.

Further, let the motion of the investigated object be described by

$$\Gamma_\theta \mathbf{x} = A_\theta \mathbf{x} + \mathbf{b}_\theta, \tag{1}$$

where a matrix

$$A_\theta = \begin{pmatrix} \cos(\theta t) & -\sin(\theta t) \\ \sin(\theta t) & \cos(\theta t) \end{pmatrix}$$

sets the rotation by the angle θt at the moment of time t from beginning of data collecting, a vector $\mathbf{b}_\theta \in \mathbb{R}^2$ specifies the shifting.

Let function $f \in L_2(B)$ describes the state of the object at the beginning of data acquisition. Then at the moment of time t the state of the object is described by $(f \circ \Gamma_\theta)$. For more details see [16]. In the frame of the setting, the dynamic Radon transform

$$\mathcal{R}_\Gamma : L_2(B) \rightarrow L_2(Z),$$

is given by

$$(\mathcal{R}_\Gamma f)(s, \xi) = \int_B f(\Gamma_\theta \mathbf{x})\, \delta(\langle \xi, \mathbf{x}\rangle - s)\, d\mathbf{x}.$$

Then the dynamic transverse and longitudinal ray transforms

$$\mathcal{P}_\Gamma^\perp, \mathcal{P}_\Gamma : L_2(B) \to L_2(Z)$$

can be defined by

$$(\mathcal{P}_\Gamma^\perp w)(s, \xi) = \int\limits_B \langle w(\Gamma_\theta \mathbf{x}), A_\theta \xi \rangle \, \delta(\langle \xi, \mathbf{x} \rangle - s) \, d\mathbf{x},$$

$$(\mathcal{P}_\Gamma w)(s, \xi) = \int\limits_B \langle w(\Gamma_\theta \mathbf{x}), A_\theta \xi^\perp \rangle \, \delta(\langle \xi, \mathbf{x} \rangle - s) \, d\mathbf{x}.$$

The statement of the vector tomography problem in dynamic case: Let the longitudinal ray transform $\mathcal{P}_\Gamma w$ and (or) the transverse ray transform $\mathcal{P}_\Gamma^\perp w$ of a vector field w be known for all $(s, \xi) \in Z$. It is required to determine the unknown vector field $w(\mathbf{x})$, $\mathbf{x} \in B$ from these data in moment of time $t = 0$.

5   Singular Value Decomposition of the Dynamic Ray Transform Operators

To solve the problem, we construct singular value decompositions of the dynamic ray transform operators.

Theorem 1. *In the case of motion (1) there is connection between the dynamic Radon transform and the dynamic ray transforms*

$$\left(\mathcal{P}_\Gamma \, d^\perp \varphi\right)(s, \xi) = \left(\mathcal{P}_\Gamma^\perp \, d\varphi\right)(s, \xi) = \frac{\partial(\mathcal{R}_\Gamma \varphi)}{\partial s}(s, \xi), \qquad \varphi|_{\partial B} = 0.$$

Proof. Introduce the notation $\Gamma_\theta \mathbf{x} := \tilde{\mathbf{x}} = (\tilde{x}, \tilde{y})$, then coordinates of points of integrating line are defined by the following formulae

$$\tilde{x} = s \cos(\phi + \theta)t - p \sin(\phi + \theta)t + (\mathbf{b}_\theta)_1,$$

$$\tilde{y} = s \sin(\phi + \theta)t + p \cos(\phi + \theta)t + (\mathbf{b}_\theta)_2,$$

and we have

$$\frac{\partial(\mathcal{R}_\Gamma \varphi)}{\partial s}(s, \xi) = \frac{\partial}{\partial s} \int\limits_B \varphi(\Gamma_\theta \mathbf{x}) \, \delta(\langle \xi, \mathbf{x} \rangle - s) \, d\mathbf{x} = \frac{\partial}{\partial s} \int\limits_B \varphi(\tilde{x}(s, p), \tilde{y}(s, p)) dp$$

$$= \int\limits_B \frac{\partial \varphi(\tilde{x}(s, p), \, \tilde{y}(s, p))}{\partial s} dp = \int\limits_B \left(\frac{\partial \varphi}{\partial \tilde{x}} \cdot \frac{\partial \tilde{x}}{\partial s} + \frac{\partial \varphi}{\partial \tilde{y}} \cdot \frac{\partial \tilde{y}}{\partial s} \right) dp$$

$$= \int\limits_B \langle d\varphi(\Gamma_\theta \mathbf{x}), A_\theta \xi \rangle \, \delta(\langle \xi, \mathbf{x} \rangle - s) \, d\mathbf{x} = (\mathcal{P}_\Gamma^\perp d\varphi)(s, \xi).$$

On the other hand, there are equalities

$$\frac{\partial(\mathcal{R}_\Gamma \varphi)}{\partial s}(s,\xi) = \int\limits_B \left(\left(-\frac{\partial \varphi}{\partial \widetilde{y}}\right) \cdot \left(-\frac{\partial \widetilde{y}}{\partial s}\right) + \frac{\partial \varphi}{\partial \widetilde{x}} \cdot \frac{\partial \widetilde{x}}{\partial s} \right) dp$$

$$= \int\limits_B \langle \mathrm{d}^\perp \varphi(\Gamma_\theta \mathbf{x}), A_\theta \xi^\perp \rangle \, \delta(\langle \xi, \mathbf{x} \rangle - s) \, d\mathbf{x} = (\mathcal{P}_\Gamma \mathrm{d}^\perp \varphi)(s,\xi). \quad \blacksquare$$

Theorem 2. *In the case of motion (1) there are connections between the dynamic ray transforms and the static ray transforms:*

$$(\mathcal{P}_\Gamma w)(s,\xi) = (\mathcal{P}w)(s + \langle A_\theta \xi, \mathbf{b}_\theta \rangle, A_\theta \xi),$$
$$(\mathcal{P}_\Gamma^\perp w)(s,\xi) = (\mathcal{P}^\perp w)(s + \langle A_\theta \xi, \mathbf{b}_\theta \rangle, A_\theta \xi).$$

Proof. We demonstrate this for the transverse ray transforms (static and dynamic), for the longitudinal ray transforms proof is analogously. We have

$$\left(\mathcal{P}_\Gamma^\perp w\right)(s,\xi) = \int\limits_B \langle w(A_\theta \mathbf{x} + \mathbf{b}_\theta), A_\theta \xi \rangle \, \delta(\langle \xi, \mathbf{x} \rangle - s) \, d\mathbf{x}$$

$$= \int\limits_B \langle w(\mathbf{y}), A_\theta \xi \rangle \, \delta(\langle \xi, A_\theta^{-1}(\mathbf{y} - \mathbf{b}_\theta) \rangle - s) \, d\mathbf{y}$$

$$= \int\limits_B \langle w(\mathbf{y}), A_\theta \xi \rangle \, \delta(\langle (A_\theta^{-1})^T \xi, \mathbf{y} - \mathbf{b}_\theta \rangle - s) \, d\mathbf{y}$$

$$= \int\limits_B \langle w(\mathbf{y}), A_\theta \xi \rangle \, \delta\left(\langle A_\theta \xi, \mathbf{y} \rangle - (\langle A_\theta \xi, \mathbf{b}_\theta \rangle + s)\right) \, d\mathbf{y}$$

$$= (\mathcal{P}^\perp w)(s + \langle A_\theta \xi, \mathbf{b}_\theta \rangle, A_\theta \xi). \quad \blacksquare$$

Theorem 3. *In the case of motion (1) the singular value decomposition of the dynamic transverse ray transform operator*

$$\mathcal{P}_\Gamma^\perp : L_2(S^1(B)) \to L_2(Z, (1 - s^2)^{-1/2})$$

is given by

$$\{(\sigma_{kn}, u_{kn}, \widetilde{v}_{kn}) | \, k, n = 0, 1, ...\}$$

where

$$\sigma_{kn} = 2\sqrt{\frac{\pi}{k + 2n + 2}},$$

$$u_{kn}(\mathbf{x}) = \mathrm{d}\left(\sqrt{\frac{k + 2n + 2}{2\pi}} \frac{C_{n+k}^k}{n + 1}(1 - |\mathbf{x}|^2)|\mathbf{x}|^k Y_k(\mathbf{x}/|\mathbf{x}|) P_n^{(k+2,k+1)}(|\mathbf{x}|^2)\right),$$

$$\widetilde{v}_{kn}(s, \alpha) = (-1)^{n+1} \frac{\sqrt{2}}{\pi} \sqrt{1 - (\widetilde{s})^2} C_{k+2n+2}^{(1)}(\widetilde{s}) Y_k(\widetilde{\alpha}),$$

with $\widetilde{s} = s + \langle A_\theta \xi, \mathbf{b}_\theta \rangle$, $\widetilde{\alpha} = (\phi + \theta)t$.

452   A. P. Polyakova et al.

Proof. This follows from the Lemma and the Theorem 2.　　　　■

Remark that the singular value decomposition of the dynamic longitudinal ray transform operator can be obtained using the Theorems 1 and 3.

6 Conclusion

We consider the problem of dynamic vector tomography for reconstructing a vector field from its known values of the dynamic longitudinal and transverse ray transforms. The movement of the investigated object is known and is a combination of rotation and shifting. The properties of the dynamic ray transform operators are studied and their singular value decompositions are obtained.

References

1. Sharafutdinov, V.A.: Integral geometry of tensor fields. VSP, Utrecht (1994). https://doi.org/10.1515/9783110900095
2. Svetov, I.E.: Reconstruction of the solenoidal part of a three-dimensional vector field by its ray transforms along straight lines parallel to coordinate planes. Numer. Anal. Appl. **5**(3), 271–283 (2012). https://doi.org/10.1134/S1995423912030093
3. Derevtsov, E.Yu., Svetov, I.E.: Tomography of tensor fields in the plain. Eur. J. Math. Comput. Appl. **3**(2), 24–68 (2015)
4. Derevtsov, E.Yu., Kashina, I.G.: Numerical solution of a vector tomography problem with the help of polynomial bases. Siberian J. Numer. Math. **5**(3), 233–254 (2002) (in Russian)
5. Derevtsov, E.Yu., Svetov, I.E., Volkov, Yu.S., Schuster, T.: Numerical B-spline solution of emission and vector $2D$-tomography problems for media with absorbtion and refraction. In: Remy, J.-G. (eds.) Proceedings of the 2008 IEEE Region 8 International Conference on Computational Technologies in Electrical and Electronics Engineering SIBIRCON 2008, pp. 212–217 (2008). https://doi.org/10.1109/SIBIRCON.2008.4602618
6. Svetov, I.E., Derevtsov, E.Yu., Volkov, Yu.S., Schuster, T.: A numerical solver based on B-splines for 2D vector field tomography in a refracting medium. Math. Comput. Simul. **97**, 207–223 (2014). https://doi.org/10.1016/j.matcom.2013.10.002
7. Derevtsov, E.Yu., Kazantsev, S.G., Schuster, T.: Polynomial bases for subspaces of vector fields in the unit ball. Method of ridge functions. J. Inverse Ill-Posed Probl. **15**(1), 1–38 (2007). https://doi.org/10.1515/JIIP.2007.002
8. Derevtsov, E.Yu., Efimov, A.V., Louis, A.K., Schuster, T.: Singular value decomposi-tion and its application to numerical inversion for ray transforms in 2D vector tomography. J. Inverse Ill-Posed Probl. **19**(4–5), 689–715 (2011). https://doi.org/10.1515/jiip.2011.047
9. Svetov, I.E., Polyakova, A.P.: Comparison of two algorithms for the numerical solution of the two-dimensional vector tomography. Siberian Electron. Math. Rep. **10**, 90–108 (2013) (in Russian). https://doi.org/10.17377/semi.2013.10.009
10. Polyakova, A.P.: Reconstruction of a vector field in a ball from its normal radon transform. J. Math. Sci. **205**(3), 418–439 (2015). https://doi.org/10.1007/s10958-015-2256-1

11. Polyakova, A.P., Svetov, I.E.: Numerical solution of the problem of reconstructing a potential vector field in the unit ball from its normal Radon transform. J. Appl. Ind. Math. **9**(4), 547–558 (2015). https://doi.org/10.1134/S1990478915040110

12. Rieder, A., Schuster, T.: The approximate inverse in action III: 3D-Doppler tomography. Numer. Math. **97**(2), 353–378 (2004). https://doi.org/10.1007/s00211-003-0512-7

13. Schuster, T.: Defect correction in vector field tomography: detecting the potential part of a field using BEM and implementation of the method. Inverse Prob. **21**(1), 75–91 (2005). https://doi.org/10.1088/0266-5611/21/1/006

14. Svetov, I.E., Maltseva, S.V., Polyakova, A.P.: Numerical solution of 2D-vector tomography problem using the method of approximate inverse. In: Ashyralyev, A., Lukashov, A. (eds.) ICAAM 2016, AIP Conference Proceedings, vol. 1759, p. 020132. AIP Publishing LLC, Melville (2016). https://doi.org/10.1063/1.4959746

15. Derevtsov, E.Yu., Louis, A.K., Maltseva, S.V., Polyakova, A.P., Svetov, I.E.: Numerical solvers based on the method of approximate inverse for 2D vector and 2-tensor tomography problems. Inverse Prob. **33**(12), 124001 (2017). https://doi.org/10.1088/1361-6420/aa8f5a

16. Hahn, B.: Efficient algorithms for linear dynamic inverse problems with known motion. Inverse Prob. **30**(3), 035008 (2014). https://doi.org/10.1088/0266-5611/30/3/035008

17. Hahn, B.: Dynamic linear inverse problems with moderate movements of the object: ill-posedness and regularization. Inverse Prob. Imaging **9**(2), 395–413 (2015). https://doi.org/10.3934/ipi.2015.9.395

18. Hahn, B.: Null space and resolution in dynamic computerized tomography. Inverse Prob. **32**(2), 025006 (2016). https://doi.org/10.1088/0266-5611/32/2/025006

Numerical Investigation into a Combined Ecologically Pure Plasma-Vortex Generator of Hydrogen and Thermal Energy

Denis Porfiriev[1,2]([✉]) [ID], Svetlana Kurushina[1] [ID], Nonna Molevich[2] [ID], and Igor Zavershinsky[1] [ID]

[1] Samara University, Moskovskoe shosse 34, Samara 443086, Russia
dporfirev@gmail.com
[2] Lebedev Physical Institute (Samara Branch),
Novo-Sadovaya 221, Samara 443011, Russia

Abstract. One of the promising approaches in clean energy is hydrogen production using the hydration reaction of a metal micropowder, stimulated by plasma formations. In present work, we carried out a series of numerical simulations of the turbulent three-dimensional swirling flow with chemical reactions in plasma vortex reactor (PVR) to provide further insights into use of such apparatus. The modelling has demonstrated that desirable behavior with heat transferred mostly downstream can be ensured by use of pipe-like electrode placed downstream from cylindrical one. Then, using exact experimental geometry and flow composition, optimized electrode system and simplified kinetic scheme of plasma-chemical reactions, we tested several operating modes of the PVR and obtained time dependent flow characteristics, which, in turn, will be used for further adjustment of the scheme and overall apparatus.

Keywords: Swirling flow · Numerical simulation · Plasma-chemical reactions · Plasma-vortex reactor

1 Introduction

For a number of important tasks and problems in the power engineering, road transport, aviation, and agriculture, mobile sources of energy are needed that would ensure the production of hydrogen and thermal energy in remote locations, especially in the northern territories. The high transportation cost of conventional hydrocarbon fuels to such territories makes it necessary to look for new ways to produce heat and electrical energy in the North. On the other hand, one of the most crucial problems of the modern world is the prevention of environmental pollution. International standards on the emission of harmful substances are constantly becoming tougher, and to ensure the competitiveness of products (and manufacturers) on the global market, it is necessary to develop new

© Springer Nature Switzerland AG 2020
Y. D. Sergeyev and D. E. Kvasov (Eds.): NUMTA 2019, LNCS 11974, pp. 454–461, 2020.
https://doi.org/10.1007/978-3-030-40616-5_43

environmentally friendly and energy-efficient technologies for the production of energy and harmless fuels. Many developed countries consider hydrogen as one of the most promising alternative fuels and foresee its strategic use [1–3]. Still, plasma technologies are generally not considered as an economically competitive technology for hydrogen generation [1], because target production cost of 1–2 US dollars seemed to be unachievable. But some plasma-based methods reach energy yield target of 60 g [H2]/kWh, which could ensure desirable pricing [4]. However, those approaches have to deal with questions of high production rate and reliability.

The technical solution of these problems can be achieved by production of cheap hydrogen in plasma-chemical reactors, including hydrogen generators, based on the use of stimulated by plasma formations chemical reactions of metal particles (aluminum, nickel, etc.) in water vapor. The advantages of this approach are simplicity, low cost, the possibility of mobile efficient production of hydrogen and thermal energy in laboratory and field conditions at high speed, an order of magnitude higher than that in traditional chemical technologies. The corresponding technology [5] currently provides the minimum price for hydrogen production among the well-known plasma-chemical methods. Still, the creation of efficiently operating PVR requires a series of experimental and theoretical studies devoted to determining the structure of swirling flow with plasma-chemical reactions, the mechanisms and possible paths of these reactions. In present work, we carried out a series of numerical simulations of the turbulent three-dimensional swirling flow in plasma vortex reactor to provide further insights into use of such apparatus: determining the optimal configuration of electrode system and calculating corresponding distributions of flow parameters.

2   Methods

First task was solved using schematic geometry of principal parts of experimental setup (Fig. 1): flow tube, swirler, and electrode system. For the second one, we employed exact experimental geometry [5] (Fig. 1) with explicit modeling of quartz glass walls and construction elements made of aluminum. Experimental setup, which was chosen based on results of calculations for model system, has following features. The cathode was merged with nozzle system. Anode has cylindrical form with radius of 5 mm and length of 66 mm. The interelectrode distance was set equal to 60 mm, the length of the quartz glass duct is 407 mm. Heat generation in gas discharge was modeled using a heat source of constant power. In what follows, we investigate the problem for water vapor flow close to the experimental conditions [5], corresponding to the total mass flow rate $m_t = 2\,\text{g/s}$, the heat source power $N = 200\,\text{W}$ and the static gas pressure $P = 10^5\,\text{Pa}$.

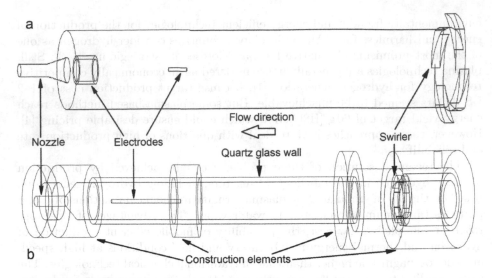

Fig. 1. Computational geometry used for task of electrode system optimization (a) and reacting flow modelling (b).

Reynolds averaged Navier-Stokes equations were used to describe the flow with chemical reactions.

$$\frac{\partial \rho}{\partial t} + \frac{\partial (\rho v_i)}{\partial x_i} = 0,$$

$$\frac{\partial (\rho v_i)}{\partial t} + \frac{\partial (\rho v_i v_j)}{\partial x_i} = -\frac{\partial P}{\partial x_i}$$
$$+ \frac{\partial}{\partial x_i}\Big[\mu_{eff}\Big(\frac{\partial v_j}{\partial x_i} + \frac{\partial v_i}{\partial x_j} - \frac{2}{3}\delta_{ij}\frac{\partial v_k}{\partial x_k}\Big)\Big] + \frac{\partial}{\partial x_i}[-\overline{\rho v_i' v_j'}],$$

$$\frac{\partial (\rho E)}{\partial t} + \frac{\partial [v_i(\rho E + P)]}{\partial x_i} = \frac{\partial}{\partial x_j}\Big[\Big(\kappa + \frac{c_P \mu_t}{Pr_t}\Big)\frac{\partial T}{\partial x_j} + v_i(\tau_{ij})_{eff}\Big]$$
$$+ \nabla\Big(\sum_{n=1}^{m} h_n \boldsymbol{J}_n\Big) + N,$$

$$\frac{\partial \rho Y_n}{\partial t} + \nabla(\rho \boldsymbol{v} Y_n) = -\nabla \boldsymbol{J}_n + R_n + S_n,$$

$$(\tau_{ij})_{eff} = \mu_{eff}\Big(\frac{\partial v_j}{\partial x_i} + \frac{\partial v_i}{\partial x_j}\Big) - \frac{2}{3}\mu_{eff}\delta_{ij}\frac{\partial v_k}{\partial x_k},$$

$$\boldsymbol{J}_n = -\Big(\rho D_t + \frac{\mu_{eff}}{Sc_t}\Big)\nabla Y_n - D_t\frac{\nabla T}{T},$$

$$E = h - \frac{P}{\rho} + \frac{v^2}{2}, P = \frac{\rho T}{M},$$

where $[-\overline{\rho v_i' v_j'}]$ are the Reynolds stresses which must be modeled to close the set of equations, v_i, v_i', ρ, T, P, E, and h are the mean and fluctuating velocity components, density, temperature, pressure, total energy, and enthalpy, respectively; N is the volumetric density of energy source power, c_P is the molar specific heat capacity at constant pressure, μ_t, μ_{eff} are the turbulent and effective viscosity coefficients, κ is the thermal conductivity coefficient, M is the molar mass, Pr_t is the turbulent Prandtl number, R_n and S_n are the terms responsible for the generation and destruction of the nth component during chemical reactions, Sc_t is turbulent Shmidt number, the value of which was set to 0.7 by default, D_t is turbulent diffusion coefficient. For the first step (optimization of electrode system), we excluded reactions from consideration. That results in significant simplification of equation system, namely, balance equation for components of the flow with mass fractions Y_n was not solved and corresponding diffusion flux J_n was eliminated from energy equation.

Mass flow rate for the inlet of the system was fixed. The pressure for the outlet was set equal to ambient one (1 atm). The no-slip velocity conditions were imposed along the tube, nozzle, electrode surfaces etc. For the of experimental geometry, temperature of outer surfaces of all internal construction elements was equal to 300 K. For all the internal boundaries separating the gas flow with electrodes, a glass tube, a nozzle and all internal boundaries between the solid elements of the setup, the conducting boundary condition was used without additional heat generation on the dividing interior wall.

To close equation set, we used Spalart-Allmaras turbulence model as one of the most adequate and computationally modest tools [6]. System of equation was solved by finite volume method using the ANSYS FLUENT 15.0 program package. Density, momentum, energy and turbulent quantities was discretized using a second-order upwind scheme. The use of higher-order schemes has not changed the flow parameters considerably. The diffusion terms are central-differenced and second-order accurate. The PRESTO! scheme for pressure treatment was used because of the strong swirl nature of the flow. For transient terms, we used the fully implicit scheme of the second-order accuracy. The SIMPLE scheme as the least resource intensive was employed for pressure-velocity coupling. The computational grid consists of about $3...4 \cdot 10^6$ hexahedral cells. The skewness and orthogonality metrics were far from critical values. Calculations with different values of time step from $1 \cdot 10^{-4}$ s to $1 \cdot 10^{-6}$ s were carried out. The time step was set equal to $5 \cdot 10^{-5}$ s since use of lower values did not improve convergence of procedure and led only to an increase in the computational time.

Full kinetic modeling of reactions in water vapor plasma in the presence of heat source is a very challenging and computationally demanding task. In our work, we used a simplified kinetic scheme of plasma-chemical reactions, which takes into account neutral particles: water vapor H_2O, hydroxyl radical OH, atomic hydrogen H; charged particles: electrons, H_2O^+, H_3O^+ ions, hydroxyl radical ions OH^-, atomic hydrogen ions H. The reactions between them are presented in Table 1, where M stands for any neutral molecule. The current values of the standard enthalpies of formation of components (at temperature $T = 298.15$ K) which are absent in the ANSYS FLUENT substance database, such as the H_2O^+ and H_3O^+ ions, were taken from the database of the Argon National Laboratory.

Table 1. Kinetic scheme used in calculations.

Reaction	Reaction rate constant
$H_2O + e \rightarrow H_2O^+ + e + e$	$1.8 \cdot 10^{-11}$ cm$^3 \cdot$ s^{-1}
$H_2O + e \rightarrow e + OH + H$	$6.0 \cdot 10^{-9}$ cm$^3 \cdot$ s^{-1}
$H_2O + e \rightarrow OH + H^-$	$5.0 \cdot 10^{-11}$ cm$^3 \cdot$ s^{-1}
$e + H_2O^+ + H_2O \rightarrow H + OH + H_2O$	$2.6 \cdot 10^{-23} \cdot (0.026)^{7/2} \cdot T^{-2} \cdot T_e^{-3/2}$ cm$^6 \cdot$ s^{-1}
$e + H_3O^+ + H_2O \rightarrow 2H + OH + H_2O$	$2.6 \cdot 10^{-23} \cdot (0.026)^{7/2} \cdot T^{-2} \cdot T_e^{-3/2}$ cm$^6 \cdot$ s^{-1}
$e + H_2O^+ \rightarrow H + OH$	$2.0 \cdot 10^{-6} \cdot (0.026/1.5)^{1/2} \cdot (300/T)$ cm$^3 \cdot$ s^{-1}
$e + H_3O^+ \rightarrow 2H + OH$	$2.0 \cdot 10^{-6} \cdot (0.026/1.5)^{1/2} \cdot (300/T)$ cm$^3 \cdot$ s^{-1}
$OH^- + H \rightarrow H_2O + e$	$1.0 \cdot 10^{-9}$ cm$^3 \cdot$ s^{-1}
$H_2O^+ + H_2O \rightarrow H_3O^+ + OH$	$1.7 \cdot 10^{-9}$ cm$^3 \cdot$ s^{-1}
$OH^- + H_2O^+ \rightarrow OH + H_2O$	$9.0 \cdot 10^{-8} \cdot (300/T)^{1/2}$ cm$^3 \cdot$ s^{-1}
$OH^- + H_2O^+ + M \rightarrow OH + H_2O + M$	$1.0 \cdot 10^{-37} \cdot (300/T)^{5/2}$ cm$^6 \cdot$ s^{-1}
$H^- + H_2O^+ \rightarrow H + H_2O$	$2.8 \cdot 10^{-7}(300/T)^{1/2}$ cm$^3 \cdot$ s^{-1}
$H^- + H_2O^+ + M \rightarrow H + H_2O + M$	$1.0 \cdot 10^{-37}(300/T)^{5/2}$ cm$^6 \cdot$ s^{-1}
$H^- + H_3O^+ \rightarrow 2H + H_2O$	$2.8 \cdot 10^{-7}(300/T)^{1/2}$ cm$^3 \cdot$ s^{-1}
$H^- + H_3O^+ + M \rightarrow 2H + H_2O + M$	$1.0 \cdot 10^{-37}(300/T)^{5/2}$ cm$^6 \cdot$ s^{-1}
$OH^- + H_3O^+ \rightarrow OH + H + H_2O$	$9.2 \cdot 10^{-8}(300/T)^{1/2}$ cm$^3 \cdot$ s^{-1}
$OH^- + H_3O^+ + M \rightarrow OH + H + H_2O + M$	$1.0 \cdot 10^{-37}(300/T)^{5/2}$ cm$^6 \cdot$ s^{-1}
$H^- + H_2O \rightarrow OH^- + H_2$	$3.8 \cdot 10^{-9}$ cm$^3 \cdot$ s^{-1}
$H^- + H \rightarrow H_2 + e$	$1.3 \cdot 10^{-9}$ cm$^3 \cdot$ s^{-1}
$H + H + M \rightarrow H_2 + M$	$4.8 \cdot 10^{-33}$ cm$^6 \cdot$ s^{-1}
$H + OH + M \rightarrow H_2O + M$	$4.2 \cdot 10^{-31}$ cm$^6 \cdot$ s^{-1}

3   Results

For the first task, the calculations were carried out with several electrode systems composed from electrodes of different size and shape, such as solid and hollow, thick and thin cylinders. That modeling revealed crucial features of flow and allowed us to propose configuration which maximizes energy output. Such desirable behavior can be achieved if the central zone of counterflow will not intersect with discharge region. That zone can be of two types: bounded by the thick electrode (Fig. 2a) and fading away due to a gradual decrease of angular momentum of the fluid and a consequent fall of the swirl number below the critical value (Fig. 2b). However, use of two thick solid electrodes creates interelectrode zone of stagnating flow (Fig. 2c). If anode is thick and cathode is thin, the counterflow zone does not prolong into interelectrode area, but cathode, although being thin, still disrupts to some extent downstream flow of energy. It cannot be made thinner due to experimental limitations. Those facts suggest that there are two potentially promising configurations. The first one is thick anode and pipe-like cathode. That design ensured downstream energy flow because of the pressure gradient along the symmetry axis which draws out the hot gas from the interelectrode area. Another possibility is to place thin anode and pipe-like cathode downstream from the end of the counterflow zone where the flow is completely downstream. Analogous problem for argon flow was solved in [7]. As can be seen, there is no principal difference in the flow structure, and we can extrapolate our conclusion to different flow compositions.

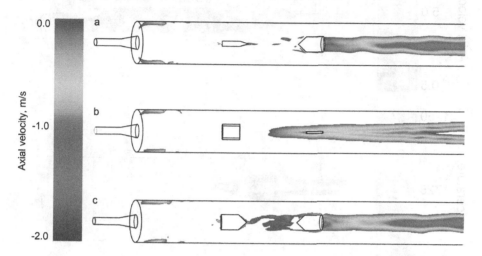

Fig. 2. Counterflow structures for different types of electrodes: a - thick anode that limits counterflow zone, b - hollow cathode placed downstrem from the end of counterflow zone, c - two thick electrodes which create stagnation zone. Limited range of values was taken for illustrative purposes.

The second problem of calculating of flow parameters distributions was solved using experimentally adjusted setup where cathode was merged with nozzle. The results are shown in Fig. 3. Aforementioned mechanism defines the flow structure and hot gas is sucked towards nozzle (Fig. 3a) because of the pressure difference. In addition, the nonstationary nature of the swirling flow under study [6] is again demonstrated. Such feature results in axial asymmetry of the heated region (Fig. 3b). Nevertheless, inertia and pressure forces still prevent hot gas leakage in the upstream region. The mass fractions of the reacting components of the mixture, except for water vapor, have maximum values in the discharge region (see Fig. 3c for hydrogen as an example) and can be controlled by modifying discharge power which leads to temperature change. Those species, the most valuable among them are atomic and molecular hydrogen, are carried away by the flow and can be utilized after proper separation.

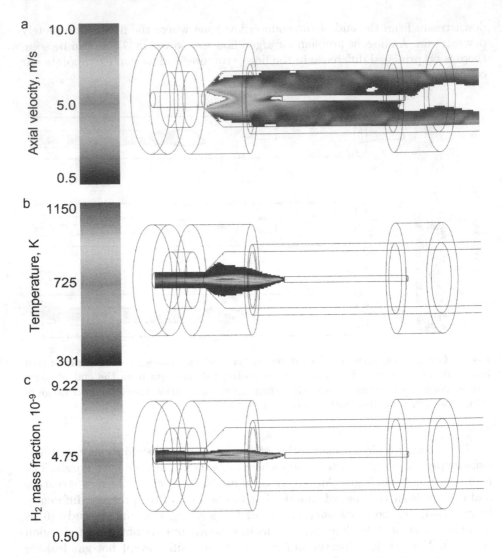

Fig. 3. Flow parameters distributions calculated for experimentally adjusted setup: a - axial velocity, b - temperature, c - molecular hydrogen mass fraction. Limited range of values was taken for illustrative purposes.

4   Conclusions

Our work results in concept of electrode configuration which allows one to minimize undesirable counterflow from discharge zone. Solid thin or thick anode placed upstream limits the zone of reverse flow. Whereas pipe-like cathode makes possible direct flow from the discharge zone towards outlet due to the pressure gradients. Such design, adjusted by experimental group by merging cathode into the exit nozzle system, seems to be the promising approach to the development of industrial samples of plasma-vortex reactor. However, the construction of effective heat and hydrogen generator requires simulation with proper model of plasma-chemical reactions. In comparison with previous research [7,8], we made the next step in present work and modelled chemically reacting flow, which allowed us to predict localization of species of interest and estimate its mass fractions. Although, in our research, we obtained good qualitative agreement with experimental data, verification of the kinetic scheme is still open question and needs further research.

Acknowledgements. The study was supported in part by the Ministry of Education and Science of Russia under the public contract with educational and research institutions within the project 3.1158.2017/4.6.

References

1. Randolph, K.: U.S. DOE. Hydrogen production. Annual Merit Review and Peer Evaluation Meeting. 16 May 2013. http://www.hydrogen.energy.gov/pdfs/review13/pd000_randolph_2013_o.pdf. Accessed 9 Oct 2019
2. Bleischwitz, R., Bader, N.: Policies for the transition towards a hydrogen economy: the EU case. Energy Pol. **38**, p5388–5398 (2010)
3. Le Duigou, A., et al.: Hydrogen pathways in France: results of the HyFrance3 project. Energy Pol. **62**, 1562–1569 (2013)
4. Mizeraczyk, J., Jasiński, M.: Plasma processing methods for hydrogen production. EPJ Appl. Phys. **75**(2), 24702 (2016)
5. Belov, N.K., Zavershinskii, I.P., Klimov, A.I., Molevich, N.E., Porfiriev, D.P., Tolkunov, B.N.: High effective heterogeneous plasma vortex reactor for production of heat energy and hydrogen. J. Phys.: Conf. Ser. **980**(1), 012040 (2018)
6. Gorbunova, A.O., et al.: Precessing vortex core in a swirling wake with heat release. Int. J. Heat Fluid Flow **59**, 100–108 (2016)
7. Kazanskii, P.N., Klimov, A.I., Molevich, N.E., Porfiriev, D.P., Zavershinskii, I.P.: Numerical simulation of an argon swirling flow in the presence of a DC discharge. J. Phys.: Conf. Ser. **980**(1), 012010 (2018)
8. Klimov, A.I., Kurushina, S.E., Molevich, N.E., Porfiriev, D.P., Zavershinskii, I.P.: Numerical simulation of argon flow structure in plasma vortex reactor. J. Phys.: Conf. Ser. **1112**(1), 012020 (2018)

Computer Modeling of Electrochemical Impedance Spectra for Defected Phospholipid Membranes: Finite Element Analysis

Tomas Raila[1]([✉]), Tadas Meškauskas[1], Gintaras Valinčius[2], Marija Jankunec[2], and Tadas Penkauskas[2]

[1] Institute of Computer Science, Vilnius University, Vilnius, Lithuania
tomas.raila@mif.vu.lt
[2] Life Sciences Center, Vilnius University, Vilnius, Lithuania

Abstract. This study deals with application of finite element method to model electrochemical impedance spectra of phospholipid membranes containing defects. Practical issues of choosing mesh and solver parameters are investigated in order to obtain the best combination of solution accuracy and computation times for the given problem. A simple mesh generation strategy suitable for membrane models with various randomly generated defect distributions is presented. Models with varying mesh densities were solved with direct and iterative solvers and solution accuracy was evaluated in terms of EIS spectral features. Computation times of models with various mesh sizes and solver configurations were also measured in two different computing environments.

Keywords: Finite element analysis · Computer modeling · Electrochemical impedance · Phospholipid membranes

1 Introduction

Tethered bilayer lipid membranes (tBLM) are versatile experimental platforms for studying a wide range of biophysical processes involving protein-membrane interactions [2]. One of alternating current (AC) techniques used to assess dielectric properties of tBLMs is electrochemical impedance spectroscopy (EIS) [7]. While this method is useful for determining macroscopic properties of bilayers, it provides no direct information on structural properties of membranes containing defects. Such cases often require more complex microscopy techniques, such as atomic force microscopy (AFM). However, recent studies showed that modeling EIS spectra either analytically [6] or numerically [3] can provide insight into important structural properties of defected tBLMs.

While EIS spectra can be obtained analytically for membrane models with homogeneous defect distributions [6], such cases are rarely encountered in practice. Modeling membranes with various irregular defect distributions usually involve finding approximate numerical solutions from which EIS spectra are constructed. Finite element method (FEM) is one of such numerical techniques for

© Springer Nature Switzerland AG 2020
Y. D. Sergeyev and D. E. Kvasov (Eds.): NUMTA 2019, LNCS 11974, pp. 462–469, 2020.
https://doi.org/10.1007/978-3-030-40616-5_44

finding approximate solutions of partial differential equations [8] and is widely used in various fields, such as aerodynamics, fluid flow, heat transfer, electromagnetism. The core concept of this method is dividing modeled domain into discrete elements, using this discretization to assemble a sparse system of linear equations and solving it to obtain an approximate solution.

An important practical consideration of using FEM is determining the right balance between computation time and solution accuracy. It is influenced by various factors, such as the choice of linear solver algorithm and its parameters, mesh generation strategy and computing environment. This study concentrates on these technical details of using FEM to effectively solve a large number of generated tBLM membrane models with various defect sizes, counts and densities, as presented by authors in the earlier work [4]. All computation experiments were carried out using COMSOL Multiphysics FEM software (version 5.4) with ACDC module.

2   EIS Spectra of Phospholipid Membranes

EIS is a versatile experimental method that provides information about physical and chemical properties of electrochemical cell. This technique works by applying an alternating current in certain frequency range to the system under study and measuring its response. While EIS data can be represented and analyzed in different ways, this study focuses on admittance phase versus frequency plots. The most important spectral features obtained from such plots is the minimum value of admittance Y (defined in Sect. 3) phase ($argY(f_{min})$) and the corresponding frequency (f_{min}) in that point (Fig. 1). It has been shown that these features are informative in determining some properties of membrane defects, such as defect density and size [4].

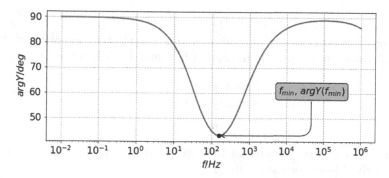

Fig. 1. Example of modeled EIS spectrum and its spectral features

3   Membrane Model

Three-dimensional model of tBLM is defined as a hexagonal prism, consisting of four layers: solution, membrane, submembrane layer and Helmholtz layer (Fig. 2, left). Each membrane defect is represented as a cylinder, intersecting membrane and submembrane layers and having the same conductivity properties as the solution layer.

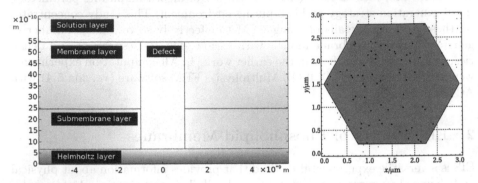

Fig. 2. Left: cross-section of three-dimensional membrane model in the vicinity of defect. Right: top view of model domain with randomly distributed defects.

The basis of this model is solving Laplace's equation for complex voltage Φ:

$$\nabla \cdot (\tilde{\sigma}(x,y,z)\nabla\Phi(x,y,z)) = 0. \tag{1}$$

Complex conductivity depends on layer properties σ and ε (parameter values are listed in [4]) and AC frequency f:

$$\tilde{\sigma}(x,y,z) = \sigma(x,y,z) + j\,2\pi f\varepsilon(x,y,z), \tag{2}$$

Boundary conditions are defined assuming $1\,\mathrm{V}$ electric potential at the top of model domain and perfectly insulating hexagonal prism sides:

$$\Phi(x,y,h_{hex}) = 1 \tag{3}$$

$$\Phi(x,y,0) = 0 \tag{4}$$

$$n \cdot J = 0 \tag{5}$$

Admittance (Y) is evaluated at the top of hexagonal prism by using current density derived from complex voltage:

$$J(x,y,z) = -\tilde{\sigma}(x,y,z)\,\nabla\Phi(x,y,z) \tag{6}$$

$$Y = \frac{\displaystyle\iint_{(x,y)\in\Gamma_{hex}} -n \cdot J(x,y,h_{hex})\,dx\,dy}{S_{hex}} \times \frac{1}{\Phi(x,y,h_{hex})} \tag{7}$$

4   Finite Element Analysis

4.1   Model Parameters

Membrane models used in experiments were constructed with randomly generated defect distributions, where X and Y coordinates of each defect are independently drawn from uniform distribution. Side length of hexagonal prism depends on predefined defect density which was set to $10\,\mu m^{-2}$ in all cases. Frequency range in logarithmic scale from $10^{-2}\,Hz$ to $10^{6}\,Hz$ is used, with 10 points per decade (81 frequency values in total). After solutions are obtained for each frequency independently, the resulting EIS spectrum is constructed by cubic spline interpolation and EIS spectral features (f_{min} and $argY(f_{min})$) are subsequently derived. Precision of these solutions is considered by one decimal place for f_{min} and two decimal places for $argY(f_{min})$.

4.2   Mesh Generation

We considered two types of mesh elements for three-dimensional membrane models - tetrahedral elements and triangular prisms. Both types of meshes were generated using built-in COMSOL mesh generation functionality. Assumption was made that current flux is the most intense inside and close to defects, so mesh generation parameters for areas inside and outside defects were set separately (Table 1), ensuring that defect areas are meshed significantly more densely. Both tetrahedral and prismatic meshes were generated in several density levels depending on the following parameters:

- k_d - ratio between defect radius and mesh element size inside defect
- k_h - ratio between hexagonal prism side length and maximum mesh element size outside defects
- k_s - number of swept mesh layers for defect and its surrounding submembrane and membrane layers (prismatic mesh only).
- r_0 - defect radius
- l_h - hexagonal prism side length.

Ratio k_d was varied, while k_h was fixed and set to 20. All defects had the same radius r_0 of 1 nm. Hexagon side length l_h was also fixed in all cases and

Table 1. COMSOL mesh generation settings.

Element size parameter	Value (defect areas)	Value (other areas)
Maximum element size	r_0/k_d	l_h/k_h
Minimum element size	r_0/k_d	l_h/k_h
Maximum element growth rate	1.7	1.7
Curvature factor	0.5	0.5
Resolution of narrow regions	0.5	0.5

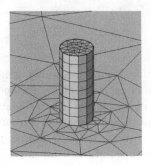

Fig. 3. Example of prism mesh elements at density level #4 in and around defect.

Table 2. DoF dependency on mesh type and k_d ratio when model contains 100 randomly distributed defects.

#	Ratio (k_d)	Swept layers (prisms)	DoF (prisms)	DoF (tetrahedra)
1	0.5	2	3.15E+05	5.05E+05
2	1.0	4	6.46E+05	7.87E+05
3	1.5	6	1.04E+06	9.99E+05
4	2.0	8	1.38E+06	1.27E+06
5	2.5	10	2.03E+06	1.82E+06
6	3.0	12	1.78E+06	2.38E+06
7	3.5	14	3.95E+06	3.25E+06

derived from defect count and density during defect distribution generation. Table 2 shows dependency between mesh generation parameters k_d and k_s and degrees of freedom (DoF) of the resulting models.

4.3   Solver Settings

Models with varying mesh densities were solved by using both direct and iterative solvers, implemented in COMSOL software package. MUMPS (MUltifrontal Massively Parallel Sparse) [1] direct solver and GMRES (Generalized Minimal Residual Method) iterative solver [5] were selected. Both solvers were used with default settings provided by COMSOL, with the exception of relative tolerance parameter for iterative solver, which was varied. Also, in both cases COMSOL was configured to parallelize computations among multiple CPU cores by distributing frequency values.

5   Experiments

5.1   Direct Solver

In order to estimate the effect of mesh density level (Table 2) on the solution accuracy expressed in terms of EIS spectral features, experiments were performed with direct solver, both mesh element types (prisms and tetrahedra) and varying mesh densities. Minimum mesh element quality in terms of element skewness was also analyzed. In all modeling cases the same model geometry having 100 randomly scattered defects was used. Results (Table 3) indicate that for both tetrahedral and prismatic meshes increasing their density past level #3 ($k_d = 1.5$) does not result in significant changes of $argY(f_{min})$ values, although f_{min} still shows decreasing trend in case of tetrahedral meshes. An anomaly at mesh density level #6 of prismatic meshes can be observed, which might possibly be attributed to COMSOL mesh generation algorithm issue. Prismatic meshes also show a clear advantage over tetrahedrals in terms of minimum element quality which is consistently orders of magnitude higher.

Table 3. Solution dependency on mesh density levels for tetrahedral and prismatic meshes where direct solver was used.

Mesh density	Tetrahedra			Prisms		
	f_{min}	$argY(f_{min})$	Min. quality	f_{min}	$argY(f_{min})$	Min. quality
1	160.22	43.071	1.1E-03	160.58	43.096	0.361
2	159.96	43.062	2.1E-03	160.12	43.076	0.341
3	159.95	43.060	1.5E-03	160.12	43.074	0.318
4	159.94	43.060	1.2E-03	160.12	43.074	0.318
5	159.91	43.061	8.2E-04	160.12	43.073	0.310
6	159.90	43.060	1.2E-04	160.44	43.088	0.058
7	159.90	43.060	8.3E-04	160.15	43.074	0.307

5.2   Iterative Solver

GMRES iterative solver was used with different relative tolerance for each mesh density level. Relative tolerance values were distributed logarithmically from 10^{-3} to 10^{-5} with 4 points per decade and from 10^{-6} to 10^{-8} with 1 point per decade. Table 4 indicates iteration counts and solutions for each selected relative tolerance value, when prismatic mesh with density level #3 was used (approximately 10^6 DoF), according to observations made in previous experiment.

Table 4. Solution and iteration count dependency on relative tolerance for fixed density ($k_d = 1.5$) prismatic mesh and iterative solver.

#	Relative tolerance	f_{min}	$argY(f_{min})$	Iteration count
1	1.0E-03	215.44	42.421	133
2	5.6E-04	193.26	42.530	173
3	3.2E-04	163.81	42.710	201
4	1.8E-04	162.34	42.995	275
5	1.0E-04	160.97	43.141	336
6	5.6E-05	159.23	43.074	392
7	3.2E-05	161.07	43.055	467
8	1.8E-05	160.39	43.071	549
9	1.0E-05	160.89	43.077	634
10	1.0E-06	160.18	43.074	1046
11	1.0E-07	160.11	43.074	1545
12	1.0E-08	160.12	43.074	2109

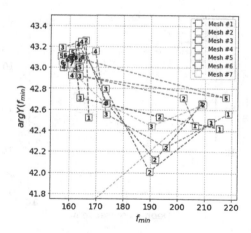

Fig. 4. Solution dependency on mesh density level and relative tolerance index. Colors represent mesh density levels, numbers indicate relative tolerance index from Table 4, limited to level #7 (Color figure online).

Results (Table 4) show that relative tolerance greater than or equal to 10^{-4} corresponds to low solution accuracy compared to results obtained using direct solver. Visually comparing solutions among different mesh densities (Fig. 4) also show a similar tendency independent on specific mesh density level. However, decreasing relative tolerance below 10^{-5} results in sharp increase of iteration counts and does not make significant accuracy improvements for $argY(f_{min})$, although for f_{min} this starts to take effect below relative tolerance of 10^{-6}.

5.3   Computation Times

In order to estimate computation time dependency on model size (expressed in terms of defect count), several models with different defect counts (10, 25, 50, 100, 200 and 500) were solved, using both types of meshes (prismatic and tetrahedral) and both solvers (direct and iterative). Mesh density level was fixed and set to #3, while relative tolerance for iterative solver was set to 10^{-5}, according to previous results. Experiments were conducted in two different computing environments: a standalone workstation and HPC cluster environment. Workstation was equipped with Intel Core i5-8600K 3.60 GHz CPU (6 cores), 64 GB of RAM and Ubuntu Linux 18.04 OS. Each cluster node had 2 x Intel Xeon X5650 2.66 GHz CPUs (6 cores each), 24 GB of RAM and Debian GNU/Linux 9 OS. To effectively use all available CPU cores, COMSOL was run in parallel mode, using 1 instance and 6 processes in workstation environment and 10 instances (one per node) in cluster environment, with 12 processes per each instance.

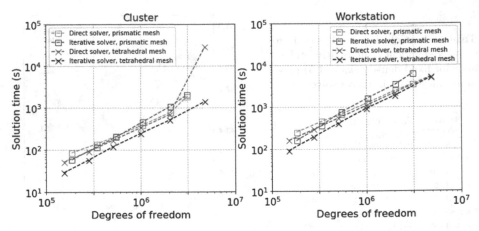

Fig. 5. Solution time dependency on defect count, solver type and mesh type in different computing environments (Color figure online).

Results (Fig. 5) indicate linear dependency in log-log scale between model degrees of freedom and solution time. A sharp increase of computation time (green curve) in cluster environment can be attributed to MUMPS solver switching to out-of-core mode due to limited amount of RAM. An overall trend of prismatic meshes taking more time to compute is visible, although its effect varies depending on defect count and, subsequently, degrees of freedom of the model. Using iterative solver with specified parameters instead of direct solver does not provide much benefit in terms of computation times and even the opposite can be observed in some cases with higher DoF. Choosing cluster environment resulted in approximately 3 times lower computation times compared to workstation.

6 Conclusions

In this work different practical aspects of applying FEM to model EIS spectra for defected phospholipid membranes were investigated. Several mesh density levels were defined and tested experimentally. Also, we experimented with different linear solvers and their parameters.

These results (see Sect. 5) allowed us to find optimal configuration (employed in computer modeling experiments presented in [4]) of FEM in terms of EIS spectral feature precision versus computation times. Note, that in this setup (defect count, density and size) both direct and iterative solvers showed similar times (Fig. 5), although in general iterative solvers could be preferred for their significantly lower memory usage.

Acknowledgements. This work was supported by Research Council of Lithuania (Project QAPHOMEDA).

References

1. Amestoy, P.R., Duff, I.S., Koster, J., L'Excellent, J.Y.: A fully asynchronous multifrontal solver using distributed dynamic scheduling. SIAM J. Matrix Anal. Appl. **23**, 15–41 (2001)
2. Cornell, B.A., et al.: A biosensor that uses ion-channel switches. Nature **387**, 580–583 (1997)
3. Kwak, K.J., et al.: Formation and finite element analysis of tethered bilayer lipid structures. Langmuir **26**, 18199–18208 (2010)
4. Raila, T., Penkauskas, T., Jankunec, M., Dreižas, G., Meškauskas, T., Valinčius, G.: Electrochemical impedance of randomly distributed defects in tethered phospholipid bilayers: finite element analysis. Electrochim. Acta **299**, 863–874 (2019)
5. Saad, Y., Schultz, M.H.: Gmres: a generalized minimal residual algorithm for solving nonsymmetric linear systems. SIAM J. Sci. Stat. Comput. **7**(3), 856–869 (1986)
6. Valinčius, G., Meškauskas, T., Ivanauskas, F.: Electrochemical impedance spectroscopy of tethered bilayer membranes. Langmuir **28**, 977–990 (2012)
7. Valinčius, G., Mickevičius, M.: Tethered phospholipid bilayer membranes: an interpretation of the electrochemical impedance response. In: Iglič, A., Kulkarni, C.V., Rappolt, M. (eds.) Advances in Planar Lipid Bilayers and Liposomes, vol. 21, chap. 2, pp. 27–61. Academic Press (2015)
8. Zienkiewicz, O.C., Taylor, R.L., Zhu, J.Z.: The Finite Element Method: Its Basis and Fundamentals, 7th edn. Butterworth-Heinemann, Oxford (2013)

A Compact Filter Regularization Method for Solving Sideways Heat Equation

Ankita Shukla and Mani Mehra[✉][iD]

Department of Mathematics, IIT Delhi, Hauz Khas 110016, New Delhi, India
ankita.shukla1509@gmail.com, mmehra@maths.iitd.ac.in

Abstract. In this paper, the stable approximate solution of the sideways heat equation is numerically investigated. The problem is severely ill-posed because if the solution exists, it does not depend continuously on the data. We introduce the compact filter regularization as a new, simple and convenient regularization method. Furthermore, the numerical implementation of the method is discussed. The numerical example shows that the proposed method is efficient and feasible.

Keywords: Sideways heat equation · Compact filter regularization · Ill-posed problem

1 Introduction

The sideways heat equation (SHE) often occurs in many industrial and engineering applications. The sideways heat equation is a model of a situation where one wishes to find out the surface temperature, but the surface itself is inaccessible for measurement [1]. Practically, the data available to us is measured data as it is based on the observations. This kind of problem is severely ill-posed since any small disruption in the observed data may lead to a drastic change in the solution. The ill-posed problems are susceptible to measurement and computational error, due to which the numerical recovery of the temperature is a tough task. Moreover, any existing numerical methods for classical PDEs cannot be directly applied to ill-posed problems to obtain a stable solution.

To overcome such difficulties, some special regularization methods [2,3] are required for stabilizing computations. Such as, Tikhonov method [4], quasi-reversibility method [5,6], optimal filtering method [7], wavelet and wavelet-Galerkin method [8], wavelet and Fourier method [9], wavelet and spectral regularization method [10,11], and so on.

A set of high order compact explicit filtering scheme is discussed in [12] to filter the high-frequency phenomena. Generally, if we use any regularization by filtering approach, the filtering is done in the Fourier space. Compact filter regularization (CFR) is a very simple and efficient method since filtering is done in the physical domain rather than other domain. The contribution of our paper can be seen as follows:

© Springer Nature Switzerland AG 2020
Y. D. Sergeyev and D. E. Kvasov (Eds.): NUMTA 2019, LNCS 11974, pp. 470–477, 2020.
https://doi.org/10.1007/978-3-030-40616-5_45

- In this paper, the key objective is to apply the compact filter as a regularization method to obtain a stable solution of SHE.
- Compact filters are a smooth filter which is preferred as compared to sharp filter.
- We do not have to go back and forth each time between physical and other domain, unlike other filtering methods.
- The proposed method is easy to implement and more efficient as all the computations are done in the physical domain only.
- A numerical example is discussed and compared with the Fourier method in terms of CPU time.

2   Problem Formulation

In a one-dimensional setting, the ill-posed problem for the heat equation in the bounded domain can be modeled as the following problem:

$$
\begin{aligned}
u_t &= \nu u_{xx}, \quad x \in (0, L), \quad t \in [0, 2\pi], \\
u(x, 0) &= 0, \quad x \in (0, L), \\
u(L, t) &= \phi(t), \quad t \in [0, 2\pi], \quad u|_{x \to \infty} \text{ bounded},
\end{aligned}
\tag{1}
$$

where the constant $L > 0$ and ν is thermal conductivity. The aim is to determine the distribution of surface temperature for $x \in (0, L)$ for the given Cauchy data $u(L, t) = \phi(t)$. It is assumed that the function $\phi(.)$, $u(x, .)$ and other functions which will appear in the paper vanish for $t < 0$. This problem is referred to as the sideways heat equation. Physically, the data ϕ can only be measured, which results in some measurement errors. The exact data ϕ and measured data ϕ^δ satisfy $\|\phi - \phi^\delta\|_2 \le \delta$, where $\|.\|_2$ denotes the L_2-norm in $[0, 2\pi]$ and $\delta > 0$ stand for the noise level.

The ill-posedness of the problem can be identified if we solve them in the Fourier domain. Let the exact solution u for the problem (1) can be represented in the Fourier expansion as

$$
u(x, t) = \sum_{k=-\infty}^{\infty} u_k(x) e^{ikt},
\tag{2}
$$

where

$$
u_k = \langle u(x, t), e^{-ikt} \rangle = \frac{1}{2\pi} \int_0^{2\pi} u(x, t) e^{-ikt} dt,
\tag{3}
$$

and $\langle ., . \rangle$ is the inner product in $L_2(0, 2\pi)$. The solution of the problem (1) in the frequency domain is

$$
u_k(x) = e^{(L-x)\nu\sqrt{ik}} \phi_k,
\tag{4}
$$

or, equivalently,

$$
u(x, t) = \sum_{k=-\infty}^{\infty} e^{(L-x)\nu\sqrt{ik}} \phi_k e^{ikt},
\tag{5}
$$

where the principle value of $\sqrt{ik}$ is given by

$$\sqrt{ik} = \begin{cases} (1+i)\sqrt{\frac{|k|}{2}}, & k \geq 0 \\ (1-i)\sqrt{\frac{|k|}{2}}, & k < 0 \end{cases}. \tag{6}$$

We note that $e^{\sqrt{ik}}$ tends to infinity as k tends to infinity since the real part of $\sqrt{ik}$ is positive. For our solution $u_k(x)$ to be in $L_2(\mathbb{R})$, (4) implies that the exact data ϕ_k must decay rapidly as $k \to \infty$. Practically, we have measured data ϕ_k^δ instead of ϕ_k, for which such decay is not attainable in the Fourier domain. To deal with this problem, we must use some regularization methods to filter away the high frequencies.

3   Compact Filter Regularization

High order compact filtering [12] is an explicit spatial filtering technique which filters the high unwanted wavenumber in the spatial domain. Filters are either smooth filter (obtained from the sharp filter by smoothing the sharp vertical edge at a cut off wavenumber) or sharp filter (whose value is either zero or one depending on whether the wavenumber modes are greater or below than some cutoff wavenumber). Compact filters are smooth filters rather than the sharp filter. If $y_f(t)$ denotes the filtered value of function $y(t)$ for $t \in (0, T)$, then general high order compact filter for the interior domain is expressed as

$$\beta y_f(t - 2\tau) + \alpha y_f(t - \tau) + y_f(t) + \alpha y_f(t + \tau) + \beta y_f(t + 2\tau) = ay(t) + \frac{b}{2}(y(t+\tau)$$

$$+ y(t-\tau)) + \frac{c}{2}(y(t+2\tau) + y(t-2\tau)) + \frac{d}{2}(y(t+3\tau) + y(t-3\tau)), \tag{7}$$

where τ is the step length. Through Fourier analysis, the transfer function $F_\alpha(\omega)$ corresponding to the (7) is given by

$$F_\alpha(\omega) = \frac{a + b\cos(\omega) + c\cos(2\omega) + d\cos(3\omega)}{1 + 2\alpha\cos(\omega) + 2\beta\cos(2\omega)}. \tag{8}$$

Here, $\omega = 2\pi k\tau/T$ is the modified wavenumber and its value lies in $(0, \pi)$. For the filters, the necessary criteria required is to eliminate all the phenomena arising at wavenumber $\omega = \pi$, i.e., $F_\alpha(\pi) = 0$. For different formal accuracy, the coefficients in (7) are obtained by matching the coefficients of Taylor series of various order along with the required condition for the transfer function $F_\alpha(\pi) = 0$. Here we use a fourth-order compact filter which can be obtained by solving the following set of coefficients

$$a = \frac{1}{8}(5 + 6\alpha), \ b = \frac{1}{2}(1 + 2\alpha), \ c = -\frac{1}{8}(1 - 2\alpha), \tag{9}$$

with $\beta = 0, d = 0$ (tridiagonal scheme). Putting the values of a, b and c from (9) to (8), we obtain

$$F_\alpha(\omega) = \frac{(5 + 6\alpha) + (4 + 8\alpha)\cos(\omega) - (1 - 2\alpha)\cos(2\omega)}{8(1 + 2\alpha\cos(\omega))}. \tag{10}$$

The coefficients a, b, and c are given in terms of single variable α, which is called the control variable. The range of α usually varies from 0 to 0.5. Values of α close to 0.5 correspond to filtering of only high wavenumber modes.

The interior compact filters given in (7) cannot be utilized for non-periodic boundary points due to relatively large symmetric stencils. Hence, for boundary filters, a one-sided fourth-order compact filter is used which is explicitly given as

$$y_f(0) = \frac{15}{16}y(0) + \frac{1}{16}\left(4y(\tau) - 6y(2\tau) + 4y(3\tau) - y(4\tau)\right),$$
$$y_f(\tau) = \frac{3}{4}y(\tau) + \frac{1}{16}\left(y(0) + 6y(2\tau) - 4y(3\tau) + y(4\tau)\right),$$

$$(11)$$

with exact filtering of wavenumber $\omega = \pi$. The similar expression can be written for $y_f(T-\tau)$ and $y_f(T)$. We use (11) for boundary point and (7) for any interior value of t to obtain filtered function $y_f(t)$ from the function $y(t)$. The plot of filtering transfer function $F_\alpha(\omega)$ (given in (10)) versus wavenumber ω for different values of α is shown in Fig. 1.

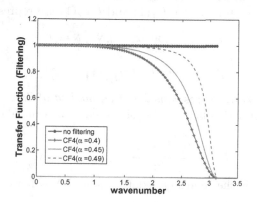

Fig. 1. Transfer function $F_\alpha(\omega)$ versus wavenumber ω for no filtering and fourth-order compact filtering with $\alpha = 0.4$, 0.45 and 0.49.

From Fig. 1, we can see that as we reduce the value of α from 0.5, not only high wavenumber mode, but a wide range of wavenumber spectrum is also filtered. It is also clear from the figure that $F_\alpha(\pi) = 0$ for all α.

The compact filter regularized solution for problem (1) with measured data $\phi^\delta(.)$ can be presented as

$$u_f^\delta(x,t) = \sum_{k=-M}^{M} e^{(L-x)\nu\sqrt{ik}} F_\alpha(\omega)\phi_k^\delta e^{ikt},$$

$$(12)$$

where $\omega = k\tau$ as $t \in [0, 2\pi]$. The step length τ plays the role of regularization parameter and M is taken as $M = [\frac{\pi}{\tau}]$ where $[.]$ denoted the greatest integer part of a real number.

4 Numerical Implementation of CFR Method

To obtain the stable numerical solution, we will use the method of lines to solve the Cauchy problem. Let t_n be the coordinate of the node after time discretization with n denotes the indices for discrete time. Consider the equidistant grid $0 = t_1 < t_2 < \ldots < t_{N+1} = 2\pi$ where $1 \leq n \leq N+1$ and $N = \frac{2\pi}{\tau}$ with $M = N/2$. The block operator equation of the discretized problem (1) using compact filter with $L = 1$ can be rewritten as

$$\begin{bmatrix} U_f(x) \\ \nu U_{f,x}(x) \end{bmatrix}_x = \begin{bmatrix} 0 & \nu^{-1} I \\ D & 0 \end{bmatrix} \begin{bmatrix} U_f(x) \\ \nu U_{f,x}(x) \end{bmatrix}, \quad x \in (0,1),$$

$$\begin{bmatrix} U_f(1) \\ U_{f,x}(1) \end{bmatrix} = [\phi_f^{\delta,1}, \ldots \phi_f^{\delta,N+1}, \rho_f^{\delta,1}, \ldots, \rho_f^{\delta,N+1}]^T \quad (13)$$

$$u_f(x, t_1) \approx u_f^1 = 0, \quad x \in (0,1),$$

where D is the time discretization matrix, and $U_f(x)$ is the vector containing the values u_f^n. We will use (7) at a given discrete node t_n to obtain the filtered value u_f^n of u. The vectors $U_f(x)$, $U_f(1)$ and $U_{f,x}(1)$ are represented as

$$U_f(x) = [u_f^1, u_f^2 \ldots u_f^N, u_f^{N+1}]^T,$$

$$U_f(1) = [\phi_f^{\delta,1}, \phi_f^{\delta,2} \ldots \phi_f^{\delta,N}, \phi_f^{\delta,N+1}]^T, \quad U_{f,x}(1) = [\rho_f^{\delta,1}, \rho_f^{\delta,2} \ldots \rho_f^{\delta,N}, \rho_f^{\delta,N+1}]^T.$$

The problem defined by (13) is considered as *method of lines*.

The time discretization matrix D is obtained using central difference approximation, i.e.,

$$D_n = \frac{u_f^{n+1} - u_f^{n-1}}{2\tau}, \quad 2 \leq n \leq N, \quad (14)$$

where, as before, τ is the time step length. At $n = N+1$, we use linear extrapolation, i.e.,

$$D_{N+1} = D_N + (D_N - D_{N-1}) = \frac{1}{2\tau} (1 \quad -2 \quad -1 \quad 2) \begin{bmatrix} u_f^{N-2} \\ u_f^{N-1} \\ u_f^N \\ u_f^{N+1} \end{bmatrix}.$$

Finally, to obtain the stable numerical solution of sideways heat equation using a compact filter regularization method, the necessary steps of the algorithm are given as follows:

1. Discretize the problem and obtain the vector $\rho = U_x(1)$ using $U(1)$.
2. For the given vector of exact data $[\phi, \rho]$, construct the vector with noisy data $[\phi^\delta, \rho^\delta]$.
3. Obtain the compact filtered vector $[\phi_f^\delta, \rho_f^\delta]$ using **Compact Filter** function.
4. Compute the central difference approximation for time derivative using (14).
5. Solve the ODE using the method of lines as discussed in (4) with filtered vector and go to next step.

6. Use the function **Compact Filter** after a fixed interval, say p to obtain the compact filter regularized solution at time t.
7. Repeat the above step 3 until the final solution $U_f(x)$ at given t is obtained.

4.1 Numerical Experiment

In this section, we present a numerical example of the sideways heat equation to illustrate the stability and effectiveness of our proposed compact filter regularization method. The method of lines is used to solve the discretized version of the sideways heat equation as given by (13) for numerical experiments. Central difference approximation is used for time derivative, and then the Runge-Kutta-Fehlberg method (ode45 in Matlab) is applied to perform the space marching. In all the experiments, the required accuracy in the R-K method was 10^{-4}.

To obtain the noisy data $\phi^\delta(t)$ and $\rho^\delta(t)$, we introduce some random noise η, i.e., $\phi^\delta(t) = \phi(t) + \eta$, $\rho^\delta(t) = \rho(t) + \eta$, where $\eta = \delta \times \text{rand}(t) \times ||.||_\infty$, $\text{rand}(t)$ is the random vector obtained using MATLAB function 'rand' and δ is the noise level of the measured data. For comparing the numerical and exact solution, relative error at fixed x is computed as

$$e_2(u) = \frac{||u_f^\delta(x,.) - u(x,.)||_2}{||u(x,.)||_2},$$

where $u_f^\delta(x,.)$ stands for the numerical regularized solution with noisy data and $u(x,.)$ stands for exact solution of the problem considered.

Example. It is easy to verify that $u(x,t) = \begin{cases} \frac{x+1}{t^{3/2}} \exp\left(-\frac{(x+1)^2}{4t}\right), & t \in (0, 2\pi), \\ 0, & t \leq 0, \end{cases}$

is the exact solution of the problem (1) with data

$$\phi(t) = \begin{cases} \frac{2}{t^{3/2}} \exp\left(-\frac{1}{t}\right), & t \in (0, 2\pi). \\ 0, & t \leq 0, \end{cases}$$

with $u(x, 0) = 0$. The exact solution at $x = 0$ is $h(t) := u(0, t) = t^{-3/2} \exp\left(-\frac{1}{4t}\right)$.

To obtain the numerical results, the noise level is fixed at 10^{-3}, and the constant coefficient of the differential operator is set at $\nu = 1$. We choose $N = \frac{2\pi}{\tau}$ and corresponding $M = N/2$ for ML solution. The step size for x is taken as $1/128$. The exact and regularized solution for different x is computed using the method of lines and shown in Fig. 2. From the figure, we find that the computed results using our proposed regularization method is stable and effective. In Table 1, the relative error at different x for two noise level δ is displayed from which we can observe that the computed regularized solution are better for large x. Furthermore, it can also be seen that the regularized solution becomes worse as we increase the noise level δ, which is obvious. To compare our proposed method in terms of CPU time, we have solved our problem using the Fourier method [9] where filtering is done in the Fourier domain with $k = M$. The CPU time for both the method is given in Table 2 from which it is clear that proposed method is more efficient.

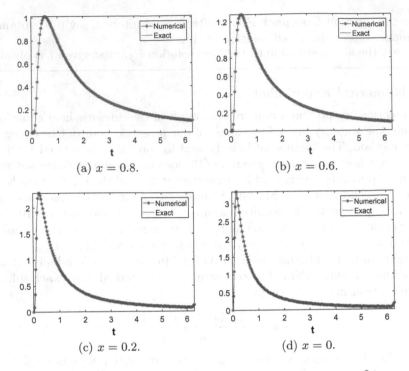

Fig. 2. Plot of exact and regularized solution with $\delta = 10^{-3}$.

Table 1. Relative error for test problem 1 with noise level $\delta_1 = 10^{-3}$, $\delta_2 = 10^{-2}$.

x	0.8	0.6	0.4	0.2	0
e_{δ_1}	0.0062	0.0089	0.0214	0.0382	0.0507
e_{δ_2}	0.031	0.0583	0.061	0.102	0.461

Table 2. Comparison of CPU time for test problem 1 with proposed method and Fourier method [9].

x	0.8	0.6	0.4	0.2	0
CPU ([9])	0.34	1.18	1.63	2.41	4.38
CPU (CF4)	0.22	0.86	1.12	1.30	2.53

5   Conclusion

In this paper, the sideways heat equation is solved by the compact filter regularization method. The numerical experiment is done to obtain a stable solution of the SHE. We conclude that the numerical implementation of the proposed method is simple and efficient. Numerical results are presented and compared with the existing one to demonstrate that the proposed method is effective and works well. In the future, we would like to extend the proposed method for the solution of two-dimensional SHE and also some other non-linear ill-posed problems.

Acknowledgement. The second author also acknowledge the support provided by the Department of Science and Technology, India, under the grant number MTR/2018/000371.

References

1. Beck, J.V., Blackwell, B., Clair, S.R.: Inverse Heat Conduction. Ill-Posed Problems. Wiley, New York (1985)
2. Engl, H.W., Hanke, M., Neubauer, A.: Regularizaton of Inverse Problems. Kluwer Academics, Dordrecht (1996)
3. Vogel, C.: Computational Methods for Inverse Problems. SIAM, Philadelphia (2002)
4. Carasso, A.: Determining surface temperatures from interior observations. SIAM J. Appl. Math. **42**(3), 558–574 (1982)
5. Eldén, L.: Approximations for a Cauchy problem for the heat equation. Inverse Probl. **3**(2), 263–273 (1987)
6. Liu, J.C., Wei, T.: A quasi-reversibility regularization method for an inverse heat conduction problem without initial data. Appl. Math. Comput. **219**(23), 10866–10881 (2013)
7. Seidman, T., Eldén, L.: An optimal filtering method for the sideways heat equation. Inverse Probl. **6**(4), 681–696 (1990)
8. Regińska, T., Eldén, L.: Solving the sideways heat equation by a wavelet-Galerkin method. Inverse Probl. **13**(4), 1093–1106 (1997)
9. Eldén, L., Berntsson, F., Regińska, T.: Wavelet and Fourier methods for solving the sideways heat equation. SIAM J. Sci. Comput. **21**(6), 2187–2205 (2000)
10. Fu, C.L., Xiong, X.T., Li, H.F., Zhu, Y.B.: Wavelet and spectral regularization methods for a sideways parabolic equation. Appl. Math. Comput. **160**(3), 881–908 (2005)
11. Xiong, X.T., Fu, C.L.: A spectral regularization method for solving surface heat flux on a general sideways parabolic. Appl. Math. Comput. **197**(1), 358–365 (2008)
12. Lele, S.K.: Compact finite difference schemes with spectral-like resolution. J. Comput. Phys. **103**(1), 16–42 (1992)

Acceleration of Global Search by Implementing Dual Estimates for Lipschitz Constant

Roman Strongin[iD], Konstantin Barkalov[(✉)][iD], and Semen Bevzuk[iD]

Lobachevsky State University of Nizhni Novgorod, Nizhny Novgorod, Russia
konstantin.barkalov@itmm.unn.ru

Abstract. The paper considers global optimization problems with a black-box objective function satisfying the Lipschitz condition. Efficient algorithms for this class of problems require reliable estimates of the Lipschitz constant to be introduced. Various approaches have been proposed to take into account both global and local properties of the objective function. In particular, algorithms using local estimates of the Lipschitz constant have shown their potential. The new approach presented in this paper is based on simultaneous use of two estimates: one is substantially larger than the other. The larger estimate ensures global convergence and the smaller one reduces the total number of trials needed to find the global optimizer. Results of numerical experiments on the random sample of multidimensional functions demonstrate the efficiency of the approach proposed by the authors.

Keywords: Global optimization · Multiextremal problems · Lipschitz constant estimates

1 Introduction

The paper considers global optimization problems of the form

$$\varphi(y^*) = \min\left\{\varphi(y) : y \in D\right\}, \tag{1}$$

$$D = \left\{y \in R^N : a_i \leq y_i \leq b_i,\ a_i, b_i \in R,\ 1 \leq i \leq N\right\}, \tag{2}$$

where the objective function is a black-box function and it is assumed to satisfy the Lipschitz condition

$$|\varphi(y_1) - \varphi(y_2)| \leq L \left\| y_1 - y_2 \right\|,\ y_1, y_2 \in D,$$

This research was supported by the Russian Science Foundation, project No. 16-11-10150.

Y. D. Sergeyev and D. E. Kvasov (Eds.): NUMTA 2019, LNCS 11974, pp. 478–486, 2020.
https://doi.org/10.1007/978-3-030-40616-5_46

with the constant L, $L < \infty$, unknown a priori.

The assumption of the objective function to be Lipschitzian is typical of many approaches to the development of the global optimization algorithms [3,11,12,17]. Moreover, the adaptive estimate of the unknown Lipschitz constant, based on the obtained search information, is one of the most important problems being solved in these algorithms. The value of the Lipschitz constant affects essentially the convergence rate of the global optimization algorithms. Therefore, the issue of its correct estimate is so important. The underestimation of the real value of this constant may result in losing the convergence of the algorithm to the global solution. At the same time, if the value of the constant estimate for the objective function is too large and does not match its real behavior, this will slow down the convergence of the algorithm to the global minimizer.

Several typical methods of adaptive estimation of the Lipschitz constant are known:

- global estimation of the constant L in the whole search domain D [7,12,17].
- local estimations of the constants L_i in different subdomains D_i of the search domain D [9,10,15].
- the choice of the estimates of the constant L from a set of possible values [4,8,13].

Each of the above approaches has its own advantages and disadvantages. For example, the use of the global estimate over the whole search domain can slow down the convergence of the algorithm to the global minimizer. The use of the local estimates to accelerate the convergence of the method requires an adequate adjustment of the algorithm parameters in order to preserve the global convergence.

In the present work, we consider a new algorithm that uses two global estimates of the Lipschitz constant. One of the two estimates is much greater than the other one. The larger estimate ensures global convergence and the smaller one reduces the total number of trials needed to find the global optimizer. The choice of one of the two estimates in the algorithm is performed adaptively during the search phase.

A rigorous substantiation of the proposed approach goes beyond the present initial publication and will be done in the forthcoming works. Here we present the results of numerical experiments that clearly demonstrate the efficiency of the new algorithm. Several hundred multiextremal test problems of various dimensionalities have been solved in the course of numerical experiments.

2   Global Search Algorithm and Dimensionality Reduction

The adaptation of the efficient algorithms that solve one-dimensional problems to solve multidimensional problems is a typical method to construct global optimization algorithms, see, for example, the diagonal partitions method in [13] or the simplicial partitions method in [18].

In this paper, we follow the approach based on the idea of reducing the dimension with the use of the Peano-Hilbert curves [16,17], which continuously and unambiguously map the unit interval $[0,1]$ onto the N-dimensional cube D from (2). By using this kind of mapping, it is possible to reduce the multidimensional problem (1) to a univariate problem

$$\varphi(y^*) = \varphi(y(x^*)) = \min\{\varphi(y(x)) : x \in [0,1]\},$$

where the function $\varphi(y(x))$ will satisfy a uniform Hölder condition

$$|\varphi(y(x_1)) - \varphi(y(x_2))| \leq H\,|x_1 - x_2|^{1/N}$$

with the Hölder constant H linked to the Lipschitz constant L by the relation $H = 2L\sqrt{N+3}$ and $y(x)$ is a Peano-Hilbert curve from $[0,1]$ onto D. Note that theoretically the Peano-Hilbert curve $y(x)$ is defined as a limit object. Therefore, in practical implementation, only some approximation to the true space-filling curve can be constructed. Some methods for constructing this type of approximations (called evolvents) are considered in [16,17]. In this case, the accuracy of the evolvent approximation to the true curve $y(x)$ depends on the density of the evolvent m (which is a parameter for constructing the evolvent) and is of the order of 2^{-m} for each coordinate.

Let us call the process of computing a function value (including the construction of the image $y = y(x)$) as a *trial*, and the pair $\{x, z = \varphi(y(x))\}$ as the outcome of the trial.

The Divide-The-Best global search algorithm used in this paper (according to [17]) can be formulated as follows. The first two trials are executed at the points $y^0 = y(0), y^1 = y(1)$. The choice of the point $y^{k+1}, k \geq 1$, for the next $(k+1)^{\text{th}}$ trial is defined by the following rules.

1. Renumber the preimages of all the points $y^i = y(x^i)$ from the trials already performed by subscripts in the increasing order of their coordinates, i.e.

$$0 = x_0 < x_1 < \cdots < x_k = 1,\tag{3}$$

and associate these with the values $z_i = \varphi(y(x_i)), 0 \leq i \leq k$, computed at these points.

2. Compute the maximum absolute value of the first divided differences

$$M = \max_{1 \leq i \leq k} \frac{|z_i - z_{i-1}|}{\Delta_i},$$

where $\Delta_i = (x_i - x_{i-1})^{1/N}$ and let

$$\mu = \begin{cases} 1, & \text{if } M = 0, \\ M, & \text{if } M \neq 0. \end{cases}\tag{4}$$

3. For each interval (x_{i-1}, x_i), $1 \leq i \leq k$, calculate the value $R(i)$ called the *characteristic* of the interval

$$R(i) = \Delta_i + \frac{(z_i - z_{i-1})^2}{r^2 \mu^2 \Delta_i} - 2\frac{z_i + z_{i-1} - 2z^*}{r\mu},\tag{5}$$

where

$$z^* = \min_{0 \le i \le k} z_i \tag{6}$$

and the real number $r > 1$ is a reliability parameter of the algorithm.

4. Select the interval (x_{t-1}, x_t) corresponding to the maximum characteristic

$$R(t) = \max_{1 \le i \le k} R(i). \tag{7}$$

5. Carry out the next trial at the point $x^{k+1} \in (x_{t-1}, x_t)$ calculated using the following formula

$$x^{k+1} = \frac{x_t + x_{t-1}}{2} - \text{sign}(z_t - z_{t-1}) \frac{1}{2r} \left[\frac{|z_t - z_{t-1}|}{\mu} \right]^N. \tag{8}$$

The algorithm terminates if the condition $\Delta_t < \epsilon$ is satisfied where t is from (7), and $\epsilon > 0$ is the predefined accuracy.

The theory of convergence of this algorithm is provided in [17]. The algorithm can be efficiently parallelized for shared and distributed memory [6] and for accelerators [1].

3   Algorithm with Dual Lipschitz Constant Estimates

The global search algorithm presented in the previous section is intended for solving the multiextremal problems, in which the objective function satisfies the Lipschitz condition. It is not necessary to define the value of the constant for the algorithm convergence. The estimation of the constant is performed in the course of global search based on available search information. According to the theorem from [17], the sequence of the trial points $\{y^k\}$ will converge to the global minimizer y^* if the condition

$$r\mu > 2^{3-1/N} L\sqrt{N+3} \tag{9}$$

is satisfied. Thus, an appropriate choice of the parameter r from (5) allows using the value $(r\mu)/(2^{3-1/N}\sqrt{N+3})$ as an estimate of the Lipschitz constant for the objective function $\varphi(y)$.

Satisfying the condition (9) will be guaranteed if we choose a large enough value of the parameter r. However, in this case the method will perform a large number of trials until the stop condition is satisfied. The choice of a small value of the parameter r (that corresponds to the lower estimate of the Lipschitz constant) would considerably reduce the number of trials but may violate the convergence to the global extremum.

An approach, in which two estimates of the Lipschitz constant are used in the rules of the algorithm, seems quite promising. This approach implies the use in the algorithm of two parameters r_{glob} and r_{loc}, where $r_{glob} > r_{loc} > 1$. When using the parameter r_{loc} we shall deal with the smaller estimate of the Lipschitz constant, and using the parameter r_{glob} will correspond to the larger one.

The rules of the algorithm with two estimates of the Lipschitz constant reproduce the ones of the global search algorithm completely except Rule 3 (the
computation of the characteristic) and Rule 4 (search for the interval with the
maximum characteristic).

The new rule for calculating the characteristic $R(i)$ of the interval (x_{i-1}, x_i)
will consist of the following operations:

– Calculate the value $R_{glob}(i)$ corresponding to the larger estimate of the Lipschitz constant

$$R_{glob}(i) - \Delta_i + \frac{(z_i - z_{i-1})^2}{r_{glob}^2 \mu^2 \Delta_i} \quad 2\frac{z_i + z_{i-1} - 2z^*}{r_{glob}\mu}.$$

– Calculate the value $R_{loc}(i)$ corresponding to the smaller estimate of the Lipschitz constant

$$R_{loc}(i) = \Delta_i + \frac{(z_i - z_{i-1})^2}{r_{loc}^2 \mu^2 \Delta_i} - 2\frac{z_i + z_{i-1} - 2z^*}{r_{loc}\mu}.$$

– Determine the characteristic $R(i)$ as

$$R(i) = \max\{\rho R_{loc}(i), R_{glob}(i)\}, \text{where } \rho = \left(\frac{1 - 1/r_{glob}}{1 - 1/r_{loc}}\right)^2, \quad (10)$$

The new rules for finding the interval with the maximum characteristic will
be as follows:

– Select the interval (x_{t-1}, x_t) corresponding to the maximum characteristic
$R(t) = \max_{1 \le i \le k} R(i)$.
– Fix the value $r = r_{loc}$ if $\rho R_{loc}(t) > R_{glob}(t)$, otherwise fix $r = r_{glob}$.
– Use this value of r in Rule 5 of the algorithm in the computing of the next
trial point.

This method for computing the characteristic can be substantiated as follows.
Each search iteration will yield an interval with the current minimum value z^*
from (6) at one of its boundaries. In the final phase of the search, this interval
will correspond to the interval containing the global minimum, i.e. it will be the
best one in terms of conducting further trials within it.

Let the current minimum value of z^* from (6) be achieved at the left point
of the ith interval, i.e. $z^* = z_{i-1}$. As proven in [17], according to the rule (5) the
following inequality will be true:

$$R(i) \ge \Delta_i (1 - 1/r)^2.$$

Therefore, for the estimates of the characteristics $R_{loc}(i)$ and $R_{glob}(i)$ calculated
with different parameters r_{loc} and r_{glob}, the following relation will hold:

$$\Delta_i (1 - 1/r_{glob})^2 > \Delta_i (1 - 1/r_{loc})^2.$$

Thus, when choosing the largest of the characteristics in accordance with

$$R(i) = \max\{R_{loc}(i), R_{glob}(i)\}$$

the characteristic $R_{glob}(i)$ corresponding to the higher estimate of the Lipschitz constant will be chosen; the lower estimate (which speeds up the process of refining the current solution) will not be used. However, if we multiply the lower estimate for the characteristic $R_{loc}(i)$ by the coefficient ρ in accordance with (10), then such lower estimates will be equal, thus the choice of the characteristic $R_{loc}(i)$ corresponding to the lower estimate of the Lipschitz constant will become more likely.

4   Numerical Experiments

A numerical comparison of the algorithms was carried out by solving several series of problems generated by the GKLS generator [5]. This generator of mul-tiextremal functions is widely used to compare global optimization methods (see, for example, [2,13,14]). In this study, six series each containing 100 problems of dimensions $N = 3, 4, 5$ were solved. For each dimension, *Simple* and *Hard* problems were generated, differing in the size of the attraction regions for local extremums and global extremum. The problem was considered solved if the method conducted the trial at a point that was in the δ-neighborhood of the global minimizer y^*, i.e. $\|y^k - y^*\| < \delta \|b - a\|$, where a and b are the boundaries of the search domain D.

Each series of problems has been solved by the original global search algorithm (GSA) and by the method with two estimates of the Lipschitz constant (GSA-DL). The evolvent constructed using the parameter $m = 10$ was used for the dimensionality reduction in both algorithms. The relative accuracy of the solution search was $\delta = 0.01$. The maximum allowable number of iterations per problem was $K_{max} = 10^6$.

The averaged numbers of iterations performed by the algorithms are presented in Table 1. For the GSA method the values of the parameter $r = 4.8$ when solving the problems of the *Simple* class and $r = 5.6$ when solving the problems of the *Hard* class were used. These values are the minimum ones (with the accuracy 0.1), at which all problems have been solved successfully. When solving the problems from the above classes by the GSA-DL method, the value of the parameter r specified above was selected for the upper estimate of the Lipschitz constant, i.e. the value $r_{glob} = r$ was set, which was complemented by the values $r_{loc} = 1.8$, $r_{loc} = 2.1$ and $r_{loc} = 2.4$. The number of unsolved problems is specified in brackets.

Table 1. Average number of iterations

	N = 3		N = 4		N = 5	
	Simple	*Hard*	*Simple*	*Hard*	*Simple*	*Hard*
AGS	2444	5345	28415	77470	25220(1)	126138(4)
AGS-DL, $r_{loc} = 1.8$	1372	2632	13273	37715	12702	94296(1)
AGS-DL, $r_{loc} = 2.1$	1502	2805	14826	38843	15213	90792(2)
AGS-DL, $r_{loc} = 2.4$	1567	2868	19447	40342	18239	106438(2)

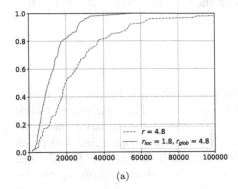
(a)

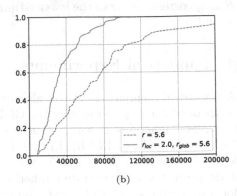
(b)

Fig. 1. Operational characteristics for GKLS *Simple* (a) and *Hard* (b) classes, $N = 4$.

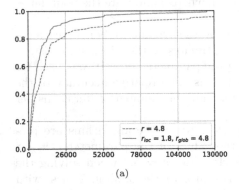
(a)

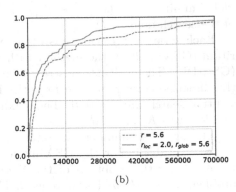
(b)

Fig. 2. Operational characteristics for GKLS *Simple* (a) and *Hard* (b) classes, $N = 5$.

The advantages of the GSA-DL algorithm over its prototype are also confirmed by the operational characteristics of the algorithms as well. Assume a series of test problems to be solved. The results of solving the series can be presented by a function $p(k)$ featuring the fraction of the total number of problems solved in k iterations. Such a function will be called the *operational characteristic* of the algorithm.

The operational characteristics for the GSA and GSA-DL methods obtained when solving the *Simple* and *Hard* problem series with the dimensionalities

$N = 4$ and $N = 5$ are presented in Figs. 1 and 2, respectively. The values of the parameters r used for estimating the Lipschitz constant are given in the figures.

The lower curves in Figs. 1 and 2 feature the characteristics of the GSA method whereas the upper ones, those of the GSA-DL. Such relative positions of the curves show the algorithm with two estimates of the Lipschitz constant is much faster on average when solving the problem series than the algorithm using a single estimate of the constant. Note that to solve problems of all classes (except for the *Hard* class at $N = 5$) the GSA-DL method requires about half as many trials as the GSA. The deterioration of the results in the case of the *Hard* class at $N = 5$ is explained by the complexity of this class' functions, which have a large attraction region of local minima and a narrow attraction region of the global minimum. To correctly solve such problems, the GSA-DL method often uses a higher estimate of the Lipschitz constant, thus reducing the speed difference of the GSA and GSA-DL algorithms.

References

1. Barkalov, K., Gergel, V., Lebedev, I.: Solving global optimization problems on GPU cluster. In: AIP Conference Proceedings, vol. 1738, p. 400006 (2016)
2. Barkalov, K., Strongin, R.: Solving a set of global optimization problems by the parallel technique with uniform convergence. J. Glob. Optim. **71**(1), 21–36 (2018)
3. Evtushenko, Y., Posypkin, M.: A deterministic approach to global box-constrained optimization. Optim. Lett. **7**, 819–829 (2013)
4. Gablonsky, J.M., Kelley, C.T.: A locally-biased form of the DIRECT algorithm. J. Glob. Optim. **21**(1), 27–37 (2001)
5. Gaviano, M., Kvasov, D.E., Lera, D., Sergeyev, Y.D.: Software for generation of classes of test functions with known local and global minima for global optimization. ACM Trans. Math. Softw. **29**(4), 469–480 (2003)
6. Gergel, V.P., Strongin, R.G.: Parallel computing for globally optimal decision making. In: Malyshkin, V.E. (ed.) PaCT 2003. LNCS, vol. 2763, pp. 76–88. Springer, Heidelberg (2003). https://doi.org/10.1007/978-3-540-45145-7_7
7. Horst, R., Tuy, H.: Global Optimization - Deterministic Approaches. Springer, Berlin (1996). https://doi.org/10.1007/978-3-662-02598-7
8. Jones, D., Perttunen, C., Stuckman, B.: Lipschitzian optimization without the Lipschitz constant. J. Optim. Theory Appl. **79**(1), 157–181 (1993)
9. Kvasov, D.E., Pizzuti, C., Sergeyev, Y.D.: Local tuning and partition strategies for diagonal GO methods. Numer. Math. **94**(1), 93–106 (2003)
10. Lera, D., Sergeyev, Y.D.: An information global minimization algorithm using the local improvement technique. J. Glob. Optim. **48**(1), 99–112 (2010)
11. Paulavičius, R., Žilinskas, J., Grothey, A.: Investigation of selection strategies in branch and bound algorithm with simplicial partitions and combination of Lipschitz bounds. Optim. Lett. **4**(2), 173–183 (2010)
12. Pinter, J.D.: Global Optimization in Action (Continuous and Lipschitz Optimization: Algorithms, Implementations and Applications). Kluwer Academic Publishers, Dordrecht (1996). https://doi.org/10.1007/978-1-4757-2502-5
13. Sergeyev, Y.D., Kvasov, D.E.: Global search based on efficient diagonal partitions and a set of Lipschitz constants. SIAM J. Optim. **16**(3), 910–937 (2006)

14. Sergeyev, Y.D., Kvasov, D.E.: A deterministic global optimization using smooth diagonal auxiliary functions. Commun. Nonlinear Sci. Numer. Simul. **21**(1–3), 99–111 (2015)
15. Sergeyev, Y.D., Mukhametzhanov, M.S., Kvasov, D.E., Lera, D.: Derivative-free local tuning and local improvement techniques embedded in the univariate global optimization. J. Optim. Theory Appl. **171**(1), 186–208 (2016)
16. Sergeyev, Y.D., Strongin, R.G., Lera, D.: Introduction to Global Optimization Exploiting Space-Filling Curves. Springer Briefs in Optimization. Springer, New York (2013). https://doi.org/10.1007/978-1-4614-8042-6
17. Strongin, R.G., Sergeyev, Y.D.: Global Optimization with Non-convex Constraints. Sequential and parallel Algorithms. Kluwer Academic Publishers, Dordrecht (2000). https://doi.org/10.1007/978-1-4615-4677-1
18. Žilinskas, J.: Branch and bound with simplicial partitions for global optimization. Math. Model. Anal. **13**(1), 145–159 (2008)

The Method of Approximate Inverse in Slice-by-Slice Vector Tomography Problems

Ivan E. Svetov[1]([⊠]) [iD], Svetlana V. Maltseva[1] [iD], and Alfred K. Louis[2] [iD]

[1] Sobolev Institute of Mathematics, Novosibirsk, Russia
{svetovie,maltsevasv}@math.nsc.ru
[2] Saarland University, Saarbrucken, Germany
louis@num.uni-sb.de

Abstract. A numerical solution of the problem of recovering the solenoidal part of three-dimensional vector field using the incomplete tomographic data is proposed. Namely, values of the ray transform for all straight lines, which are parallel to one of the coordinate planes, are known. The recovery algorithms are based on the method of approximate inverse.

Keywords: Vector tomography · Ray transform · Solenoidal vector field · Method of approximate inverse

1 Introduction

In the article we consider the following three-dimensional vector tomography problem. Let a vector field v in a bounded domain of the space $\mathbb{R}^3$ be unknown. It is required to reconstruct the field by its known values of the ray transform $[\mathcal{I}^3 v]$.

The fact that there is not a unique solution of the problem is due to the following. The potential fields ∇g with potentials g vanishing on the boundary of domain belong to the kernel of ray transform [1]. Therefore, only the solenoidal part $^s v$ of vector field v can be reconstructed by $[\mathcal{I}^3 v]$.

In the three-dimensional space the problem of reconstruction of the solenoidal part of vector field from their known values of the ray transform is overdetermined in terms of its dimension. Namely, it is necessary to recover the functions $^s v_j(x)$, $j = 1, 2, 3$, $x \in \mathbb{R}^3$, from the values of function $[\mathcal{I}^3 v]$ defined on the four-dimensional set of oriented lines. In the articles [2,3] it was proposed to consider the problem of recovering $^s v$ from the incomplete data $[\mathcal{I}^3 v]|_{M^3}$, where M^3 is some three-dimensional set of oriented straight lines. We mention the papers [4,5] devoted to the development and numerical study of the algorithms

This research was partially supported by RFBR and DFG according to the research project 19-51-12008.

Y. D. Sergeyev and D. E. Kvasov (Eds.): NUMTA 2019, LNCS 11974, pp. 487–494, 2020.
https://doi.org/10.1007/978-3-030-40616-5_47

based on the inversion formulas [3]. It should be noted, that in $\mathbb{R}^2$ the task of reconstruction of a vector field v from the values of ray transform $[\mathcal{I}^2 v]$ is not overdetermined (see, for example, [6–8]). In the space $\mathbb{R}^3$, the problem of reconstruction of the potential part of vector field v from the values of normal Radon transform $[\mathcal{R}_1^\perp v]$ is not overdetermined [9–11].

In this paper, the problem of reconstruction of the solenoidal part $^s v$ of vector field v is considered in the following statement. Let the values of ray transform $[\mathcal{I}^3 v]|_{M^3}$ be known, where M^3 is the set of all straight lines parallel to the coordinate planes. It is necessary to recover $^s v$ by these data. We propose to use the method of approximate inverse [12–15] to solve the problem. This numerical method was successfully applied, in particular, to solve problems of the Doppler tomography [16,17] in R^3 and the vector tomography in $\mathbb{R}^2$ [18,19].

2   Definitions and Statement of Problems

We give basic definitions of the vector tomography in $\mathbb{R}^n$ for arbitrary n, but in the article we consider only $n = 2$ and $n = 3$. Let $\xi \in \mathbb{S}^{n-1}$ and v be a vector field in $\mathbb{R}^n$, then *the ray transform* of v is defined by the formula

$$[\mathcal{I}^n v](\xi, x) = \sum_{j=1}^{n} \int_{\mathbb{R}} \xi_j v_j(x + t\xi) \, dt$$

where $x \in \xi^\perp$.

It is known [1], that every vector field v may be presented uniquely as the sum

$$v = {}^s v + \nabla g,$$

where vector field $^s v$ satisfies the condition $\delta\, ^s v = \sum_{j=1}^{n} \frac{\partial}{\partial x_j}\, ^s v_j = 0$, and function g is such as $g(x) \to 0$ at $|x| \to \infty$, $(\nabla g)_j = \frac{\partial}{\partial x_j} g$. A field $^s v$ is called the *solenoidal part* of vector filed v, and a field ∇g is called the *potential part* of v. Ray transform of the potential field is identically zero: $[\mathcal{I}^n (\nabla g)] = 0$ for arbitrary $g \in C^1(\mathbb{R}^n)$ such that $g(x) \to 0$ at $|x| \to \infty$. Thus we can reconstruct only the solenoidal part $^s v$ of field v by its known values of $\mathcal{I}^n v$.

Let $\pi_j = \{x_j = 0\} \subset \mathbb{R}^3$ $(j = 1, 2, 3)$ be the coordinate planes. We define the manifolds

$$M^3(\pi_j) = \{(\xi, x) \in \mathbb{R}^3 \times \mathbb{R}^3 \mid |\xi| = 1,\ \xi_j = 0,\ \langle \xi, x \rangle = 0\}, \qquad j = 1, 2, 3.$$

We introduce the coordinates (θ, s, z) on $M^3(\pi_j)$ so that

$$\xi = \cos\theta\, e_{j+1} + \sin\theta\, e_{j+2}, \qquad x = s(-\sin\theta\, e_{j+1} + \cos\theta\, e_{j+2}) + z e_j,$$

where (e_1, e_2, e_3) is the standard basis. The lower indices take values 1, 2, 3, so, for example, if $j = 2$ we have: $e_{j+1} = e_3$, $e_{j+2} = e_1$. For the three-dimensional vector field v the function $[\mathcal{I}_{(j)}^3 v] = [\mathcal{I}^3 v]|_{M^3(\pi_j)}$ is defined by the following formula

$$[\mathcal{I}_{(j)}^3 v](\theta, s, z) = \int_{\mathbb{R}} \big(\cos\theta\, v_{j+1}(x + t\xi) + \sin\theta\, v_{j+2}(x + t\xi) \big)\, dt.$$

We consider two problems:

The 2P-problem. It is required to recover the solenoidal part ^sv of vector field v by two given functions $[\mathcal{I}_{(1)}^3 v]$ and $[\mathcal{I}_{(2)}^3 v]$.

The 3P-problem. It needs to reconstruct the solenoidal part ^sv of vector field v by tree given functions $[\mathcal{I}_{(j)}^3 v]$, $j = 1, 2, 3$.

It is proved [3], that the 2P-problem has an unique solution and a solution of the 3P-problem is stable.

3   The Theoretical Background of Algorithms

Let us present the theoretical basis of algorithms proposed for a solving of the posed problems.

3.1   The Two-Dimensional Slices of Three-Dimensional Vector Field

Let $S_{(j)}$, $j = 1, 2, 3$, be an orthogonal projector of the vector field on the plane π_j. We can consider the field $(S_{(j)}v)(x)$ for the given $v(x)$ as the three-dimensional vector field with zero j-th component, or as the two-dimensional vector field

$$(S_{(j)}v)(x_{j+1}, x_{j+2})(x_j) = (v_{j+1}, v_{j+2})(x_{j+1}, x_{j+2})(x_j)$$

on the plane (x_{j+1}, x_{j+2}), with smooth dependence of x_j. The two-dimensional vector field $(S_{(j)}v)(x_{j+1}, x_{j+2})(z)$ is called the slice of three-dimensional vector field $v(x)$ on the plane $\{x_j = z\}$. Note that $[\mathcal{I}_{(j)}^3 v](\theta, s, x_j) = [\mathcal{I}^2(S_{(j)}v)(x_j)](\theta, s)$, where $(S_{(j)}v)(x_j)$ is j-th slice of v for fixed x_j. In this way only the solenoidal parts $^s(S_{(j)}v)$ of two-dimensional vector fields $S_{(j)}v$, $j = 1, 2, 3$, may be reconstructed by their known values of $[\mathcal{I}_{(j)}^3 v]$, $j = 1, 2, 3$.

There is a connection between components of the solenoidal part of three-dimensional vector field and components of the solenoidal parts of its slices [3]. Namely, for $j = 1, 2, 3$ there are equalities

$$y_{j+1} F[^s(S_{(j)}v)]_{j+2}(y) - y_{j+2} F[^s(S_{(j)}v)]_{j+1}(y)$$
$$= y_{j+1} F[^sv]_{j+2}(y) - y_{j+2} F[^sv]_{j+1}(y). \tag{1}$$

Here $F[\cdot]$ is the three-dimensional Fourier transform, defined by the formula

$$F[g](y) = (2\pi)^{-3/2} \iiint_{\mathbb{R}^3} e^{-i\langle y, x\rangle} g(x)\, dx.$$

Here and further, we denote points in the main space by x and points in the space of the Fourier transform image by y. For the Fourier images of components of solenoidal parts the following equalities are also fulfilled

$$y_1 F[^sv]_1(y) + y_2 F_3[^sv]_2(y) + y_3 F[^sv]_3(y) = 0, \tag{2}$$
$$y_{j+1} F[^s(S_{(j)}v)]_{j+1}(y) + y_{j+2} F[^s(S_{(j)}v)]_{j+2}(y) = 0.$$

3.2   The Method of Approximate Inverse For Reconstruction of the Two-Dimensional Vector Fields

We propose to use the method of approximate inverse for reconstruction of the solenoidal parts of two-dimensional slices of three-dimensional vector field. This approach was successfully applied in [18,19].

There is a relation between the ray transform $\mathcal{I}^2$ of two-dimensional vector field and the Radon transform $\mathcal{R}$ of function f defined by the formula

$$[\mathcal{R}f](\theta, s) = \int\limits_{\mathbb{R}} f(s\xi^\perp + t\xi)dt,$$

where $\xi = (\cos\theta, \sin\theta)$, $\xi^\perp = (-\sin\theta, \cos\theta)$. Namely, there are following equalities

$$[\mathcal{R}(^su)_i](\theta, s) = \xi_i [\mathcal{I}^2 u](\theta, s), \qquad i = 1, 2.$$

It is necessary to remind the scheme of the method of approximate inverse for the function reconstruction by its Radon transform. Let $e \in L_2(\mathbb{R}^2)$ be a function with the feature $\|e\|_{L_1(\mathbb{R}^2)} = 1$. Using the operator of shifting and dilating $T_{1,\gamma}^p : L_2(\mathbb{R}^2) \to L_2(\mathbb{R}^2)$ we form the *mollifier* e_γ^p from the function e

$$e_\gamma^p(x) = T_{1,\gamma}^p e(x) = \gamma^{-2} e\left((x - p)/\gamma\right), \qquad x, p \in \mathbb{R}^2.$$

Namely, for the function $e_\gamma^p(x)$ and an arbitrary function $f \in L_2(\mathbb{R}^2)$ the following equality

$$\lim_{\gamma \to 0} \langle f, e_\gamma^p \rangle_{L_2(\mathbb{R}^2)} = f(p)$$

holds.

The operator of shifting and dilating $T_{2,\gamma}^p$ at fixed $\gamma > 0$, $p \in \mathbb{R}^2$ acts on a function $g(\theta, s)$ according to the formula

$$T_{2,\gamma}^p g(\theta, s) = \gamma^{-2} g\left(\theta, (s - \langle p, \xi^\perp \rangle)/\gamma\right)$$

and is connected with $T_{1,\gamma}^p$ by the equality $\mathcal{R}^* T_{2,\gamma}^p = T_{1,\gamma}^p \mathcal{R}^*$. Here $\mathcal{R}^*$ is an adjoined operator for the Radon transform $\mathcal{R}$ acting on a function $g(\theta, s)$ by the following formula

$$(\mathcal{R}^* g)(x) = \int\limits_0^{2\pi} g(\theta, \langle x, \xi^\perp \rangle)\, d\theta.$$

Let e belong to the range of operator $\mathcal{R}^*$ and ψ be a solution of the equation $\mathcal{R}^*\psi = e$, then for fixed γ and p the function

$$\psi_\gamma^p = T_{2,\gamma}^p \psi \qquad (3)$$

is a solution of the equation $\mathcal{R}^* \psi_\gamma^p = e_\gamma^p$. At small γ we obtain

$$f(p) \approx \langle f, e_\gamma^p \rangle_{L_2(\mathbb{R}^2)} = \langle f, \mathcal{R}^* \psi_\gamma^p \rangle_{L_2(\mathbb{R}^2)} = \langle \mathcal{R}f, \psi_\gamma^p \rangle_{L_2(Z)},$$

where $Z = \{(\theta, s) \in \mathbb{R}^2 : \theta \in [0, 2\pi), s \in \mathbb{R}\}$. Thus formulas for the approximate inverse of the ray transform operator $\mathcal{I}^2$ have the form (at small γ)

$$^su_i(p) \approx \langle \xi_i [\mathcal{I}^2 u], \psi_\gamma^p \rangle_{L_2(Z)}, \qquad i = 1, 2. \qquad (4)$$

3.3   About Solution of the 2P- and 3P-Problems

We introduce the notations $w = F[{}^s v]$,

$$\nu_{(j)}(y) = y_{j+1}F[{}^s(S_{(j)}v)]_{j+2}(y) - y_{j+2}F[{}^s(S_{(j)}v)]_{j+1}(y), \qquad (5)$$

and rewrite the equalities (1) for each j

$$y_2 w_3(y) - y_3 w_2(y) = \nu_{(1)}(y), \qquad (6)$$
$$y_3 w_1(y) - y_1 w_3(y) = \nu_{(2)}(y), \qquad (7)$$
$$y_1 w_2(y) - y_2 w_1(y) = \nu_{(3)}(y). \qquad (8)$$

In the statement of $3P$-problem we know the values of three functions $[\mathcal{I}_{(j)}v]$, $j = 1, 2, 3$. Thus we can reconstruct the solenoidal parts ${}^s(S_{(j)}v)$ of three slices of vector field v, therefore we have three functions $\nu_{(j)}$, $j = 1, 2, 3$. It is not difficult to obtain the solution of system (2), (6)–(8) for each $y \neq 0$,

$$w_1(y) = |y|^{-2}\big(y_3 \nu_{(2)}(y) - y_2 \nu_{(3)}(y)\big), \qquad (9)$$
$$w_2(y) = |y|^{-2}\big(y_1 \nu_{(3)}(y) - y_3 \nu_{(1)}(y)\big), \qquad (10)$$
$$w_3(y) = |y|^{-2}\big(y_2 \nu_{(1)}(y) - y_1 \nu_{(2)}(y)\big). \qquad (11)$$

In its turn, in the $2P$-problem the values of two functions $[\mathcal{I}_{(1)}v]$, $[\mathcal{I}_{(2)}v]$ are known. In this way, there are two functions $\nu_{(1)}$ and $\nu_{(2)}$, and it is necessary to solve the system (2), (6), (7) for finding the vector field w. When $y_3 \neq 0$ the solution has the form

$$w_1(y) = |y|^{-2}y_3^{-1}\big(y_1 y_2 \nu_{(1)}(y) + (y_2^2 + y_3^2)\nu_{(2)}(y)\big), \qquad (12)$$
$$w_2(y) = -|y|^{-2}y_3^{-1}\big((y_1^2 + y_3^2)\nu_{(1)}(y) + y_1 y_2 \nu_{(2)}(y)\big), \qquad (13)$$

the component $w_3(y)$ should be found by the formula (11).

4   The Schemes of Algorithms for Solving the 2P- and 3P-Problems

4.1   Solving the 3P-Problem

Let three functions $[\mathcal{I}_{(j)}v]$, $j = 1, 2, 3$ be given for a vector field v. It is necessary to realize the following steps for recovery of the solenoidal part ${}^s v$ of field v by these data.

The step 1. Reconstruction of the components ${}^s(S_{(j)}v)_{j+i}$, $i = 1, 2$ of solenoidal parts of slices $S_{(j)}v$, $j = 1, 2, 3$ of field v at fixed x_j using the formulas (4) of the method of approximate inverse. For the mollifiers construction we use the Gauss function

$$e_G(x) = (2\pi)^{-1}\exp(-|x|^2/2).$$

The solution of equation $\mathcal{R}^*\psi_G = e_G$ has the form

$$\psi_G(s) = (2\pi^2)^{-1}\left(1 - \sqrt{2}\,s\,\mathrm{D}(s/\sqrt{2})\right),$$

where $\mathrm{D}(t) = \exp(-t^2)\int_0^t \exp(z^2)dz$ is the Dawson integral. The values of functions $\psi_{G,\gamma}^p(\theta, s)$ are calculated using (3).

The step 2. Finding of the functions $\nu_{(j)}(y)$, $j = 1, 2, 3$ using (5).

The step 3. Calculating $w(y) = F[^sv](y)$ by the formulas (9)–(11).

The step 4. Finding $^sv(x)$, applying the three-dimensional inverse Fourier transform to $w(y)$.

4.2   Solving the 2P-Problem

Let two functions $[\mathcal{I}_{(1)}v]$ and $[\mathcal{I}_{(2)}v]$ be given for a vector field v. It is necessary to reconstruct the solenoidal part ^sv of field v.

The algorithm for the 2P-problem differs from the one for the 3P-problem by that the first and the second steps are realized for $j = 1, 2$, and at the third step the components of field $w(y) = F[^sv](y)$ are found by the formulas (11)–(13).

5   Simulations

We demonstrate the results of the numerical experiment aimed to finding of the value of γ being optimal for the chosen discretization of the input data with respect to parameters θ, s. As "optimal value" of γ we mention such value of it (among several values under consideration) at which the relative error (in percents) for the test field takes the smallest value. Discretization of the ray transform with respect to z is 64, discretization with respect to θ, s is changing and takes values 100, 200, 300, 400. Values of the parameter γ of the approximate inverse method are equal 0.005, 0.01, 0.02, 0.03, 0.04. Test solenoidal vector field $v = v(x)$ is given by the equality

$$v(x) = \begin{cases} 4\left(0.49 - r^2\right)\left(x_3 - (x_2 + 0.1), \frac{x_1 - 0.2}{0.8} - x_3, (x_2 + 0.1) - \frac{x_1 - 0.2}{0.8}\right), & r^2 < 0.49, \\ (0,0,0), & \text{otherwise,} \end{cases}$$

where $r^2 = \frac{(x_1 - 0.2)^2}{0.64} + (x_2 + 0.1)^2 + x_3^2$.

Values of the relative error of the reconstruction of the field v in solving 2P- and 3P-problems at different discretization of the ray transform and different values of γ are shown in the Table 1. Discretization of the ray transform is changing in rows, values of γ and the type of the problem (2P or 3P)—in columns. The smallest value of the error which may be achieved at the selected set for the parameter γ in solving 2P and 3P-problems and at the fixed discretization of the data is bold. The results show the following relative error behavior. As γ increases from 0.005 to 0.05 we see the error decreasing to the some smallest value depending from the type of the problem and the discretization and after this the error is increasing. Thus we conclude that optimal value of γ at discretization 100 is 0.03, at discretization 200—0.02, at discretization 400—0.01 for both types of problems. Also we obtain that at discretization 300 optimal value of γ is 0.01 for 2P-problem and 0.02 for 3P-problem.

Table 1. The relative error at different discretization of the ray transform and different values of γ

γ	0.005		0.01		0.02		0.03		0.04		0.05	
discr. \ Pr	2P	3P	2P	3P	2P	3P	2P	3P	2P	3P	2P	3P
100	>100	>100	>100	>100	14.62	10.17	**13.62**	**7.25**	14.27	8.07	15.25	9.5
200	>100	>100	14.69	10.61	**12.99**	**6.39**	13.41	6.83	14.1	7.76	15.06	9.17
300	63.07	72.86	**12.83**	6.49	12.94	**6.32**	13.37	6.75	14.05	7.67	15.01	9.07
400	16.27	13.57	**12.72**	**6.23**	12.92	6.3	13.34	6.72	14.02	7.62	14.98	9.03

6 Conclusion

We consider the problem of reconstruction of the solenoidal part $^s v$ of vector field v in $\mathbb{R}^3$ from the incomplete data. Namely, it is required to recover $^s v$ by the values of ray transform $[\mathcal{I}^3 v]|_{M^3}$, where M^3 is the set of all lines parallel to the coordinate planes. We use the denotation $[\mathcal{I}^3_{(j)} v]$ for the values of ray transform known for all lines parallel to the coordinate plane $\pi_j = \{x_j = 0\}$. Two statements of the problem are considered:

- *The 2P-problem.* It is required to recover the solenoidal part $^s v$ of vector field v by two given functions $[\mathcal{I}^3_{(1)} v]$ and $[\mathcal{I}^3_{(2)} v]$.
- *The 3P-problem.* It needs to reconstruct the solenoidal part $^s v$ of vector field v by tree given functions $[\mathcal{I}^3_{(j)} v]$, $j = 1, 2, 3$.

For a numerical solution of the problems we develop the algorithms based on the method of approximate inverse.

References

1. Sharafutdinov, V.A.: Integral Geometry of Tensor Fields. VSP, Utrecht (1994). https://doi.org/10.1515/9783110900095
2. Denisjuk, A.: Inversion of the x-ray transform for 3D symmetric tensor fields with sources on a curve. Inverse Prob. **22**(2), 399–411 (2006). https://doi.org/10.1088/0266-5611/22/2/001
3. Sharafutdinov, V.: Slice-by-slice reconstruction algorithm for vector tomography with incomplete data. Inverse Prob. **23**(6), 2603–2627 (2007). https://doi.org/10.1088/0266-5611/23/6/021

4. Svetov, I.E.: Reconstruction of the solenoidal part of a three-dimensional vector field by its ray transforms along straight lines parallel to coordinate planes. Numer. Anal. Appl. **5**(3), 271–283 (2012). https://doi.org/10.1134/S1995423912030093

5. Svetov, I.: Slice-by-slice numerical solution of 3D-vector tomography problem. In: Kosmas, T., Vagenas, E., Vlachos, D. (eds.) Journal of Physics: Conference Series, IC-MSQUARE 2012, vol. 410, p. 012042. IOP Publishing Ltd., Bristol (2013). https://doi.org/10.1088/1742-6596/410/1/012042

6. Derevtsov, E.Y., Efimov, A.V., Louis, A.K., Schuster, T.: Singular value decomposition and its application to numerical inversion for ray transforms in 2D vector tomography. J. Inverse Ill-posed Probl. **19**(4–5), 689–715 (2011). https://doi.org/10.1515/jiip.2011.047

7. Svetov, I.E., Derevtsov, E.Y., Volkov, Y.S., Schuster, T.: A numerical solver based on B-splines for 2D vector field tomography in a refracting medium. Math. Comput. Simul. **97**, 207–223 (2014). https://doi.org/10.1016/j.matcom.2013.10.002

8. Derevtsov, E.Y., Svetov, I.E.: Tomography of tensor fields in the plain. Eurasian J. Math. Comput. Appl. **3**(2), 24–68 (2015)

9. Polyakova, A.: Reconstruction of potential part of 3D vector field by using singular value decomposition. In: Kosmas, T., Vagenas, E., Vlachos, D. (eds.) IC-MSQUARE 2012, Journal of Physics: Conference Series, vol. 410, p. 012015. IOP Publishing Ltd., Bristol (2013). https://doi.org/10.1088/1742-6596/410/1/012015

10. Polyakova, A.P.: Reconstruction of a vector field in a ball from its normal Radon transform. J. Math. Sci. **205**(3), 418–439 (2015). https://doi.org/10.1007/s10958-015-2256-1

11. Polyakova, A.P., Svetov, I.E.: Numerical solution of the problem of reconstructing a potential vector field in the unit ball from its normal Radon transform. J. Appl. Indu. Math. **9**(4), 547–558 (2015). https://doi.org/10.1134/S1990478915040110

12. Louis, A.K.: Inverse und schlecht gestellte Probleme. Vieweg+Teubner Verlag, Stuttgart (1989).https://doi.org/10.1007/978-3-322-84808-6

13. Louis, A.K., Maass, P.: A mollifier method for linear operator equations of the first kind. Inverse Prob. **6**(3), 427–440 (1990). https://doi.org/10.1088/0266-5611/6/3/011

14. Louis, A.K.: Approximate inverse for linear and some nonlinear problems. Inverse Prob. **12**(2), 175–190 (1996). https://doi.org/10.1088/0266-5611/12/2/005

15. Rieder, A., Schuster, T.: The approximate inverse in action with an application to computerized tomography. SIAM J. Numer. Anal. **37**(6), 1909–1929 (2000). https://doi.org/10.1137/S0036142998347619

16. Rieder, A., Schuster, T.: The approximate inverse in action III: 3D-Doppler tomography. Numer. Math. **97**(2), 353–378 (2004). https://doi.org/10.1007/s00211-003-0512-7

17. Schuster, T.: Defect correction in vector field tomography: detecting the potential part of a field using BEM and implementation of the method. Inverse Prob. **21**(1), 75–91 (2005). https://doi.org/10.1088/0266-5611/21/1/006

18. Svetov, I., Maltseva, S., Polyakova, A.: Numerical solution of 2D-vector tomography problem using the method of approximate inverse. In: Ashyralyev, A., Lukashov, A. (eds.), ICAAM 2016, AIP Conference Proceedings, vol. 1759, p. 020132. AIP Publishing LLC, Melville (2016). https://doi.org/10.1063/1.4959746

19. Derevtsov, E.Y., Louis, A.K., Maltseva, S.V., Polyakova, A.P., Svetov, I.E.: Numerical solvers based on the method of approximate inverse for 2D vector and 2-tensor tomography problems. Inverse Prob. **33**(12), 124001 (2017). https://doi.org/10.1088/1361-6420/aa8f5a

Computational Fluid Dynamics Methods for Wind Resources Assessment

Sabine Upnere[1]([✉])[ID], Valerijs Bezrukovs[1,2], Vladislavs Bezrukovs[1][ID], Normunds Jekabsons[1], and Linda Gulbe[1]

[1] Ventspils University of Applied Science, Ventspils, Latvia
upnere@protonmail.com
[2] Institute of Physical Energetics, Riga, Latvia

Abstract. The use of already existing infrastructure for mounting of wind speed sensors could be a promising way of how to assess wind resources instead to install the new meteorological mast. One part of this study is devoted to exploring the impact of the mast on the flow field around it. Computational Fluid Dynamics (CFD) is chosen to predict airflow using Reynolds-Averaged Navier-Stokes equations. In the second part of this research, the typical topology near the Baltic Sea is selected to evaluate numerically the turbulent airflow over coastal terrain. The lidar images are utilized to describe the topology of the interested area. Digital Surface Model is used to generate the ground surface which is applied as the input to develop the high-resolution computational mesh of the terrain. Computational domain parallelization and the computational cluster is applied due to the complexity of the numerical simulations. Obtained results are compared with experimentally measured data from wind speed sensors located on the telecommunication mast.

Keywords: CFD · Flow modelling · RANS equations · Wind resources

1 Introduction

Increasing wind resource applicability leads to the interest of the wind turbine installations in areas with more uniform topography. Special meteorological mast installation to obtain measurements is the expensive way how to determine the suitability of the area for the wind power generation. The alternative could be to use already existing infrastructures (such as telecommunications masts) for the mounting of measurement sensors.

The one part of this research is related to the numerical evaluation of the telecommunication mast induced distortions in the flow field around it. The estimation of changes in the field distribution allows to chose better locations for wind speed sensors. In the scientific literature is available numbers of researches where the numerical evaluation of the flow pattern around masts and the flow distortion effect induced by the mast on the sensors measurements is analysed. For example, atmospheric flow around tubular and lattice meteorological masts

© Springer Nature Switzerland AG 2020
Y. D. Sergeyev and D. E. Kvasov (Eds.): NUMTA 2019, LNCS 11974, pp. 495–502, 2020.
https://doi.org/10.1007/978-3-030-40616-5_48

has been investigated in [1]. An experimental and numerical research of flow around triangular lattice mast with anemometer booms is described in [2]. The impact of top-mounted sensors is studied in [3] and [4]. An analytical estimation of the center-line wind speed deficit as function of the mast solidity is available in IEC standard [5].

The second part of this study is about the flow field prediction over coastal terrain. In this research, the typical topology near the Baltic Sea is analysed. The telecommunication mast on which are placed wind speed sensors is located approximately 2 km from the coast. A feasibility study described [6] has been carried out in which the coastal topology was described with a smooth surface using orthophoto. In this study, data derived from lidar measurements were used to obtain the digital model of the ground surface, providing a significantly more detailed description. Obtained results are compared with experimentally measured wind speed in the corresponding area.

2   Numerical Approach

The numerical method is based on the Computational Fluid Dynamics (CFD) approach. To carry out a CFD analysis, the partial differential equations are solved using the Finite Volume discretization. A computational study of the airflow around the communication mast and over the coastal terrain has been realised by the open source CFD tool OpenFOAM 2.4.x [7].

The flow is assumed to be incompressible and isothermal therefore the equation of energy is not solved. The three-dimensional equations describing the conservation of the mass and momentum are calculated applying the time-averaged approximation – Reynolds-Averaged Navier-Stokes (RANS) method. To obtain the closure of the system of equations due to the turbulence, two additional transport equations (one for the turbulence kinetic energy, k and one for the rate of dissipation of turbulence energy, ε) have been added. The SIMPLE algorithm has been selected using OpenFOAM solver *simpleFoam* for solving pressure-correction equation from the momentum and mass balance equations [8].

2.1   Implementation of the Mast

The analysed telecommunication mast is a triangular guyed mast which consists of 1 m long segments. The total length of the mast is 92 m but modelled is only four mast segments. The construction of the cellular communication mast has been modelled as a rigid object.

The mast model consists of three vertical tubular tubes (7.6 cm diameter) which are connected with tubular cross-arms (3.8 cm diameter). One segment of the mast is 1.76 m high with a leg-to-leg distance of 1.2 m. The digital representation of the mast contains also cables and stairs. Structural elements are created as STL surfaces which are generated by mesh generator NetGen [9].

The computational domain for the flow field simulations around the mast is defined as a rectangular area with the following dimensions – 30 m in the flow direction, 9 m in the perpendicular direction of the flow and 5 m in the vertical direction. The scheme of the domain is shown in Fig. 1.

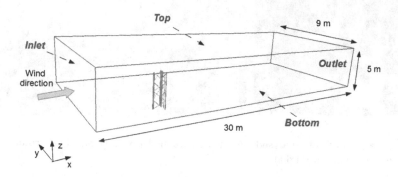

Fig. 1. Computational domain for calculations of the flow field around the communication mast.

The computational mesh is developed using built-in OpenFOAM mesh utilities. Around the mast construction is generated as unstructured mesh using *snappyHexMesh* utility. *blockMesh* is used to obtain structured mesh in the remaining part of the computational domain.

2.2   Implementation of the Terrain

The flow simulations over coastal terrain are done using four rectangular computational domains. Dimensions of the domains are summarized in Table 1.

Table 1. Dimensions of the computational domains.

Domain	Streamwise (x)	Spanwise (y)	Vertical (z)
D1	5000 m	5000 m	1000 m
D2	5000 m	5000 m	600 m
D3	2000 m	2000 m	600 m
D4	400 m	800 m	600 m

Digital Surface Model (DSM), employed for mesh development, is obtained from the lidar data. Surface roughness coefficients to each DSM pixel are assigned according to the land cover classes. The classification is performed using supervised k nearest neighbours (kNN) classifier. A sample set for kNN covers 3000 m × 3000 m area around the telecommunication mast. Each of the sample points is classified in 6 classes: water, sand, buildings and asphalt, grasslands and agricultural lands, forest and marsh. Resultant land cover maps are prepared at 1 m per pixel spatial resolution (see the left side of Fig. 2).

Fig. 2. The map of the analysed area with classified pixels (left). The ground of the computational domain (right).

In the next step, the obtained data is converted to STL format using Matlab script. The STL file describes smaller part (see Table 1) of terrain with the maximum difference of the height – 93.11 m. To obtain ground surface applicable for OpenFOAM the *snappyHexMesh* utility is used. The ground level of the domain can be seen in the right side of Fig. 2. Simulation domains contain approximately from $12.2 \cdot 10^6$ to $34.4 \cdot 10^6$ hexahedral cells.

2.3   Boundary Conditions

Atmospheric boundary layer (ABL) library is used to define boundary conditions for both simulations – to predict the flow around the mast and the flow over the terrain. The thermal stratification and Coriolis effects are not taken into account. It is assumed that the turbulence intensity is 10% and turbulence length scale is 10% of the computational domain height.

Inlet velocity is defined by logarithmic law:

$$U(z) = \frac{u_\tau}{\kappa} ln((z + z_0)/z_0), \tag{1}$$

where U is the time-averaged horizontal velocity, z is the vertical coordinate, κ is the von Karman constant, z_0 is the physical roughness height and the friction velocity u_τ is defined as:

$$u_\tau = \sqrt{\tau_\omega/\rho}, \tag{2}$$

where τ_ω is the wall shear stress and ρ is the density.

Other boundary conditions are represented in Table 2.

3   Results and Discussion

At the beginning of the computational simulations, the preliminary study was realized to investigate the effect of the turbulence model, the size of the computational domain, the mesh resolution and boundary conditions. The solutions

Table 2. Boundary conditions of the computational domain

Boundary	Description
Inlet (incoming flow)	Dirichlet conditions are applied for U, k and ε using ABL libraries with Richardson and Hoxey expressions [10]. Neumann zero gradient condition is used for p
Outlet (outflow)	A constant pressure, $p = 0$ is set. For all other variables Neumann zero gradient condition is considered
Sky (top of the domain)	Slip condition (zero for normal component of a vector, zero gradient for tangential and for any scalar) is used
Sides (parallel to the flow)	Slip conditions are considered
Ground (for the terrain)	Standard OpenFOAM wall functions have been applied for ν_t, k and ε. Velocity on surface walls is equal to zero. Neumann zero gradient condition is used for p
Bottom (for the mast)	Symmetry plane

have been assumed as converged when the sum of the residuals is kept within limits of 10^{-5}.

Two turbulence models – standard k-ε and realizable k-ε was compared. Flow velocity profile before the communication mast was chosen to evaluate the impact of the model. From obtained results follows that there is no significant difference between predicted profile. However, the standard k-ε has better convergence therefore for the future simulations this turbulence model is chosen.

The impact of the domain vertical dimension was analysed using domains D1 and D2. Results show that the domain height reduction from 1000 m to 600 m changes the velocity at 100 m from the ground by 1.68%. Therefore for the next investigation, the domain with 600 m vertical dimension was used. The effect of domain size in streamwise and spanwise direction was analysed using D2, D3 and D4. Based on the obtained results the domain D4 was chosen as optimal.

The mesh sensitivity study was realized using three different cell sizes at the ground in the vertical direction: M1 with 2 m height; M2 with 1.67 m height and M3 with 1 m height. An investigation of the vertical resolution influence showed the obtained difference is negligible and for future calculations can be used mesh M1.

To evaluate the telecommunication mast influence on the sensor measurement accuracy several series of simulations were done. Simulations of flow field distribution around segments of the mast are conducted for two different wind velocities and 4 different angles of attack, α. Analysed wind directions are illustrated in Fig. 3.

Figure 4 shows the examples of velocity distribution of a wind speed around a telecommunication mast at a distance of 2.8 m relative to the mast centre. The distance from the mast centre was chosen on the basis of the previous study where optimum boom length was select, see [11]. Flow velocities are 5.0 and 10.0 m/s and the angles of attack are 0°, 90°, 180° and 270°.

In Fig. 4, you can see that the pattern of the distortions is similar to both wind speeds. Significant structure impacts on wind speed are only observed in

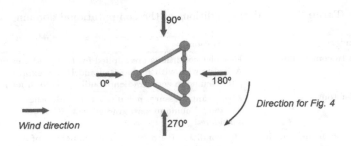

Fig. 3. Analysed angles of attack.

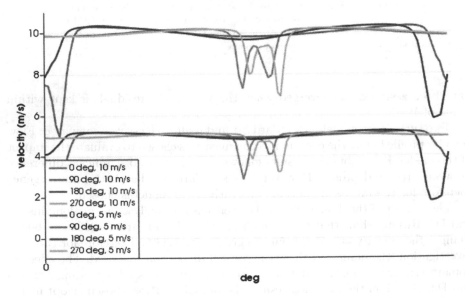

Fig. 4. Simulation results of the wind flow field around a triangular cellular communication mast at 2.8 m from the centre, for wind velocities $U = 5.0$ and $10.0\,\mathrm{m/s}$.

the tail region. Due to cable lines, the distortions are larger and nonsymmetric in corresponding places. It is recommended that you use two sensors at one height to avoid incorrect measurements obtained in the shadow of the mast.

The second group of simulations is used to model the flow over the terrain. Figure 5 demonstrates a comparison of simulated velocity profile and experimentally measured velocity at certain points. It can be seen that the largest differences are between 20 and 60 m from the ground. For example, at 40 m height simulations overpredicts velocity by 17%. It could be speculated that the main reason for the difference is the description of the terrain.

Comparing the velocity in the computational domain with and without terrain can evaluate the effect of the ground surface roughness on measured wind speed. For example, at 62 m height, the difference in the flow velocity is approximately 5%. Increasing distance from the ground, the difference decrease. It can

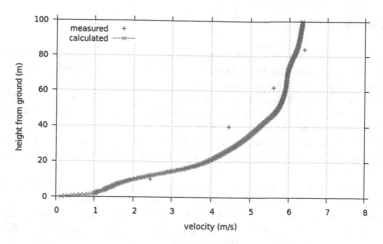

Fig. 5. The comparison of the velocity profile predicted in the computational domain with the terrain and experimentally measured values on site.

be concluded that the flat terrain affects the airflow only by a few percentages if measurements is taken above 60 m.

4   Conclusions

The obtained results show that the presence of cable lines inside a lattice cellular communication mast slows down wind flow speed and it is the main disadvantage of telecommunication masts comparing to meteorological masts. However, it can be concluded that the telecommunications masts of the type considered may be used to install wind sensors at the selected distance from the mast centre if two sensors in one level are used.

The CFD simulations of flow field over coastal terrain were done to analyse changes of the velocity profile due to the terrain. Using the described methodology the largest differences between experimentally measured values and numerically predicted were from 20 m to 60 m from the ground. It can be guessed that this is related to the representation of the terrain. Calculation results show that the flat terrain effect on the wind speed above 60 m from the ground is less than 5%.

Acknowledgements. The work is carried out within the project New European Wind Atlas (NEWA), ENER/FP7/618122/NEWA ERA-NET PLUS, supported by the EUROPEAN COMMISSION under the 7th Framework Programme for Research, Technological Development and Demonstration.

References

1. Tusch, M., et al.: Modeling of turbulent atmospheric flow around tubular and lattice meteorological masts. J. Solar Energy Eng. **133**(1), 011011 (2011)
2. Stickland, M., et al.: Measurement and simulation of the flow field around a triangular lattice meteorological mast. J. Energy Power Eng. **7**, 1934–1939 (2013)
3. Perrin, D., et al.: The effect of meteorological tower on its top-mounted anemometer. Appl. Energy **84**(4), 413–424 (2007)
4. Barlow, J.F., et al.: A wind-tunnel study of flow distortion at a meteorological sensor on top of the BT Tower, London, UK. J. Wind Eng. Ind. Aerodyn. **99**, 899–907 (2011)
5. Wind Turbines - Part 12–1: Power Performance Measurements of Electricity Producing Wind Turbines. IEC 61400–12-1, Geneva, Switzerland
6. Upnere, S., et al.: Simulation of the flow field over the coastal terrain. In: Abdel Wahab, M. (eds.) Proceedings of the 1st International Conference on Numerical Modelling in Engineering, pp. 400–407 (2018)
7. OpenFOAM Homepage. https://openfoam.org/release/2-4-0/. Accessed 22 May 2019
8. Wilcox, D.C.: Turbulence modeling for CFD. DCW Industries, La Canada, CA (2006)
9. NetGen Homepage. https://ngsolve.org/. Accessed 22 May 2019
10. Richards, P.J., Hoxey, R.P.: Appropriate boundary conditions for computational wind engineering models using the k-epsilon turbulence model. J. Wind Eng. Ind. Aerodyn. **46 & 47**, 145–153 (1993)
11. Bezrukovs, V., et al.: The assessment of wind speed distortions in a simulated flow around a lattice cellular communication mast. In: IEEE 2017 European Conference on Electrical Engineering and Computer Science (2017)

On Stationary Points of Distance Depending Potentials

Alexei Uteshev[ID] and Marina Goncharova[✉][ID]

Faculty of Applied Mathematics, St. Petersburg State University,
7–9 Universitetskaya nab., St. Petersburg 199034, Russia
{a.uteshev,m.goncharova}@spbu.ru

Abstract. We continue investigations started in the previous publications by the authors (*LNCS*, volumes 8136 (2013) and 9570 (2016)). The structure of stationary point sets is established for the family of functions given as linear combinations of an exponent L of Euclidean distances from a variable point to the fixed points in 2D and 3D spaces. We compare the structure of the stationary point sets for several values of the exponent L, focusing ourselves mainly onto the cases of Coulomb potential and Weber facility location problem. We develop the analytical approach to the problem aiming at finding the exact number of stationary points and their location in relation to the parameters involved.

Keywords: Stationary points · Coulomb potential · Weber problem

1 Introduction

Given the coordinates of $K \geq 3$ points $\{P_j\}_{j=1}^{K} \subset \mathbb{R}^n$, the problem is to find the exact number and the coordinates of the stationary points for the function

$$F(P) = \sum_{j=1}^{K} m_j |PP_j|^L, \quad P = (x_1, \ldots, x_n) \in \mathbb{R}^n. \tag{1}$$

Here $\{m_j\}_{j=1}^{K}$ are assumed to be real non-negative numbers, the exponent $L \in \mathbb{R}$ is nonzero while $|\cdot|$ stands for the Euclidean distance.

For $L = 2$, the problem has a well-known solution: the center of mass (barycenter) $P_* = \sum_{j=1}^{K} m_j P_j / \sum_{j=1}^{K} m_j$ provides the global minimum for the function $F(P)$. For $L = 1$, the problem is know as the generalized Fermat–Torricelli or the Weber problem, and it is the origin of the branch of Operation Research known as Facility Location. For this case, a unique stationary point might exist for (1), and for the specialization $n = 2, K = 3$, its coordinates can be expresses by radicals via the parameters of the function [5]. For $L = -1$ and $n = 3$, the problem can be viewed to as a classical electrostatics one with the function (1) representing the Coulomb potential of the charges $\{m_j\}_{j=1}^{K}$ placed at fixed positions $\{P_j\}_{j=1}^{K}$ in the space. In despite of

© Springer Nature Switzerland AG 2020
Y. D. Sergeyev and D. E. Kvasov (Eds.): NUMTA 2019, LNCS 11974, pp. 503–510, 2020.
https://doi.org/10.1007/978-3-030-40616-5_49

its classical looking formulation, the problem has not been given a systematic exploration — with the exception of some special configurations [4]. The difficulty of the problem can be acknowledged also from the state of the art with its part known as

Maxwell's Conjecture [3]. The total number of stationary points of any configuration with K charges in $\mathbb{R}^3$ never exceeds $(K-1)^2$.

After more than a century and a half from its formulation, this conjecture has recently attracted attention in [2]. It remains still open even for the case of $K = 3$ charges [6].

Aside from the stated problem of localization of stationary points corresponding to minimum (or *stable* equilibria for the potential), we also treat the *parameter synthesis problem*. Namely, we look for the largest possible domain $\mathbb{P}$ in the parameter space $\mathbb{R}^K$ such that for any specialization of parameter vector $(m_1, \ldots, m_K)$ from this domain, there exists at least one local minimum for (1). On the other hand, we look for the set $\mathbb{S}$ in coordinate space $\mathbb{R}^n$ where every point might be made a stationary minimum one for (1) by a suitable specialization of $(m_1, \ldots, m_K) \in \mathbb{P}$. Every set $\mathbb{S}$ or $\mathbb{P}$ will be hereinafter referred to as the **stability domain** in the corresponding space.

Although the stated problem hardly expect the closed form analytical solution, the latter can be suggested for sufficiently wide class of functions (1). To illuminate this statement, we tackle the case of function (1) where $K = n + 1$.

2 Multidimensional Case

Stationary points of the function

$$F(P) = \sum_{j=1}^{n+1} m_j |PP_j|^L \qquad (2)$$

are given by the system

$$\partial F/\partial x_1 = 0, \ldots, \partial F/\partial x_n = 0. \qquad (3)$$

These equations are linear homogeneous ones with respect to $\{m_j\}_{j=1}^{n+1}$. Their elimination from this system leads to the following

Theorem 1. *Let the points* $\{P_j = (x_{j1}, \ldots, x_{jn})\}_{j=1}^{n+1}$ *be chosen such that the condition*

$$V = \begin{vmatrix} 1 & 1 & \ldots & 1 \\ x_{11} & x_{21} & \cdots & x_{n+1,1} \\ \vdots & \vdots & & \vdots \\ x_{1n} & x_{2n} & \cdots & x_{n+1,n} \end{vmatrix} > 0 \qquad (4)$$

is satisfied. Denote by V_j *the determinant obtained on replacing the jth column of (4) by the column[1]* $[1, x_1, \ldots, x_n]^\top$. *Any solution to the system (3) is a solution to the system*

---

[1] Hereinafter $^\top$ denotes transposition.

$$m_1 : m_2 : \cdots : m_{n+1} = |PP_1|^{2-L}V_1 : |PP_2|^{2-L}V_2 : \cdots : |PP_{n+1}|^{2-L}V_{n+1}. \quad (5)$$

Since a solution to system (3) is invariant under substitution $\{m_j \rightarrow cm_j\}_{j=1}^{n+1}$, $c \neq 0$, it is possible to assume that at this solution the following relations

$$\{m_j = |PP_j|^{2-L}V_j\}_{j=1}^{n+1} \quad (6)$$

are valid up to a common numerical factor. This permits one to find the boundary for stability domain $\mathbb{S}$. Indeed, this boundary is defined by the condition for the Hessian $\mathcal{H}(P)$ of (2) to lose the property of positive definiteness, and this happens when

$$\det \mathcal{H}(P) = 0 \quad (7)$$

at some stationary point P_* of $F(P)$. Joint fulfillment of the equalities (3) and (7) at this point for some specialization of parameters means then that it is a multiple one for the system (3). Namely, its appearance is due to collision of (generically) two non-degenerate stationary points of the function $F(P)$ when the parameter vector $(m_1, \ldots, m_{n+1})$ tends to a bifurcation point lying at the boundary of the stability domain $\mathbb{P}$ in the parameter space.

Since at this point the relationships (6) are valid, one can eliminate the parameters $\{m_j\}_{j=1}^{n+1}$ from (7). This results in an equation for the manifold in $\mathbb{R}^n$ yielding the boundary of the stability domain $\mathbb{S}$ provided the latter is not empty. We detail the structure of this manifold for $n \in \{2, 3\}$ in the foregoing sections, and restrict ourselves here with the following condition for the emptiness of $\mathbb{S}$.

Theorem 2. *For* $2 - n \leq L < 0$, *none of the stationary point of (2) is a point of minimum.*

For $L > 0$ or $L < 2 - n$, the principal existence of the point of minimum is confirmed by the following

Example 1. Let the points $\{P_j\}_{j=1}^{n+1}$ compose a regular n-dimensional simplex in $\mathbb{R}^n$ centered at $P_* = \mathbb{O}$. For instance, for $n = 3$, one may take

$$P_1 = \begin{pmatrix} -1/3 \\ -\sqrt{2}/3 \\ -\sqrt{2}/3 \end{pmatrix}, P_2 = \begin{pmatrix} -1/3 \\ \sqrt{8}/3 \\ 0 \end{pmatrix}, P_3 = \begin{pmatrix} -1/3 \\ -\sqrt{2}/3 \\ \sqrt{2}/3 \end{pmatrix}, P_4 = \begin{pmatrix} 1 \\ 0 \\ 0 \end{pmatrix}. \quad (8)$$

For the function $F(P) = \sum_{j=1}^{n+1} |PP_j|^L$, the point $P_* = \mathbb{O} \in \mathbb{R}^n$ is evidently a stationary one. Hessian $\mathcal{H}(\mathbb{O})$ possesses a single eigenvalue (of the multiplicity n) equal (up to a positive factor) to $L(L + n - 2)$. Therefore, $\mathbb{O}$ is the point of minimum iff $L > 0$ or $L < 2 - n$. $\quad\square$

As for the stability domain $\mathbb{P}$ in the parameter space, its boundary can be obtained on elimination of the variables $x_1, \ldots, x_n$ from equation (7) using system (3). Compared with the previously considered procedure of elimination of parameters, this time one cannot expect even the algebraicity of the procedure. Only for the case of the rationality of the exponent L, the system (3) can be reduced to an algebraic one. Elimination of variables from such a system can be organized with the aid of the multivariate resultant computation or via the Gröbner basis construction [1].

3   The 2D Case

The planar counterpart of (5) is as follows

$$m_1 : m_2 : m_3 = |PP_1|^{2-L}S_1 : |PP_2|^{2-L}S_2 : |PP_3|^{2-L}S_3. \tag{9}$$

Here $\{P_j = (x_j, y_j)\}_{j=1}^3, P = (x, y)$ and

$$S_1(x, y) = \begin{vmatrix} 1 & 1 & 1 \\ x & x_2 & x_3 \\ y & y_2 & y_3 \end{vmatrix}, \quad S_2(x, y) = \begin{vmatrix} 1 & 1 & 1 \\ x_1 & x & x_3 \\ y_1 & y & y_3 \end{vmatrix}, \quad S_3(x, y) = \begin{vmatrix} 1 & 1 & 1 \\ x_1 & x_2 & x \\ y_1 & y_2 & y \end{vmatrix}.$$

Theorem 3. *Let the points P_1, P_2, P_3 be noncollinear and be counted counter-clockwise. If $L \geq 1$ then the stability domain $\mathbb{S}$ in the coordinate plane coincides with the interior of the triangle $P_1P_2P_3$, i.e. any point $P_* = (x_*, y_*)$ inside the triangle is the point of minimum for the function*

$$F_*(P) = \sum_{j=1}^3 m_j^* |PP_j|^L \quad \text{where} \quad \{m_j^* = |P_*P_j|^{2-L}S_j(x_*, y_*)\}_{j=1}^3. \tag{10}$$

If $L < 1, L \neq 0$ then the boundary of the stability domain $\mathbb{S}$ is given by the equation

$$\widetilde{\Phi}_L(x, y) := \frac{S_1(x, y)S_2(x, y)S_3(x, y)}{|PP_1|^2|PP_2|^2|PP_3|^2} \sum_{j=1}^3 S_j(x, y)|PP_j|^2 + \frac{L-1}{(L-2)^2}S^2 = 0 \tag{11}$$

where

$$S = S_1 + S_2 + S_3 \equiv \begin{vmatrix} 1 & 1 & 1 \\ x_1 & x_2 & x_3 \\ y_1 & y_2 & y_3 \end{vmatrix}.$$

Proof. The characteristic polynomial of the Hessian at any stationary point P_* of the function $F(P)$ can be represented as follows [5]:

$$\det(\lambda I - \mathcal{H}(P_*)) = \lambda^2 - L^2 S \lambda + L^2(L-2)^2 \widetilde{\Phi}_L(x_*, y_*). \tag{12}$$

It turns out that for $L < 1, L \neq 0$, the inequality $\widetilde{\Phi}_L(x, y) > 0$ provides a nonempty domain inside the triangle $P_1P_2P_3$, and if the point P_* lies within this domain then all the zeros of (12) are positive and therefore P_* is point of minimum for (10).   □

As for the localization of stability domain in the parameter space, this problem is rather more difficult since the variables x and y to be eliminated from (9) are involved in it in a highly nonlinear manner.

Example 2. Let $P_1 = (1, 1), P_2 = (5, 1), P_3 = (2, 6)$. For the Coulomb potential $F(P) = 1/|PP_1| + m_2/|PP_2| + m_3/|PP_3|$, stability domain $\mathbb{P}$ in the (m_2, m_3)-plane has a boundary which is obtained via resultant computation for the equations

$$m_2^2 S_1^2 |PP_1|^6 - S_2^2 |PP_2|^6 = 0, \quad m_2^2 S_3^2 |PP_3|^6 - m_3^2 S_2^2 |PP_2|^6 = 0$$

accomplished with the condition (7); this time the variables x and y are to be eliminated. This results in the algebraic equation $\Psi(m_2, m_3) = 0$ where $\Psi(m_2, m_3) \in \mathbb{Z}[m_2, m_3], \deg \Psi = 48$ and the coefficients of the magnitude of up to 10^{80}. The details of computation and the image of both domains $\mathbb{S}$ and $\mathbb{P}$ are presented in [6]. □

Representation (9) permits one to trace the curve of stationary points under variation of an extra parameter of the considered function, namely the exponent L.

Theorem 4. *For any specialization of $\{m_j, P_j\}_{j=1}^3$, the stationary points of the function (10) lie in the curve*[2]

$$(\log|PP_2| - \log|PP_3|) \log \frac{S_1}{m_1} + (\log|PP_3| - \log|PP_1|) \log \frac{S_2}{m_2}$$

$$+ (\log|PP_1| - \log|PP_2|) \log \frac{S_3}{m_3} = 0 \tag{13}$$

Proof. From (9) it follows that

$$\begin{cases} \log S_2/m_2 - \log S_1/m_1 - (2 - L)(\log|PP_1| - \log|PP_2|) = 0, \\ \log S_3/m_3 - \log S_1/m_1 - (2 - L)(\log|PP_1| - \log|PP_3|) = 0. \end{cases} \tag{14}$$

Elimination of L results in (13). □

Though the curve (13) is not an algebraic one, its depiction does not cause trouble.

Example 3. For $m_1 = m_2 = m_3 = 1$ and $P_1 = (1,1), P_2 = (5,1), P_3 = (2,6)$, the curve (13) is plotted in Fig. 1 (a).

It passes through nearly all the significant points of the triangle $P_1 P_2 P_3$, namely its vertices ($L \to 1$), the midpoints of the sides ($L \to -\infty$), centroid ($L = 2$), circumcenter ($L \to +\infty$ and $L \to -\infty$), Fermat–Torricelli point ($L = 1$). The two extra points in the curve

$$\left(\frac{8}{3} \pm \frac{\sqrt{-6 + \sqrt{61}}}{3}, \frac{8}{3} \mp \frac{\sqrt{6 + \sqrt{61}}}{3}\right) \approx \{(3.1151, \ 1.4279); (2.2181, 3.9054)\}$$

corresponding to $L \to 0$ are the stationary points of the logarithmic potential $\log|PP_1| + \log|PP_2| + \log|PP_3|$ (or, equivalently, for $|PP_1| \cdot |PP_2| \cdot |PP_3|$).

The bifurcation values for the exponent L are obtained from the condition that Eq. (14) have a multiple zero with respect to x and y. This is equivalent to vanishment of their Jacobian. It turns out that the latter equals $(L-2)^2 \widetilde{\Phi}_L(x,y)/(S_1 S_2 S_3)$ with $\widetilde{\Phi}_L(x,y)$ defined by (11). Resolving the obtained non-algebraic system yields two bifurcation values for L, namely $L_1 \approx -13.5023$ and $L_2 \approx 0.7948$ (Fig. 1 (b)). An extra bifurcation value is $L_3 = 1$. When L

---

[2] With the logarithm considered to an arbitrary positive base.

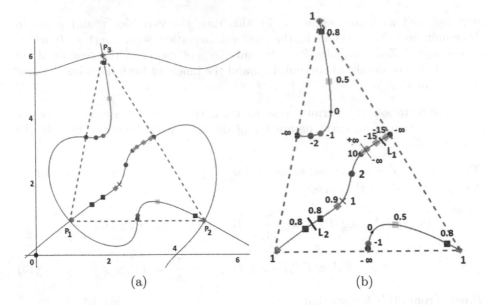

Fig. 1. Example 3. Stationary points for different values of the exponent L.

approaches L_1 from the left the two stationary points tend to collision point at $\approx (3.3099, 3.3907)$; when L approaches L_2 from the right the two stationary points tend to collision at $\approx (1.8354, 1.6141)$. When L approaches 1 from the left, the three stationary points tend to P_1, P_2 and P_3. Bifurcation values L_1, L_2, L_3 separate the intervals in the L-axis corresponding to distinct numbers of stationary points for the function (10). The latter possesses four stationary points if $L < L_1$ or $L_2 < L < 1$, two points if $L_1 < L < L_2$, and a single point if $L \geq 1$.   □

4   The 3D Case

We next treat the case $n = 3$, i.e.,

$$F(P) = \sum_{j=1}^{4} m_j \, |PP_j|^L, \quad P = (x, y, z), \ \{P_j = (x_j, y_j, z_j)\}_{j=1}^{4} \subset \mathbb{R}^3 \quad (15)$$

In the following result we give a corrected version of one erroneous statement from [5].

Theorem 5. *Let the points $\{P_j\}_{j=1}^{4}$ satisfy the assumptions of Theorem 1. If $L \geq 2$ then the stability domain $\mathbb{S}$ for the function (15) in the coordinate space coincides with the interior of the simplex $P_1 P_2 P_3 P_4$. If $L \in [-1; 0]$ then the domain is empty (v. Theorem 2). Else the boundary for the domain of stability is given by the equation*

$$\widetilde{\Phi}_L(x, y, z) = (L-1)V^3 + (L-2)^2 V t_2(x, y, z) + (L-2)^3 t_3(x, y, z) = 0 \quad (16)$$

where

$$t_2(x, y, z) = \sum_{1 \leq j < k \leq 4} V_j V_k \frac{\det(M_{jk} \cdot M_{jk}^{\mathsf{T}})}{|P_* P_j|^2 |PP_k|^2}, \quad M_{jk} = \begin{bmatrix} x - x_j & y - y_j & z - z_j \\ x - x_k & y - y_k & z - z_k \end{bmatrix};$$

and

$$t_3(x, y, z) = \frac{V_1 V_2 V_3 V_4}{|PP_1|^2 |PP_2|^2 |PP_3|^2 |PP_4|^2} \sum_{j=1}^{4} V_j |PP_j|^2.$$

Proof. For the function (15) with $\{m_j = m_j^*\}_{j=1}^4$ defined by (6), compute the characteristic polynomial of the Hessian matrix $\mathcal{H}(P_*)$:

$$\lambda^3 - L(L+1)V\lambda^2 + L^2 \left[(2L-1)V^2 + (L-2)^2 t_2(x_*, y_*, z_*)\right]\lambda - L^3 \widetilde{\Phi}_L(x_*, y_*, z_*).$$

Both expressions t_2 and t_3 are non-negative if the point P_* lies inside the simplex. For $L \in [-1; 0]$, the coefficient of λ^2 is not negative, therefore, at least one of the eigenvalues of the Hessian should be non-positive. From this follows the first statement of the theorem. On the contrary, for $L \geq 2$, the three variations in sign in the sequence of the coefficients of the characteristic polynomial certify, due to Descartes rule of signs, that all the eigenvalues of $\mathcal{H}(P_*)$ are positive. Wherefrom follows the last statement of the theorem. $\qquad\square$

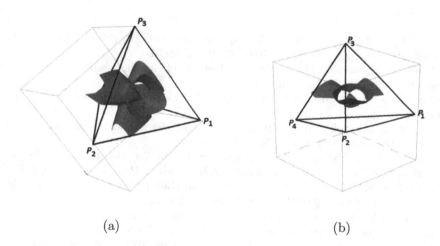

(a)   (b)

Fig. 2. Example 4. Stability domain in the coordinate space (in red). (Color figure online)

Example 4. For the simplex (8) and the function $\sum_{j=1}^{4} m_j / |PP_j|^2$, the surface (16) is given by the equation

$$81(x^2 + z^2 + y^2)^4 - 144(-2x^2 + 2\sqrt{2}xy - y^2 + 3z^2))(x + \sqrt{2}y)(x^2 + z^2 + y^2)^2$$

$$+ \cdots - 304(-2x^2 + 2\sqrt{2}xy - y^2 + 3z^2)(x + \sqrt{2}y) - 132(x^2 + z^2 + y^2) + 1 = 0.$$

Stability domain $\mathbb{S}$ is bounded by a closed tetrahedron-looking part of this surface surrounding the origin (Fig. 2).

5   Conclusion

We have treated the problem of structure specification for the set of stationary points of the function (1). We have also keened in establishing the influence of the involved parameters onto this set. The suggested solution results in the pair of sets. The one in parameter space absorbs all the bifurcation values while the other in the space of variables contains all the possible locations for the points of minimum. Both sets has been represented analytically by *algebraic* equations or inequalities. This opportunity is certainly granted only for the case of functions with rational exponents L. Even for this case, the computational complexity of the algorithm grows drastically with the increase of the dimension from $n = 2$ to $n = 3$. The hope to overcome this obstacle is connected with a counterpart in $\mathbb{R}^3$ of Theorem 4.

Acknowledgements. The present work is supported by RFBR according to the projects No **17-29-04288** (A. Uteshev) and No **18-31-00413** (M. Goncharova).

References

1. Cox, D., Little, J., O'Shea, D.: Ideals, Varieties, and Algorithms. Springer, New York (2007). https://doi.org/10.1007/978-0-387-35651-8
2. Gabrielov, A., Novikov, D., Shapiro, B.: Mystery of point charges. Proc. Lond. Math. Soc. **95**, 443–472 (2007). (series 3)
3. Maxwell, J.: A Treatise on Electricity and Magnetism, vol. 1. Dower, New York (1954)
4. Tsai, Y.L.: Maxwell's conjecture on three point charges with equal magnitudes. Physica D: Nonlinear Phenom. **309**, 86–98 (2015)
5. Uteshev, A.Y., Yashina, M.V.: Stationary points for the family of Fermat–Torricelli–Coulomb-like potential functions. In: Gerdt, V.P., Koepf, W., Mayr, E.W., Vorozhtsov, E.V. (eds.) CASC 2013. LNCS, vol. 8136, pp. 412–426. Springer, Cham (2013). https://doi.org/10.1007/978-3-319-02297-0_34
6. Uteshev, A.Y., Yashina, M.V.: On Maxwell's conjecture for coulomb potential generated by point charges. In: Gavrilova, M.L., Tan, C.J.K. (eds.) Transactions on Computational Science XXVII. LNCS, vol. 9570, pp. 68–80. Springer, Heidelberg (2016). https://doi.org/10.1007/978-3-662-50412-3_5

A Partition Based Bayesian Multi-objective Optimization Algorithm

Antanas Žilinskas$^{(\boxtimes)}$ (iD) and Linas Litvinas

Institute of Data Science and Digital Technologies,
Vilnius University, Akademijos Street 4, 08663 Vilnius, Lithuania
antanas.zilinskas@mii.vu.lt, linas.litvinas@mif.vu.lt

Abstract. The research is aimed at coping with the inherent computational intensity of Bayesian multi-objective optimization algorithms. We propose the implementation which is based on the rectangular partition of the feasible region and circumvents much of computational burden typical for the traditional implementations of Bayesian algorithms. The included results of the solution of testing and practical problems illustrate the performance of the proposed algorithm.

Keywords: Non-convex optimization · Multi-objective optimization · Bayesian approach

1 Introduction

Applied optimization problems, especially those in engineering design, frequently are multi-objective and non-convex. The class of non-convex objective functions is non-homogeneous from the point of view of design of optimization algorithms. We address the black-box problems differently from the problems, where objective functions and constraints are described by mathematical formulas. Methods for the latter problems generalize classical mathematical programming methods [8,10]. Metaheuristic methods are popular for black-box problems [5]. However, the metaheuristic methods frequently are not appropriate for the expensive black-box problems because of the limited budget of the computations of the objective functions.

The optimization of expensive black-box objective functions can be considered as a sequence of decisions under uncertainty, and the ideas of the theory of rational decision making seem most appropriate for the development of the corresponding algorithms. Gaussian random fields (GRF) normally are used as models representing uncertainty about aimed objective functions. Bayesian algorithms are designed maximizing a criterion of average utility with respect to the random field chosen for a model. The most frequently used criteria are: the maximum average improvement and the maximum improvement probability. The research and applications of the single-objective Bayesian methods is booming during last years. The Bayesian approach has recently extended to multi-objective optimization. The idea of the maximization of the multi-objective improvement

© Springer Nature Switzerland AG 2020
Y. D. Sergeyev and D. E. Kvasov (Eds.): NUMTA 2019, LNCS 11974, pp. 511–518, 2020.
https://doi.org/10.1007/978-3-030-40616-5_50

probability is implemented in [17]. A popular alternative method is based on the reduction to single objective optimization where the hyper-volume of a sub-region of the objective space, bounded by the approximation of the Pareto front, is maximized; see e.g. [6,7]. For the relevant single-objective black-box optimization methods we refer to [11–14].

The inherent computational burden of Bayesian multi-objective algorithms bounds their application area similarly to the single-objective case. In the present paper we propose coping with such a challenge by the implementation based on the rectangular partition of the feasible region. We merge the ideas of [17] (where the multi-objective P-algorithm was proposed), and of [3] (where the partition based single-objective P-algorithm was implemented). The developed algorithm was applied to the optimization of a biochemical process modeling of which is computationally extremely intensive.

2   The Proposed Algorithm

A black-box multi-objective minimization problem is considered

$$\min F(x), \ x \in A \subset \mathbb{R}^d, \ F(x) = (f_1(x), \ldots, f_m(x))^T; \tag{1}$$

we assume that A is of simple structure, e.g., a hyper-rectangular. We assume that objective functions are computationally expensive.

We start from a brief introduction of the original single-objective ($m = 1$) P-algorithm. For a review of related methods we refer to [15]. A GRF $\xi(x)$, $x \in A$, is accepted for a model of an objective function. The results of k function evaluations $y_i = f(x_i)$, $i = 1, \ldots, k$, are available and can be taken into account for planning a current evaluation point. The P-algorithm selects for the evaluation of the objective function the point of maximum of conditional improvement probability

$$x_{k+1} = \arg \max_{x \in A} \mathbb{P}\{\xi(x) < y_{ok} \,|\, \xi(x_i) = y_i, \ i = 1, \ldots, k\}, \tag{2}$$

where $y_{ok} = \min\{y_1, \ldots, y_k\} - \varepsilon_k$, $\varepsilon_k > 0$ is an improvement threshold.

A generalization of the P-algorithm to the multi-objective case was proposed in [17]. A vector valued GRF $\Xi(x) = (\xi_1(x), \ldots, \xi_m(x))^T$, $m > 1$, $x \in A$, is accepted for a model of objective functions. Let $Y_i = F(x_i)$, $i = 1, \ldots, k$, denote the vectors of objective function values evaluated in previous iterations, and Y_{ok} be a reference point. The P-algorithm computes a current vector of the objectives at the point

$$x_{k+1} = \arg \max_{x \in A} \mathbb{P}\{\Xi(x) < Y_{ok} \,|\, \Xi(x_i) = Y_i, \ i = 1, \ldots, k\}. \tag{3}$$

The main disadvantage of the standard implementation of the multi-objective P-algorithm is its computational burden where the maximization problem (3) is multimodal, and the computation of $\mathbb{P}(\cdot)$ involves inverting large ill conditioned matrices. The computational burden in a single-objective case can be substantially reduced by the partition based implementation as shown in [3]. We will generalise that implementation, in the present paper, for the multi-objective case.

Let the feasible region A be a hyper-rectangle. The algorithm is designed as a sequence of subdivisions by means of the bisection. The selected hyper-rectangle is bisected by a hyper-plane orthogonal to its longest edges. The values of $F(x)$ are computed at 2^{d-1} intersection points. A hyper-rectangle is selected for the subdivision according to a criterion which is an approximation of (3). The criterion of a hyper-rectangle is computed as the conditional improvement probability at its center x_c. The computational complexity for that probability is defined by the complexity of the computations of the conditional mean $\mu(x_c \,|\, \cdot)$ and conditional variance $\sigma^2(x_c \,|\, \cdot)$ of $\Xi(x_c)$. We approximate the conditional probability in (3) by restricting information used in the definition of $\mu(x_c \,|\, \cdot)$ and $\sigma^2(x_c \,|\, \cdot)$ with function values at the vertices of the considered rectangle. Thus the computational complexity of $\mu(\cdot)$ and $\sigma(\cdot)$ in the proposed implementation is lower than in the standard implementation thanks to the restriction of the involved information. Further, the expressions of $\mu(\cdot)$ and $\sigma(\cdot)$ are replaced with their asymptotic expressions obtained by shrinking the hyper-rectangle to a point [3]:

$$\mu(x_c \,|\, \cdot) \sim \frac{1}{|I|} \sum_{i \in I} Y_i, \;\; \sigma^2(x_c \,|\, \cdot) \sim V, \tag{4}$$

where I denotes the set of indices of the vertices, and V denotes the hyper-volume of the considered hyper-rectangle. These simplifications imply the following expression of the selection criterion

$$\frac{V}{\left\| \sum_{i \in I} Y_i - Y_{ok} \right\|}.$$

The high asymptotic convergence of the bi-objective version of that algorithm is proved in [4]. The partition of the feasible region at the initial iterations is quite uniform, thus it is rational with respect to the modest information about the considered objective functions at the initial iterations [16]. The accumulated information guides the selection of hyper-rectangles at later iterations towards the set of efficient decisions. The computational complexity of the proposed algorithm at a current iteration t can be evaluate similarly to the single-objective optimization algorithm [20] since the same operations are performed to manage the accumulated data. Thus the computational complexity of iteration t is $T(n) = O(n \times m \times \log(n \times m))$, where n is the number of evaluations of $F(x)$ made at previous iterations. The complexity of computations at a current t iteration of the standard implementation of Bayesian algorithms, e.g. of the algorithm of average improvement, is much higher than $T(n)$ (here $t = n$). The iteration t includes inverting of, generally speaking, m $n \times n$ matrices the time-complexity of which is $O(m \times n^3)$. Moreover, the inverting of the considered matrices is challenging since their condition numbers typically are very large. The other serious computational complexity of a standard implementation is the maximization of the average improvement which is a non-convex optimization problem.

3 Comparison with the Standard Implementation of the P-Algorithm

The proposed algorithm is a simplified version of its predecessor, i.e., of the standard implementation of the P-algorithm. It is interesting to compare their performance. The standard implementation is described in detail in [17] where several test problems are solved to illustrate its performance. We use the same test problems.

A non-convex problem, proposed in [9], is quite frequently used for testing multi-objective algorithms; see e.g., [5]. The objective functions are

$$f_1(\mathbf{x}) = 1 - e^{-\sum_{i=1}^{d}(x_i - 1/\sqrt{d})^2},$$
$$f_2(\mathbf{x}) = 1 - e^{-\sum_{i=1}^{d}(x_i + 1/\sqrt{d})^2}, \tag{5}$$

$d = 2$, and the feasible region is $\mathbf{A} : -4 \leq x_1, x_2 \leq 4$. The next test problem is composed of two Shekel functions which are frequently used for testing of single-objective global optimization algorithms:

$$f_1(\mathbf{x}) = -\frac{0.1}{(0.1 + (x_1 - 0.1)^2 + 2(x_2 - 0.1)^2)}$$
$$-\frac{0.1}{(0.14 + 20((x_1 - 0.45)^2 + (x_2 - 0.55)^2))},$$
$$f_2(\mathbf{x}) = -\frac{0.1}{(0.15 + 40((x_1 - 0.55)^2 + (x_2 - 0.45)^2))}$$
$$-\frac{0.1}{(0.1 + (x_1 - 0.3)^2 + (x_2 - 0.95)^2)}. \tag{6}$$

The visualization of these test problems including graphs of the Pareto fronts and sets of Pareto optimal decisions is presented, e.g. in [5,10].

Several metrics are used for the quantitative assessment of the precision of a Pareto set approximation. For the comparison of the proposed implementation of the P-algorithm with the standard one we apply the metrics which were used in recent publications related to the standard implementation. The generational distance (GD) is used to estimate the distance between the found approximation and the true Pareto front [5]. GD is computed as the maximum of distances between the found non-dominated solutions and their closest neighbors from the Pareto set. The epsilon indicator (EI) is a metric suggested in [21] which integrates measures of the approximation precision and spread: it is the min max distance between the Pareto front and the set of the found non-dominated solutions

$$\text{EI} = \max_{1 \leq i \leq K} \min_{1 \leq j \leq N} ||Z_i - F_j^*||, \tag{7}$$

where $F_j^*, j = 1, \ldots, N$, are the non-dominated solutions found by the considered algorithm, and $\{Z_i, i = 1, \ldots, K\}$ is the set of points well representing the Pareto set, i.e. Z_i are sufficiently densely and uniformly distributed over the Pareto set.

The algorithms were stopped after 100 computations of $F(x)$. Although the P-algorithm theoretically is deterministic, its version implemented in [17] is randomised because of a stochastic maximization method used for (3). Therefore, test problem were solved 100 times. We present the mean values and standard deviations of the considered metrics from [10] in two columns of Table 1. The proposed algorithm is deterministic; thus its results occupy single column for each test problem. The results of the proposed algorithm for the test problem (5) are even better than the results of the standard version of the P-algorithm. However, the standard version outperforms the proposed algorithm in solving (6). The set of optimal decisions of the latter problem consists of three disjoint subsets, and the diameter of one of subsets is relatively small. A considerable number of partitions of the feasible region was needed to indicate the latter subset. The experimentally measured solution time was 8.6 ms for the proposed algorithm, and 3.9 s for the algorithm of [17].

Table 1. Performance criteria of the standard and partition based implementations of the P-algorithm for Problems (5) and (6)

Implementation	Standard				Partition based	
Problem	Problem (5)		Problem (6)		Problem (5)	Problem (6)
NN	9.87	1.4	15.7	2.0	27	18
GD	0.015	0.0061	0.070	0.051	0.015	0.21
EI	0.20	0.034	0.13	0.053	0.092	0.25

4 Performance Evaluation on a Real World Problem

Quite many optimization problems in biotechnology can be characterized as black-box expensive ones. For example, optimal design of bio-sensors and bio-reactors requires solving optimization problems the objective functions of which are defined by computer models of high complexity [1,2]. The micro bio-reactors are computationally modeled by a two-compartment model based on reaction-diffusion equations containing a nonlinear term related to the Michaelis-Menten enzyme kinetics; we refer to [1,2] for the description and substantiation of the model. The computation of one objective function value of such problems take up to 10 min, and for some special cases possibly more. However, we are optimistic about these challenges since the problems in question are low dimensional.

We consider an optimization problem related to the optimal design of a micro bio-reactor, which well represents real world problems for which the proposed algorithm is potentially appropriate. Specific technical aspects of the computer model of the micro bio-reactor, and the substantiation of the objective functions are addressed in [18].

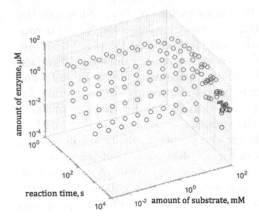

Fig. 1. Pareto front of the set of objective vectors computed by the proposed and genetic algorithms. (Color figure online)

Optimal design of a micro bioreactor is formulated as a three-objective optimization problem of four variables. The objectives are: the time of the reactions; the amount of the substrate per volume unite converted to the product; the total amount of the enzyme used per volume unit of the reactor; first and third objectives are minimized, and the second one is maximized. The variables are: two constructive parameters of a bio-reactor, and the concentrations of the enzyme and of the substrate. The computation time depends on the parameters of the bio-reactor; the average time of computation of a single value of the first objective function is 4.32 min (a computer with Intel Xeon X5650 2.66 GHz processor was used). A long reaction time means that the bioreactor with such parameters is not appropriate; correspondingly, the simulation was interrupted if it exceeded 10 min.

We demonstrate the performance of the proposed algorithm under the conditions typical to a real world applications. The following optimization problem is considered where $F(x)$ is available as a function in C:

$$\min_{x \in A} F(x), \ F(x) = (f_1(x), \ f_2(x), \ f_3(x))^T,$$

$$A = \{x : \ -4 \leq x_1, \ x_2 \leq -3, \ -8 \leq x_3 \leq -4, \ -5 \leq x_4 \leq -1\}. \tag{8}$$

We focus here on black-box optimization without discussing its applied aspects. The problem will be presented from the point of view of applied optimal designed in the next paper co-authored with experts in biotechnology. The proposed algorithm was applied to the solution of (8) with the predefined budget of evaluations of $F(x)$ equal to 1000; the solution time was 72 h and 14 min. The genetic algorithm (GA) from the MATLAB Optimization Toolbox was also applied to the solution of the considered problem. The following parameters of GA were chosen: the population size was 50, the crossover fraction was 0.8, Pareto fraction was 0.35, and the other parameters were chosen as predefined in the Optimization Toolbox. The termination condition of GA was the same: budget of the objective function evaluations equal to 1000. GA is a randomized algorithm, thus results of a single run are not sufficiently reliable from the point of view of statistical testing. Indeed, the use of real world problems with expensive objective functions for testing randomised algorithms is challenging especially where the experimentation time is limited.

The Pareto front approximation computed by the proposed algorithm consisted of 124 vectors, and the approximation computed by GA consisted of 15 vectors. None vector of the of first approximation was dominated by a vector of the second approximation. Only four vectors of the second approximation were not dominated by the vectors of the first approximation. The hypervolume indicators of both approximations were equal to 42.3 and 12.4 correspondingly. Thus the proposed algorithm clearly outperforms GA. The non-dominated vectors (in the commonly used physical units) of the union of both approximations are presented in Fig. 1 where circles denote the points of the first approximation, and the red rectangles denote the points of the second one. This figure shows only a general shape of the Pareto front. User oriented interfaces and visualization methods are available to experts in biotechnology to aid selecting an appropriate Pareto optimal decision; the advantages of visualising not only Pareto optimal solutions but also Pareto optimal decisions are argued in [18].

The performance of the proposed method is appropriate for low dimensional black-box expensive problems. Extensions to higher dimensionality can be challenging because of large number of computations of the objective function values (2^{d-1}) at a current iteration. We plan the investigation of the simplicial partition based Bayesian algorithms to cope with that challenge; note that such a partition proved efficient in Lipschitzian optimization [11]. The hybridization of the proposed global search algorithm with a local one seems promising. The good performance of hybrids of the single-objective Bayesian algorithms with local ones [19,20] gives hope that similar synergy of the global and local search strategies will be achieved also in the multi-objective case.

5 Conclusions

A rectangular partition based implementation of a Bayesian multi-objective method is proposed where the typical for Bayesian algorithms inherent computational burden is avoided. The proposed algorithm is appropriate for solving practical problems characterized as expensive black-box problems of modest dimensionality. The prospective directions of further research are: increasing dimensionality of efficiently solvable problems, and studying efficiency of other partition strategies.

Acknowledgments. This work was supported by the Research Council of Lithuania under Grant No. P-MIP-17-61. We thank the reviewers for their valuable remarks.

References

1. Baronas, R., Ivanauskas, F., Kulys, J.: Mathematical modeling of biosensors based on an array of enzyme microreactors. Sensors **6**(4), 453–465 (2006)
2. Baronas, R., Kulys, J., Petkevičius, L.: Computational modeling of batch stirred tank reactor based on spherical catalyst particles. J. Math. Chem. **57**(1), 327–342 (2019)
3. Calvin, J., Gimbutienė, G., Phillips, W., Žilinskas, A.: On convergence rate of a rectangular partition based global optimization algorithm. J. Global Optim. **71**, 165–191 (2018)

4. Calvin, J., Žilinskas, A.: On efficiency of a single variable bi-objective optimization algorithm. Optim. Lett. **14**(1), 259–267 (2020). https://doi.org/10.1007/s11590-019-01471-4
5. Deb, K.: Multi-Objective Optimization Using Evolutionary Algorithms. Wiley, Chichester (2009)
6. Emmerich, M., Deutz, A.H., Yevseyeva, I.: On reference point free weighted hypervolume indicators based on desirability functions and their probabilistic interpretation. Proc. Technol. **16**, 532–541 (2014)
7. Feliot P., Bect J., Vazquez E.: User preferences in Bayesian multi-objective optimization: the expected weighted hypervolume improvement criterion. arXiv:1809.05450v1 (2018)
8. Floudas, C.: Deterministic Global Optimization: Theory, Algorithms and Applications. Kluwer, Dodrecht (2000)
9. Fonseca, C.M., Fleming, P.J.: On the performance assessment and comparison of stochastic multiobjective optimizers. In: Voigt, H.-M., Ebeling, W., Rechenberg, I., Schwefel, H.-P. (eds.) PPSN 1996. LNCS, vol. 1141, pp. 584–593. Springer, Heidelberg (1996). https://doi.org/10.1007/3-540-61723-X_1022
10. Pardalos, P., Žilinskas, A., Žilinskas, J.: Non-Convex Multi-Objective Optimization. Springer, Cham (2017). https://doi.org/10.1007/978-3-319-61007-8
11. Paulavičius, R., Žilinskas, J.: Simplicial Global Optimization. Springer, New York (2014). https://doi.org/10.1007/978-1-4614-9093-7
12. Paulavičius, R., Sergeyev, Y., Kvasov, D., Žilinskas, J.: Globally-biased DISIMPL algorithm for expensive global optimization. J. Global Optim. **59**, 545–567 (2014)
13. Sergeyev, Y., Kvasov, D., Mukhametzhanov, M.: On the efficiency of nature-inspired metaheuristics in expensive global optimization with limited budget. Sci. Rep. **453**, 1–8 (2018)
14. Sergeyev, Y., Kvasov, D.: Deterministic Global Optimization: An Introduction to the Diagonal Approach. Springer, New York (2017). https://doi.org/10.1007/978-1-4939-7199-2
15. Žilinskas, A., Zhigljavsky, A.: Stochastic global optimization: a review on the occasion of 25 years of informatica. Informatica **27**(2), 229–256 (2016)
16. Žilinskas, A.: On the worst-case optimal multi-objective global optimization. Optim. Lett. **7**(8), 1921–1928 (2013)
17. Žilinskas, A.: A statistical model-based algorithm for black-box multi-objective optimisation. Int. J. Syst. Sci. **45**(1), 82–92 (2014)
18. Žilinskas, A., Baronas, R., Litvinas, L., Petkevičius, L.: Multi-objective optimization and decision visualization of batch stirred tank reactor based on spherical catalyst particles. Nonlinear Anal. Model. Control **24**(6), 1019–1033 (2019)
19. Žilinskas, A., Žilinskas, J.: A hybrid global optimization algorithm for non-linear least squares regression. J. Global Optim. **56**(2), 265–277 (2013)
20. Žilinskas, A., Gimbutienė, G.: A hybrid of Bayesian approach based global search with clustering aided local refinement. Commun. Nonlinear Sci. Numer. Simul. **78**, 104785 (2019)
21. Zitzler, E., Thiele, L., Lauman, M., Fonseca, C., Fonseca, V.: Performance measurement of multi-objective optimizers: an analysis and review. IEEE Trans. Evol. Comput. **7**(2), 117–132 (2003)

Problem Statement for Preparing a Single Batch of End Product Under Uncertainty

Anna Zykina[✉][iD], Olga Kaneva[iD], Vladimir Savkin[iD], and Tatyana Fink[iD]

Omsk State Technical University, Mira Avenue 11, 644050 Omsk, Russia
avzykina@mail.ru

Abstract. Oil refining is a key industry of the world economy. Growing hydrocarbon production cost and global competition in the oil market encourage the oil refining industry to optimize the production scheme. The evolution of mathematical tools of automated enterprise control systems is closely connected with the systems development at each level of control. Mathematical models for organizational and economic control of the enterprise and process control models are widely presented in publications and implemented in the enterprise information systems. The management of operational scheduled and dispatching production is one of the most complex problems. The paper deals with the problem of finding an optimal ratio for the components from the tanks to obtain an oil product of the required amount and quality in a commercial tank. The peculiarity of the mathematical models proposed for solving the problem is that only the boundaries for each quality indicator of petroleum product are known. To formalize the emerging uncertainty, models utilizing the interval approach are proposed.

Keywords: Oil refining · Mathematical tools of automated systems · Production processes optimization · Calendar task · Dispatch schedule · Interval optimization

1 Introduction

The increase in the crude hydrocarbons recovery costs and the global competition on the oil products market induce the oil processing industry to optimize petroleum refineries. Production control systems used at oil refineries should promptly respond to increasing dynamics of perturbances such as volumes and quality of the feed as well as orders for the manufactured products volumes in terms of the constraints imposed by shipment logistics. Inexhaustible opportunities of solving this kind of problems lie in enhancing a petroleum refinery control system, its main tool being automation.

Developing mathematical tools of an integrated automated control system for an enterprise is closely connected with constructing systems for every control

The study was carried out with the financial support of RFBR and Omsk region in the framework of the scientific project No. 18-41-550003.

level and is performed due to priorities from the top-down and bottom-up. The questions of constructing and investigating mathematical models of the enterprise for the business management purposes are widely discussed in the literature [1]. Operational-calendar and dispatch production control is one of the most complex control forms. That is the reason why this stage of operational planning of continuous operation has not been sufficiently addressed in the literature and has no ready-made practical solutions [2].

The article aims to develop an approach to mathematical modelling of dispatch control to optimize the processes in layered control systems of a petroleum refinery.

In mathematical programming models, to which the tasks of planning, design, and control are reduced, some or all of the parameters (characteristics) of the quality indicators and constraints may appear uncertain or random. In some cases, experience, statistics, and the study of the processes, determining the change in the source data and forming the conditions under which the plan, project or control system is implemented, allow you to set certain probabilistic characteristics of the task parameters. In other cases, there is no information about the features of the phenomena that can change the expected values of the task conditions parameters. Both situations are the scope of the study for stochastic programming [3–5].

The practical problems of choosing solutions with conflicting conditions or under risk and uncertainty often describe the study subject and the investigated system more adequately than those under solvability and certainty. This indicates the relevance of developing and implementing algorithms to solve problems with incomplete information, including controversial ones. The effective solution of such problems requires not only the development of new algorithms to find solutions but also fundamentally new approaches and new mathematical designs to model and determine the concept of solving such problems [6, 7].

2   Problem Statement and Mathematical Model

Consider the production problem of preparing an end product. For a given range, weight and quality restrictions of the end product based on the calendar task, one needs to calculate the corresponding shipment operations and the full schedule of the pumping equipment operation in commodity production for the entire period of the dispatch schedule so as to ensure a minimum components arrival deviation from the formulation specified in the calendar task. The deviation is calculated for each component as the difference between the value in the calendar job and the sum of the component arrival operations in the dispatch schedule. The scheme of objects interaction in preparing the end product from the components is as follows. There are multiple component containers $R_1, R_2, ..., R_r$, multiple pump units $U_1, U_2, ..., U_m$ and a commercial tank where the blending of the components, i.e. the end product P manufacture, occurs. Each of the components can be stored in several component tanks.

The task of the dispatch schedule is to find an optimal ratio of the components from the component tanks to obtain the required amount of the product, with the specified quality indicators and with minimal deviations from the specified formulation, for a given period in the commercial tank.

To formalize the proposed conditions, the following notations are introduced:

n is the number of components involved in blending, $i = 1, ..., n$;

m is the number of pumping units, $p = 1, ..., m$;

r is the number of component tanks, $j = 1, ..., r$, $n \leq r$, $r \leq m$;

g is the number of types of quality indicators, $k = 1, ..., g$;

v_{ij} is the available amount of the i-th component in the j-th tank;

ρ_{ij} is the density of the i-th component in the j-th tank;

θ_i is the required amount of the i-th component due to the formulation;

q_{ki}^j is the value of the k-th quality indicator for the i-th component in the j-th container;

V^E is the remaining amount of the end product in the commercial tank;

q_k^E is the value of the k-th indicator for the quality of the end product remaining in the commercial tank;

V^P is the required amount of the end product;

q_k^P is the value of the k-th indicator of the end product quality;

C_p is the maximum capacity of the p-th pumping unit at the specified period for the arrival of the components to prepare the end product;

c_p is the minimum capacity of the p-th pumping unit at the specified period for the arrival of the components to prepare the end product.

Let X be the matrix, whose elements are values x_{ij}, i.e. the amount of the i-th component taken from the j-th container; the matrix X is the solution to the set problem.

Let us introduce the graph G, which is a tree, whose root is a commercial tank, inner nodes are pumping units, and the leaves are component tanks (corresponding to the elements of the matrix X), which form the set $L(G)$. Let G_p be a subgraph G that is a tree whose root is the pump unit with number p, where $p = 1, ..., m$. Denoting through $L(G_p)$ the set leaves of the rooted tree G_p, as $L(G_p) \subseteq L(G)$, the set $I(G_p)$ is introduced for the indices of the leaves of the rooted tree G_p:

$$I(G_p) = (i, j) | x_{ij} \in L(G_p), \quad p = 1, ..., m.$$

Since the task of minimizing the deviation of the components i arrival from the given formulation θ_i, $i = 1, ..., n$ is set, then we consider:

$$F(x) = \sum_{i=1}^{n} \left(\sum_{j=1}^{r} x_{ij} - \theta_i \right)^2 \tag{1}$$

of the optimization problem which $F(x)$ as the target function.

Let us formulate the constraints of the problem.

The amount of the used i-th component from the j-th tank satisfies the non-negativity conditions and does not exceed the available amount v_{ij}:

$$0 \le x_{ij} \le v_{ij}, \quad i = 1, ..., n, \quad j = 1, ..., r. \tag{2}$$

The total number of all used components $i = 1, ..., n$ from all tanks $j = 1, ..., r$ is equal to the required amount of end product:

$$\sum_{i=1}^{n}\sum_{j=1}^{r} x_{ij} = V^P. \tag{3}$$

The constraints on the capacity of the pump unit with number p are:

$$c_p \le \sum_{i,j \in I(G_p)} \rho_{ij} x_{ij} \le C_p, \quad p = 1, ..., m. \tag{4}$$

The constraints on the quality indicators of the end product in terms of the remaining end product V^E in the commercial tank are:

$$\sum_{i=1}^{n}\sum_{j=1}^{r} q_{ki}^{j} x_{ij} = \left(V^P + V^E\right) q_k^P - V^E q_k^E, \quad k = 1, ..., g. \tag{5}$$

To form restrictions (5) a linear dependence of the end product quality indicators on their components quality indicators was used (considering V^E, i.e. the remaining end product in the commercial tank)

$$q_k^P(X) = \frac{\left(\sum_{i=1}^{n}\sum_{j=1}^{r} q_{ki}^{j} x_{ij} + q_k^E V^E\right)}{\left(V^P + V^E\right)}, \quad k = 1, ..., g.$$

The studies of this dependence were carried out based on the initial data from the laboratory system of the enterprise and demonstrated satisfactory results [8].

3   Interval Problem Statement

Forming the dispatch schedule according to the calendar task based on the model (1)–(5) has a significant drawback consisting in the following. If permissible ranges for the values $[c_p, C_p]$, $p = 1, ..., m$ are set for the capacity constraints of the pump unit with number $p = 1, ..., m$ in the production process, the quality indicators of the end product q_k^P, $k = 1, ..., g$ are constants. But the quality of the components differs in individual batches of filling the component tanks $j = 1, ..., r$, while the production process includes no measurement of the quality indicators values for the blended components $i = 1, ..., n$. According to dependence (5), the resulting quality indicators q_k^P, $k = 1, ..., g$ of the end product, in terms of the remaining end product V^E in the commercial tank, depend on the quality indicators values q_{ki}^{j}, $j = 1, ..., r$, $i = 1, ..., n$, of the blended components, and constraints (5) are incorrect if the quality values for the components are set incorrectly. To formalize the obtained uncertainty, various approaches are used [9,10].

Let the boundaries of the quality indicators $k = 1, ..., g$ be known (or defined in the calendar task) for the end product. Let us denote the lower and upper limits of each quality indicator as $\underline{q_k^P}$ and $\overline{q_k^P}$, correspondingly. Consequently, problem statement under interval uncertainty results.

To solve such problems, there are specially developed methods of solving the interval linear programming [11], as well as the methods of the interval global optimization [12].

The advantage of interval approach is the possibility to reduce the nondeterministic problem to solving a pair of the corresponding deterministic problems (lower boundary problem and upper boundary problem), in which uncertain parameters are determined by the lower and upper boundaries of each of them. The peculiarity of the method is that the solution result is presented in the same form as the uncertain parameters, that is, in the form of the intervals for possible values. The disadvantage of the interval approach to the optimization problem is the possible insolubility of the lower and/or upper boundary problem.

Let us construct the lower and upper boundary problems for tasks (1)–(5), replacing the parameters q_k^P in constraint (5) with the corresponding values $\underline{q_k^P}$ and $\overline{q_k^P}$, $k = 1, ..., g$.

The lower boundary problem for (1)–(5) is one in which target function (1) and constraints (2)–(4) remain the same, while constraints (5) are replaced by the constraints of the form:

$$\sum_{i=1}^{n} \sum_{j=1}^{r} q_{ki}^j x_{ij} = \left(V^P + V^E \right) \underline{q_k^P} - V^E q_k^E, \quad k = 1, ..., g. \tag{6}$$

Its solution will be denoted by $\underline{X}$.

The upper boundary problem for (1)–(5), correspondingly, is one in which target function (1) and constraints (2)–(4) remain the same, while constraints (5) are replaced by the constraints of the form:

$$\sum_{i=1}^{n} \sum_{j=1}^{r} q_{ki}^j x_{ij} = \left(V^P + V^E \right) \overline{q_k^P} - V^E q_k^E, \quad k = 1, ..., g. \tag{7}$$

Its solution will be denoted by $\overline{X}$.

In the conducted numerical experiments, the sets of the parameters providing the solvability of the boundary problems were formed. Table 1 shows an example of this set of parameters $\underline{q_k^P}$ and $\overline{q_k^P}$, $k = 1, ..., g$, for the upper and lower boundary problems. The initial data were taken from the laboratory system of the enterprise. In this experiment the number of quality indicators is equal to 4, the number of the components involved in blending is 5, and the number of the component tanks is equal to 5. Table 2 provides the solutions $\underline{X}$ and $\overline{X}$ for the corresponding upper and lower boundary problems. The solutions set the intervals $[\underline{x}_{ij}, \overline{x}_{ij}]$ for the solutions x_{ij} of the initial task (1)–(5) under the conditions of interval uncertainty.

Table 1. Lower and upper boundary problem.

k	Quality indicator	The lower boundary	The upper boundary
1	Octane number (research method)	99	101
2	Density	0,73	0,75
3	Mass fraction of sulphur	3,87	3,94
4	Volume fraction of benzene	0,31	0,32

Table 2. Solutions $\underline{X}$ and $\overline{X}$ for both upper and lower boundary problems.

x_{ij}	The available amount	The lower problem	The upper problem
x_{11}	69	1,1	23,5
x_{22}	780,5	148,6	325,3
x_{33}	1391,7	519,2	171,7
x_{44}	37,5	37,5	37,5
x_{55}	∞	63,6	212

4   Interval Approach to Controversial Problem

Problem (1)–(5) is a convex quadratic programming problem: constraints (2)–(5) are given by linear functions, the Hesse matrix of target function (1) is positively semidefinite (which, due to the sufficient convexity condition, provides the convexity of the target function). In the deterministic statement, in the case of the consistency of the constraints, this problem has an optimal solution. Indeed, constraints (2) provide compactness of tolerance region (2)–(5), hence quadratic programming problem (1)–(5) has an optimal solution under the compatibility of constraint system (2)–(5).

The production process and the calendar task provide the consistency of constraint (3) by the required amount of the end product and interval constraints (4) by the capacity for the pumping units with constraints (2) by filling the component tanks. Therefore, constraints (2)–(4) are consistent and specify a non-empty compact domain. Let us denote this set D.

Interval constraints (5) (or (6) and (7) for the lower and upper boundary problems, respectively) depend on the set of the quality indicators values q_k^P, $k = 1, ..., g$, and at the boundary values $\underline{q_k^P}$ and $\overline{q_k^P}$, $k = 1, ..., g$, constraints (5) (or (6), or (7), respectively) may be incompatible with the compact set D specified by the set constraints (2)–(4).

To determine the boundary values $\underline{q_k^P}$ and $\overline{q_k^P}$ ensuring the solvability of the upper and lower boundary problems, respectively, let us note the following two problems.

Table 3. Boundary value sets providing the solvability of the lower and upper boundary problems.

k	Quality indicator	The lower boundary	The upper boundary
1	Octane number (research method)	97,168	106,36
2	Density	0,68	0,83
3	Mass fraction of sulphur	3,41	3,83
4	Volume fraction of benzene	0,28	0,35

Table 4. Solutions $\underline{X}$ and $\overline{X}$ to both upper and lower solvable boundary problems.

x_{ij}	The available amount	The lower problem	The upper problem
x_{11}	69	0	64,3
x_{22}	780,5	0	689,2
x_{33}	1391,7	275,5	0
x_{44}	37,5	0	16,4
x_{55}	∞	494,5	0

Problem 1 (for the lower boundary problem):

$$\sum_{k=1}^{g} \alpha_k \underline{q_k^P} \rightarrow min. \qquad (8)$$

under the constraints (2), (3), (4), (6), $\underline{q_k^P} \geq 0$, $k = 1, ..., g$.

Problem 2 (for the upper boundary problem):

$$\sum_{k=1}^{g} \overline{\alpha_k} \overline{q_k^P} \rightarrow max. \qquad (9)$$

under the constraints (2), (3), (4), (7), $\overline{q_k^P} \geq 0$, $k = 1, ..., g$.

Problems 1 and 2 are linear optimization problems. Optimized unknown parameters in these problems are sets of the boundary values $\underline{q_k^P}$ and $\overline{q_k^P}$, respectively. The coefficients in the linear objective functions (8) and (9) are $\alpha_k \geq 0$ and $\overline{\alpha_k} \geq 0$ and they set the weight value (normalization) of the desired parameters $\underline{q_k^P}$ and $\overline{q_k^P}$, respectively.

Table 3 provides the results of the numerical experiments for Problems 1 and 2. The sets of the boundary values $\underline{q_k^P}$ and $\overline{q_k^P}$, providing the solvability of the lower and upper boundary problems, respectively, are obtained. Table 4 gives the solutions $\underline{X}$ and $\overline{X}$ for the corresponding solvable lower and upper boundary tasks. These solutions determine the tolerance intervals $\left[\underline{x_{ij}}, \overline{x_{ij}}\right]$ ensuring the solvability for the solutions x_{ij} of the initial tasks (1)–(5). These values can be recommended to form the dispatch schedule.

5   Conclusion

The article studies mathematical models of a petroleum refinery. The problem of finding an optimal ratio for the components from the tanks to obtain the oil product of required amount and quality has been considered. The result of the work is the developed approaches to calculating the boundary values of the end product quality indicators at filling the tanks in the case when the production process of the petroleum refining includes no measurement of the quality indicators values. The scientific significance of the article lies in the use of mathematical modelling apparatus under uncertainty and inconsistency to construct mathematical models for optimizing a petroleum refinery. Numerical studies of the constructed models are of practical importance: a set of calculated boundary values for the quality indicators of the end product provides solvability, and hence the guarantee of the practical feasibility of the production process under the specified parameters.

References

1. Huseynov, I.A., Melikov, E.A., Khanbutaeva, N.A., Efendiyev, I.R.: Models and algorithms of a multi-level control system for installations of primary oil refining. Izvestiya RAS. Theory Control Syst. **1**, 83–91 (2012)
2. Zykina, A.V., Savelev, M.Yu., Fink, T.Yu.: Multi-level management of oil refining production. Requir. Res. Tasks. Omsk Sci. Bull. **162**(6), 271–274 (2018)
3. Yudin, D.B.: Mathematical methods of control in the conditions of incomplete information: problems and methods of stochastic programming. KRASAND, Moscow (2010)
4. Taha, H.A.: Operations Research: An Introduction. Pearson Prentice Hall, Upper Saddle River (2007)
5. Kibzun, A.I., Kan, Y.S.: Stochastic Programming Problems with Probability and Quantile Functions. Wiley, Chichester (1996)
6. Zykina, A., Kaneva, O.: Stochastic optimization models in transportation logistics. In: Selected Papers of the 1-st International Scientific Conference Convergent Cognitive Information Technologies on CEUR Workshop proceedings, Moscow, vol. 1763, pp. 251–255 (2016)
7. Zykina, A.V., Kaneva, O.N.: Algorithm for solving stochastic transportation problem. In: Proceedings of the VIII International Conference on Optimization and Applications on CEUR Workshop Proceedings, Moscow, vol. 1987, pp. 598–603 (2017)
8. Savkin, V.N., Motkov, A.M., Saveliev, M.Y.: Comparison of approaches to the calculation of quality indicators of petrols. In: Informational Bulletin of the Omsk Scientific and Educational Center Omsk State Technical University and the Institute of Mathematics and Siberian Branch of the Russian Academy of Sciences in the Field of Mathematics and Computer Science, vol. 2, no. 1, pp. 94–97 (2018)
9. Diligensky, N.V., Dymova, L.G., Sevastyanov, P.V.: Fuzzy modeling and multi-criteria optimization of production systems under uncertainty: technology, economics, ecology. Mashinostroenie, Moscow (2004)
10. Giles, M.B.: Multilevel Monte Carlo methods. Acta Numerica **24**, 259–328 (2015)

11. Fiedler, M., Nedoma, J., Ramik, J., Rohn, J., Zimmermann, K.: Linear Optimization Problems with Inexact Data. Springer, New York (2006). https://doi.org/10.1007/0-387-32698-7
12. Hansen, E., Walster, G.W.: Global Optimization Using Interval Analysis. Marcel Dekker, New York (2004)

Author Index

Printed in the United States
By Bookmasters